CAMBRIDGE STUDIES IN ADVANCED MATHEMATICS 218

MULTIPLICATIVE NUMBER THEORY II: PRIMES AND SIEVES

This long-anticipated work shares the aims of its celebrated companion: namely, to provide an introduction for students and a reference for researchers to the techniques, results, and terminology of multiplicative number theory. This volume builds on Volume I (which served as an introduction to basic, classical results) and focuses on sieve methods. From the mid-1960s this area has witnessed a number of major advances. Central to many of these has been the Bombieri–Vinogradov theorem, which is an application of the large sieve and can act as a substitute for the Generalized Riemann Hypothesis. A further stimulus was the discovery of a simple combinatorial identity for treating primes with oscillatory terms. Many of the advances are included as applications. Amongst the more recent is the proof of Gelfond's conjecture on the distribution of the sum of digits of primes to a given base, and the establishment of infinitely many bounded gaps between primes. Despite the fact that the book can serve as an entry to contemporary mathematics, it remains largely self-contained, with appendices containing background or material more advanced than undergraduate mathematics. Again, exercises, of which there is a profusion, illustrate the theory or indicate ways in which it can be developed. Each chapter ends with a thorough set of references, which will be essential for all analytic number theorists.

Hugh L. Montgomery is Professor Emeritus in the Department of Mathematics at the University of Michigan, Ann Arbor.

Robert C. Vaughan, FRS, was formerly Professor of Pure Mathematics at Imperial College London, and is currently Professor of Mathematics in the Department of Mathematics at Pennsylvania State University, University Park.

CAMBRIDGE STUDIES IN ADVANCED MATHEMATICS

All the titles listed below can be obtained from good booksellers or from Cambridge University Press. For a complete series listing, visit www.cambridge.org/mathematics.

Already Published

181 A. Agrachev, D. Barilari & U. Boscain *A Comprehensive Introduction to Sub-Riemannian Geometry*
182 N. Nikolski *Toeplitz Matrices and Operators*
183 A. Yekutieli *Derived Categories*
184 C. Demeter *Fourier Restriction, Decoupling and Applications*
185 D. Barnes & C. Roitzheim *Foundations of Stable Homotopy Theory*
186 V. Vasyunin & A. Volberg *The Bellman Function Technique in Harmonic Analysis*
187 M. Geck & G. Malle *The Character Theory of Finite Groups of Lie Type*
188 B. Richter Category *Theory for Homotopy Theory*
189 R. Willett & G. Yu *Higher Index Theory*
190 A. Bobrowski *Generators of Markov Chains*
191 D. Cao, S. Peng & S. Yan *Singularly Perturbed Methods for Nonlinear Elliptic Problems*
192 E. Kowalski *An Introduction to Probabilistic Number Theory*
193 V. Gorin *Lectures on Random Lozenge Tilings*
194 E. Riehl & D. Verity *Elements of ∞-Category Theory*
195 H. Krause *Homological Theory of Representations*
196 F. Durand & D. Perrin *Dimension Groups and Dynamical Systems*
197 A. Sheffer *Polynomial Methods and Incidence Theory*
198 T. Dobson, A. Malnič & D. Marušič *Symmetry in Graphs*
199 K. S. Kedlaya *p-adic Differential Equations*
200 R. L. Frank, A. Laptev & T. Weidl *Schrödinger Operators:Eigenvalues and Lieb–Thirring Inequalities*
201 J. van Neerven *Functional Analysis*
202 A. Schmeding *An Introduction to Infinite-Dimensional Differential Geometry*
203 F. Cabello Sánchez & J. M. F. Castillo *Homological Methods in Banach Space Theory*
204 G. P. Paternain, M. Salo & G. Uhlmann *Geometric Inverse Problems*
205 V. Platonov, A. Rapinchuk & I. Rapinchuk *Algebraic Groups and Number Theory, I (2nd Edition)*
206 D. Huybrechts *The Geometry of Cubic Hypersurfaces*
207 F. Maggi *Optimal Mass Transport on Euclidean Spaces*
208 R. P. Stanley *Enumerative Combinatorics, II (2nd Edition)*
209 M. Kawakita *Complex Algebraic Threefolds*
210 D. Anderson & W. Fulton *Equivariant Cohomology in Algebraic Geometry*
211 G. Pineda Villavicencio *Polytopes and Graphs*
212 R. Pemantle, M. C. Wilson & S. Melczer *Analytic Combinatorics in Several Variables (2nd Edition)*
213 A. Yadin *Harmonic Functions and Random Walks on Groups*
214 Y. Kawamata *Algebraic Varieties: Minimal Models and Finite Generation*
215 J. Gillespie *Abelian Model Category Theory*
216 L. Anderson *Oriented Matroids*
217 Y. Motohashi *Essays in Classical Number Theory*
218 H. L. Montgomery & R. C. Vaughan *Multiplicative Number Theory II*
219 E. Rijke *Introduction to Homotopy Type Theory*
220 N. Berestycki & E. Powell *Gaussian Free Field and Liouville Quantum Gravity*

Multiplicative Number Theory
II. Primes and Sieves

HUGH L. MONTGOMERY
University of Michigan, Ann Arbor

ROBERT C. VAUGHAN
Pennsylvania State University, University Park

CAMBRIDGE
UNIVERSITY PRESS

CAMBRIDGE
UNIVERSITY PRESS

Shaftesbury Road, Cambridge CB2 8EA, United Kingdom

One Liberty Plaza, 20th Floor, New York, NY 10006, USA

477 Williamstown Road, Port Melbourne, VIC 3207, Australia

314–321, 3rd Floor, Plot 3, Splendor Forum, Jasola District Centre, New Delhi – 110025, India

Cambridge University Press is part of Cambridge University Press & Assessment, a department of the University of Cambridge.

We share the University's mission to contribute to society through the pursuit of education, learning and research at the highest international levels of excellence.

www.cambridge.org
Information on this title: www.cambridge.org/9781009445054
DOI: 10.1017/9781009445030

When citing this work, please include a reference to the DOI 10.1017/9781009445030

First published 2026

A catalogue record for this publication is available from the British Library

A Cataloging-in-Publication data record for this book is available from the Library of Congress

ISBN 978-1-009-44505-4 Hardback

Dedicated with love to our partners:

Anita B. Bosky
Michele M. MacFarlane

with thanks for their support and patience.

Contents

Preface

We reiterate that our objective is to introduce the interested student to the techniques, results, and terminology of multiplicative number theory. Whilst it is not intended that our discussion will always reach the research frontier, it is hoped that the material here will prepare the student for tackling the more advanced research literature. As far as possible, the topics of this volume are either self-contained or build on material from Volume I. We continue to assume that the reader has some acquaintance with the fundamentals of elementary number theory, abstract algebra, measure theory, complex analysis, and classical harmonic analysis. More specialized or advanced background material in analysis is provided in the appendices. It should be noted that as we build on the earlier volume and develop the more advanced material, there is often also increased complexity of detail and this requires greater stamina in the reader. The average chapter length in this volume is about 50 pages, compared with 30 or so for Volume I.

The relationship of exercises to the material developed in a given section varies widely. Some exercises are designed to illustrate the theory directly, whilst others are intended to give some idea of the ways in which the theory can be extended, or developed, or paralleled in other areas. The reader is cautioned that papers cited in exercises do not necessarily contain a solution.

The years since our first volume appeared have witnessed many developments, especially in sieves and gaps between primes, and very recently on large values of Dirichlet polynomials and zero density estimates. As happened with the first volume, we again have too much material for one volume, so we are emphasizing sieves in this volume, and postpone such topics as Vinogradov's method of exponential sums, the wider zero-free region for the zeta function, mean and large values of Dirichlet polynomials, zero density theorems, Linnik's theorem, probabilistic number theory, and pair correlation of zeta zeros for the next volume.

While it is to be expected that we will be building on the first volume, there are three topics that might have appeared minor there but which, perhaps rather surprisingly, make significant appearances in Volume II. (1) The Ramanujan sum, as discussed in §4.1, will turn up repeatedly. (2) The function $\psi(x, y)$, which counts the integers $n \le x$ all of whose prime factors are $\le y$ was discussed in §7.1, where we found that it is asymptotic to $\rho(u)x$ with $u = (\log x)/\log u$. Here $\rho(u)$ is the Dickman function. (3) The quantity $\Phi(x, y)$, which is defined to be the number of integers $n \le x$ all of whose prime factors are $\ge y$. In §7.2 we found that $\Phi(x, y) \sim (w(u)x - y)/\log y$ when u is bounded. Here $w(u)$ is the Buchstab function.

Many people have assisted us in this work, including P. T. Bateman, E. Bombieri, T. Chan, J. B. Conrey, H. G. Diamond, T. Estermann, J. B. Friedlander, S. M. Gonek, S. W. Graham, A. Granville, H. Halberstam, D. R. Heath-Brown, H. Iwaniec, H. Maier, G. G. Martin, D. W. Masser, A. M. Odlyzko, G. Peng, C. Pomerance, H.-E. Richert, K. Soundararajan, and U. M. A. Vorhauer. In particular, our doctoral students, and their students also, have been most helpful in detecting errors of all types. We are grateful to them all. We would be most happy to hear from any reader who detects a misprint, or might suggest improvements.

Finally, we thank our loved ones and friends for their long-term support, David Tranah at Cambridge University Press for his encouragement and forbearance, and all who have helped with the production of this volume, including Sam Chow, Ayla Gafni, Steve Gonek, Anna Johansson, and Amita Malik who have helped with the proofreading, and especially the copyeditor Eleanor Bolton for her unbounded patience.

Notation

We continue to use notation defined in Volume I, mostly without repeating mention of them here. Some symbols are used in more than one way. The intended interpretation should be clear from the context in which it arises.

Symbol	Meaning
A^*	The adjoint of the matrix A. See page 398.
$B(z)$	Beurling's function. See page 338.
$C(\mathbb{T})$	The set of continuous functions with period 1. See §E.1.
$c_q(n)$	Ramanujan's sum. See Theorem 4.1.
$D(N,\alpha)$	$= Z(N,\alpha) - N\alpha$. Discrepancy. See page 373.
$D^\star(N)$	Discrepancy. See page 373.
$D(N)$	Discrepancy. See page 377.
$\deg P$	The degree of the polynomial P.
$e(x)$	$= e^{2\pi i x}$; the complex exponential with period 1. See page 335.
E^σ	Entire functions of exponential type σ. See page 338.
$E_0(\chi)$	$= 1$ if $\chi = \chi_0$, $= 0$ otherwise. See page 19.
$\widehat{f}(t)$	The Fourier transform of f. See page 335.
$n_2(p)$	The least positive quadratic non-residue of p. See page 161.
P_k	An almost prime; i.e., a product of at most k primes. See page 182.
$s(x)$	The sawtooth function. See page 378.
$s(n)$	Sum of the binary digits of n. See page 86.
$S_\pm(x)$	Selberg's functions. See page 341.
$\mathrm{si}(x)$	The sine integral. See page 354.
$\mathrm{sgn}(x)$	$= x/\lvert x\rvert$ for $x \neq 0$; $\mathrm{sgn}(0) = 0$. The sign or signum function.
$\left\{ n \atop k \right\}$	Stirling number of the second kind. See page 142.

Symbol	Meaning
$w(u)$	The Buchstab function, used to approximate $\Phi(x, y)$. See §7.2.
$Z(N, \alpha)$	The number of n, $1 \le n \le N$, such that $\{u_n\} \le \alpha$. See page 369.
δ	Dirac delta measure. See page 371.
$\Delta(x)$	The error term in the Dirichlet Divisor Problem. See page 45.
$\Delta(n)$	Hooley's function. See (21.125).
$\Delta_N(x)$	The Fejér kernel. See (E.2).
λ	Lebesgue measure. See page 371.
$\nu(A)$	Numerical radius of the square matrix A. See page 403.
$\rho(A)$	Spectral radius of the square matrix A. See page 403.
$\rho(y)$	$= \limsup_{x\to\infty} \pi(x + y) - \pi(x)$. See page 133.
$\overline{\rho}(N)$	$= \max_M \sum^{M+N}_{\substack{n=M+1 \\ p\mid n \implies p>N}} 1$. See page 133.
$\rho(u)$	The Dickman function, used to estimate $\psi(x, y)$. See §7.1.
$\sigma_N(x)$	A Cesàro partial sum of a Fourier series. See page 335.
$\Phi(x, y)$	The number of $n \le x$ composed entirely of primes $p \ge y$. See §7.2.
$\sum^\star_\chi$	A sum over primitive characters modulo q. See page 174.
$\psi(x, y)$	The number of $n \le x$ composed entirely of primes $p \le y$. See §7.1.
$\psi'(x,\chi) = \psi(x,\chi) - E_0(\chi)x$	See §20.2, page 193.
$\mathsf{X}_{i=1}^n S_i$	A Cartesian product of sets. See page 382.
$\lfloor x \rfloor$	The floor of x, which is the unique integer n such that $n \le x < n + 1$; formerly denoted by $[x]$.
$\lceil x \rceil$	The ceiling of x, which is the unique integer n such that $n - 1 < x \le n$.
$\widehat{f}(n)$	A Fourier coefficient of f. See page 335.
$\widehat{f}(t)$	The Fourier transform of f. See page 337.
$\Vert \boldsymbol{x} \Vert$	Norm of the vector $\boldsymbol{x}$. See page 397.
$\Vert \alpha \Vert$	$= \min_{n\in\mathbb{Z}} \lvert \alpha - n \rvert$. See page 5.
$\Vert A \Vert$	The operator norm of the matrix A. See page 397.

16

Exponential Sums I: Van der Corput's Method

16.1 Exponential Integrals

We are interested in non-trivial bounds for sums of the form

$$\sum_{n=1}^{N} e(f(n)),$$

where $f(x)$ is a smooth, real-valued function. In this chapter, we develop methods whereby one may show that such a sum is indeed $o(N)$. The quality of the results depends on the finer properties of f. In some simple cases the estimates are best possible, but in most situations the bounds we achieve fall far short of what we suppose to be the truth. We begin with the simpler continuous analogue. This provides motivation, and the results we obtain are also useful when dealing with the discrete case.

We seek bounds for integrals of the form $\int_a^b r(t)e^{i\theta(t)}\,dt$ in terms of the behaviour of the real-valued functions $r(t)$ and $\theta(t)$. We begin by generalizing the obvious inequality

$$\left|\int_a^b e^{i\alpha t}\,dt\right| \le \min\left(b-a, \frac{2}{|\alpha|}\right). \tag{16.1}$$

Theorem 16.1 *Let $r(t)$ and $\theta(t)$ be real-valued functions on $[a, b]$ for which $r(t)$ is continuous on $[a, b]$, $\theta(t)$ is differentiable on $[a, b]$ (where if necessary we take the right and left derivatives at a and b, respectively), $\theta'(t)$ is continuous on $[a, b]$, and $\theta'(t) \ne 0$. Suppose that λ satisfies $Var_{[a,b]} r(t)/\theta'(t) \le 2\lambda$ and $|r(t)/\theta'(t)| \le \lambda$ when $a \le t \le b$. Then*

$$\left|\int_a^b r(t)e^{i\theta(t)}\,dt\right| \le 4\lambda.$$

In many interesting cases $r(t)/\theta'(t)$ is monotonic, and then the bound on r/θ' implies the bound on the variation.

Proof Let $\rho(t) = r(t)/\theta'(t)$. We integrate by parts, using the Riemann–Stieltjes integral as developed in Appendix A. Thus

$$\begin{aligned}\int_a^b r(t)e^{i\theta(t)}\,dt &= -i\int_a^b \rho(t)de^{i\theta(t)} \\ &= \Big[-i\rho(t)e^{i\theta(t)}\Big|_a^b + i\int_a^b e^{i\theta(t)}\,d\rho(t).\end{aligned} \tag{16.2}$$

Hence

$$\left|\int_a^b r(t)e^{i\theta(t)}\,dt\right| \le |\rho(a)| + |\rho(b)| + \int_a^b |d\rho(t)| \le 4\lambda.$$

□

It is instructive to view the above argument geometrically. When $a \le t \le b$, let $Z(t) = \int_a^t r(u)e^{i\theta(u)}\,du$. These points describe a curve in the complex plane with tangent vector $r(t)e^{i\theta(t)}$. Thus $Z(t)$ is moving with speed $|r(t)|$, and the argument of the tangent vector is changing at a rate $\theta'(t)$. Hence the curve has curvature $\kappa = |\theta'(t)/r(t)|$. Consequently, the radius of curvature at time t is $|\rho(t)|$, and $C(t) = Z(t) + i\rho(t)e^{i\theta(t)}$ is the centre of the osculating circle. One may reach $Z(b)$ from the origin by following the path $Z(t)$. Alternatively, to reach $Z(b)$ one may first move along the line segment from 0 to $C(a)$, then follow the path $C(t)$ to $C(b)$, and finally pass along the line segment from $C(b)$ to $Z(b)$. These two alternatives are expressed in the identity (16.2). When $\rho(t)$ is differentiable we find that $C'(t) = i\rho'(t)e^{i\theta(t)}$. Thus the tangent vector $C'(t)$ to the curve $C(t)$ is at all times perpendicular to the tangent vector $Z'(t)$ to the curve $Z(t)$, and $C(t)$ moves with a speed equal to the rate of change of the radius of curvature. Suppose for simplicity that $\rho(t)$ is positive and decreasing. Then the curve $Z(t)$ spirals inward, in the sense that the osculating circles are nested. To see this, observe that if $a \le t_1 \le t_2 \le b$, then

$$\big|C(t_1) - C(t_2)\big| = \left|i\int_{t_1}^{t_2} e^{i\theta(t)}\,d\rho(t)\right| \le \int_{t_1}^{t_2} |d\rho(t)| = \rho(t_1) - \rho(t_2).$$

In particular, the circle with centre $C(a)$ and radius $\rho(a)$ passes through the point $Z(a) = 0$, whilst $Z(b)$ falls within the circle. Hence $|Z(b)| \le 2\rho(a)$ in this case.

In many cases we do not need the full generality of Theorem 16.1, and the following special case suffices.

Corollary 16.2 *Let $r(t)$ and $\theta(t)$ be real-valued functions on $[a, b]$ for which $r(t)$ is continuous on $[a, b]$, $\theta(t)$ is differentiable on $[a, b]$ (where, if necessary,* *we take the right and left derivatives at a and b, respectively),* $\theta'(t)$ *is continuous on* $[a,b]$*, and* $\theta'(t) \neq 0$*. Put* $\rho(t) = r(t)/\theta'(t)$*. If* ρ *is monotonic and* λ *is a number such that* $-\lambda \le \rho(t) \le \lambda$ *for* $a \le t \le b$*, then*

$$\left| \int_a^b r(t)e^{i\theta(t)}\,dt \right| \le 4\lambda.$$

If $\theta'(t)$ vanishes at some point of the interval $[a,b]$, then Theorem 16.1 does not apply, but we can still obtain a bound when $\theta''(t)$ exists and is not too small.

Theorem 16.3 *Suppose that* $r(t)$ *and* $\theta(t)$ *are real valued and continuous on* $[a,b]$*, that* $0 < r(t) \le M$*, that* $\theta(t)$ *is twice differentiable on* $[a,b]$ *(where if necessary we take the right and left derivatives at a and b, respectively), that* $\theta'(t)/r(t)$ *is monotonic, and that* $0 < \mu \le \theta''(t)$ *when* $a \le t \le b$*. Then*

$$\left| \int_a^b r(t)e^{i\theta(t)}\,dt \right| \le \frac{8M}{\sqrt{\mu}}.$$

Theorem 16.3 often suffices in applications. If necessary, a more precise approximation can be derived, say via the more elaborate Theorem 16.19 below. However, generally the above bound is of the correct order of magnitude. For example, in the case $r(t) \equiv 1$ and $\theta(t) = ct^2$ with $c > 0$ we have $\theta''(t) = 2c$ and

$$\int_{-\infty}^{\infty} e^{ict^2}\,dt = e(1/8)\sqrt{\pi/c}. \tag{16.3}$$

(A proof of this is outlined in Exercise 9.3.1.5.) If we were to apply Theorem 16.3 to the integral above, we would find that it is $\ll 1/\sqrt{c}$, which is to say we would obtain a bound of the correct order of magnitude. In Figure 16.1 we depict the curve $Z(t) = \int_{-\infty}^{t} e^{iu^2}\,du$, which spirals tightly except near the inflection point at $t = 0$.

Proof Let $\delta > 0$ be a parameter at our disposal. Since $\theta''(t) > 0$, we know that $\theta'(t)$ is increasing, and hence if there are t for which $|\theta'(t)| \le \delta\mu$, then such t comprise an interval, say I_0. If I_0 is a proper subinterval of $[a,b]$, then the complement of I_0 consists of one or two intervals, say $I_{\pm 1}$. The length of I_0 is at most 2δ since $\theta''(t) \ge \mu$. Hence

$$\left| \int_{I_0} r(t)e^{i\theta(t)}\,dt \right| \le 2M\delta.$$

For $t \in I_{\pm 1}$ we have $|\theta'(t)| \ge \delta\mu$. Thus, by Theorem 16.1 with $\lambda = M\mu^{-1}\delta^{-1}$, we deduce that

$$\left| \int_{I_{\pm 1}} r(t)e^{i\theta(t)}\,dt \right| \le \frac{4M}{\delta\mu}.$$

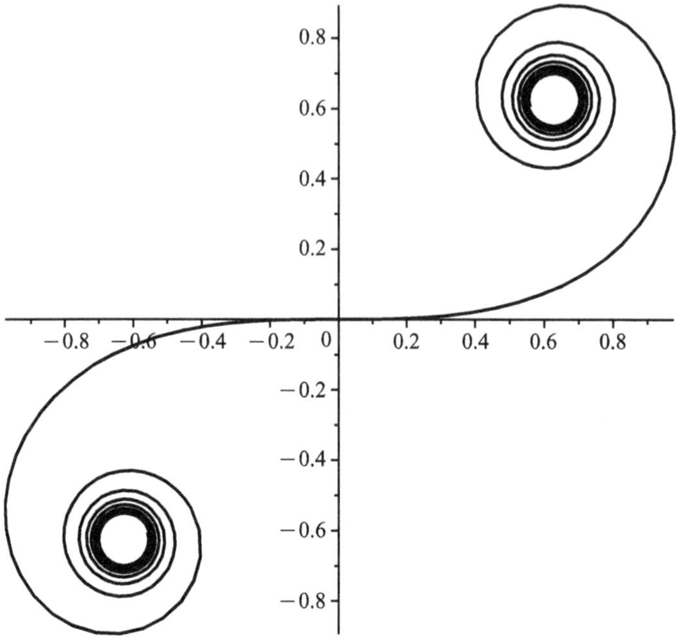

Figure 16.1 Graph of $z(t) = \int_0^t e^{iu^2}\,du$ for $-7 \le t \le 7$.

Hence altogether,

$$\left|\int_a^b r(t)e^{i\theta(t)}\,dt\right| \le 2M\delta + \frac{8M}{\delta\mu},$$

and the desired bound follows on taking $\delta = 2\mu^{-1/2}$. □

16.1.1 Exercises

1. Suppose that $k \ge 2$, that $f\colon [a,b] \to \mathbb{R}$ is k times differentiable on $[a,b]$, and that there is a positive number λ_k such that for each x in (a,b) we have $f^{(k)}(x) \ge \lambda_k$. Show that

$$\left|\int_a^b e(f(x))\,dx\right| \le k2^k\lambda_k^{-1/k}.$$

2. Suppose that $\alpha_1, \alpha_2, \ldots, \alpha_k$ are real and let

$$I(t;\boldsymbol{\alpha}) = \int_0^t e(\alpha_1 u + \alpha_2 u^2 + \cdots + \alpha_k u^k)\, du.$$

Show that for any positive number t,

$$I(t;\boldsymbol{\alpha}) \ll \frac{t}{(1 + |\alpha_1|t + |\alpha_2|t^2 + \cdots + |\alpha_k|t^k)^{1/k}}.$$

3. (Talmage 2022) Suppose that $k \ge 2$ and $\theta_1, \ldots, \theta_k$, β, and γ are real numbers with $\theta_k\gamma \ne 0$, $(k+1)/(k+2) \le \beta \le 1$, and write $\rho = \beta + i\gamma$. Suppose further that X is a real number with $X \ge 1$, and put

$$I(X;\boldsymbol{\theta},\rho) = \int_0^X e(\theta_1 t + \cdots + \theta_k t^k) t^{\rho-1} dt.$$

Show that

$$I(X;\boldsymbol{\theta},\rho) \ll \frac{X^\beta}{(1 + X|\theta_1| + \cdots + X^k|\theta_k| + |\gamma|)^{1/(k+1)}}.$$

16.2 Elementary Estimates

We now derive discrete analogues of the estimates of §16.1. Corresponding to the estimate (16.1) we have the following lemma.

Lemma 16.4 *Let $\|\alpha\|$ denote the distance from the real number α to the nearest integer, $\|\alpha\| = \min_{n\in\mathbb{Z}} |\alpha - n|$. Then*

$$\left|\sum_{n=1}^{N} e(n\alpha)\right| \le \min\left(N, \frac{1}{2\|\alpha\|}\right). \tag{16.4}$$

Proof The sum has N summands, each of them unimodular, so by the triangle inequality we see that N is an upper bound for the modulus of the sum. Now suppose that α is not an integer. Then $e(\alpha) \ne 1$, so by the formula for the sum of a geometric progression we see that the modulus of the sum is

$$\left|\frac{e((N+1)\alpha) - e(\alpha)}{e(\alpha) - 1}\right| \le \frac{2}{|e(\alpha) - 1|} = \frac{1}{|\sin \pi\alpha|} \le \frac{1}{2\|\alpha\|}.$$

□

As an analogue of Theorem 16.1 we have the following.

Theorem 16.5 (Kusmin–Landau) *Let $\alpha_1, \alpha_2, \ldots, \alpha_N$ be real numbers, and for $1 \le n < N$ put $\delta_n = \alpha_{n+1} - \alpha_n$. Suppose that Δ is a positive real number and that $\Delta \le \delta_1 \le \delta_2 \le \cdots \le \delta_{N-1} \le 1 - \Delta < 1$. Then*

$$\left| \sum_{n=1}^{N} e(\alpha_n) \right| \le \cot \frac{\pi\Delta}{2}.$$

Proof Let $z_n = e(\alpha_n)$, $w_n = z_{n+1}/z_n = e(\delta_n)$, and $\rho_n = 1/(1 - w_n)$. Then

$$\sum_{n=1}^{N} e(\alpha_n) = \sum_{n=1}^{N-1} \rho_n(z_n - z_{n+1}) + z_N.$$

Hence, by partial summation,

$$\sum_{n=1}^{N} e(\alpha_n) = \rho_1 z_1 + \sum_{n=2}^{N-1} (\rho_n - \rho_{n-1}) z_n + (1 - \rho_{N-1}) z_N. \tag{16.5}$$

Therefore, by the triangle inequality

$$\left| \sum_{n=1}^{N} e(\alpha_n) \right| \le |\rho_1| + \sum_{n=2}^{N-1} |\rho_n - \rho_{n-1}| + |1 - \rho_{N-1}|.$$

If $\rho = 1/(1 - w)$ and $w = e(\delta)$ with $0 < \delta < 1$, then $\rho = (1 + i \cot \pi\delta)/2$ and $|\rho| = |1 - \rho| = 1/(2 \sin \pi\delta)$. Hence the left hand side of the above is

$$\begin{aligned}
&\le \frac{1}{2 \sin \pi\delta_1} + \frac{1}{2} \sum_{n=2}^{N-1} \left(\cot \pi\delta_{n-1} - \cot \pi\delta_n \right) + \frac{1}{2 \sin \pi\delta_{N-1}} \\
&= \frac{1}{2} \left(\frac{1}{\sin \pi\delta_1} + \frac{1}{\tan \pi\delta_1} - \frac{1}{\tan \pi\delta_{N-1}} + \frac{1}{\sin \pi\delta_{N-1}} \right) \\
&\le \frac{1}{\sin \pi\Delta} + \frac{1}{\tan \pi\Delta} \\
&= \cot \frac{\pi\Delta}{2},
\end{aligned}$$

and the proof is complete. □

The above argument can be interpreted geometrically as follows. Let s_n denote the nth partial sum of the sum in Theorem 16.5, and for $1 < n < N$, put $c_n = s_n + z_{n+1}\rho_n$. Then $c_n = s_{n-1} + z_n\rho_n = s_{n+1} - z_{n+1}(1 - \rho_n)$. Thus c_n is the centre of the circle that passes through the three points s_{n-1}, s_n, s_{n+1}, and the radius of this circle is $|\rho_n|$. Hence ρ_n corresponds to the function $\rho(t)$ introduced in the proof of Theorem 16.1. One may construct a polygonal path from 0 to s_N whose vertices are the partial sums s_n. Alternatively, we may construct such a path that goes from 0 to c_1, then to c_2, and so on, and finally from c_{N-1} to s_N. This suggests writing s_N as a telescoping sum

$$s_N = c_1 + \sum_{n=2}^{N-1} (c_n - c_{n-1}) + (s_N - c_{N-1}).$$

Since $c_n - c_{n-1} = z_n(\rho_n - \rho_{n-1})$, this is precisely the identity (16.5).

In most applications, the α_n are values of a function with continuous derivatives, as follows.

Corollary 16.6 *Let $f(x)$ be a real-valued function continuous on $[a, b]$, with a continuous derivative on (a, b), and such that $f'(x)$ is increasing. Suppose further that M_1 is a positive real number such that $\|f'(x)\| \geq M_1$ for all $x \in (a, b)$. Then*

$$\Big| \sum_{a \leq n \leq b} e(f(n)) \Big| \leq \frac{2}{\pi M_1}.$$

Proof If there were an integer k such that for some x and y in (a, b) with $x \neq y$ we had $f'(x) \leq k \leq f'(y)$, it would follow from the intermediate value theorem that there is a $z \in (a, b)$ such that $f'(z) = k$ and $\|f'(z)\| = 0$. Hence we may suppose that there is an integer k such that for every $x \in (a, b)$ we have $k < f'(x) < k + 1$ and so $k + M_1 \leq f'(x) \leq k + 1 - M_1$. If we replace $f(x)$ by $f(x) - kx$, then as $kn \in \mathbb{Z}$ the sum is unchanged and $M_1 \leq f'(x) \leq 1 - M_1$, which allows us to apply Theorem 16.5 with $\alpha_n = f(n)$ and $\Delta = M_1$. By the mean value theorem for derivatives we know that if $[n, n + 1] \subseteq [a, b]$, then there is a $\xi_n \in (n, n + 1)$ such that $\delta_n = f(n + 1) - f(n) = f'(\xi_n)$. Thus the hypotheses of Theorem 16.5 are satisfied and it remains only to note that $\cot u < 1/u$ when $0 < u \leq \pi/2$. □

The bounds provided in Theorem 16.5 and Corollary 16.6 are quite sharp (see Exercise 16.2.1.1). The partial sums spiral tightly in intervals in which $\|f'(x)\|$ is large, but the terms tend to pull in one direction when $f'(x)$ is near an integer. For example, consider $f(x) = x^2/1600$ with $a = 0$, $b = 800$. Then $f'(a) = 0$ and $f'(b) = 1$, but $f'(x)$ is increasing and $\|f'(x)\| \geq 1/50$ when $16 \leq x \leq 784$, so that

$$\left| \sum_{n=16}^{784} e\left(\frac{n^2}{1600}\right) \right| \leq \frac{100}{\pi} < 31.831.$$

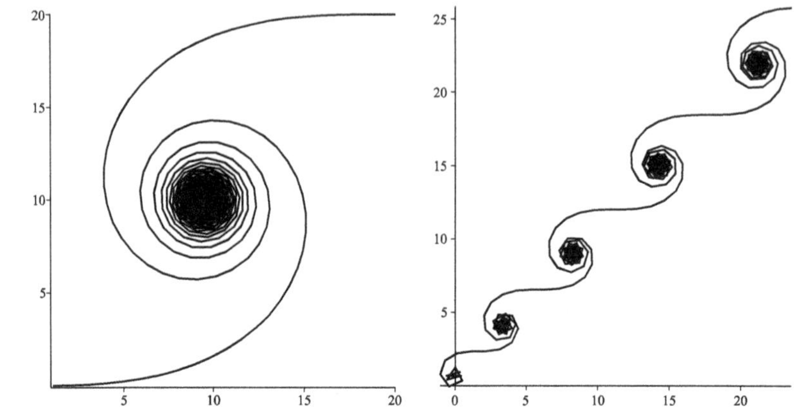

(a) (b)

Figure 16.2 (a) $\sum_{n=1}^{800} e(n^2/1600)$; (b) $\sum_{n=1}^{300} e((n/3)^{3/2}) = 25.56 + 25.81i$.

By combining this with the trivial bound for the contribution of the first 15 and last 16 terms we find that

$$\left|\sum_{n=1}^{800} e\left(\frac{n^2}{1600}\right)\right| < 62.831.$$

The exact value of this sum is $20 + 20i$, as we see from Corollary 9.16.

In general when $f'(b) - f'(a)$ is large but $f''(x)$ is small we may obtain a useful bound by treating separately the subintervals in which $\|f'(x)\|$ is small or large.

Theorem 16.7 *Let N be a positive integer with $a \le b \le a + N$ and suppose that f is twice differentiable on $[a, b]$ and that when $a \le x \le b$ we have $0 < M_2 \le f''(x) \le AM_2$ for some real numbers A and M_2. Then*

$$\sum_{a \le n \le b} e(f(n)) \ll_A M_2^{1/2} N + M_2^{-1/2}.$$

If instead we have $-AM_2 \le f''(x) \le -M_2$, then the same bound applies, as we see by taking complex conjugates. This remark also applies later to the corresponding derivatives in Theorems 16.11, 16.12, and 16.20 and Corollary 16.21.

If $M_2 \ge 1$, then the bound given by Theorem 16.7 is trivial, as it must be, since $f(x)$ may be increasing so rapidly that all the numbers $f(n)$ are integers (consider the case $f(x) = x(x + 1)/2$). If $M_2 \le N^{-2}$, then again the bound is trivial, because $f(x)$ may be essentially constant throughout the interval in question (here consider $f(x) = (x/(2N))^2$ on the interval $[a, b] = [0, N]$). If

$N^{-2} \le M_2 \le N^{-1}$, then the bound provided is likely to be of the correct order of magnitude, unless Theorem 16.5 is applicable. If $N^{-1} \le M_2 \le 1$, then it may be possible to obtain a sharper estimate by using Theorem 16.11. We could estimate how the implicit constant depends on A, but in practice one should cut the interval into subintervals so that A is bounded in each application. For example, suppose that we wish to estimate

$$\sum_{n=1}^{N} e\big((n/3)^{3/2}\big). \tag{16.6}$$

We take $f(x) = (x/3)^{3/2}$, and note that we may take $M_2 \asymp a^{1/2}$ and $A \ll 1$ when $a \le x \le b \le 2a$. Then Theorem 16.5 gives the estimate

$$\sum_{a<n<b} e\big((n/3)^{3/2}\big) \ll a^{3/4}.$$

On summing over dyadic blocks, we deduce that the sum in (16.6) is $\ll N^{3/4}$, which is best possible (see Exercise 16.3.1.3). In Figure 16.2(b) one may note that the partial sums resemble a number of copies of the curve in Figure 16.1, one for each solution of $f'(x) \in \mathbb{Z}$. If $f'(x_\nu) = \nu \in \mathbb{Z}$, then we obtain a copy of the curve of Figure 16.1, scaled by a factor $\asymp f''(x_\nu)^{-1/2}$, and rotated by $2\pi(f(x_\nu) - \nu x_\nu)$. In the case under consideration we find that $x_\nu = 12\nu^2$, and hence $f(x_\nu) - \nu x_\nu = -4\nu^3 \in \mathbb{Z}$, so that these contributions all pull in the same direction. More typically, in general the $f(x_\nu)$ are not integers and one is led to consider a new exponential sum of the form $\sum_\nu e(f(x_\nu) - \nu x_\nu)$. The transformation from the original sum to this new sum is achieved by means of an analytic technique that we develop in the next section.

Proof of Theorem 16.7 We have already noted that the bound is trivial when $M_2 \gg 1$. Thus we suppose that $M_2 \le 1/4$. Since $f'(b) - f'(a) = (b - a)f''(\xi) \le AM_2(b-a)$ and f' is increasing, we see that the interval $f'([a,b])$ contains $\ll AM_2(b-a) + 1$ integers. Let λ be a positive parameter at our disposal. Then the set of $x \in [a,b]$ such that $\|f'(x)\| \ge \lambda$ can be partitioned into at most $\ll AM_2(b-a)+1$ intervals and, likewise, so can the set $x \in [a,b]$ such that $\|f'(x)\| < \lambda$ and in the latter case each interval is of length at most $\ll \lambda M_2^{-1}$. By Corollary 16.6, the contribution to the sum from the terms with n in a subinterval of the first kind is $\ll \lambda^{-1}$, and trivially the contribution from such n in a subinterval of the second kind is $\ll \lambda M_2^{-1} + 1$. Hence the sum in question is

$$\ll_A (M_2 N + 1)\big(\lambda^{-1} + \lambda M_2^{-1} + 1\big),$$

and the choice $\lambda = M_2^{1/2}$ gives the stated bound. □

As a further application of Theorem 16.7, we consider the trigonometric polynomial

$$P(\alpha) = \sum_{n=1}^{N} e(n \log n + n\alpha). \tag{16.7}$$

Take $f(x) = x \log x + \alpha x$. Then we find that $f''(x) = 1/x$, and Theorem 16.7 gives the estimate

$$\sum_{a \le n \le b} e(n \log n + n\alpha) \ll a^{1/2}$$

when $a \le b \le 2a$. On summing over dyadic blocks we deduce that

$$P(\alpha) \ll N^{1/2}. \tag{16.8}$$

This is best possible, at least for some α, since by Parseval's identity we have

$$\int_0^1 |P(\alpha)|^2 \, d\alpha = N.$$

Thus $P(\alpha)$ is an example of a trigonometric polynomial with unimodular coefficients and such that $\|P\|_2 \asymp \|P\|_\infty$.

We have noted that Corollary 16.6 is useless when $f'(x)$ is large, but that Theorem 16.7 provides a substitute when $f''(x)$ is small. If $f''(x)$ is large, then Theorem 16.7 is useless, but we may still obtain non-trivial estimates if $f'''(x)$ (or some higher derivative) is small. To derive bounds that depend on higher derivatives we introduce an important new idea.

Lemma 16.8 (van der Corput) *Let $z_1, z_2, \ldots, z_N$ be arbitrary complex numbers. Then for any integer H with $1 \le H \le N$ we have*

$$H^2 \left| \sum_{n=1}^{N} z_n \right|^2 \le H(N + H - 1) \sum_{n=1}^{N} |z_n|^2 + 2(N + H - 1) \sum_{h=1}^{H-1} (H - h) \left| \sum_{n=1}^{N-h} z_{n+h} \bar{z}_n \right|.$$

Proof To simplify the ranges of summation, we suppose that $z_n = 0$ when $n < 1$ or $n > N$. Then

$$H \sum_{n=1}^{N} z_n = \sum_{0 \le r < H} \sum_{0 < n < N+H} z_{n-r} = \sum_{0 < n < N+H} \sum_{0 \le r < H} z_{n-r}.$$

Hence, by Cauchy's inequality we see that

$$H^2 \left| \sum_{n=1}^{N} z_n \right|^2 \le (N + H - 1) \sum_{0 < n < N+H} \left| \sum_{0 \le r < H} z_{n-r} \right|^2. \tag{16.9}$$

On multiplying out the square, and inverting the order of summation, we see that this is

$$= (N + H - 1) \sum_{\substack{0 \le r < H \\ 0 \le s < H}} \sum_n z_{n-r}\overline{z_{n-s}}.$$

The inner sum depends only on $r - s$, and a given value h of $r - s$ occurs for $H - |h|$ different pairs r, s. Thus the above is equal to

$$H(N + H - 1) \sum_{n=1}^{N} |z_n|^2 + 2(N + H - 1) \operatorname{Re} \sum_{h=1}^{H-1} (H - h) \sum_n z_{n+h}\overline{z_n}, \quad (16.10)$$

and the desired result now follows. □

In applications, it is likely that some cancellation has been discarded when Cauchy's inequality is applied. That is, there may be some loss in the inequality (16.9). Similarly, in passing from (16.10) to the final result by means of the triangle inequality, some further cancellation may have been lost.

The van der Corput lemma has an immediate application to Weyl's criterion (Theorem F.1) concerning the distribution of a sequence u_n modulo 1, as discussed in §F.1: We take $z_n = e(ku_n)$ in the above, and thus find that

$$\sum_{n=1}^{N} e(ku_n) \ll NH^{-1/2} + N^{1/2} \left(\frac{1}{H} \sum_{h=1}^{H} \left| \sum_{n=1}^{N-h} e(k(u_{n+h} - u_n)) \right| \right)^{1/2}. \quad (16.11)$$

Suppose that the sequence $u_{n+h} - u_n$ is uniformly distributed, for each fixed positive h. Then by Weyl's criterion (Theorem F.1) the inner sum over n on the right-hand side is $o(N)$. Hence the entire term containing this sum is $o(N)$. Since H may be taken to be arbitrarily large, it follows that

$$\sum_{n=1}^{N} e(ku_n) = o(N)$$

as $N \to \infty$, for any fixed non-zero integer k. Thus, by a second application of Weyl's criterion we have

Theorem 16.9 (van der Corput) *Let $\{u_n\}$ be a sequence of real numbers with the property that, for each positive integer h, the sequence $\{u_{n+h} - u_n\}$ is uniformly distributed. Then the sequence $\{u_n\}$ is uniformly distributed.*

From the example $u_n = n\theta$ with θ irrational, we see that the converse is false.

Corollary 16.10 (Weyl) *Let $P(x) = \sum c_j x^j$ be a polynomial with real coefficients. If there is a $j > 0$ for which the coefficient c_j is irrational, then the sequence $\{P(n)\}$ is uniformly distributed modulo 1.*

The constant term c_0 may be rational or irrational since it only causes the fractional parts to be translated. The converse is obvious, for if the coefficients c_j were to be rational for all $j > 0$, then the sequence $\{P(n)\}$ would be periodic and then the numbers $P(n)$ would not even be dense in $\mathbb{T}$.

Proof We first prove the assertion by induction under the stronger hypothesis that the leading coefficient is irrational. If $\deg P = 1$, then the result follows by Theorem F.2. If $\deg P = d > 1$ and the leading coefficient c_d is irrational, then for any positive integer h the polynomial $P(x+h) - P(x)$ has an irrational leading coefficient hdc_d. Hence the numbers $P(n+h) - P(n)$ are uniformly distributed by the inductive hypothesis. This establishes the result when the leading coefficient is irrational.

Now suppose that $P(x)$ has an irrational coefficient (other than the constant term), which may or may not be the leading coefficient. Write $P(x) = P_i(x) + P_r(x)/q$, where all the non-zero coefficients of P_i are irrational and all the coefficients of P_r are integers. Then $P_i(x)$ has positive degree. Moreover, for any integer a the polynomial $P_i(qx + a)$ has positive degree and an irrational leading coefficient. Hence the sequence $P_i(qn + a)$ is uniformly distributed. On the other hand, the sequence $P_r(qn + a)/q$ is constant modulo 1. Hence the sequences $P(qn + a)$ are uniformly distributed. It follows at once from the definition of uniform distribution that the sequence $P(n)$ is also uniformly distributed. □

We now use the van der Corput lemma (Lemma 16.8) to derive bounds for the exponential sum $\sum e(f(n))$ that depend on higher derivatives of f.

Theorem 16.11 *Let N be a positive integer and suppose that $a \le b \le a + N$ and $0 < M_3 \le f'''(x) \le AM_3$ when $a \le x \le b$. Then*

$$\sum_{a\le n\le b} e(f(n)) \ll_A N\big(M_3^{1/6} + N^{-1/4} + N^{-3/4}M_3^{-1/4}\big).$$

If $M_3 < N^{-3}$ or $M_3 > 1$, then the bound is trivial, for then the second factor on the right is larger than 1. Of the three terms in parentheses on the right, we see that the first one is largest when $N^{-3/2} \le M_3 \le 1$, the second is largest when $N^{-2} \le M_3 \le N^{-3/2}$, and the third is largest when $N^{-3} \le M_3 \le N^{-2}$.

Proof In view of the remarks above, we may suppose that $N^{-3} \le M_3 \le 1$. Suppose that $0 < h \le b - a$, and let $f_h(x) = f(x+h) - f(x)$ for $a \le x \le b - h$. By the van der Corput lemma we see that

$$\sum_{a\le n\le b} e(f(n)) \ll NH^{-1/2} + N^{1/2}\left(\frac{1}{H}\sum_{h=1}^{H}\left|\sum_{n} e(f_h(n))\right|\right)^{1/2}. \tag{16.12}$$

Since $f_h''(x) = f''(x+h) - f''(x) = hf'''(\xi) \asymp hM_3$, it follows from Theorem 16.7 that the inner sum is $\ll_A h^{1/2}M_3^{1/2}N + h^{-1/2}M_3^{-1/2}$. On inserting this estimate, we obtain a bound

$$\ll_A NH^{-1/2} + M_3^{1/4}H^{1/4}N + M_3^{-1/4}H^{-1/4}N^{1/2}.$$

If $N^{-3/2} \le M_3 \le 1$, then we take $H = \left[M_3^{-1/3}\right]$, and the first two terms are the same size and the third is smaller. If $N^{-2} \le M_3 \le N^{-3/2}$, then we take $H = \left[M_3^{-1}N^{-1}\right]$, whence the second and third terms are the same size and the first is smaller. In both these cases the chosen value of H satisfies the requirement that $1 \le H \le N$. Finally, if $N^{-3} \le M_3 \le N^{-2}$, then we take $H = N$, and the third term is the largest. □

We note that if the innermost sum on the right in (16.12) is estimated trivially, then the bound obtained for the left-hand side is trivial, but no worse. Consequently, a non-trivial estimate for the inner sum on the right will yield a non-trivial estimate for the sum on the left. Thus the Weyl–van der Corput inequality is a very useful tool, although (as we have already noted) it may be expected to involve some loss of quantitative precision. One may attempt to avoid some of this loss by constructing estimates for two-dimensional exponential sums, i.e., sums of the form $\sum_{h,n} e(f(h,n))$. Such estimates may then be applied to the double sum in (16.10), thereby avoiding the appeal to the triangle inequality in the last step of the proof of the lemma.

If $f'''(x)$ is large, then the estimate of Theorem 16.11 is trivial, but if $f^{(4)}(x)$ is small, then we may still obtain a useful estimate by applying the van der Corput lemma and Theorem 16.11, in the same way that we derived Theorem 16.11 from Theorem 16.7. Continuing by induction, we obtain the following general result, of which Theorems 16.7 and 16.11 are the first two cases.

Theorem 16.12 *Let N be a positive integer, and let r be an integer with $r \ge 2$. Suppose that $a \le b \le a+N$ and that $0 < M_r \le f^{(r)}(x) \le AM_r$ when $a \le x \le b$. Put $R = 2^r$. Then*

$$\sum_{a\le n\le b} e(f(n)) \ll_{A,r} N\big(M_r^{1/(R-2)} + N^{-2/R} + (N^rM_r)^{-2/R}\big).$$

Proof Since we have already established this for $r = 2$ and $r = 3$, we may suppose that $r \ge 4$, and that the estimate has been established for $r - 1$. We may also suppose that $N^{-r} \le M_r \le 1$, for otherwise the bound is trivial. We apply the van der Corput lemma, as in the proof of Theorem 16.11, to obtain the estimate (16.12). As $f_h^{(r-1)}(x) = f^{(r-1)}(x+h) - f^{(r-1)}(x) = hf^{(r)}(\xi) \asymp hM_r$, we deduce from the inductive hypothesis that

$$\sum_{a\le n\le b-h} e(f_h(n)) \ll_{A,r} N\big((hM_r)^{2/(R-4)} + N^{-4/R} + (N^{r-1}hM_r)^{-4/R}\big).$$

Inserting this in (16.12), we find that the sum in question is

$$\ll_{A,r} N\big(H^{-1/2} + (HM_r)^{1/(R-4)} + N^{-2/R} + (N^{r-1}HM_r)^{-2/R}\big).$$

If M_r is not very small, say $N^{-2+4/R} \le M_r \le 1$, then we take $H = \big[M_r^{-2/(R-1)}\big]$. Then the first two terms are the same size, and the remaining terms are smaller. If M_r is extremely small, say $N^{-r} \le M_r \le N^{-r+1}$, then we take $H = N$. Then the last term is largest. In the intermediate range $N^{-r+1} \le M_r \le N^{-2+4/R}$ we have some freedom in our choice of H, because it suffices to choose H so that the first, second, and fourth terms are majorized by the third term. That is, we take H to be an integer such that $H \gg N^{4/R}$, $H \ll M_r^{-1}N^{-2+8/R}$, $H \gg M_r^{-1}N^{2-r}$, and of course $1 \le H \le N$. To complete the proof it suffices to verify that the lower bounds for H are indeed smaller than the upper bounds when M_r is in the interval under consideration. □

We now consider what our estimates yield when they are applied to sums of the form $\sum_{a\le n\le b} n^{-it}$. By Corollary 16.6 with $f(x) = \frac{-t}{2\pi}\log x$, we see that if $\tau \le a \le b \le 2a$, then

$$\sum_{a\le n\le b} n^{-it} \ll \frac{a}{\tau}, \tag{16.13}$$

where t is real. Similarly, by Theorem 16.7 we find that if $\tau^{2/3} \le a \le \tau$ and $a \le b \le 2a$, then

$$\sum_{a\le n\le b} n^{-it} \ll \tau^{1/2}. \tag{16.14}$$

This bound also holds for $\tau^{1/2} \le a \le \tau^{2/3}$, but for such smaller a we obtain a better bound from Theorem 16.11: If $\tau^{1/3} \le a \le \tau^{2/3}$ and $a \le b \le 2a$, then

$$\sum_{a\le n\le 2a} n^{-it} \ll a^{1/2}\tau^{1/6}. \tag{16.15}$$

Further such estimates can be derived for smaller values of a, but they become successively weaker. Our very first estimate, given by (16.13), is the correct order of magnitude, but is flawed because we can derive a much more precise statement about such sums by using the following theorem.

Theorem 16.13 *Suppose that $0 < \delta \le 1/2$, that f is continuous and monotonic on $[a,b]$, and that $-1+\delta \le f'(x) \le 1-\delta$ for $a \le x \le b$. Then*

$$\sum_{a\le n\le b} e(f(n)) = \int_a^b e(f(x))\,dx + O_\delta(1).$$

This is a precursor to the more elaborate Theorem 16.18 (which we shall prove in §16.3) and may be viewed as an instance of a Riemann sum approximation to an integral, but with an error term that is much smaller than would normally be the case due to the special shape of the integrand.

Proof First, assume that f' is increasing. By Riemann–Stieltjes integration we see that the left-hand side above is

$$\int_{a^-}^b e(f(x))\,d\lfloor x\rfloor = \int_a^b e(f(x))\,dx - \int_{a^-}^b e(f(x))\,d\{x\}.$$

Thus our only task is to bound this last integral, which is

$$\begin{aligned}
&= \int_{a^-}^b e(f(x))\,d(\{x\}-1/2)\\
&= [e(f(x))(\{x\}-1/2)|_{a^-}^b - \int_a^b (1/2-\{x\})\,de(f(x))\\
&= 2\pi i \int_a^b (\{x\}-1/2)e(f(x))f'(x)\,dx + O(1). \qquad (16.16)
\end{aligned}$$

In (E.14) we define the sawtooth function $s(x)$ to be $s(x) = \{x\}-1/2$ when $x \notin \mathbb{Z}$, and $s(x) = 0$ when $x \in \mathbb{Z}$ (see also Lemma D.1). Thus we can switch from $\{x\}-1/2$ to $s(x)$ in the above integral without altering its value. In Appendix D we determined the Fourier series of $s(x)$, showed that it is boundedly convergent to $s(x)$, and even established this in a sharp quantitative form

$$s(x) = -\sum_{0<|k|\le K} \frac{e(kx)}{2\pi ik} + O\left(\min\left(1, \frac{1}{K\|x\|}\right)\right). \qquad (16.17)$$

We can now see why the integral in (16.16) is so small: $s(x)$ is essentially a linear combination of functions of the form $e(kx)$, each one of which is turning quite quickly, while $e(f(x))$ is turning comparatively slowly. Thus the product $e(f(x))e(kx)$ is turning at approximately the same speed as $e(kx)$, and so we can estimate the contribution of this term by appealing to Theorem 16.1. We take $r(x) = f'(x)$, $\theta(x) = 2\pi(kx+f(x))$. Thus

$$\frac{r(x)}{\theta'(x)} = \frac{f'(x)}{2\pi(k+f'(x))}. \qquad (16.18)$$

Now it is familiar that a function of the form $\frac{au+b}{cu+d}$ is linear if $c = 0$, but if $c \ne 0$ it has a simple pole at $-d/c$, and is monotonic on both the intervals $(-\infty, -d/c)$, $(-d/c, \infty)$. Moreover, on both these intervals the function is increasing, constant, or decreasing, according to the sign of $ad - bc$. In the present case, the point $-d/c = -k$ lies outside the interval $[-1+\delta, 1-\delta]$ and $ad - bc = 2\pi k$, so if $k \ge 1$ the expression is increasing and lies in the interval

$$\left[\frac{-1+\delta}{2\pi(k-1+\delta)}, \frac{1-\delta}{2\pi(k+1-\delta)}\right].$$

Thus the expression (16.18) has absolute value not exceeding

$$\frac{1-\delta}{2\pi(k-1+\delta)} \le \frac{1}{k-1+\delta} \le \frac{1}{k\delta}.$$

A similar argument applies when $k \le -1$, so by Theorem 16.1 it follows that

$$\int_a^b e(kx + f(x))f'(x)\,dx \ll \frac{1}{|k|\delta}$$

for all non-zero integers k. Thus, when the sawtooth function in (16.16) is replaced by the two terms in (16.17), the first term contributes an amount $\ll \delta^{-1}\sum_{k=1}^{\infty} k^{-2} \ll 1/\delta$. Clearly

$$\int_0^1 \min\left(1, \frac{1}{K\|x\|}\right)\,dx \ll \frac{\log K}{K},$$

so the contribution to (16.16) of the second term in (16.17) is $\ll (b+1-a)(\log K)/K$, and this can be made arbitrarily small by taking K to be large. Thus we have the result when f' is increasing. If f' is decreasing, then $-f'$ is increasing, so we have the result for $-f$, and we obtain the result for f by taking complex conjugates. □

By taking $f(x) = \frac{-t}{2\pi}\log x$ in the above, we see immediately that if $\tau \le x \le y$, then

$$\sum_{x<n\le y} n^{-it} = \frac{y^{1-it} - x^{1-it}}{1-it} + O(1). \tag{16.19}$$

This allows us to establish a further useful result.

Lemma 16.14 *If $\sigma \ge 0$, $s \ne 1$, and $\tau \le x \le y$, then*

$$\sum_{x<n\le y} n^{-s} = \frac{y^{1-s} - x^{1-s}}{1-s} + O(x^{-\sigma}). \tag{16.20}$$

Proof If $\sigma = 0$, then this is just (16.19), so we assume that $\sigma > 0$. In what follows, we consider t to be fixed. Put

$$A(u) = \sum_{x<n\le u} n^{-it}, \qquad B(u) = \frac{u^{1-it} - x^{1-it}}{1-it}.$$

We see easily that

$$\int_x^y u^{-\sigma}\, dA(u) = \sum_{x<n\le y} n^{-s}, \qquad \int_x^y u^{-\sigma}\, dB(u) = \frac{y^{1-s} - x^{1-s}}{1-s}.$$

Put $R(u) = A(u) - B(u)$. The difference between the two main terms in (16.20) is

$$\int_x^y u^{-\sigma}\, dR(u) = R(y)y^{-\sigma} - R(x)x^{-\sigma} + \sigma \int_x^y \frac{R(u)}{u^{\sigma+1}}\, du.$$

By (16.19) we know that $R(u) \ll 1$. Hence the above is $\ll x^{-\sigma}$, and we have the stated result. □

On future occasions, we may dismiss an argument of the above type by simply saying. 'By integration by parts it follows that . . ..' However, it is worth noting that the integration by parts is simpler if one first removes the main term (as we did above) before integrating.

Theorem 16.15 *Suppose that $\sigma > 0$, that $s \neq 1$, and that $x \ge \tau$. Then*

$$\zeta(s) = \sum_{n\le x} n^{-s} + \frac{x^{1-s}}{1-s} + O\big(x^{-\sigma}\big). \tag{16.21}$$

It follows, in particular, that if $\sigma > 0$, $s \neq 1$, and $\tau \le x \le C\tau$, then

$$\zeta(s) = \sum_{n\le x} n^{-s} + O\big(\tau^{-\sigma}\big). \tag{16.22}$$

Proof We quote Theorem 1.12, which asserts that

$$\zeta(s) = \sum_{n\le y} n^{-s} + \frac{y^{1-s}}{s-1} + \frac{\{y\}}{y^s} - s\int_y^\infty \{u\}u^{-s-1}\, du. \tag{16.23}$$

We briefly outline the proof of this. We suppose first that $\sigma > 1$, write

$$\zeta(s) = \sum_{n\le y} n^{-s} + \int_y^\infty u^{-s}\, d\lfloor u\rfloor,$$

add and subtract $y^{1-s}/(s-1) = \int_y^\infty u^{-s}\, du$, and integrate the resulting integral by parts. Then we observe that the resulting integral is analytic for $\sigma > 0$. This gives (16.23).

The integral in (16.23) is $\ll |s|/y^{\sigma}$. We choose y to be so large that $|s|/y^{-\sigma} \le x^{-\sigma}$. Then we subtract (16.20) from both sides to obtain the result. □

We know (recall Corollary 1.17) that $\zeta(1+it) \ll \log\tau$ for $|t| \ge 1$. We also know (recall Corollary 10.5) that $|\zeta(it)| \asymp |\zeta(1+it)|\tau^{1/2}$ for $t \ge 1$. It follows by convexity (recall Exercise 10.1.1.19(c)) that $\zeta(s) \ll \tau^{(1-\sigma)/2}\log\tau$ for $0 \le \sigma \le 1$ and, in particular, that $\zeta(1/2+it) \ll \tau^{1/4}\log\tau$. We now derive a subconvex bound for $\zeta(1/2+it)$.

Theorem 16.16 *Let $\tau = |t| + 4$. Then for any real t,*

$$\zeta(1/2+it) \ll \tau^{1/6}\log\tau.$$

Proof We first show that if $1 \le a \le b \le 2a \le 2\tau$, then

$$\sum_{a\le n\le b} n^{-it} \ll a^{1/2}\tau^{1/6}. \tag{16.24}$$

To do this, we consider a in several ranges. First, suppose that $a \le \tau^{1/3}$. We argue trivially that

$$\sum_{a\le n\le b} n^{-it} \ll a = a^{1/2}a^{1/2} \le a^{1/2}\tau^{1/6}.$$

Second, if $\tau^{1/3} \le a \le \tau^{2/3}$, we use (16.15), which gives precisely the desired estimate. Finally, if $\tau^{2/3} \le a \le \tau$, then by (16.14),

$$\sum_{a\le n\le b} n^{-it} \ll \tau^{1/2} = \tau^{1/3}\tau^{1/6} \le a^{1/2}\tau^{1/6}.$$

Thus (16.24) is established. Next we show that if $x \le \tau$, then

$$\sum_{n\le x} n^{-it} \ll x^{1/2}\tau^{1/6}. \tag{16.25}$$

To do this, we cut the interval $[1,x]$ into dyadic blocks, and apply the bound (16.24) to each block. The bounds grow exponentially, so the size of the sum of all of them is the size of the largest term, which is $x^{1/2}\tau^{1/6}$. From (16.25) it follows by integrating by parts that

$$\sum_{n\le\tau} n^{-1/2-it} \ll \tau^{1/6}\log\tau.$$

The stated result now follows by combining this with (16.22). □

By cutting the interval $[1, x]$ into dyadic blocks and appealing to (16.13)–(16.15), we find that

$$\sum_{n \le x} n^{-it} \ll \begin{cases} x & x \le \tau^{1/3}, \\ x^{1/2}\tau^{1/6} & \tau^{1/3} \le x \le \tau^{2/3}, \\ \tau^{1/2}\log\tau & \tau^{2/3} \le x \le \tau. \end{cases} \tag{16.26}$$

Thus we see that

$$\sum_{n \le x} n^{-it} \ll \tau^{1/2}\log\tau \tag{16.27}$$

whenever $x \le \tau$. This bound is reminiscent of the Pólya–Vinogradov inequality (Theorem 9.18), which asserts that if χ is a non-principal character modulo q, then

$$\sum_{n=M+1}^{M+N} \chi(n) \ll q^{1/2}\log q.$$

We now establish a hybrid bound that includes both of these estimates, although only for initial sums, not sums over arbitrary intervals. To ease the insertion of a contribution that occurs only when a character is principal, we set

$$E_0(\chi) = \begin{cases} 1 & \chi = \chi_0, \\ 0 & \text{otherwise}. \end{cases} \tag{16.28}$$

Theorem 16.17 *Let χ be a Dirichlet character* (mod q). *Then*

$$\sum_{n \le x} \chi(n)n^{-it} = E_0(\chi)\frac{\varphi(q)}{q} \cdot \frac{x^{1-it}}{1-it} + O\big((q\tau)^{1/2}\log q\tau\big). \tag{16.29}$$

Proof Suppose first that $q = 1$. If $x \le \tau$, then it suffices to appeal to (16.27). If $x > \tau$, then we treat the range from 1 to τ using (16.27), and the range from τ to x by appealing to (16.19).

We use the case $q = 1$ to treat principal characters to moduli $q > 1$.

$$\begin{aligned}\sum_{\substack{n\le x\\(n,q)=1}} n^{-it} &= \sum_{n\le x} n^{-it} \sum_{d\mid(n,q)} \mu(d) = \sum_{d\mid q}\mu(d)\sum_{\substack{n\le x\\ d\mid n}} n^{-it}\\ &= \sum_{d\mid q}\mu(d)d^{-it}\sum_{m\le x/d} m^{-it}\\ &= \sum_{d\mid q}\mu(d)d^{-it}\left(\frac{(x/d)^{1-it}}{1-it} + O\big(\tau^{1/2}\log\tau\big)\right)\\ &= \left(\sum_{d\mid q}\frac{\mu(d)}{d}\right)\frac{x^{1-it}}{1-it} + O\big(d(q)\tau^{1/2}\log\tau\big).\end{aligned}$$

Here the sum over d is $\varphi(q)/q$, and $d(q) \le 2q^{1/2}$, so we have the result for χ_0 modulo q.

Now suppose that χ is a primitive character modulo q, $q > 1$. From Corollary 9.8 we know that

$$\chi(n) = \frac{1}{\tau(\overline{\chi})}\sum_{a=1}^{q}\overline{\chi}(a)e(an/q)$$

for all n. Here $\tau(\chi) = \sum_{a=1}^{q}\chi(a)e(a/q)$ is the Gauss sum of χ, which is not to be confused with our standard notation $\tau = |t| + 4$, which we also employ here, and we know by Theorem 9.7 that $|\tau(\chi)| = q^{1/2}$. Thus

$$\sum_{n\le x}\chi(n)n^{-it} = \frac{1}{\tau(\overline{\chi})}\sum_{a=1}^{q-1}\overline{\chi}(a)\sum_{n\le x} e(an/q)n^{-it}. \tag{16.30}$$

We show that if $q \nmid a$, then

$$\sum_{n\le x} e(an/q)n^{-it} \ll \tau^{1/2}\left(\frac{1}{\|a/q\|} + \log q\tau\right). \tag{16.31}$$

This bound suffices, for then the right-hand side of (16.30) is

$$\begin{aligned}\ll q^{-1/2}\tau^{1/2}\sum_{a=1}^{q-1}\left(\frac{1}{\|a/q\|} + \log q\tau\right) &\ll (q\tau)^{1/2}\left(\log q\tau + \sum_{1\le a\le q/2}\frac{1}{a}\right)\\ &\ll (q\tau)^{1/2}\log q\tau.\end{aligned}$$

To prove (16.31) put $f(u) = au/q - t(\log u)/(2\pi)$. Then $f'(u) = a/q - t/(2\pi u)$. Let u_0 be determined by the equation

$$\frac{|t|}{2\pi u_0} = \frac{1}{2}\left\|\frac{a}{q}\right\|.$$

If $u_0 \le x$, then by Corollary 16.6 we see that

$$\sum_{u_0 \le n \le x} e(an/q)n^{-it} \ll \frac{1}{\|a/q\|}.$$

To treat the sum over n in the interval $1 \le n \le \min(x, u_0)$, we divide this interval into dyadic blocks. Since $\|a/q\| \ge 1/q$, we know that $u_0 \ll q\tau$, and hence the number of dyadic blocks is $\ll \log q\tau$. We note that $f''(u) = t/(2\pi u^2)$. By Theorem 16.7 it follows that

$$\sum_{U \le n \le 2U} e(an/q)n^{-it} \ll |t|^{1/2} + U|t|^{-1/2}.$$

When this is summed over the dyadic blocks, the total is

$$\ll |t|^{1/2} \log q\tau + u_0|t|^{-1/2} \ll |t|^{1/2} \log q\tau + \frac{|t|^{1/2}}{\|a/q\|}.$$

On combining these last two estimates we see that we have established (16.31), so the desired result is proved for primitive non-principal characters.

Finally, suppose that χ is a non-principal character (mod q) that is induced by a primitive character $\chi^\star$ modulo d, for some $d|q$. Put $r = q/d$. Then $\chi(n) = \chi^\star(n)$ if $(n, r) = 1$, and $\chi(n) = 0$ otherwise. Thus

$$\sum_{n \le x} \chi(n)n^{-it} = \sum_{\substack{n \le x \\ (n,r)=1}} \chi^\star(n)n^{-it} = \sum_{n \le x} \chi^\star(n)n^{-it} \sum_{k|(n,r)} \mu(k)$$

$$= \sum_{k|r} \mu(k) \sum_{\substack{n \le x \\ k|n}} \frac{\chi^\star(n)}{n^{it}} = \sum_{k|r} \frac{\mu(k)\chi^\star(k)}{k^{it}} \sum_{m \le x/k} \frac{\chi^\star(m)}{m^{it}}.$$

Here the outer sum has $d(r) \le 2r^{1/2}$ summands, and the inner sum is $\ll (d\tau)^{1/2} \log d\tau$ by what we have already proved for primitive characters. Hence the above is $\ll (q\tau)^{1/2} \log q\tau$, so the proof is complete. □

16.2.1 Exercises

1. Let M, K be positive integers with $K \le (M-1)/2$, and define $N = 2M$, $\Delta = (K + 1/2)/M$, and $\alpha_n = \Delta n$ if $1 \le n \le M$ and $\alpha_n = (1 - \Delta)(n - M) + \Delta M$ if $M < n \le N$. Further, let $\delta_n = \alpha_{n+1} - \alpha_n$ when $1 \le n < N$ and $S = \sum_{n=1}^N e(\alpha_n)$. Show that

$$\Delta \le \delta_1 \le \delta_2 \le \cdots \le \delta_{N-1} \le 1 - \Delta,$$

that $|S| = 2 \cot \pi\Delta$, and that if M/K is large, then

$$|S| \sim \cot \frac{\pi\Delta}{2}.$$

2. Suppose that the sequence u_n is weakly increasing, that $u_{n+1} - u_n$ is weakly decreasing to 0, and that $\lim_{n\to\infty} n(u_{n+1} - u_n) = \infty$. (Note that the sequence $u_n = \log n$ satisfies only the first two of these hypotheses, not the third, and Exercise F.1.1.4(a,b) show that this sequence is not uniformly distributed modulo 1. Also, Exercise F.1.1.4(c) shows that the Weyl criterion fails for this sequence.)

   (a) Show that $\lim_{n\to\infty} u_n/\log n = \infty$.

   (b) Use the Kusmin–Landau inequality to show that u_n is uniformly distributed (mod 1).

   (c) (Fejér) Suppose that $f(x)$ is a real-valued function defined on the positive real numbers such that f is weakly increasing, that f' decreases weakly to 0, and that $xf'(x) \to \infty$ as $x \to \infty$. Show that the sequence $f(n)$ is uniformly distributed (mod 1).

3. Let $P(\alpha) = \sum_{n=1}^{N} e(n^2/N + n\alpha)$.

   (a) Show that $P(\alpha) \ll N^{1/2}$ uniformly in α.

   (b) Show that
$$\int_0^1 |P(\alpha)|^2 \, d\alpha = N.$$

   (c) Deduce that there is an α such that $|P(\alpha)| \geq N^{1/2}$.

4. For arbitrary real $c > 0$, prove that
$$\sum_{n=N}^{2N} e(c/n^2) \ll c^{1/2}N^{-1} + c^{-1/2}N^2.$$

5. Show that
$$\sum_{n=N}^{2N} e\left(\frac{n^2}{6N}\right) \ll 1,$$
but that
$$\left|\sum_{n=N}^{2N} e\left(\frac{n^2}{3N}\right)\right| \asymp N^{1/2}.$$

6. Prove that if $M_3 \leq f'''(x) \leq AM_3$ and $f''(0) = 0$, then
$$\sum_{n=1}^{N} e(f(n)) \ll_A M_3^{-1/3} + N^{3/2}M_3^{1/2}.$$

7. Prove that if $1/N \leq c \leq 2/N$, then $\sum_{n=1}^{N} e\left(cn^3\right) \ll N^{5/6}$. Better still, show that this bound can be replaced by $N^{3/4+\varepsilon}$.

8. (a) By writing $n = mp^2 + h$, show that

$$\sum_{n=1}^{p^3} e\left(\frac{n^3}{p^3}\right) = \sum_{h=1}^{p^2} e\left(\frac{h^3}{p^3}\right) \sum_{m=1}^{p} e\left(\frac{3mh^2}{p}\right).$$

(b) Deduce that if $p \neq 3$, then the expression in part (a) is equal to p^2.

(c) By writing $n = 3m + h$, show that

$$\sum_{n=1}^{27} e\left(\frac{n^3}{27}\right) = \sum_{h=1}^{3} e\left(\frac{h^3}{27}\right) \sum_{m=1}^{9} e\left(\frac{h^2m}{3}\right) = 9.$$

9. In many applications, such as in treating the sum $\sum_{n=N}^{2N} n^{it}$, we find that $M_{r+1} \asymp M_r/N$. Show that when this is the case, the best estimate from Theorem 16.12 is obtained by taking r so that

$$N^{-2+4/R} \ll_r M_r \ll_r N^{-1+2/R},$$

and that the estimate is then

$$\ll_r NM_r^{1/(R-2)}.$$

10. Let $f(x)$ be real-valued with $k+1$ continuous derivatives, and put

$$P(x) = \sum_{r=0}^{k} \frac{f^{(r)}(0)}{r!} x^r.$$

Show that for $k \geq 1$,

$$\sum_{n=1}^{N} e(f(n)) \ll S^*\left(1 + MN^{k+1}\right),$$

where

$$M = \max_{0<x<N} \frac{|f^{(k+1)}(x)|}{(k+1)!}, \qquad S^* = \max_{X \leq N} \sum_{n \leq X} e(P(n)).$$

11. Let $z_1, z_2, \ldots, z_N$ be N arbitrary complex numbers, H be an integer with $1 \leq H \leq N$, and define

$$S(\alpha) = \sum_{n=1}^{N} z_n e(\alpha n), \quad T(\alpha) = \sum_{m=1}^{N+H} e(\alpha m), \quad K(\alpha) = \sum_{h=1}^{H} e(\alpha h).$$

(a) Prove that

$$H \sum_{n=1}^{N} z_n = \int_0^1 S(\alpha)K(\alpha)T(-\alpha)\,d\alpha.$$

(b) Prove that

$$\int_0^1 |S(\alpha)|^2|K(\alpha)|^2 d\alpha = H\sum_n |z_n|^2 + 2\,\mathrm{Re}\sum_{h=1}^{H}(H-h)\sum_n z_{n+h}\overline{z}_n.$$

(c) Use parts (a) and (b) to derive van der Corput's lemma (Lemma 16.8).

16.3 Van der Corput's Method

By means of the van der Corput lemma, Lemma 16.8, we may reduce the problem of estimating one exponential sum to that of estimating some other sums. We now use the Poisson summation formula to establish a second, quite different, transformation of the initial sum.

Theorem 16.18 *Let $f(x)$ be real valued, and suppose that $f'(x)$ is continuous and increasing on the interval $[a,b]$. Put $f'(a) = \alpha$ and $f'(b) = \beta$. Then*

$$\sum_{a\le n\le b} e(f(n)) = \sum_{\alpha-1\le \nu\le\beta+1} \int_a^b e(f(x)-\nu x)\,dx + O\big(\log(2+\beta-\alpha)\big). \tag{16.32}$$

Proof Let N be an integer such that $|N-(\alpha+\beta)/2| \le 1/2$. If we replace $f(x)$ by $f(x)-Nx$, then the terms in the sum on the left are unchanged, $f'(x)$ is still continuous and increasing, and the sum on the right is unchanged, although the indexing of the terms has been translated, as α has been replaced by $\alpha' = \alpha - N$, and β has been replaced by $\beta' = \beta - N$. We note that $\alpha' + \beta' = \alpha + \beta - 2N$, so that $|\alpha' + \beta'| \le 1$. Thus by making a change of variable of this sort, we may suppose that $|\alpha+\beta| \le 1$.

Let $F(x) = e(f(x))$ for $a \le x \le b$, and put $F(x) = 0$ otherwise. Then $F \in L^1(\mathbb{R})$ and F has bounded variation on $\mathbb{R}$, so by the Poisson summation formula (Theorem D.3),

$$\sum_n \frac{1}{2}(F(n^+)+F(n^-)) = \lim_{K\to\infty}\sum_{k=-K}^{K}\widehat{F}(k).$$

Since $F(x)$ is continuous apart from possible jump discontinuities at a or b, the left-hand side here is within $O(1)$ of the left-hand side in (16.32). The integral on the right in (16.32) is simply $\widehat{F}(\nu)$, so to complete the proof it suffices to show that

$$\sum_{\substack{|k|\le K\\ k\notin[\alpha-1,\beta+1]}} \widehat{F}(k) \ll \log(2+\beta-\alpha) \tag{16.33}$$

for all sufficiently large K. Integrating by parts, we find that

$$\widehat{F}(k) = \frac{e(f(a) - ka)}{2\pi i k} - \frac{e(f(b) - kb)}{2\pi i k} + \frac{1}{k}\int_a^b f'(x)e(f(x) - kx)\,dx.$$

If $k > \beta$, then $f'(x)/(f'(x) - k)$ is monotonic, so by Theorem 16.1 this integral is $\ll \beta/(k - \beta)$. We note that

$$\sum_{k>\beta+1} \frac{\beta}{k(k-\beta)} \asymp \log(2+\beta).$$

We treat $k < \alpha$ similarly, and find that the left-hand side of (16.33) is

$$\frac{e(f(a))}{2\pi i} \sum_{\substack{0<|k|\le K \\ k\notin[\alpha-1,\beta+1]}} \frac{e(-ka)}{k} - \frac{e(f(b))}{2\pi i} \sum_{\substack{0<|k|\le K \\ k\notin[\alpha-1,\beta+1]}} \frac{e(-kb)}{k} + O(\log(2+\beta-\alpha)).$$

Since $|\alpha+\beta| \le 1$, we may pair each k in these sums with $-k$, except for at most one k, whose contribution is bounded. Hence the above is

$$= e(f(b)) \sum_{\beta+1<k\le K} \frac{\sin 2\pi k b}{\pi k} - e(f(a)) \sum_{\beta+1<k\le K} \frac{\sin 2\pi k a}{\pi k}$$
$$+ O(\log(2+\beta-\alpha)).$$

That these sums are bounded can be seen from Theorem D.1, but we find the following direct argument to be instructive. It suffices to bound the first sum, which is an odd function of b with period 1. Hence it suffices to bound this sum when $0 < b \le 1/2$. For those k (if there are any) for which $k \le 1/b$, we use the inequality $\sin u \le u$ to see that the summand is $\le 2b$. Since the number of such k is $\ll 1/b$, it follows that the total contribution of such terms is $\ll 1$. By taking the imaginary part of (16.4) we see that

$$\sum_{u\le k\le v} \sin 2\pi k b \ll \frac{1}{b}.$$

By summation by parts it follows that if $u > 0$, then

$$\sum_{u\le k\le v} \frac{\sin 2\pi k b}{\pi k} \ll \frac{1}{ub}.$$

Since $u \ge 1/b$ in our application, this contribution is also bounded, and the proof is complete. □

Suppose we apply Theorem 16.18 to a function $f(x)$ such that $f''(x) \ge M_2 > 0$. By Theorem 16.3, the integrals on the right-hand side are $\ll M_2^{-1/2}$. The number of terms in the sum on the right is $f'(b) - f'(a) + O(1)$. If we suppose that $f''(x) \le AM_2$, then the number of terms is $\ll_A (b-a)M_2 + 1$, and thus the right-hand side is $\ll_A (b-a)M_2^{1/2} + M_2^{-1/2}$. This provides a second (more complicated) proof of Theorem 16.7, but now we are in a position to determine whether there is any cancellation in the sum on the right in (16.32). To this end we must first derive a more precise estimate for the integrals in (16.32). Suppose that $g(x)$ is a real-valued function on $[a, b]$, that there is a point $x_0 \in [a, b]$ such that $g'(x_0) = 0$, and also that

$$0 < M_2 \le g''(x) \tag{16.34}$$

for $x \in [a, b]$. Let $q(x)$ be the quadratic polynomial $q(x) = g(x_0) + \frac{1}{2}g''(x_0)(x - x_0)^2$. We expect that $q(x)$ provides a good approximation to $g(x)$, at least when x is near x_0. Consider first the idealized situation in which $g(x)$ is exactly equal to $q(x)$. By (16.3) we see that

$$\int_{-\infty}^{\infty} e(q(x))\,dx = e(g(x_0) + 1/8)g''(x_0)^{-\frac{1}{2}}.$$

As $q'(x)$ is increasing and $q'(x) \ge M_2(b - x_0)$ for $x \ge b$, we see from Theorem 16.1 that

$$\int_b^{\infty} e(q(x))\,dx \ll M_2^{-1}(b - x_0)^{-1}.$$

This estimate is weak if x_0 is close to b, in which case we use Theorem 16.3 instead and obtain

$$\int_b^{\infty} e(q(x))\,dx \ll M_2^{-\frac{1}{2}}.$$

We may treat $\int_{-\infty}^{a} e(q(x))\,dx$ similarly, and thus we find that

$$\int_a^b e(q(x))\,dx = e\big(g(x_0) + 1/8\big)g''(x_0)^{-\frac{1}{2}} + O(R_1) \tag{16.35}$$

where

$$\begin{aligned} R_1 &= \min\big(M_2^{-1}(x_0 - a)^{-1}, M_2^{-\frac{1}{2}}\big) \\ &\quad + \min\big(M_2^{-1}(b - x_0)^{-1}, M_2^{-\frac{1}{2}}\big). \end{aligned} \tag{16.36}$$

In the general case, $g(x)$ is not a quadratic polynomial, but if the higher derivatives of g are not too large, then the expression above provides a good approximation to the integral in question.

Theorem 16.19 *Let $g(x)$ be a thrice continuously differentiable real-valued function on $[a, b]$. Suppose that there is an $x_0 \in [a, b]$ such that $g'(x_0) = 0$,* *and that* (16.34) *holds throughout this interval. If* $|g'''(x)| \le M_3$ *for* $x \in [a, b]$, *then*

$$\int_a^b e(g(x))\,dx = e\big(g(x_0) + 1/8\big)g''(x_0)^{-1/2} + O(R_1) + O(R_2), \qquad (16.37)$$

where R_1 *is given by* (16.36) *and*

$$R_2 = M_2^{-1}M_3^{1/3}. \qquad (16.38)$$

If, additionally, $g^{(4)}(x)$ *exists, is continuous, and* $|g^{(4)}(x)| \le M_4$ *for* $x \in [a, b]$, *then we may take*

$$R_2 = (b-a)M_2^{-2}M_4 + (b-a)M_2^{-3}M_3^2. \qquad (16.39)$$

If instead of (16.34) we have

$$g''(x) \le -M_2 < 0, \qquad (16.40)$$

then we apply the theorem to $-g(x)$ and take complex conjugates in (16.37). This gives a similar result, but the main term in (16.37) must be replaced by

$$e\big(g(x_0) - 1/8\big)|g''(x_0)|^{-1/2}. \qquad (16.41)$$

Proof By (16.34) and Theorem 16.3 we know that the integral in (16.37) is $\ll M_2^{-1/2}$. Thus if $a \le x_0 < a + M_2^{-1/2}$ or $b - M_2^{-1/2} < x_0 \le b$, then there is nothing further to be done, in view of the error term R_1. Thus in continuing, we may assume that

$$a + M_2^{-1/2} \le x_0 \le b - M_2^{-1/2}. \qquad (16.42)$$

We multiply both sides of (16.37) by $e(-g(x_0))$ to reduce to the case $g(x_0) = 0$. Similarly, we may translate the coordinates so that $x_0 = 0$. We take $q(x)$, as above, to be the Taylor approximation of order 2. Then $g(x) = q(x) + r(x)$ where the remainder term $r(x)$ may be written explicitly as

$$r(x) = \frac{1}{2}x^3\int_0^1 (1-u)^2 g^{(3)}(xu)\,du. \qquad (16.43)$$

Similarly, $q'(x)$ is the Taylor approximation of order 1 to $g'(x)$, so the remainder term $r'(x)$ can be written as

$$r'(x) = x^2\int_0^1 (1-u)g^{(3)}(xu)\,du. \qquad (16.44)$$

In view of (16.34), it suffices to show that

$$\int_a^b e(q(x))(e(r(x)) - 1)\,dx \ll R_1 + R_2. \qquad (16.45)$$

Let δ be a parameter at our disposal, and let $I = [c, d]$ denote the portion of the interval $[a, b]$ for which $|x| \le \delta$, and let $J = [a, b] \setminus I$. The set J may be empty, but if it is not, then it consists of one or two intervals. By (16.34) we see that $|g'(x)| \ge \delta M_2$ for all $x \in J$. Hence by Theorem 16.1 we find that

$$\int_J e(g(x))\,dx \ll \delta^{-1} M_2^{-1}.$$

Since $q''(x) = g''(0) \ge M_2$, a similar argument applies to $q(x)$, and so

$$\int_J e(q(x))(e(r(x)) - 1)\,dx \ll \delta^{-1} M_2^{-1}. \tag{16.46}$$

We now consider the integral (16.45), restricted to the interval I. Because the expression $e(q(x))/(2\pi i g''(0))$ is an antiderivative of $xe(q(x))$, we integrate by parts to see that the integral is

$$\begin{aligned}&= \left[\frac{e(q(x))(e(r(x)) - 1)}{2\pi i g''(0)x}\right]_c^d \\ &\quad - \frac{1}{g''(0)} \int_I e(q(x)) \left(\frac{e(r(x))r'(x)}{x} - \frac{e(r(x)) - 1}{2\pi i x^2}\right) dx.\end{aligned} \tag{16.47}$$

Since $d = \min(b, \delta)$, it follows that $1/d \le 1/b + 1/\delta$. Thus the upper endpoint contributes an amount

$$\ll b^{-1} M_2^{-1} + \delta^{-1} M_2^{-1} \ll R_1 + \delta^{-1} M_2^{-1}.$$

The lower endpoint is treated similarly. By (16.43) we see that $r(x) \ll |x|^3 M_3$, and by (16.44) we find that $r'(x) \ll x^2 M_3$. Using the inequality $|e(u) - 1| \le 2\pi|u|$, we deduce that the integrand is $\ll |x| M_3$, and hence the second term in (16.47) is $\ll \delta^2 M_2^{-1} M_3$. On comparing this with (16.46), we discover that the choice $\delta = M_3^{-1/3}$ is optimal. This gives (16.45) with R_2 given by (16.38). Our choice of δ is plausible since (16.43) allows us to show that $r(x)$ is small precisely when $x \in I$.

It remains to derive (16.37) with the refined error term (16.39). We integrate by parts as above, but take $I = [a, b]$. Since $d = b$, the upper endpoint now contributes an amount $\ll b^{-1} M_2^{-1} \ll R_1$. The lower endpoint is treated similarly. Write the integral in (16.47) as $T_1 + T_2$ where T_1 arises from the first term in brackets, and T_2 from the second. Let $h(x) = r'(x)x^{-2}$ and $j(x) = g'(x)x^{-1}$. Then

$$\begin{aligned}T_1 &= \int_a^b \frac{h(x)}{j(x)} e(g(x)) g'(x)\,dx \\ &= \left[\frac{h(x)e(g(x))}{j(x)2\pi i}\right]_a^b - \int_a^b \frac{d}{dx}\left(\frac{h(x)}{j(x)}\right) \frac{e(g(x))}{2\pi i}\,dx.\end{aligned}$$

Since $h(x)$ is the integral in (16.44), we see that $h(x) \ll M_3$. By differentiating this integral with respect to x, we find also that $h'(x) \ll M_4$. Similarly $j(x) = \int_0^1 g''(xu)\,du \geq M_2$ by (16.34), and $j'(x) = \int_0^1 ug'''(xu)\,du \ll M_3$. Hence

$$\frac{d}{dx}\left(\frac{h(x)}{j(x)}\right) = \frac{h'(x)}{j(x)} - \frac{h(x)j'(x)}{j(x)^2} \ll \frac{M_4}{M_2} + \frac{M_3^2}{M_2^2},$$

so that

$$T_1 \ll M_2^{-1}M_3 + M_2^{-1}M_4(b-a) + M_2^{-2}M_3^2(b-a).$$

To bound the integral T_2 we follow the method used to derive the estimate (16.38). We let I and J be defined as before. Put $k(x) = r(x)x^{-3}$. By (16.43) we see that $k(x) \ll M_3$, and that $k'(x) \ll M_4$. Set $m(x) = (e(x)-1)/x$. Then $m(x) \ll 1$ and $m'(x) \ll 1$. The contribution of the interval I to T_2 is

$$\begin{aligned}\int_c^d & e(q(x))xm(r(x))k(x)\,dx \\ &= \left[\frac{e(q(x))}{2\pi i g''(0)}m(r(x))k(x)\right|_c^d \\ &\quad - \int_c^d \frac{e(q(x))}{2\pi i g''(0)}(m'(r(x))r'(x)k(x) + m(r(x))k'(x))\,dx \\ &\ll M_2^{-1}M_3 + M_2^{-1}M_3^2(d-c)^3 + M_2^{-1}M_4(d-c).\end{aligned}$$

In the second factor we use the inequality $d - c \leq \delta$, but in the third factor we use instead $d - c \leq b - a$. Thus we find that

$$\int_I e(q(x))\left(\frac{e(r(x))-1}{2\pi i x^2}\right)dx \ll M_2^{-1}M_3 + M_2^{-1}M_3^2\delta^3 + M_2^{-1}M_4(b-a).$$

As for the set J, we consider separately the integrals $\int_J e(g(x))x^{-2}\,dx$ and $\int_J e(q(x))x^{-2}\,dx$. Applying Theorem 16.1 to the first of these integrals, we are led to consider the function $g'(x)x^2$. This quantity has absolute value $\geq M_2\delta^3$, and the expression is monotonic since its derivative is $g''(x)x^2 + 2g'(x)x > 0$. Thus by Theorem 16.1, $\int_J e(g(x))x^{-2}\,dx \ll M_2^{-1}\delta^{-3}$. Similarly, as $q'(x)x^2 =$

$g''(0)x^3$ is monotonic, $\int_J e(q(x))x^{-2}\,dx \ll M_2^{-1}\delta^{-3}$. On combining these estimates, we conclude that

$$T_2 \ll M_2^{-1}M_3 + M_2^{-1}M_3^2\delta^3 + M_2^{-1}M_4(b-a) + M_2^{-1}\delta^{-3}.$$

To optimize this estimate we again take $\delta = M_3^{-1/3}$. We combine this with our estimate for T_1 to see that the integral in (16.47) is

$$\begin{aligned}&\ll R_1 + M_2^{-1}(T_1 + T_2)\\ &\ll R_1 + M_2^{-2}M_3 + M_2^{-2}M_4(b-a) + M_2^{-3}M_3^2(b-a).\end{aligned}$$

Put $U = M_2^{-1}(b-a)^{-1}$. The second term above is the geometric mean of U and the fourth term. By (16.42) we deduce that $U \ll R_1$, so the second term is majorized by the maximum of the first and fourth terms, and therefore may be omitted. Thus we have (16.37) with the error term (16.39), and the proof is complete. □

Theorem 16.20 *Let N be a positive integer and $a \le b \le a + N$, and suppose that f is thrice continuously differentiable on $[a, b]$ and that*

$$0 < M_2 \le f''(x) \le AM_2, \qquad |f'''(x)| \le M_3.$$

Let $\alpha = f'(a)$, $\beta = f'(b)$, and for each integer ν in $[\alpha, \beta]$ let x_ν be defined by $f'(x_\nu) = \nu$. Then

$$\sum_{a\le n\le b} e(f(n)) = \sum_{\alpha\le\nu\le\beta} \frac{e\big(f(x_\nu) - \nu x_\nu + 1/8\big)}{\sqrt{f''(x_\nu)}} + O_A(E_1 + E_2), \tag{16.48}$$

where

$$E_1 = \log(2 + M_2N) + M_2^{-1/2}$$

and

$$E_2 = M_3^{\frac{1}{3}}N. \tag{16.49}$$

If, moreover, $f^{(4)}(x)$ exists, is continuous, and satisfies $|f^{(4)}(x)| \le M_4$ on $[a, b]$, then (16.49) *may be replaced by*

$$E_2 = \frac{M_4}{M_2}N^2 + \frac{M_3^2}{M_2^2}N^2. \tag{16.50}$$

If instead

$$0 < M_2 \le -f''(x) \le AM_2,$$

then (16.48) holds with $\alpha = f'(b)$, $\beta = f'(a)$, $1/8$ replaced by $-1/8$, and the $-f''(x_\nu)$ in the sum on the right-hand side replaced by $-f''(x_\nu)$.

Proof We may suppose that $M_2 \geq N^{-2}$, for otherwise the conclusion is trivial since the number of terms on the left-hand side is at most $N+1$ and $E_1 \gg M_2^{-1/2}$. By Theorem 16.18,

$$\sum_{a\leq n\leq b} e(f(n)) = \sum_{\alpha-1\leq\nu\leq\beta+1} \int_a^b e(f(x) - \nu x)\, dx + O(\log(2+\beta-\alpha)).$$

By Theorem 16.3,

$$\int_a^b e(f(x) - \nu x)\, dx \ll M_2^{-\frac{1}{2}} \tag{16.51}$$

uniformly in ν, and

$$\beta - \alpha = (b-a)f''(\xi) \ll_A M_2 N. \tag{16.52}$$

Hence

$$\sum_{a\leq n\leq b} e(f(n)) = \sum_{\alpha\leq\nu\leq\beta} \int_a^b e(f(x) - \nu x)\, dx + O_A(E_1).$$

If $\beta-\alpha \leq 1$, then by (16.51) we are done. Thus we may suppose that $\beta-\alpha > 1$, and then by (16.52) the sum on the right-hand side is non-empty and the number of terms is

$$\sum_{\alpha\leq\nu\leq\beta} 1 \asymp_A M_2 N. \tag{16.53}$$

By Theorem 16.19 we may replace each integral on the right by

$$\frac{e\big(f(x_\nu) - \nu x_\nu + 1/8\big)}{\sqrt{f''(x_\nu)}}$$

with an error

$$\begin{aligned}\ll & M_2^{-1}M_3^{\frac{1}{3}} + \min\big(M_2^{-1}(x_\nu - a)^{-1}, M_2^{-1/2}\big)\\ &+ \min\big(M_2^{-1}(b-x_\nu)^{-1}, M_2^{-1/2}\big).\end{aligned} \tag{16.54}$$

By (16.53) the first term contributes a total amount $\ll M_3^{\frac{1}{3}}N = E_2$. To treat the second term we observe that $\nu - \alpha = f'(x_\nu) - f'(a) = (x_\nu - a)f''(\xi) \leq AM_2(x_\nu - a)$ and so the second term is bounded by

$$\min\left(\frac{A}{\nu-\alpha}, M_2^{-1/2}\right).$$

Thus the total contribution from the second term is

$$\ll_A M_2^{-1/2} + \sum_{\alpha+1\leq\nu\leq\beta} \frac{1}{\nu-\alpha} \ll E_1.$$

Likewise, the same upper bound holds for the contribution from the third term. The first part of the theorem now follows.

For the second part of the theorem we appeal to the concomitant part of Theorem 16.19. Then the term $M_2^{-1}M_3^{1/3}$ in (16.54) is replaced by

$$(b-a)M_2^{-2}M_4 + (b-a)M_2^{-3}M_3^2,$$

and so by (16.53) the total contribution is

$$\ll \frac{M_4}{M_2}N^2 + \frac{M_3^2}{M_2^2}N^2.$$

□

Corollary 16.21 *Suppose that I is a subinterval of $[N, 2N]$, f has four continuous derivatives on I, and that there are positive real numbers A, λ, θ such that*

$$\begin{aligned} 0 < \lambda N^{-\theta-1} &\le f^{(2)}(x) \le A\lambda N^{-\theta-1}, \\ |f^{(3)}(x)| &\le A\lambda N^{-\theta-2}, \\ |f^{(4)}(x)| &\le A\lambda N^{-\theta-3} \end{aligned}$$

for $x \in I$. Then the error term in (16.48) *is*

$$\ll_A \log\left(2 + \lambda N^{-\theta}\right) + \lambda^{-1/2}N^{(\theta+1)/2}.$$

The proof is immediate on observing that the contribution from E_2, given by (16.50), is $\ll 1$, which can be absorbed in the logarithmic term.

The conditions of Corollary 16.21 are those which are very largely met in applications.

We now have two essentially different lines of approach for dealing with a given exponential sum. In each of these we begin by transforming the sum into a new one. The first of these is via the Weyl–van der Corput lemma (Lemma 16.8). The second is via Theorem 16.20 (or, usually more conveniently, via Corollary 16.21). With each of these processes, the presumption is that the transformed sum is one about which we already have information. The normal requirement is that the function f behaves somewhat like that considered in the corollary. To this end we define the following class of functions.

Definition 16.1 Let N, λ, θ, ε be positive real numbers, let r be a positive integer, and let I be a subinterval of $[N, 2N]$. Let

$$\phi(x) = \begin{cases} \frac{\lambda x^{1-\theta}}{1-\theta} & \text{when } \theta \ne 1, \\ \lambda \log x & \text{otherwise.} \end{cases} \tag{16.55}$$

We define $\mathscr{F}(N, I, \lambda, \theta, r, \varepsilon)$ to be the set of functions f that are r-times continuously differentiable on I and which for each s with $1 \le s \le r$ and $x \in I$ satisfy

$$\left|f^{(s)}(x) - \phi^{(s)}(x)\right| < \varepsilon\left|\phi^{(s)}(x)\right|. \tag{16.56}$$

We are now in a position to define precisely what we mean by exponent pairs.

Definition 16.2 An *exponent pair* is a pair (k, l) of real numbers k and l satisfying

$$0 \le k \le \frac{1}{2} \le l \le 1 \tag{16.57}$$

and such that for every $\theta > 0$ there is an integer $r = r(k, l, \theta) \ge 2$ and an $\varepsilon = \varepsilon(k, l, \theta)$ satisfying $0 < \varepsilon < 1/2$ for which for every $N > 0$, $\lambda > 0$, $I \subseteq [N, 2N]$, and $f \in \mathscr{F}(N, I, \lambda, \theta, r, \varepsilon)$ we have

$$\sum_{n \in I} e(f(n)) \ll \left(\lambda N^{-\theta}\right)^k N^l + \lambda^{-1} N^{\theta}. \tag{16.58}$$

We now make a number of observations concerning exponent pairs.

1. In establishing that a particular pair is an exponent pair we may suppose that

$$\lambda N^{-\theta} \ge 1, \tag{16.59}$$

for otherwise the inequality always holds. To see this we consider two cases. First, if $\lambda N^{-\theta} < 1/2$, then by Corollary 16.6 we have

$$\sum_{n \in I} e(f(n)) \ll \frac{N^{\theta}}{\lambda}.$$

Second, if $1/2 \le \lambda N^{-\theta} < 1$, then by Theorem 16.7, we have

$$\sum_{n \in I} e(f(n)) \ll N^{\frac{1}{2}} \ll \left(\lambda N^{-\theta}\right)^k N^l$$

since $l \ge 1/2$. Henceforward we always assume that (16.59) holds.

2. By examining some special functions f we can explain why we have imposed the conditions (16.57) on the ordered pairs. Let $M = \lfloor N \rfloor$, let $\lambda = \operatorname{lcm}(1, 2, \ldots, 2M)$, and let

$$f(x) = -\frac{\lambda}{x}.$$

Thus $f(x)$ is the function $\phi(x)$ defined in (16.55) with $\theta = 2$, and $f(n) \in \mathbb{Z}$ for $1 \le n \le 2N$ so that

$$\sum_{M+1 \le n \le 2M} e(f(n)) = M \gg N.$$

Now $\lambda = \exp(\psi(2M))$, and so $\lambda = \exp(2N + o(N))$ by the Prime Number Theorem. Since λ is exponentially large, we deduce that if the estimate

$$N \ll (\lambda N^{-2})^k N^l$$

is to hold, then $k \geq 0$, and if $k = 0$, then $l \geq 1$. In particular, the only exponent pair of the form $(k, 1)$ is $(0, 1)$.

3. Suppose we have an exponent pair with $l > 1$. In view of (16.59) the bound (16.58) would then be worse than $(k, 1)$, and this in turn would be worse than the trivial pair $(0, 1)$. This explains why we have imposed the condition $l \leq 1$ in (16.57).
4. Consider the expression

$$\int_{\Lambda}^{2\Lambda} \left| \sum_{M+1}^{2M} e(-\lambda n^{-1}) \right|^2 d\lambda,$$

where $M = \lfloor N \rfloor$. The numbers $1/n$ with $M+1 \leq n \leq 2M$ are spaced at least $\frac{1}{2M(2M-1)}$ apart. Let $S_-(x)$ be the function of Theorem E.3 with $\alpha = \Lambda$, $\beta = 2\Lambda$, and $\delta = \frac{1}{2M(2M-1)}$. Then the integral is

$$\begin{aligned} &\geq \int_{-\infty}^{\infty} S_-(\lambda) \left| \sum_{M+1}^{2M} e(-\lambda n^{-1}) \right|^2 d\lambda \\ &= \sum_{m=M+1}^{2M} \sum_{n=M+1}^{2M} \widehat{S}_-(1/m - 1/n) \\ &= \widehat{S}_-(0)M = (\Lambda - 2M(2M-1))M. \end{aligned}$$

Thus we see that if $\Lambda = 4M(2M-1)$, then there is a $\lambda \in [\Lambda, 2\Lambda]$ such that

$$\left| \sum_{n=M+1}^{2M} e(f(n)) \right| \gg N^{1/2},$$

where $f(x) = -\lambda/x$. Now $f'(x) = \lambda x^{-2}$, and so if (k, l) is an exponent pair we would have

$$N^{1/2} \ll (\lambda N^{-2})^k N^l + N^2 \lambda^{-1} \ll N^l$$

since we have already seen that $k \geq 0$. Thus it is also necessary that $\frac{1}{2} \leq l$ when (k, l) is an exponent pair.
5. By Theorem 16.7 we see that $(1/2, 1/2)$ is an exponent pair, and we have already seen in observation 4. that, of necessity, $\frac{1}{2} \leq l$. Thus an exponent pair (k, l) with $k > 1/2$ would give a bound that is inferior to that provided by $(1/2, 1/2)$. Thus we can happily restrict our attention to $k \leq \frac{1}{2}$, as in (16.57).

6. Next, we show that if $(k, 1/2)$ is an exponent pair, then perforce $k = 1/2$. Let H be an arbitrary positive integer, and define λ to be the positive number with $\lambda^2 = \operatorname{lcm}\{1, 2, \ldots, H\}$, so that $\nu|\lambda^2$ for any positive integer with $\nu \le H$. Now let $N = \lambda^2 H^{-2}$ and $f(x) = 2\lambda x^{\frac{1}{2}}$, and suppose that $(k, 1/2)$ is an exponent pair, so that $0 \le k \le 1/2$. Then, by Definitions 16.1 and 16.2,

$$\sum_{N<n\le 2N} e(f(n)) \ll \left(\lambda N^{-1/2}\right)^k N^{1/2} + \lambda^{-1} N^{1/2}$$

and $\lambda N^{-\frac{1}{2}} = H$ so that

$$\sum_{N<n\le 2N} e(f(n)) \ll H^k N^{1/2} + H^{-1} \ll H^k N^{1/2}.$$

By Corollary 16.21 we have

$$\sum_{a\le n\le b} e(f(n)) = \sum_{\alpha\le\nu\le\beta} \frac{e\left(f(x_\nu) - \nu x_\nu - 1/8\right)}{\sqrt{-f''(x_\nu)}} + O_A\left(\log(2 + H) + N^{1/2} H^{-1/2}\right), \tag{16.60}$$

where $\alpha = H/\sqrt{2}$, $\beta = H$, $x_\nu = \lambda^2\nu^{-2}$. Thus $-f''(x_\nu) = \frac{1}{2}\nu^3\lambda^{-2} \ll HN^{-1}$ and $f(x_\nu) - \nu x_\nu = \lambda^2\nu^{-1} \in \mathbb{Z}$.

Hence by (16.60),

$$\left|\sum_{N<n\le 2N} e(f(n))\right| \gg H^{1/2} N^{1/2}.$$

Thus, of necessity,

$$l = \frac{1}{2} \implies k = \frac{1}{2}. \tag{16.61}$$

7. The set of exponent pairs forms a convex set since given any two exponent pairs (k', l'), (k'', l''), we have (assuming (16.59), of course)

$$\sum_{n\in I} e(f(n)) \ll \min((\lambda N^{-\theta})^{k'} N^{l'}, (\lambda N^{-\theta})^{k''} N^{l''}),$$

and for any η with $0 \le \eta \le 1$ we can replace this by

$$(\lambda N^{-\theta})^k N^l$$

with $k = k'\eta + k''(1 - \eta)$, $l = l'\eta + l''(1 - \eta)$. In particular, the ordered pairs $(\frac{1}{2}, \frac{1}{2})$ and $(0, 1)$ with $\eta = 2k$ shows that each of the pairs

$$(k, 1 - k) \text{ with } \left(0 \le k \le \frac{1}{2}\right)$$

are exponent pairs. Moreover, given any pair above this line, there will always be one on the line which gives superior bounds. Thus, in practice, the main interest lies in finding suitable exponent pairs below this line.

We now show that when we apply the van der Corput lemma, the parameters describing the functions arising in the transformed sums are related to those of the original function.

Lemma 16.22 *Suppose that f is in the class*

$$\mathscr{F}(N, [a, b], \lambda, \theta, r, \varepsilon)$$

of functions defined in Definition 16.1, *and that*

$$1 \leq h \leq \min\left(b - a, \frac{2\varepsilon N}{r + \theta}\right).$$

Let $\mathscr{J} = [a, b - h]$ and $f_1(x) = f(x; h) = f(x) - f(x + h)$. Then

$$f_1 \in \mathscr{F}(N, \mathscr{J}, \lambda\theta h, \theta + 1, r - 1, 3\varepsilon).$$

Proof This is a simple verification. Let

$$\phi_1(x) = \phi_1(x; h) = \phi(x) - \phi(x + h),\ \psi(x) = -\lambda\theta h x^{-\theta}.$$

The latter of these two functions plays the same role for f_1 that ϕ does for f. For $1 \leq s \leq r - 1$, we have

$$f_1^{(s)}(x) - \phi_1^{(s)}(x) = -\int_x^{x+h} \left(f^{(s+1)}(y) - \phi^{(s+1)}(y)\right) dy,$$

and in modulus this does not exceed

$$\int_x^{x+h} \varepsilon|\phi^{(s+1)}(y)|\, dy = \varepsilon|\phi_1^{(s)}(x)|.$$

We also have $h\phi'(x) = -\psi(x)$, so that

$$\begin{aligned}\phi_1^{(s)}(x) - \psi^{(s)}(x) &= -\int_x^{x+h} \left(\phi^{(s+1)}(y) - \phi^{(s+1)}(x)\right) dy \\ &= -\int_x^{x+h} \left(\int_x^y \phi^{(s+2)}(z)\, dz\right) dy,\end{aligned}$$

and in modulus this does not exceed

$$\frac{1}{2}h^2\left|\phi^{(s+2)}(x)\right| = \frac{1}{2}h\left|\psi^{(s+1)}(x)\right| \leq \varepsilon\left|\psi^{(s)}(x)\right|.$$

Combining inequalities, we have

$$\big|\phi_1^{(s)}(x)\big| < (1+\varepsilon)\big|\psi^{(s)}(x)\big|,$$

and

$$\big|f_1^{(s)}(x) - \psi^{(s)}(x)\big| < \varepsilon\big(\big|\phi_1^{(s)}(x)\big| + \big|\psi^{(s)}(x)\big|\big) < \big(2\varepsilon + \varepsilon^2\big)\big|\psi^{(s)}(x)\big|. \qquad \square$$

We now formulate the precise terms of 'Process A'.

Theorem 16.23 (Process A) *Suppose that (k,l) is an exponent pair. Then so also is*

$$(k',l') = A(k,l) = \left(\frac{k}{2k+2}, \frac{k+l+1}{2k+2}\right).$$

Proof We first check that $0 \le k' \le \frac{1}{2} \le l' \le 1$. We have $0 \le \frac{k}{2k+2} < \frac{k+1}{2k+2} = \frac{1}{2}$ and $\frac{1}{2} \le \frac{1}{2} + \frac{l}{2k+2} = \frac{k+l+1}{2k+2} \le \frac{1}{2} + \frac{1}{2k+2} \le 1$. Suppose that $N \le a \le b \le 2N$ and $\mathcal{I} = [a,b]$, and suppose that $r' \ge 1 + r(k,l,\theta)$, $0 < \epsilon' \le \frac{1}{3}\varepsilon(k,l,\theta)$, and

$$f \in \mathcal{F}(N, \ell, \lambda, \theta, r', \epsilon).$$

Note that then we have $f \in \mathcal{F}(N, \mathcal{I}, \lambda, \theta, r(k,l,\theta), \epsilon(k,l,\theta))$. We will show that

$$\sum_{n\in\mathcal{I}} e(f(n)) \ll (\lambda N^{-\theta})^{k'} N^{l'}.$$

We observe that

$$l' = \frac{1}{2} + \frac{l}{2k+2} \ge \frac{1}{2} + \frac{1/2}{2\cdot\frac{1}{2}+2} = \frac{2}{3}.$$

As usual, we may assume (16.59). Hence we may suppose that

$$|\mathcal{I}| > N^{2/3}, \tag{16.62}$$

for otherwise the conclusion is immediate. When $1 \le \lambda N^{-\theta} \le N^{\frac{1}{6}}$ we have, by Theorem 16.7,

$$\sum_{n\in\mathcal{I}} e(f(n)) \ll (\lambda N^{-\theta-1})^{1/2} N + (\lambda^{-1} N^{\theta+1})^{1/2} \ll N^{2/3},$$

which is more than sufficient. Thus we may also suppose that

$$\lambda N^{-\theta} \ge N^{\frac{1}{6}}. \tag{16.63}$$

Let

$$S = \sum_{n\in\mathcal{I}} e(f(n)).$$

By the Weyl–van der Corput lemma (Lemma 16.8), with N replaced by $2N$ and $z_n = e(f(n))$ when $n \in \mathcal{J}$ and 0 otherwise, we have

$$|S|^2 \ll N^2H^{-1} + NH^{-1} \sum_{1\le h\le H} |S_1(h)|,$$

where

$$S_1(h) = \sum_{a<n\le b-h} e(f_1(n;h)),$$

and we suppose that

$$1 \le H \le \min\left(b-a, \frac{2\varepsilon' N}{r'+\theta}\right). \tag{16.64}$$

Here H is otherwise at our disposal. Let $\mathcal{J} = [a, b-h]$. Then by Lemma 16.22,

$$f_1 \in \mathcal{F}(N, \mathcal{J}, \lambda\theta h, \theta+1, r'-1, 3\varepsilon'),$$

and by the choices made for r' and ε' above we see that the exponent pair (k,l) applies to f_1. Thus

$$\begin{aligned}|S|^2 &\ll N^2H^{-1} + NH^{-1} \sum_{1\le h\le H} \left((h\lambda N^{-\theta-1})^k N^l + h^{-1}\lambda^{-1}N^{\theta+1}\right)\\ &\ll N^2H^{-1} + N^{l+1-k(\theta+1)}\lambda^k H^k + N^{\theta+2}\lambda^{-1}H^{-1}\log N.\end{aligned} \tag{16.65}$$

By (16.63) the last term is bounded by the first. The good choice for H would be given by

$$H^{k+1} = N^{-l+1+k(\theta+1)}\lambda^{-k} \tag{16.66}$$

provided that this does not violate (16.64), and this leads to the bound

$$S \ll (\lambda N^{-\theta})^{\frac{k}{2k+2}} N^{\frac{k+l+1}{2k+2}},$$

as required. If (16.66) violates (16.64), then we take

$$H = \min\left(b-a, \frac{2\varepsilon' N}{r'+\theta}\right).$$

In this case, the first term on the left-hand side of (16.65) will dominate the second. Hence by (16.62) we have

$$S \ll NH^{-1/2} \ll N^{2/3},$$

and the theorem follows once more. □

We now come to 'Process B'. This corresponds to applying the Poisson summation formula as embodied in Corollary 16.21. For a suitable function f we need to understand how the function $f(x(y)) - yx(y)$ behaves when x and y are related by

$$f'(x(y)) = y. \tag{16.67}$$

Let

$$g(y) = yx(y) - f(x(y)). \tag{16.68}$$

The function $x(y)$ is the inverse function of f', so we have

$$x'(y) = 1/f''(x(y)) \tag{16.69}$$

and

$$g'(y) = x(y) + yx'(y) - f'(x(y))x'(y) = x(y). \tag{16.70}$$

In the special case that

$$f(x) = \phi(x),$$

we have

$$f'(x) = \lambda x^{-\theta}, \quad x(y) = \lambda^{1/\theta} y^{-1/\theta}.$$

Let

$$\psi(y) = \begin{cases} \frac{\lambda^{1/\theta} y^{1-1/\theta}}{1-1/\theta} & \text{when } \theta \neq 1, \\ \lambda \log y & \text{when } \theta = 1. \end{cases} \tag{16.71}$$

Then, in general, we can expect that if f is close to ϕ, then g is close to ψ. We need to show that our concept of close in terms of the first r derivatives of f and ϕ carries through to g and ψ. We have

$$g''(y) = \frac{1}{f''(x(y))}, \tag{16.72}$$

and it is an easy induction on s to show that for $3 \leq s \leq r$ there are coefficients $c_s(\boldsymbol{t})$ which depend only on s and $\boldsymbol{t}$ such that

$$g^{(s)}(y) = \frac{1}{f''(g'(y))^{2s-3}} \sum_{\substack{t_1=2 \\ t_1+\cdots+t_{s-2}=3s-6}}^{s} \cdots \sum_{t_{s-2}=2}^{s} c_s(\boldsymbol{t}) f^{(t_1)}(g'(y)) \cdots f^{(t_{s-2})}(g'(y)), \tag{16.73}$$

and with an obvious convention for an empty sum of products this also holds when $s = 2$.

Lemma 16.24 *Suppose that*

$$f \in \mathscr{F}(N, [a, b], \lambda, \theta, r, \varepsilon),$$

let $\alpha = f'(b)$, $\beta = f'(a)$, *and let* g, ψ *be defined as (16.68) and (16.71), respectively. Then there is a positive number* $C = C(\theta, r)$ *such that*

$$|g^{(s)}(y) - \psi^{(s)}(y)| < C\varepsilon|\psi^{(s)}(y)|$$

whenever $1 \le s \le r$ *and* $y \in [\alpha, \beta]$.

Proof We have $\alpha \ge (1-\varepsilon)\lambda(2N)^{-\theta}$ and $\beta \le (1+\varepsilon)\lambda N^{-\theta}$. Also, for $x \in [N, 2N]$ we have $\phi'(x) \le \lambda N^{-\theta}$, and for $y \in [\alpha, \beta]$ we have, by (16.71),

$$\psi'(y) \le \lambda^{1/\theta}\alpha^{-1/\theta} \le (1-\varepsilon)^{-1/\theta}2N$$

and $\psi'(y) \ge \lambda^{1/\theta}\beta^{-1/\theta} \ge (1+\varepsilon)^{-1/\theta}N$.

By (16.70) and the facts that $x(y)$ is the inverse function of f' and ψ' is the inverse function of ϕ' we have

$$\phi'(g'(y)) - f'(g'(y)) = \phi'(g'(y)) - y = \phi'(g'(y)) - \phi'(\psi'(y)),$$

and by the first mean value of the differential calculus this is

$$= (g'(y) - \psi'(y))\phi''(\xi)$$

for some ξ between $g'(y)$ and $\psi'(y)$. Moreover $g'(y) \in [N, 2N]$ so that by (new number above) we have $\xi^{-1} \ge (1-\varepsilon)^{1/\theta}(2N)^{-1}$, whence

$$|\phi''(\xi)| = \theta\lambda\xi^{-\theta-1} \ge \theta\lambda(1-\varepsilon)^{1+1/\theta}(2N)^{-\theta-1}.$$

Thus

$$|\phi'(x(y)) - f'(x(y))| \ge |g'(y) - \psi'(y)|\theta\lambda(1-\varepsilon)^{1+1/\theta}(2N)^{-\theta-1}.$$

Hence

$$\begin{aligned}|g'(y) - \psi'(y)| &\le \theta^{-1}\lambda^{-1}(1-\varepsilon)^{-1-1/\theta}(2N)^{\theta+1}\varepsilon\phi'(x(y))\\ &\le \theta^{-1}(1-\varepsilon)^{-1-1/\theta}2^{\theta+1}\varepsilon N\\ &\le \theta^{-1}(1-\varepsilon)^{-1-1/\theta}2^{\theta+1}(1+\varepsilon)^{1/\theta}\varepsilon\psi'(y).\end{aligned}$$

This settles the first derivative. To deal with higher derivatives we use (16.73), both as stated and in the special case $f' = \phi'$ (and so $g' = \psi'$). Consider the effect first of all on a single monomial term in the sum (16.73) of replacing f' by ϕ'. We have an expression of the general shape

$$F(z_1, \dots, z_k) = cz_1^{-m}z_2 \cdots z_k.$$

Moreover,

$$F(z_1, \dots, z_k) - F(w_1, \dots, w_k)$$
$$= \sum_{j=1}^{k} \big(F(z_1, \dots, z_j, w_{j+1}, \dots, w_k) - F(z_1, \dots, z_{j-1}, w_j, \dots, w_k)\big),$$

and by the mean value theorem of the differential calculus, provided z_1 and w_1 have the same sign, the general term here is of the form

$$(z_j - w_j)F_j(z_1, \dots, z_{j-1}, \xi_j, w_{j+1}, \dots, w_k),$$

where ξ_j lies between z_j and w_j. Thus in considering $g^{(s)}(y) - \psi^{(s)}(y)$ the difference $z_j - w_j$ becomes an expression of the form $f^{(t)}(g'(y)) - \phi^{(t)}(\psi'(y)) = f^{(t)}(g'(y)) - \phi^{(t)}(g'(y)) + \phi^{(t)}(g'(y)) - \phi^{(t)}(\psi'(y))$. The first difference here is bounded by $\varepsilon|\phi^{(t)}(g'(y))|$, and to the second we may apply the mean value theorem once more to obtain $(g'(y) - \psi'(y))\phi^{(t+1)}(\xi)$ and to this we can apply the first derivative bound obtained above. Thus,

$$|f^{(t)}(g'(y)) - \phi^{(t)}(\psi'(y))| \le \varepsilon|\phi^{(t)}(g'(y))| + C'\varepsilon|\psi'(y)||\phi^{(t+1)}(\xi)|.$$

A straightforward calculation now completes the argument. □

Theorem 16.25 (Process B) *Suppose that (k, l) is an exponent pair. Then so is*

$$(k', l') = B(k, l) = (l - 1/2, k + 1/2).$$

Proof It is immediate that if (k, l) is an exponent pair, then $0 \le l - \frac{1}{2} \le \frac{1}{2} \le k + \frac{1}{2}$. Also, we know that $(0, 1)$ and $(\frac{1}{2}, \frac{1}{2})$ are exponent pairs and that there are no others with $l = \frac{1}{2}$. Hence we may suppose that $l > 1/2$.

Choose $r' \ge \max(3, r(k, l, 1/\theta))$ and let $C = C(\theta, r')$ be as in Lemma 16.24. Then choose ε' so small that

$$0 < \varepsilon' \le \min(1, C^{-1})\varepsilon(k, l, 1/\theta).$$

Let $f \in \mathscr{F}(N, I, \lambda, \theta, r', \varepsilon)$. Choose a, b so that $I = [a, b]$ and define $\alpha = f'(b)$, $\beta = f'(a)$. Now suppose that $J = [M, M'] \subseteq [\alpha, \beta]$ with $M' \le 2M$. Then the function g defined by (16.68) certainly includes J in its support, and so by Lemma 16.24, $g \in \mathscr{F}(M, J, \lambda^{1/\theta}, 1/\theta, r', \varepsilon')$. Hence

$$\sum_{n \in J} e(-g(n)) \ll (\lambda^{1/\theta} M^{-1/\theta})^k M^l + \lambda^{-1/\theta} M^{1/\theta}.$$

We have $\lambda N^{-\theta} \ll \alpha \le \beta \ll \lambda N^{-\theta}$, and as usual we are assuming (16.59).

Summing over $M = \alpha, 2\alpha, \ldots$, we see that for any interval $K = [\alpha, \gamma]$ with $\alpha \le \gamma \le \beta$, we have

$$\sum_{n\in\mathcal{K}} e(-g(n)) \ll N^k(\lambda N^{-\theta})^l + N^{-1}.$$

Moreover, $-f''(x(n)) \asymp \lambda N^{-1-\theta}$ where $x(y)$ is given by (16.57), and since $r \ge 3$, f'' is monotonic. Hence, by partial summation,

$$\sum_{n\in[\alpha,\beta]} \frac{e(-g(n))}{\sqrt{-f''(x(n))}} \ll N^k(\lambda N^{-\theta})^l \lambda^{-\frac{1}{2}} N^{\frac{1}{2}+\frac{\theta}{2}} + N^{-1}\lambda^{-\frac{1}{2}} N^{\frac{1}{2}+\frac{\theta}{2}}$$
$$\ll (\lambda N^{-\theta})^{l-\frac{1}{2}} N^{k+\frac{1}{2}} + \lambda^{-\frac{1}{2}} N^{\frac{1}{2}+\frac{\theta}{2}}.$$

Thus, by Corollary 16.21,

$$\sum_{n\in I} e(f(n)) \ll (\lambda N^{-\theta})^{l-\frac{1}{2}} N^{k+\frac{1}{2}} + \log(1+\lambda N^{-\theta}) + \lambda^{-\frac{1}{2}} N^{\frac{1}{2}+\frac{\theta}{2}}.$$

By (16.59) and the fact that $k \ge 0$, the second term is easily seen to be dominated by the first. Likewise, the third term is bounded by $N^{1/2}$, which is also dominated by the first term. This completes the proof of the theorem. □

We can now compute some exponent pairs (see Table 16.1). It is normal to start from the trivial exponent pair $(0, 1)$. This is equivalent to taking the trivial bound for an exponential sum at the final stage.

If one takes the rational points listed in Table 16.1, adjoins the further point $(1/2, 1)$, and takes the convex hull, then we obtain a set all of whose members are exponent pairs. However, the entries on the second and third rows are in the

Pair	Operation	Pair	Operation
$(0,1)$		$(\frac{1}{2}, \frac{1}{2})$	B
$(\frac{1}{6}, \frac{2}{3})$	AB	$(\frac{1}{6}, \frac{2}{3})$	BAB
$(\frac{1}{14}, \frac{11}{14})$	A^2B	$(\frac{2}{7}, \frac{4}{7})$	BA^2B
$(\frac{1}{9}, \frac{13}{18})$	ABA^2B	$(\frac{2}{9}, \frac{11}{18})$	$BABA^2B$
$(\frac{1}{20}, \frac{33}{40})$	A^2BA^2B	$(\frac{13}{40}, \frac{11}{20})$	BA^2BA^2B
$(\frac{1}{30}, \frac{13}{15})$	A^3B	$(\frac{11}{30}, \frac{8}{15})$	BA^3B
$(\frac{11}{82}, \frac{57}{82})$	ABA^3B	$(\frac{8}{41}, \frac{26}{41})$	$BABA^3B$
$(\frac{11}{186}, \frac{25}{31})$	A^2BA^3B	$(\frac{19}{62}, \frac{52}{93})$	BA^2BA^3B
$(\frac{4}{49}, \frac{75}{98})$	$ABABA^3B$	$(\frac{13}{49}, \frac{57}{98})$	$BABABA^3B$
$(\frac{1}{62}, \frac{57}{62})$	A^4B	$(\frac{13}{31}, \frac{16}{31})$	BA^4B

Table 16.1 *Some exponent pairs.*

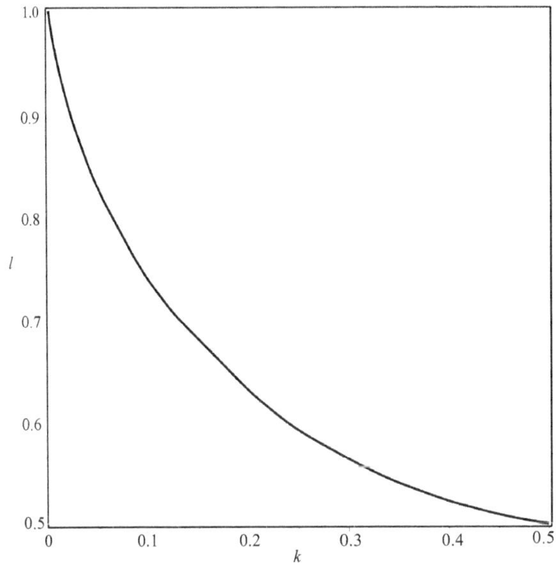

Figure 16.3 Polygonal path determined by 386 exponent pairs.

interior of this convex polygon. As we form longer words, the polygon becomes larger, and it is to be expected that most of the pairs listed will eventually lie in the interior. On the other hand, a new pair constructed with a longer word does not necessarily enlarge the polygon. For example, the operations ABA^4B and $BABA^4B$ produce points that lie in the interior of the present polygon. In many applications one needs to minimize $k+l$. For that purpose, the best of the pairs in Table 16.1 are $(\frac{11}{82}, \frac{57}{82})$ and $(\frac{8}{41}, \frac{26}{41})$.

We now return to the question of bounding the Riemann zeta function on the $\frac{1}{2}$-line.

Theorem 16.26 *Let $\tau = |t| + 4$ and let (k, l) be an exponent pair. Then for any real t,*

$$\zeta(1/2 + it) \ll \tau^{(k+l)/2-1/4} \log \tau.$$

Proof The pattern has already been set in Theorem 16.16, where, in retrospect, we see that that conclusion follows from the exponent pair $(\frac{1}{6}, \frac{2}{3})$. Following the proof there, we see that it suffices to show that when $a \le b \le 2a$ and $a \le \tau^2$, we have

$$\sum_{a \le n \le b} n^{it} \ll a^{\frac{1}{2}} \tau^{\eta},$$

where

$$\eta = \eta(k,l) = \frac{k}{2} + \frac{l}{2} - \frac{1}{4}.$$

Again, as in Theorem 16.16, this is immediate from Corollary 16.6 when $\tau < a \le \tau^2$. By the exponent pairs (k,l) and $B(k,l) = (l - \frac{1}{2}, k + \frac{1}{2})$, we see that

$$\sum_{a \le n \le b} n^{it} \ll \min\left(\left(\tau a^{-1}\right)^k a^l, \left(\tau a^{-1}\right)^{l-1/2} a^{k+1/2}\right) + \tau^{-1} a$$
$$\ll a^{1/2} \min\left(a^{l-k-1/2}\tau^k, a^{k-l+1/2}\tau^{l-1/2}\right) + 1.$$

We replace the minimum of $a^{l-k-1/2}\tau^k$ and $a^{k-l+1/2}\tau^{l-1/2}$ by their geometric mean to obtain the desired conclusion. □

The following corollary is immediate from the exponent pair $\left(\frac{11}{82}, \frac{57}{82}\right)$.

Corollary 16.27 *Let $\tau = |t| + 4$. Then for any real t,*

$$\zeta(1/2 + it) \ll \tau^{27/164} \log \tau.$$

Many questions in analytic number theory can be rephrased in terms of the sawtooth function $s(x)$, which is defined in (E.14) and which we have already used in (16.17).

It is natural to approximate this function by trigonometric polynomials and thereby relate the original question to the theory of exponential sums.

Theorem 16.28 *Suppose that (k,l) is an exponent pair and that $\theta > 0$. Let $r = r(k,l,\theta)$, $\varepsilon = \varepsilon(k,l,\theta)$, $N > 0$, $\lambda > 0$, $I \subseteq [N, 2N]$, and $f \in \mathcal{F}(N, I, \lambda, \theta, r, \varepsilon)$. Then*

$$\sum_{n \in I} s(f(n)) \ll \left(\lambda N^{-\theta}\right)^{k/(k+1)} N^{(k+l)/(k+1)} + \lambda^{-1} N^{\theta}.$$

Proof By Exercise E.2.1.4, for any given positive integer J there are trigonometric polynomials

$$T_{\pm}(x) = \sum_{j=-J}^{J} \widehat{T}_{\pm}(j) e(jx)$$

with period 1 and degree at most J such that

$$T_{-}(x) \le s(x) \le T_{+}(x)$$

for all x, $\widehat{T}_\pm(0) = \pm 1/(2J+2)$, and $\widehat{T}_\pm(j) \ll 1/|j|$ for $j \ne 0$. Hence

$$\sum_{n \in I} s(f(n)) \le \frac{N}{2J+2} + \sum_{0<|j|\le J} \widehat{T}_+(j) \sum_{n\in I} e(jf(n))$$
$$\le \frac{N}{2J+2} + C\sum_{j=1}^{J} \frac{1}{j}\left|\sum_{n\in I} e(jf(n))\right|.$$

Similarly,

$$\sum_{n\in I} s(f(n)) \ge \frac{-N}{2J+2} - C\sum_{j=1}^{J}\frac{1}{j}\left|\sum_{n\in I} e(jf(n))\right|.$$

Moreover, for $f \in \mathscr{F}(N, I, \lambda, \theta, r, \varepsilon)$ we have $|j|f \in \mathscr{F}(N, I, |j|\lambda, \theta, r, \varepsilon)$, and so the exponent pair (k, l) applies to each of the sums

$$\sum_{n\in\mathcal{J}} e(jf(n)).$$

Thus

$$\sum_{n\in I} s(f(n)) \ll \frac{N}{J+1} + \sum_{j=1}^{J}\frac{1}{j}\big((j\lambda N^{-\theta})^k N^l + j^{-1}\lambda^{-1}N^{\theta}\big)$$
$$\ll \frac{N}{J+1} + (J\lambda N^{-\theta})^k N^l + \lambda^{-1}N^{\theta}.$$

We take

$$J = \left[\lambda^{-k/(k+1)} N^{(1+k\theta-l)(k+1)}\right],$$

and this gives the desired conclusion when $k \ne 0$. When $k = 0$ we have $l = 1$ and the conclusion is trivial. □

One obvious application of the above is to the Dirichlet divisor problem.

Theorem 16.29 *Let $\Delta(x) = \sum_{n\le x} d(n) - x\log x - (2C_0 - 1)x$, and suppose that (k, l) is an exponent pair. If $(k, l) \ne (\frac{1}{2}, \frac{1}{2})$, then*

$$\Delta(x) \ll x^{\frac{k+l}{2k+2}}.$$

Proof From the initial steps of the proof of Theorem 2.3 we see that

$$\sum_{n\le x} d(n) = 2\sum_{n\le\sqrt{x}} \frac{x}{n} - 2\sum_{n\le\sqrt{x}} s(x/n) - \left[\sqrt{x}\right]^2 - \left[\sqrt{x}\right], \tag{16.74}$$

where $s(y)$ is as in (E.14), and from the initial steps of the proof of (1.26) we have

$$\sum_{n\le y}\frac{1}{n} = \log y + C_0 - \frac{s(y)}{y} - \int_y^\infty \frac{s(u)}{u^2}\,du.$$

We observe that $\int_y^u s(v)\,dv \ll 1$, and so by integrating the last term by parts it follows that it is $\ll y^{-2}$. Hence

$$\sum_{n\le y}\frac{1}{n} = \log y + C_0 - \frac{s(y)}{y} + O\left(1/y^2\right).$$

We also have

$$x - \left[\sqrt{x}\right]^2 - \left[\sqrt{x}\right] = 2\sqrt{x}s(\sqrt{x}) + O(1).$$

Inserting these two expressions into (16.74) gives

$$\Delta(x) = -2\sum_{n\le\sqrt{x}} s(x/n) + O(1).$$

We now divide the interval of summation into subintervals of the form $[N, N']$, with $N' \le 2N$ and $N \le \sqrt{x}$, and appeal to Theorem 16.28 with $\theta = 2$. The contribution from a typical such subinterval is

$$\ll (xN^{-2})^{k/(k+1)}N^{(k+l)/(k+1)} + x^{-1}N^2 \ll x^{k/(k+1)}N^{(k-l)/(k+1)} + x^{-1}N^2.$$

Since $(k, l) \ne (\frac{1}{2}, \frac{1}{2})$ we have $l > k$. Hence on summing the contribution from the different subintervals we obtain the bound

$$X^{(k+l)/(2k+2)} + 1 \ll x^{(k+l)/(2k+2)},$$

as required. □

For completeness we observe that in the case of the exponent pair $(\frac{1}{2}, \frac{1}{2})$ the proof gives an extra factor of $\log x$ in the conclusion. More interestingly, one can observe that $(k', l') = A(k, l)$ satisfies

$$k' = \frac{k}{2k+2}, \qquad l' = \frac{k+l+1}{2k+2},$$

and so the exponent of x in the conclusion is $k' + l' - \frac{1}{2}$. With the exponent pairs obtained by the A and B operations there is symmetry in the line $l = k + \frac{1}{2}$ between those in which the last operation is an A and those in which the last operation is a B. Thus, just as in Theorem 16.26, we are interested in exponent pairs (k', l') in which $k' + l'$ is minimal. Amongst those listed above $(k, l) = (\frac{11}{30}, \frac{8}{15})$ (which gives $(k', l') = (\frac{11}{82}, \frac{57}{82})$, of course) gives the following corollary.

Corollary 16.30 *Let $\Delta(x)$ denote the error term in the divisor problem, as defined in Theorem 16.29. Then*

$$\Delta(x) \ll x^{\frac{27}{82}}.$$

16.3.1 Exercises

1. Let $I(\alpha) = \int_0^1 e(\alpha x + \log\log e/x)\,dx$. Show that

$$I(\alpha) = \frac{1}{2\pi i\alpha}\big(e(\alpha) - e(\log\log\alpha)\big) + o(1/\alpha)$$

as $\alpha \to \infty$. Note that this is larger than $f''(x_0)^{-1/2}$. Why?

2. Let $I(\alpha) = \int_0^1 e\left(\alpha x + \log\log\frac{1}{x(1-x)}\right)\,dx$. Show that

$$|I(\alpha)| \asymp \frac{1}{\alpha\sqrt{\log\alpha}}$$

as $\alpha \to +\infty$.

3. Show that if N is a positive integer, then

$$\sum_{n=1}^{N} e\big((n/3)^{3/2}\big) = 2^{1/4}3^{-3/2}N^{3/4} + O\big(N^{1/4}\big).$$

4. Let $E(n)$ denote the number of words using the two letters A and B with the property that the last letter is B, and the word does not contain a pair of consecutive B's. Among such words, let $A(n)$ be the number in which the leftmost letter is A, and $B(n)$ the number in which the leftmost letter is B. Thus $E(n) = A(n) + B(n)$. Note that $A(1) = 0$, $B(1) = 1$, $A(2) = 1$, $B(2) = 0$. Let F_n denote the nth Fibonacci number, as defined by the relations $F_0 = 0$, $F_1 = 1$, $F_{n+1} = F_n + F_{n-1}$.

   (a) Show that if n is an integer such that both $A(n) = F_{n-1}$ and $B(n) = F_{n-2}$, then $A(n+1) = F_n$ and $B(n+1) = F_{n-1}$.

   (b) Deduce that $E(n) = F_n$ for all positive integers n.

   (c) Suppose that you start with $(0, 1)$, and use words of length from 1 to n to generate new exponent pairs. Show that, in total, you have F_{n+2} exponent pairs. (Ignore the fact that the pairs generated are not guaranteed to bc distinct. In fact they are not all distinct. In Table 16.1 we see that AB and BAB generate the same point.)

5. (a) Let $\chi(d) = \left(\frac{-4}{d}\right)$ be the non-principal character modulo 4, and let

$$S(y) = \sum_{n\le y} \chi(n).$$

Show that

$$S(y) = \frac{1}{2} - s\left(\frac{y-1}{4}\right) + s\left(\frac{y-3}{4}\right),$$

and that

$$\sum_{n\le y}\frac{\chi(n)}{n}=\frac{\pi}{4}+\frac{S(y)-\frac{1}{2}}{y}+O(1).$$

(b) Let

$$r(n)=4\sum_{d|n}\chi(d),$$
$$R(x)=\sum_{n\le x}r(n)-\pi x,$$
$$T(y;a,b)=\sum_{n\le y}s\left(\frac{x-a}{4n+b}\right).$$

Show that

$$\frac{1}{4}R(x)=T(\sqrt{x};0,1)-T(\sqrt{x};0,3)$$
$$+T(\sqrt{x};3,0)-T(\sqrt{x};1,0)+O(1).$$

(c) Suppose that (k,l) is an exponent pair other than $\left(\frac{1}{2},\frac{1}{2}\right)$. Show that

$$R(x)\ll x^{\frac{k+l}{2k+2}},$$

and in particular that

$$R(x)\ll x^{\frac{27}{82}}.$$

6. (a) Let $Q(x,h)$ denote the number of square-free numbers q with $x-h<q\le x$. Suppose that $1\le h\le\frac{x}{2}$ and that $\sqrt{h}\le z\le\sqrt{x}$. Show that

$$Q(x,h)=\frac{6h}{\pi^2}+O((R+S)\log x+\sqrt{h}),$$

where

$$R=\sup_{a\le z}\sup_{b\le 2a}\sup_{x-h\le y\le x}\left|\sum_{a\le n\le b}s\left(\frac{y}{n^2}\right)\right|$$

and

$$S=\sup_{a\le xz^{-2}}S(a),\quad S(a)=\sup_{b\le 2a}\sup_{x-h\le y\le x}\left|\sum_{a\le n\le b}s\left(\frac{y^{1/2}}{n^{1/2}}\right)\right|.$$

(b) Show that

$$R\ll x^{1/3}z^{-1/3}+xz^{-3},$$

and that if (k,l) is an exponent pair, then

$$S\ll x^{k/(k+1)}a^{(l-2k)/(k+1)}+x^{-1}z^{3}.$$

(c) Show that there is a positive number C such that whenever $Cx^{2/9}\log x \le h \le x$ there is a square-free number q with $x - h < q \le x$.

16.4 Notes

Section 16.1. Exponential integrals have been used and studied for centuries. The plot in Figure 16.1 is *Euler's Spiral.* Leonhard Euler (1707–1783) encountered his spiral in 1744 while investigating a problem concerning elasticity posed by Jakob Bernoulli. Euler noted then that the spiral converges to a single point, but that it is difficult to name that point. In 1781 he found the limit, which is to say that he proved (16.3). In 1818, French physicist A.-J. Fresnel (1788–1827), in the course of his seminal investigation of the diffraction of light defined the integrals

$$S(t) = \int_0^t \sin\left(\frac{\pi u^2}{2}\right)\,du, \qquad C(t) = \int_0^t \cos\left(\frac{\pi u^2}{2}\right)\,du.$$

These are now known as the *Fresnel integrals.* Here, $C(t) + iS(t) = z(t)$ as defined in the caption of Figure 16.1, but Fresnel was unaware of Euler's prior work. He spent considerable effort to compute values of his integrals, and later the French physicist M. A. Cornu (1841–1902) computed detailed tables of $z(t)$, also for the purposes of optics. Today, such calculations are done for us since common software provides the error function,

$$\operatorname{erf} z = \frac{2}{\sqrt{\pi}}\int_0^z e^{-u^2}\,du,$$

even with complex arguments, and

$$z(t) = \frac{\sqrt{\pi}}{2}\left(\frac{1}{\sqrt{2}} + \frac{i}{\sqrt{2}}\right)\operatorname{erf}\left(\left(\frac{1}{\sqrt{2}} - \frac{i}{\sqrt{2}}\right)t\right).$$

Euler's spiral (also known as Cornu's spiral), was discovered independently a third time, in 1890, for the following reason. The point $z(t)$ moves with velocity $z'(t) = e^{it^2}$; thus $|z'(t)| = 1$, so the arc length of the curve $z(t)$ for $0 \le t \le T$ is exactly T. Moreover, the acceleration on the curve is $z''(t) = 2ite^{it^2}$, which has absolute value $2t$ for $t > 0$. If a train were to pass from a straight line directly onto a circular arc, its acceleration would undergo a jump discontinuity, which is uncomfortable for passengers and hard on the equipment. So railway beds are designed to pass from a straight line onto (a suitably scaled copy of) Euler's spiral. When the acceleration has reached the desired level, the course then continues on a circular arc, and finally transitions back on a segment of such a spiral to a straight line. This technique is used also in the construction of highways and roller coasters. For a detailed account of the history of Euler's spiral, see Levien (2008).

Exercises 16.1.1.2 and 16.1.1.3 are useful in applications of the Hardy–Littlewood method. See Lemma 10.1 of Hua (1965), Theorem 7.3 of Vaughan (1997), and Theorem 3.1 of Talmage (2022).

Section 16.2. The methods developed here were first studied systematically in van der Corput (1921) and van der Corput (1922), with the main intent of applying them to the Dirichlet divisor problem. Van der Corput does acknowledge Weyl (1916) for approximations of the kind in Theorem 16.13, and one has to believe that he was greatly influenced by Weyl's paper. Landau, Hardy, Littlewood, and their colleagues and students, beginning before WWI, had organized written accounts of everything that was known concerning the Riemann zeta function and the distribution of primes, and kept it up to date as advances were made. There is an intriguing footnote on page 316 of Weyl (1916) which states 'Vgl. H. Bohr und J. E. Littlewood, The Riemann Zeta-function and the Theory of Prime Numbers (Cambridge Tracts in Mathematics and Mathematical Physics; noch nicht erschienen)'! Presumably there was already an intent to publish this material as a Cambridge Tract, but this was interrupted by WWI and perhaps also by fast-moving developments in research. The more important research appeared in papers such as Hardy & Littlewood (1916a,b). Some of it was promised but never published. See the announcement in Littlewood (1922), which presumably was overtaken by developments elsewhere. A little later the Bohr–Littlewood manuscript was divided into two and appeared as Cambridge Tracts, by Ingham (1932) and Titchmarsh (1930). The latter was expanded into the celebrated text Titchmarsh (1951, 1986) (the latter being the second edition), and was the place from which the authors of this work initially learnt the material. Many of the estimates of this section are also used extensively in harmonic analysis. See Stein (1993).

Theorem 16.5 as stated occurs in Landau (1928). In a slightly weaker form it occurs in Kusmin (1927). See Mordell (1958) for some generalizations and other references. The trigonometric polynomial (16.7) was noted by Hardy & Littlewood (1916a,b), who established the estimate (16.8). Lemma 16.8 is the Fundamental Inequality of (van der Corput, 1931, p. 407), and Theorem 16.9 is a special case of the 'Dritte Haupteigenschaft' in the same article. Theorem 16.17 is from Fujii, Gallagher, Montgomery (1976).

Section 16.3. The van der Corput method, including exponent pairs, originates in van der Corput (1921, 1922), and was developed further by Phillips (1933).

Rankin (1955) optimized the choice of exponent pairs for the purpose of estimating $\zeta(\frac{1}{2}+it)$. For further expositions of van der Corput's method, see

Graham & Kolesnik (1991) and §3.3 of Montgomery (1994). For a discussion of applications of the van der Corput method to the zeta function, see Graham & Kolesnik (1991, §4.1) or Titchmarsh (1986, §5.20).

The *exponent pair conjecture* is the conjecture that $(k, \frac{1}{2}+k)$ is an exponent pair for every k with $0 < k \le \frac{1}{2}$.

There is a more recent history of small reductions which transcend the methods described here and their two-dimensional variants. These depend on a method for treating exponential sums introduced in Bombieri & Iwaniec (1986a,b), and further refined by Iwaniec & Mozzochi (1988) and Huxley & Watt (1988). Further work by Huxley, culminating in Huxley (2000, 2003) has established that

$$\zeta\left(\tfrac{1}{2}+it\right) \ll \tau^{\phi+\varepsilon}, \quad R(x) \ll x^{\theta+\varepsilon}, \quad \Delta(x) \ll x^{\theta+\varepsilon},$$

with

$$\phi = \frac{32}{205} = 0.1560975609\ldots, \qquad \theta = \frac{131}{416} = 0.3149038461\ldots.$$

The values $\phi = 0$ and $\theta = \frac{1}{4}$ follow from the exponent pair conjecture, and these conjectural values for ϕ and θ are known to be essentially best possible. That is, there is a limitation as to how small the upper bounds can be for the Dirichlet divisor and Gauss lattice point problems. In that regard there is also a long history beginning with Hardy (1916) and culminating in Soundararajan (2003), which also contains an overview of previous work in the area.

Over the years, further exponent pairs have been found that cannot be obtained via the A and B operations. For the current state of the art on this topic, see Tao, Trudgian, and Yang (2025).

16.5 References

Bombieri, E. & Iwaniec, H. (1986a). On the order of $\zeta(1/2+it)$, *Ann. Scuola Norm. Sup. Pisa Cl. Sci.* **13**, 449–472.

(1986b). Some mean value theorems for exponential sums, *Ann. Scuola Norm. Sup. Pisa Cl. Sci.* **13**, 473–486.

van der Corput, J. G. (1921). Zahlentheoretische Abschätzungen, *Math. Ann.* **84**, 53–79.

(1922). Verschärfung der Abschätzung beim Teilerproblem, *Math. Ann.* **87**, 39–65.

(1931). Diophantische Ungleichungen. I. Zur Gleichverteilung Modulo Eins. *Acta Math.* **56** (1), 373–456.

Fujii, A., Gallagher, P. X., & Montgomery, H. L. (1976). Some hybrid bounds for character sums and Dirichlet L-series, *Topics in Number Theory* (Proc. Colloq., Debrecen, 1974), Colloq. Math. Soc. János Bolyai, Vol. 13, Amsterdam: North-Holland, pp. 41–57.

Graham, S. W., & Kolesnik, G. (1991). *Van der Corput's Method for Exponential Sums,* London Math. Soc. Lecture Notes 126, Cambridge: Cambridge University Press, iii+120 pp.

Hardy, G. H. (1916). On Dirichlet's divisor problem, *Proc. Lond. Math. Soc.* (2) **15**, 1–25.

Hardy, G. H. & Littlewood, J. E. (1916a). Some problems of Diophantine approximation: A remarkable trigonometric series, *Proc. Natl. Acad. Sci.* **2**, 583–586.

(1916b). Contributions to the theory of the Riemann zeta-function and the theory of the distribution of primes, *Acta Math.* **41**, 119–196.

Hua, L. K. (1965). *Additive Theorem of Prime Numbers*, Trans. of Math. Monographs, Volume: 13, Providence: Amer. Math. Soc., 190 pp.

Huxley, M. N. (2000). Integer points, exponential sums and the Riemann zeta function. In Bennett, M. A., Berndt, B. C., Boston, N., Diamond, H. G., Hildebrand, A. J., & Philipp, W. (eds.). *Number Theory for the Millennium, II* (University of Illinois, Urbana, May 21–26, 2000), Natick: A K Peters, pp. 275–290.

(2003). Exponential sums and lattice points III, *Proc. London Math. Soc.* **87** (3), 591–609.

Huxley, M. N. & Watt, N., (1988). Exponential sums and the Riemann zeta-function, *Proc. London Math. Soc.* **57**, 1–24.

Ingham, A. E. (1932). *The Distribution of Prime Numbers*, Cambridge Tracts in Mathematics and Mathematical Physics 30, Cambridge: Cambridge University Press, 114 pp.

Iwaniec, H. & Mozzochi, C. J. (1988). On the divisor and circle problems, *J. Number Theory* **29**, 60–93.

Kusmin, R. O. (1927). On some trigonometric inequalities. (Über einige trigonometrische Ungleichungen.), *Leningrad, J. Soc. Phys.-Math.* **1**, 233–239.

Landau, E. (1928). Über einer trigonometrische Summen, *Nachr. Akad. Wiss. Göttingen* **1928**, 21–24; *Collected Works*, Vol. 9, Essen: Thales Verlage (1987), pp. 41–44.

Levien, R. (2008). *The Euler Spiral: A Mathematical History*, Tech. Report No. UCB/EECS-2008-111, `www2.eecs.berkeley.edu/Pubs/TechRpts/2008/EECS-2008-111.html`

Littlewood, J. E. (1922). Researches in the theory of Riemann zeta-function, *Proc. London Math. Soc.* (2) **20**, xxii–xxviii. Records for 10 February 1921.

Montgomery, H. L. (1994). *Ten Lectures on the Interface Between Analytic Number Theory and Harmonic Analysis*, CBMS **84**, Providence: Amer. Math. Soc., xiii+220 pp.

Mordell, L. J. (1958). On the Kusmin–Landau inequality for exponential sums, *Acta Arith.* **4**, 3–9.

Phillips, E. (1933). The zeta-function of Riemann; further developments of van der Corput's method, *Quart. J. Math. Oxford* **4**, 209–225.

Rankin, R. A. (1955). Van der Corput's method and the theory of exponent pairs, *Quart. J. Math. Oxford* (2) **6**, 147–153.

Soundararajan, K. (2003). Omega results for the divisor and circle problems, *Int. Math. Res. Not.* **2003**, 1987–1998.

Stein, E. (1993). *Harmonic Analysis: Real-Variable Methods, Orthogonality and Oscillatory Integrals*, Princeton: Princeton University Press.

Talmage, A. (2022). Prime solutions of diagonal diophantine systems, arXiv:2209.06934 [math.NT].

Tao, T., Trudgian, T., and Yang, A. (2025). New exponent pairs, zero density estimates, and zero additive energy estimates: a systematic approach, *Math. Comp.* Nov. 2025 DOI:10.1090/mcom/4138.

Titchmarsh, E. C. (1930). *The Zeta-Function of Riemann*, Cambridge Tracts in Mathematics and Mathematical Physics 26, Cambridge: Cambridge University Press.

(1951). *The Theory of the Riemann Zeta-Function*, First edition, Oxford: Clarendon Press.

(1986). *The Theory of the Riemann Zeta-Function*, Second edition, revised by D. R. Heath-Brown, Oxford: Clarendon Press, x+412 pp.

Vaughan, R. C. (1997). *The Hardy–Littlewood Method*, Second edition, Cambridge: Cambridge University Press, xiii+232 pp.

Weyl, H. (1916). Über die Gleichverteilung von Zahlen mod. Eins, *Math. Ann.* **77** (3), 313–352.

17
Estimates for Sums over Primes

17.1 Principles of the Method

Let

$$S = \sum_{n \le N} f(n)\Lambda(n).$$

If f is monotonic, then we can estimate S by using the Prime Number Theorem and integration by parts. If f is multiplicative, then we can gain information concerning S by studying the properties of the associated Dirichlet series $\sum f(n)n^{-s}$. This has already been especially successful when f is of the form $f(n) = \chi(n)n^{-\alpha}$. We now introduce an entirely different method that is most successful when f is *far* from being multiplicative. Let $P = \prod_{p \le \sqrt{N}} p$. Vinogradov (1937a,b) had the idea of writing

$$f(1) + \sum_{\sqrt{N} < p \le N} f(p) = \sum_{\substack{1 \le n \le N \\ (n,P)=1}} f(n) = \sum_{\substack{t \mid P \\ t \le N}} \mu(t) \sum_{r \le N/t} f(rt).$$

If we can demonstrate that there is considerable cancellation in the inner sum on the right, then we can obtain a non-trivial estimate for the left-hand side. However, when t is near N in size, one expects to have little cancellation in the inner sum on the right, and indeed when $N/2 < t \le N$ the sum has only one term, and hence no cancellation at all. Thus the terms on the right must be rearranged before satisfactory estimates can be derived. This approach, known as *Vinogradov's method for prime number sums*, is rather complicated. The general aim is to express S as a linear combination of sums of the following two sorts:

$$\sum_{t \le T} a(t) \sum_{r \le N/t} f(tr) \tag{17.1}$$

and

$$\sum_{\substack{mk\le N\\ m>U\\ k>V}} b(m)c(k)f(mk), \tag{17.2}$$

where $a(t)$, $b(m)$, and $c(k)$ are certain fixed arithmetic functions (independent of f), and T, U, and V are parameters. Such sums are said to be of Type I and Type II, respectively. In the Type I sum we choose T to be small compared with N, so that we have a hope of showing that the inner sum enjoys some cancellation. Although $b(m)$ and $c(k)$ are fixed, we generally treat a Type II sum as if it were a general bilinear form. In any case, it is essential that we can avoid small values of m and small values of k. Within this framework, Vaughan (1977a,b) devised a variant known as *Vaughan's version of Vinogradov's method* (V^3M), which we now describe.

We start by expressing $\Lambda(n)$ as a linear combination of several other arithmetic functions. Put

$$F(s) = \sum_{d\le U} \Lambda(d)d^{-s}, \qquad G(s) = \sum_{k\le V} \mu(k)k^{-s}. \tag{17.3}$$

Clearly,

$$\begin{aligned} -\frac{\zeta'}{\zeta}(s) &= F(s) - \zeta(s)F(s)G(s) - \zeta'(s)G(s) \\ &\quad + (-\zeta'(s) - F(s)\zeta(s))\left(\frac{1}{\zeta(s)} - G(s)\right) \end{aligned} \tag{17.4}$$

for $\sigma > 1$. We write

$$\zeta(s) = \sum_{r=1}^{\infty} r^{-s}, \qquad -\zeta'(s) = \sum_{m=1}^{\infty} (\log m)m^{-s},$$

and calculate the coefficients of the four Dirichlet series on the right in (17.4). Thus *Vaughan's identity* asserts that

$$\Lambda(n) = c_1(n) + c_2(n) + c_3(n) + c_4(n), \tag{17.5}$$

where

$$c_1(n) = \begin{cases} \Lambda(n) & \text{if } n \le U, \\ 0 & \text{if } n > U, \end{cases}$$

$$c_2(n) = -\sum_{\substack{rdk=n\\ d\le U\\ k\le V}} \Lambda(d)\mu(k),$$

and

$$c_3(n) = \sum_{\substack{mk=n \\ k\le V}} \mu(k) \log m.$$

To calculate $c_4(n)$ we observe that in the first factor of the final product in (17.4), the coefficient of m^{-s} is

$$\log m - \sum_{\substack{d|m \\ d\le U}} \Lambda(d) = \sum_{\substack{d|m \\ d>U}} \Lambda(d).$$

Thus

$$c_4(n) = \sum_{\substack{mk=n \\ m>U,\, k>V}} \left(\sum_{\substack{d|m \\ d>U}} \Lambda(d) \right) \mu(k).$$

We multiply (17.5) through by $f(n)$ and sum to see that

$$S = S_1 + S_2 + S_3 + S_4, \tag{17.6}$$

where

$$S_i = \sum_{n\le N} f(n) c_i(n).$$

Thus

$$S_1 = \sum_{n\le U} f(n)\Lambda(n).$$

We generally estimate this sum trivially. Let

$$a(t) = -\sum_{\substack{dk=t \\ d\le U \\ k\le V}} \Lambda(d)\mu(k).$$

Then $c_2(n) = \sum_{t|n} a(t)$, and hence

$$S_2 = \sum_{t\le UV} a(t) \sum_{r\le N/t} f(tr), \tag{17.7}$$

which is a Type I sum. Since $|a(t)| \le \sum_{d|t} \Lambda(d) = \log t \le \log UV$, it follows that

$$S_2 \ll (\log UV) \sum_{t\le UV} \left| \sum_{r\le N/t} f(rt) \right|. \tag{17.8}$$

As for S_3, we find that

$$S_3 = \sum_{k\le V} \mu(k) \sum_{m\le N/k} f(km) \log m.$$

This is not quite a Type I sum, but $\log m$ is smoothly increasing, so we write $\log m = \int_1^m dw/w$ and invert the order of integration and summation to see that

$$S_3 = \int_1^N \sum_{k\le V} \mu(k) \sum_{w\le m\le N/k} f(km)\,\frac{dw}{w}$$

$$\ll (\log N) \max_{w\ge 1} \sum_{k\le V} \Bigg| \sum_{w\le m\le N/k} f(km) \Bigg|. \tag{17.9}$$

This is still not quite a Type I sum, but is instead the maximum over a family of Type I sums. However, in most cases our estimate for the sum over m is uniform in w, so for practical purposes we have a Type I sum. Let

$$b(m) = \sum_{\substack{d\mid m\\ d>U}} \Lambda(d). \tag{17.10}$$

Then

$$c_4(n) = \sum_{\substack{mk=n\\ m>U\\ k>V}} b(m)\mu(k), \tag{17.11}$$

and so

$$S_4 = \sum_{\substack{mk\le N\\ m>U\\ k>V}} b(m)\mu(k)f(mk) = \sum_{U<m\le N/V} b(m) \sum_{V<k\le N/m} \mu(k)f(mk). \tag{17.12}$$

This is a Type II sum. Suppose that $\Delta(M) = \Delta(M, N, f)$ is defined so that

$$\Bigg| \sum_{M<m\le 2M} b_m \sum_{k\le N/m} c_k f(mk) \Bigg| \le \Delta(M) \Bigg(\sum_{M<m\le 2M} |b_m|^2 \Bigg)^{1/2} \Bigg(\sum_{k\le N/M} |c_k|^2 \Bigg)^{1/2} \tag{17.13}$$

for arbitrary complex numbers b_m and c_k. By cutting the interval $U \le m \le N/V$ into $\ll \log N$ subintervals of the form $M < m \le 2M$, we deduce that

$$S_4 \ll (\log N) \max_{U\le M\le N/V} \Delta(M) \Bigg(\sum_{M<m\le 2M} b(m)^2 \Bigg)^{1/2} \Bigg(\sum_{k\le N/M} |\mu(k)|^2 \Bigg)^{1/2}.$$

Since $|b(m)| \le \log m$, the sum over m is $\ll M(\log 2M)^2$. The sum over k is $\ll N/M$, so

$$S_4 \ll N^{1/2}(\log N)^2 \max_{U\le M\le N/V} \Delta(M). \tag{17.14}$$

We interrupt our development at this point in order to assess the situation. For the purposes of discussion, in this paragraph only, we assume that $|f(n)| \le 1$ for all n. The bound $S \ll N$ is trivial, and if f is oscillatory we hope to show that $S = o(N)$. Trivially $S_1 \ll U$, so S_1 poses no problem provided that $U = o(N)$. In (17.8) the trivial bound would be that

$$S_2 \ll (\log UV) \sum_{t \le UV} \frac{N}{t} \ll N(\log UV)^2.$$

Thus in order to get a bound that is $o(N)$ we only need to demonstrate a modest amount of cancellation in the sum over r in (17.8), and even this only on average over t. We note, however, that there will be little or no cancellation if the inner sum has very few terms (a single term is the worst case). For this reason it will be necessary to choose the parameters U and V so that UV is considerably smaller than N. Similar remarks apply to (17.9) where the situation is even more favourable since the range of k in (17.9) is shorter than that of t in (17.8). To obtain a trivial bound for $\Delta(M)$ we first observe that

$$\left| \sum_{M<m\le 2M} b_m \sum_{k \le N/m} c_k f(mk) \right| \le \sum_{M<m\le 2M} |b_m| \sum_{k\le N/M} |c_k|.$$

By two applications of Cauchy's inequality, this in turn is

$$\ll \left(M \cdot N/M\right)^{1/2} \left(\sum_{M<m\le 2M} |b_m|^2 \right)^{1/2} \left(\sum_{k\le N/M} |c_k|^2 \right)^{1/2}.$$

Thus the bound $\Delta(M) \ll N^{1/2}$ is trivial. By inserting this in (17.14) we deduce that $S_4 \ll N(\log N)^2$ trivially. That is, we will be able to show that $S_4 = o(N)$ if we can obtain a bound for $\Delta(M)$ that is only a power of a logarithm smaller than trivial. In summary, it seems that we have not dug ourselves into too deep a hole, and that we can expect to show that $S = o(N)$ whenever we can derive estimates that are only moderately better than trivial. We note, however, that if f were to be unimodular and totally multiplicative, then we might obtain non-trivial estimates for S_2 and S_3, but no non-trivial estimate for $\Delta(M)$ can hold because of the possibility that $b_m = \overline{f(m)}$ and $c_k = \overline{f(k)}$. Despite this observation, we shall find in Chapter 20 that we can still use our present approach when we average over several multiplicative functions f_i.

In order to estimate $\Delta(M)$, we first observe that by Cauchy's inequality the left-hand side of (17.13) is

$$\le \left(\sum_{M<m\le 2M} |b_m|^2 \right)^{1/2} \left(\sum_{M<m\le 2M} \left| \sum_{k\le N/m} c_k f(mk) \right|^2 \right)^{1/2}.$$

Here the second sum over m is

$$= \sum_{j \le N/M} c_j \sum_{k \le N/M} \overline{c_k} \sum_{\substack{M < m \le 2M \\ m \le N/j \\ m \le N/k}} f(mj)\overline{f(mk)}. \tag{17.15}$$

By the arithmetic-geometric mean inequality we know that $|c_j c_k| \le \frac{1}{2}|c_j|^2 + \frac{1}{2}|c_k|^2$. Thus (17.15) is

$$\le \sum_{k \le N/M} |c_k|^2 \sum_{j \le N/M} \left| \sum_{\substack{M < m \le 2M \\ m \le N/j \\ m \le N/k}} f(mj)\overline{f(mk)} \right| \tag{17.16}$$

$$\le \left(\sum_{k \le N/M} |c_k|^2 \right) \left(\max_{k \le N/M} \sum_{j \le N/M} \left| \sum_{\substack{M < m \le 2M \\ m \le N/j \\ m \le N/k}} f(mj)\overline{f(mk)} \right| \right).$$

Hence

$$\Delta(M) \le \left(\max_{k \le N/M} \sum_{j \le N/M} \left| \sum_{\substack{M < m \le 2M \\ m \le N/j \\ m \le N/k}} f(mj)\overline{f(mk)} \right| \right)^{1/2}, \tag{17.17}$$

and so by (17.14) we conclude that

$$S_4 \ll N^{1/2}(\log N)^2 \times \max_{U \le M \le N/V} \max_{k \le N/M} \left(\sum_{j \le N/M} \left| \sum_{\substack{M < m \le 2M \\ m \le N/j \\ m \le N/k}} f(mj)\overline{f(mk)} \right| \right)^{1/2}. \tag{17.18}$$

Clearly our bound (17.8) for S_2 becomes better when UV is reduced. On the other hand, our bound above for S_4 becomes better when U and V are increased. In practice, we choose the parameters to balance these bounds.

Our strategy for bounding S_4 may be inferior, for two reasons. In the first place, we need to bound the double sum on the left-hand side of (17.13) not for arbitrary b_m and c_k but only in the special case that $b_m = b(m)$ and $c_k = \mu(k)$. Secondly, the double sum on the left-hand side of (17.13) is a linear function of the b_m, and is also linear in the c_k. Such an expression is known as a *bilinear form*, and in Appendix G we develop a general theory concerning bounds for bilinear forms. Indeed, we could have passed directly from (17.13) to (17.17) simply by appealing to Corollary G.4. Although we have taken a more elementary route, the general theory offers some insights. From Theorem G.1 we see that from (17.13) to (17.15) we have thrown nothing away if the bounds are to hold for arbitrary b_m and c_k. In (17.15) we again have a bilinear form, but this time the coefficient matrix is not only square, but Hermitian as well, and hence normal. Thus by Corollary G.11 the problem is to determine (or estimate) the spectral radius of this matrix. In passing from (17.15) to (17.16) we have in effect derived a bound for this spectral radius, but our bound may be considerably larger than the truth.

In S_2, which is a Type I sum, when t is large the inner sum is over a shorter interval, with the result that there may be less cancellation. In such a situation, sometimes a better estimate can be obtained by writing

$$S_2 = \sum_{t\le U} + \sum_{U<t\le UV} = S_2^{(1)} + S_2^{(2)}, \tag{17.19}$$

say. Then we treat $S_2^{(1)}$ in as we did S_1, i.e., as a Type I sum, and estimate $S_2^{(2)}$ we did S_4, i.e., as a Type II sum.

17.1.1 Exercises

1. (Linnik 1960, 1961)
   (a) Show that $|\zeta(s) - 1| < 1$ if $\sigma \ge 2$.
   (b) Show that if $\sigma \ge 2$, then
   $$\log \zeta(s) = \sum_{k=2}^{\infty} \frac{(-1)^{k-1}}{k}(\zeta(s) - 1)^k.$$
   (c) For positive integers k, let $d_k'(n) = \operatorname{card}\{(n_1, \ldots, n_k) : n_1 n_2 \cdots n_k = n,\ n_i > 1\}$. Show that
   $$(\zeta(s) - 1)^k = \sum_{n=2}^{\infty} d_k'(n) n^{-s}$$
   for $\sigma > 1$.
   (d) Deduce that
   $$\log \zeta(s) = \sum_{n=2}^{\infty} \left(\sum_{k=1}^{\infty} \frac{(-1)^{k-1}}{k} d_k'(n) \right) n^{-s}$$
   for $\sigma > 2$.
   (e) Conclude that
   $$\frac{\Lambda(n)}{\log n} = \sum_{k=1}^{\infty} \frac{(-1)^{k-1}}{k} d_k'(n)$$
   for all $n > 1$.

(f) Show that $d'_k(n) = 0$ if $n < 2^k$.

(g) Show that if $K \ge (\log n)/\log 2$, then

$$\frac{\Lambda(n)}{\log n} = \sum_{k=1}^{K} \frac{(-1)^{k-1}}{k} d'_k(n).$$

2. (a) Show that

$$\Lambda(n) = \sum_{\substack{d,k \\ dk|n}} \Lambda(d)\mu(k).$$

(b) Observe that

$$\begin{aligned}\Lambda(n) = \sum_{\substack{d,k \\ dk|n \\ d\le U}} \Lambda(d)\mu(k) + \sum_{\substack{d,k \\ dk|n \\ k\le V}} \Lambda(d)\mu(k) \\ - \sum_{\substack{d,k \\ dk|n \\ d\le U \\ k\le V}} \Lambda(d)\mu(k) + \sum_{\substack{d,k \\ dk|n \\ d>U \\ k>V}} \Lambda(d)\mu(k) \\ = C_1(n) + C_2(n) + C_3(n) + C_4(n),\end{aligned}$$

say.

(c) In the notation of Vaughan's identity (17.5), show that $C_i(n) = c_i(n)$ for $i = 1, 2, 3, 4$.

3. Suppose that $\Delta'(M) = \Delta'(M, N, V, f)$ is defined so that

$$\begin{aligned}\left|\sum_{M<m\le 2M} b_m \sum_{V<k\le N/m} c_k f(mk)\right| \\ \le \Delta'(M)\Bigg(\sum_{M<m\le 2M} |b_m|^2\Bigg)^{1/2}\Bigg(\sum_{V<k\le N/M} |c_k|^2\Bigg)^{1/2}\end{aligned} \tag{17.20}$$

for arbitrary complex numbers b_m and c_k.

(a) Show that

$$S_4 \ll N^{1/2}(\log N)^2 \max_{U\le M\le N/V} \Delta'(M). \tag{17.21}$$

(b) Deduce that

$$\Delta'(M) \ll \left(\max_{V<k\le N/M} \sum_{V<j\le N/M} \left|\sum_{\substack{M<m\le 2M \\ m\le N/j \\ m\le N/k}} f(mj)\overline{f(mk)}\right|\right)^{1/2}. \tag{17.22}$$

(c) Conclude that

$$S_4 \ll N^{1/2}(\log N)^2 \times \max_{U\le M\le N/V} \max_{V<k\le N/M} \left(\sum_{V<j\le N/M} \left| \sum_{\substack{M\le m\le 2M \\ m\le N/j \\ m\le N/k}} f(mj)\overline{f(mk)} \right| \right)^{1/2}. \tag{17.23}$$

4. Let S_2 be defined as in (17.7), and write $S_2 = S_2^{(1)} + S_2^{(2)}$, as in (17.19). Show that

$$S_2^{(2)} \ll N^{1/2}(\log N)^2 \max_{V\le M\le UV} \Delta(M). \tag{17.24}$$

5. Let $\Delta(M)$ denote the best constant in the bilinear form inequality (17.13). By appealing to an appropriate result from Appendix G, or otherwise, show also that if $|f(n)| \ge 1$ for all n, then

$$\Delta(M) \gg \max(M^{1/2}, (N/M)^{1/2}).$$

(Hence our method, as presently constituted, never gives an upper bound better than $N^{3/4}$ when f is unimodular. Also, our bound (17.14) will be trivial if M is allowed to be as small as $(\log N)^4$ or as large as $N/(\log N)^4$). Thus U and V must be at least moderately large.)

6. (Heath-Brown 1982) Show that if $s \ne 1$ and $\zeta(s) \ne 0$, then

$$-\frac{\zeta'}{\zeta}(s) = \sum_{k=1}^{K}(-1)^k\binom{K}{k}\zeta(s)^{k-1}\zeta'(s)M^k - \frac{\zeta'}{\zeta}(s)(1-\zeta(s)M)^K.$$

The above holds for any complex M, but as usual we take

$$M = M(s) = \sum_{n\le Y} \mu(n)n^{-s}.$$

We set

$$b(n) = \begin{cases} \sum_{\substack{d\mid n \\ d\le Y}} \mu(d) & (n > Y), \\ 0 & (n \le Y). \end{cases}$$

Thus $\sum_{n=1}^{\infty} b(n)n^{-s} = \zeta(s)M(s) - 1$ for $\sigma > 1$. Show that

$$\Lambda(n) = \sum_{k=1}^{K}(-1)^{k-1}\binom{K}{k}a_k(n) + s(n)$$

for all n, where

$$a_k(n) = \sum_{\substack{r_1 \cdots r_{2k} = n \\ i > k \implies r_i \le Y}} \mu(r_{k+1}) \cdots \mu(r_{2k}) \log r_1$$

and

$$s(n) = (-1)^K \sum_{d_0 \cdots d_K = n} \Lambda(d_0) b(d_1) \cdots b(d_K).$$

Note that $s(n) = 0$ if $n \le Y^K$, so we obtain only Type I sums in this range.

7. (Montgomery & Vaughan 1981) Let $G(s)$ be defined as in (17.3). From the identity

$$\frac{1}{\zeta(s)} = 2G(s) - G(s)^2\zeta(s) + \left(\frac{1}{\zeta(s)} - G(s)\right)(1 - \zeta(s)G(s)), \quad (17.25)$$

or otherwise, show that

$$\mu(n) = a_0(n) + a_1(n) + a_2(n),$$

where

$$a_0(n) = \begin{cases} 2\mu(n) & n \le V, \\ 0 & n > V, \end{cases}$$

$$a_1(n) = -\sum_{\substack{dem = n \\ d \le V \\ e \le V}} \mu(d)\mu(e),$$

$$a_2(n) = -\sum_{\substack{dk = n \\ d > V \\ k > V}} \mu(d) \left(\sum_{\substack{e \mid k \\ e \le V}} \mu(e) \right).$$

8. Show that if $1 \le V \le N$, then

$$\sum_{n=1}^{N} \mu(n) f(n) = T_0 + T_1 + T_2,$$

where

$$T_0 = 2 \sum_{n \le V} \mu(n) f(n),$$

$$T_1 = -\sum_{m \le V^2} b_m \sum_{n \le N/m} f(mn)$$

with

$$b_m = \sum_{\substack{de=m \\ d,e\le V}} \mu(d)\mu(e), \tag{17.26}$$

and

$$T_2 = -\sum_{V<m\le N/V}\ \sum_{V<k\le N/m} \mu(m)c_k f(mk)$$

with

$$c_k = \sum_{\substack{d|k \\ d\le V}} \mu(d). \tag{17.27}$$

9. With the T_i defined as in the previous exercise, show that

$$T_0 \ll \sum_{n\le V} |f(n)|, \tag{17.28}$$

$$T_1 \ll \sum_{m\le V^2} d(m) \left| \sum_{k\le N/m} f(mk) \right|, \tag{17.29}$$

and

$$T_2 \ll N^{1/2}(\log N)^{5/2}$$
$$\times \max_{V\le M\le N/V}\ \max_{j\le N/M} \left(\sum_{k\le N/M} \left| \sum_{\substack{M<m\le 2M \\ m\le N/k \\ m\le N/j}} f(mj)\overline{f(mk)} \right| \right)^{1/2}. \tag{17.30}$$

10. Let $\Lambda_2(n) = \Lambda(n)\log n + \sum_{bc=n} \Lambda(b)\Lambda(c)$, as in Theorem 8.3.

(a) Show that

$$\frac{\zeta''}{\zeta}(s) = \left(\frac{\zeta'}{\zeta}(s)\right)' + \left(\frac{\zeta'}{\zeta}(s)\right)^2.$$

(b) Show that

$$\frac{\zeta''}{\zeta}(s) = \sum_{n=1}^{\infty} \Lambda_2(n) n^{-s}$$

for $\sigma > 1$.

(c) Let $G(s)$ be defined as in (17.3), and put $H(s) = \sum_{n\le U} \Lambda_2(n)$. Observe that

$$\frac{\zeta''}{\zeta}(s) = H(s) - \zeta(s)G(s)H(s) + \zeta''(s)G(s)$$
$$+ \left(\frac{\zeta''}{\zeta}(s) - H(s)\right)(1-\zeta(s)G(s)).$$

(d) Define arithmetic functions $a_i(n)$ so that $\Lambda_2(n) = a_1(n) + a_2(n) + a_3(n) + a_4(n)$.

17.2 An Exponential Sum Formed with Primes

Vinogradov applied his method to the generating function $\sum_{p\le x} e(p\alpha)$, and thus showed that the generating function is small when α is not near a rational number with small denominator. This 'minor arc estimate' enabled him to show (as we shall in Theorem 18.1) that all sufficiently large odd numbers can be written as a sum of three primes. We find it simpler to work with the generating function

$$S(\alpha) = \sum_{n=1}^{N} \Lambda(n)e(n\alpha) \tag{17.31}$$

because $\Lambda(n)$ has the decomposition (17.5), which gives rise to sums for which (in many cases) we can derive non-trivial estimates.

Theorem 17.1 *Let $S(\alpha)$ be as in (17.31). If $(a,q) = 1$ and $|\alpha - a/q| \le 1/q^2$, then*

$$S(\alpha) \ll \left(Nq^{-1/2} + N^{4/5} + N^{1/2}q^{1/2}\right)(\log N)^{5/2}. \tag{17.32}$$

Proof By (16.4) we see that

$$\sum_{0<t\le T} \max_{w\ge 1} \left| \sum_{w\le r\le N/t} e(rt\alpha) \right| \ll \sum_{0<t\le T} \min\left(\frac{N}{t}, \frac{1}{\|t\alpha\|}\right). \tag{17.33}$$

To estimate the right-hand side, we write $t = hq + r$ and sum over $0 \le h \le T/q$ and $1 \le r \le q$. Let $\delta = \alpha - a/q$. We consider first the case in which $h = 0$ and $1 \le r \le q/2$. Since $|\delta| \le 1/q^2$, $\|r\alpha\|$ differs from $\|ra/q\|$ by at most $1/(2q)$. But $\|ra/q\| \ge 1/q$ for these r, and hence $\|r\alpha\| \asymp \|ra/q\|$. Consequently

$$\sum_{1\le r\le q/2} \frac{1}{\|r\alpha\|} \ll \sum_{1\le r\le q/2} \frac{1}{\|ra/q\|} \ll \sum_{1\le r\le q/2} \frac{q}{r} \ll q\log 2q.$$

For all other terms we have $hq + r \gg (h+1)q$. Thus it suffices to estimate

$$\sum_{0\le h\le T/q} \sum_{r=1}^{q} \min\left(\frac{N}{(h+1)q}, \frac{1}{\|hq\alpha + ra/q + r\delta\|}\right). \tag{17.34}$$

For any given h, the q points $hq\alpha + ra/q + r\delta$ are uniformly within $1/q$ of the equally spaced points $hq\alpha + ra/q$. Thus if $\|hq\alpha + ra/q + r\delta\| < 1/q$, then $\|hq\alpha + ra/q\| < 2/q$, and this holds for at most four values of r. For all other

r, the numbers $\|hq\alpha + ra/q + r\delta\|$ are comparable to the numbers $\|r/q\|$ for $0 < r < q$. Hence the double sum (17.34) is

$$\ll \sum_{0 \le h \le T/q} \left(\frac{N}{(h+1)q} + q \log 2q \right) \ll \frac{N}{q} \log 2T/q + T \log 2q + q \log 2q.$$

That is, we have shown that

$$\sum_{0<t\le T} \min\left(\frac{N}{t}, \frac{1}{\|t\alpha\|}\right) \ll (N/q + T + q) \log 2Tq. \tag{17.35}$$

By (17.8) we deduce that

$$S_2 \ll (N/q + UV + q)(\log 2qUV)^2.$$

Similarly, from (17.9) we see that

$$S_3 \ll (N/q + V + q)(\log 2qVN)^2.$$

By (17.18) and (16.4) we find that

$$S_4 \ll N^{1/2}(\log N)^2 \max_{U \le M \le N/V} \max_{k \le N/M} \left(\sum_{j \le N/M} \min\left(M, \frac{1}{\|(j-k)\alpha\|}\right)\right)^{1/2}.$$

Here the sum over j is

$$\ll M + \sum_{0<j\le N/M} \min\left(M, \frac{1}{\|j\alpha\|}\right) \ll M + \sum_{0<j\le N/M} \min\left(\frac{N}{j}, \frac{1}{\|j\alpha\|}\right)$$

since $M \le N/j$ for $j \le N/M$. Thus by a further application of (17.35) we deduce that

$$S_4 \ll \left(Nq^{-1/2} + NU^{-1/2} + NV^{-1/2} + N^{1/2}q^{1/2}\right)(\log 2qN)^{5/2}.$$

By taking $U = V = N^{2/5}$ we deduce that

$$S(\alpha) \ll \left(Nq^{-1/2} + N^{4/5} + N^{1/2}q^{1/2}\right)(\log 2qN)^{5/2}.$$

To complete the argument it suffices to note that we may assume that $q \le N$ since otherwise the estimate (17.32) is weaker than the trivial estimate $S(\alpha) \ll N$. □

17.2.1 Exercises

1. Show that if $|\alpha - a/q| \le 1/q^2$ and $(a, q) = 1$, then

$$\sum_{n\le N} \mu(n)e(n\alpha) \ll \left(Nq^{-1/2} + N^{4/5+\varepsilon} + N^{1/2}q^{1/2}\right)(\log N)^3. \tag{17.36}$$

2. Show that if q is a positive integer, then for any integer c,

$$e(c/q) = \sum_{\substack{d|q \\ d|c}} \frac{1}{\varphi(q/d)} \sum_{\substack{\chi \\ (\text{mod } q/d)}} \tau(\overline{\chi})\chi(c/d).$$

3. Let

$$M(x;\chi,\delta) = \sum_{n\le x} \chi(n)\mu(n)e(n\delta),$$

where χ is a Dirichlet character, x is real, and $\delta \in \mathbb{T}$. Let A and B be given positive real numbers. Show that if $\alpha = a/q + \delta$ with $(a,q) = 1$, then

$$\sum_{n\le x} \mu(n)e(n\alpha) = \sum_{d|q} \frac{\mu(d)}{\varphi(q/d)} \sum_{\substack{\chi \\ (\text{mod } q/d)}} \tau(\overline{\chi})\chi(a)M(x/d;\chi\chi_{0(d)},\delta),$$

where $\chi_{0(d)}$ denotes the principal character modulo d.

4. Let $M(x;\chi,\delta)$ be defined as in the preceding problem. Show that if χ is a character modulo q and $q \le (\log x)^A$, then

$$M(x;\chi,\delta) \ll (1 + x\|\delta\|)x(\log x)^{-B}.$$

5. (Davenport 1937a,b) Show that if $|\alpha - a/q| \le 1/q^2$, $(a,q) = 1$, and $q \le (\log x)^A$, then

$$\sum_{n\le x} \mu(n)e(n\alpha) \ll x(\log x)^{-B}. \tag{17.37}$$

By combining this with the result of Exercise 17.2.1.1, show that the estimate (17.37) holds uniformly in α.

6. (Bateman & Chowla 1963)

(a) Let $\lambda(n)$ denote the *Liouville lambda function*, which is to say that $\lambda(n) = (-1)^{\Omega(n)}$ where $\Omega(n) = \sum_{p^a\|n} a$. Show that

$$\sum_{d^2|n} \mu(n/d^2) = \lambda(n)$$

for all positive integers n.

(b) Deduce that

$$\sum_{n\le x} \lambda(n)e(n\alpha) = \sum_{d\le x^{1/2}} \sum_{m\le x/d^2} \mu(m)e(d^2 m\alpha)$$

for all $x \ge 1$ and all real α.

(c) Conclude that

$$\sum_{n\le x} \lambda(n)e(n\alpha) \ll x(\log x)^{-B}$$

uniformly in α.

(d) Let

$$f(\alpha) = \sum_{n=1}^{\infty} \frac{\mu(n)}{n} e(n\alpha), \qquad g(\alpha) = \sum_{n=1}^{\infty} \frac{\lambda(n)}{n} e(n\alpha). \quad (17.38)$$

Show that these series are uniformly convergent, and hence define continuous functions on $\mathbb{T}$.

(e) Show that

$$\sum_{a=1}^{q} f(a/q) = 0, \qquad \sum_{a=1}^{q} g(a/q) = 0 \quad (17.39)$$

for all positive integers q.

(f) By using the result of Exercise 9.2.1.1(a), or otherwise, show that if $(a,q) = 1$, then

$$\sum_{\substack{n=1 \\ (n,q)=1}}^{\infty} \frac{\lambda(n)}{n} e(an/q) = \frac{1}{\varphi(q)} \sum_{\chi \ne \chi_0} \overline{\chi}(a)\tau(\chi) \frac{L(2,\overline{\chi}^2)}{L(1,\overline{\chi})}.$$

(g) Show that if χ is an even primitive character modulo q, $q > 1$, then

$$L(2,\chi) = \frac{-\pi^2}{q^2\tau(\overline{\chi})} \sum_{a=1}^{q-1} \overline{\chi}(a)a(q-a).$$

(h) (See Figure 17.1) Show that

$$\operatorname{Re} g(1/5) = \frac{\pi^2}{10\log\left(\frac{1+\sqrt{5}}{2}\right)} = 2.05098958\ldots.$$

(i) Suppose that p_1 and p_2 are distinct primes, and that $(a, p_1p_2) = 1$. Show that

$$g\left(\frac{a}{p_1p_2}\right) = -g(a/p_2)/p_1 - g(a/p_1)/p_2 + \sum_{\substack{n=1 \\ (n,p_1p_2)=1}}^{\infty} \frac{\lambda(n)}{n} e\left(\frac{an}{p_1p_2}\right).$$

(j) Show that $\operatorname{Re} g(1/10) = 0$.

Suppose that

$$F(n) = \sum_{d|n} f(d)$$

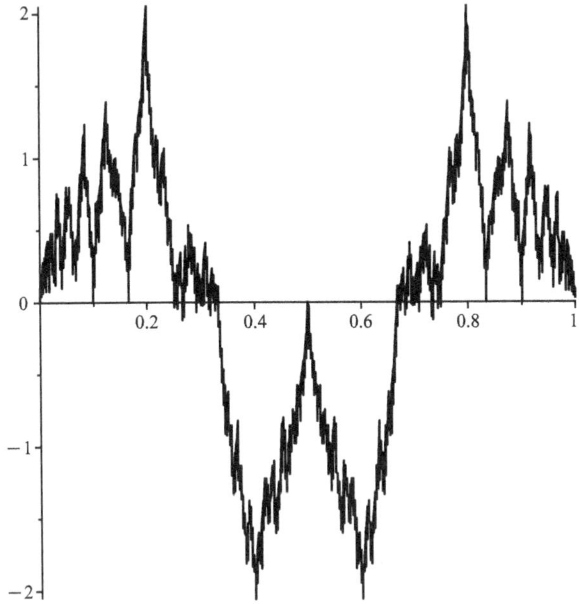

Figure 17.1 Graph of Re $g(x)$ with g defined as in (17.38).

for all n, and let $s(x)$ denote the sawtooth function with period 1, as defined in (E.14). By the Fourier series expansion of Lemma D.1 (see also §E.3), we see that possibly

$$\sum_{d=1}^{\infty} \frac{f(d)}{d} s(d\alpha) = -\sum_{d=1}^{\infty} \frac{f(d)}{d} \sum_{m=1}^{\infty} \frac{\sin 2\pi m d\alpha}{\pi m} \tag{17.40}$$

$$= -\sum_{n=1}^{\infty} \frac{F(n)}{\pi n} \sin 2\pi n\alpha \tag{17.41}$$

by grouping together those pairs m, d for which $md = n$. This is merely a *formal* argument since we have not justified the reorganization of terms in passing from (17.40) to (17.41). In the next several exercises, we treat this issue in the interesting case that $f(d) = \mu(d)$.

7. Let

$$S_D(\alpha) = \sum_{d \le D} \frac{\mu(d)}{d} s(d\alpha). \tag{17.42}$$

(a) Let N be a parameter to be chosen later such that $N > D$, and let $E_K(x)$ be defined as in Lemma D.1. Show that

$$S_D(\alpha) = \frac{-1}{\pi}\sin 2\pi\alpha + T_1(\alpha) + T_2(\alpha),$$

where

$$T_1 = \frac{1}{\pi}\sum_{D<d\le N}\frac{\mu(d)}{d}\sum_{n\le N/d}\frac{\sin 2\pi n d\alpha}{\pi n},$$

$$T_2 = \sum_{d\le D}\frac{\mu(d)}{d}E_{N/d}(\alpha).$$

(b) Show that

$$T_1 = \sum_{n\le N/D}\frac{1}{n}\sum_{D<d\le N/n}\frac{\mu(d)}{d}\sin 2\pi n d\alpha.$$

(c) Use (17.37) to show that $T_1 \ll (\log D)^{-B}(\log N/D)^2$.

(d) Explain why $E_K(0) = 0$.

(e) Show that if $(a,q) = 1$ and $q \le D$, then $T_2(a/q) \ll DN^{-1}\log 2q$.

(f) Take $N = D(\log D)^A$, and deduce that

$$S_D(\alpha) = \frac{-1}{\pi}\sin 2\pi\alpha + O\big((\log D)^{-B}\big) \tag{17.43}$$

when $\alpha = a/q$, $(a,q) = 1$, and $q \le D$.

8. Let $S_D(\alpha)$ be defined as in (17.42).

(a) Show that $S_D(\alpha)$ is piecewise linear with slope

$$M(D) = \sum_{d\le D}\mu(d)$$

and jump discontinuities at the Farey fractions of order D.

(b) Write

$$x\sum_{\substack{n\le x\\(n,q)=1}}\frac{\mu(n)}{n} = \sum_{\substack{n\le x\\(n,q)=1}}\mu(n)\lfloor x/n\rfloor + \sum_{\substack{n\le x\\(n,q)=1}}\mu(n)\{x/n\}$$
$$= \Sigma_1 + \Sigma_2,$$

say. Show that Σ_1 is the number of integers not exceeding x that are composed entirely of prime numbers that divide q. Hence deduce that $|\Sigma_1| \le x$.

(c) Explain why $|\Sigma_2| \le x$.

(d) Deduce that

$$\Bigg|\sum_{\substack{n\le x\\(n,q)=1}}\frac{\mu(n)}{n}\Bigg|\le 2$$

uniformly in x and q.

(e) Let a/q denote a Farey fraction of order D. Show that the jump discontinuity of $S_D(\alpha)$ at $\alpha = a/q$ is

$$-\sum_{\substack{d\le D\\q|d}}\frac{\mu(d)}{d}.$$

(f) Show that the sum in (e) has absolute value not exceeding $2/q$.

(g) Let $\mathscr{R}$ denote the set of numbers composed entirely of primes dividing q. Show that

$$\sum_{\substack{d|n\\d\in\mathscr{R}}}\mu(n/d)=\begin{cases}\mu(n) & \text{if } (n,q)=1,\\ 0 & \text{otherwise.}\end{cases}$$

(h) Deduce that

$$\sum_{\substack{n\le x\\(n,q)=1}}\frac{\mu(n)}{n}=\sum_{\substack{d\le x\\d\in\mathscr{R}}}\frac{1}{d}\sum_{m\le x/d}\frac{\mu(m)}{m}.$$

(i) By adapting the techniques developed in §7.1, show that if $q\le x^2$, then the number of members of $\mathscr{R}$ not exceeding x is $\ll x^{\varepsilon}$.

(j) Deduce that if $q\le x$, then

$$\sum_{\substack{n\le x\\(n,q)=1}}\frac{\mu(n)}{n}\ll \exp\big(-c\sqrt{\log x}\,\big).$$

(k) (Davenport 1937a,b) Conclude that (17.43) holds uniformly in α.

9. Let d_α denote the multiplicative function defined by

$$d_\alpha(p^k)=\binom{-\alpha}{k}(-1)^k$$

and let $f:\mathbb{N}\to\mathbb{C}$ be such that $\sum_n|f(n)|<\infty$.

(a) Prove that $0\le d_{\frac12}(n)\le 1$ and $0\le |d_{-\frac12}(n)|\le 1$.

(b) Prove that

$$\sum_{m|n}d_{\frac12}(m)d_{-\frac12}(n/m)=\begin{cases}1 & n=1,\\ 0 & n>1.\end{cases}$$

(c) Let $u, v \in \mathbb{R}$, $u, v \geq 1$, and $w = \min(u, v)$. By considering the formal identity

$$\zeta^{\frac{1}{2}} = F - \frac{1}{2}F^2G + \frac{1}{2}G\zeta - (\zeta^{\frac{1}{2}} - F)\left(\frac{1}{2}\zeta^{\frac{1}{2}}G + \frac{1}{2}FG - 1\right),$$

or otherwise, prove that

$$\sum_n d_{\frac{1}{2}}(n)f(n) = S_1 - \tfrac{1}{2}S_2 + \tfrac{1}{2}S_3 - S_4,$$

where

$$S_1 = \sum_{n\leq u} d_{\frac{1}{2}}(n)f(n),$$

$$S_2 = \sum_{l\leq u}\sum_{m\leq u}\sum_{n\leq v} d_{\frac{1}{2}}(l)d_{\frac{1}{2}}(m)d_{-\frac{1}{2}}(n)f(lmn),$$

$$S_3 = \sum_{m\leq v}\sum_{n} d_{-\frac{1}{2}}(m)f(mn),$$

$$S_4 = \sum_{m>u}\sum_{n>w} d_{\frac{1}{2}}(m)b(n)f(mn),$$

where $b(1) = 1$ and for $n > 1$,

$$b(n) = \sum_{\substack{m|n\\ m\leq v}} \frac{1}{2}d_{-\frac{1}{2}}(m)d_{\frac{1}{2}}(n/m) + \sum_{\substack{m|n\\ m\leq v\\ n/m\leq u}} \frac{1}{2}d_{-\frac{1}{2}}(m)d_{\frac{1}{2}}(n/m).$$

(d) Suppose that $\alpha \in \mathbb{R}$ and there are $a \in \mathbb{Z}$, $q \in \mathbb{N}$ with $(a, q) = 1$ such that $|\alpha - a/q| \leq q^{-2}$. Prove that

$$\sum_{n\leq x} d_{\frac{1}{2}}(n)e(\alpha n) \ll (xq^{-\frac{1}{2}} + x^{\frac{6}{7}} + x^{\frac{1}{2}}q^{\frac{1}{2}})(\log x)^3.$$

17.3 Further Applications

Before considering specific applications, we make two technical remarks. Firstly, we sometimes obtain sharper results not by treating $1 \leq n \leq N$ directly, but rather by treating $N < n \leq 2N$, and then summing the bounds obtained to treat $1 \leq n \leq N$. The point is that the parameters chosen to treat $N < n \leq 2N$ may not succeed as well for smaller n. An example of this is seen in Exercise 17.3.1.3, where the estimate for the sum over $M_1 < m \leq M_2$ would not apply if the sum were over $0 < m \leq M$.

Our second observation concerns our treatment of Type II sums, say

$$S = \sum_{\substack{M<m\le 4M}} \sum_{\substack{K<k\le 4K \\ N<mk\le 2N}} b_m c_k f(mk).$$

Our existing treatment of this gives rise to the problem of bounding

$$\sum_{K<j\le 4K} \left| \sum_{M<m\le 4M} f(mj)\overline{f(mk)} \right|.$$

If the bound we can derive for the above sum over m is smaller when $0 < |j-k| = o(K)$, then we may obtain a better final result by partitioning the interval $(K, 4K]$ into R subintervals $\mathscr{K}_r$ of equal length. By Cauchy's inequality,

$$|S|^2 \le R \sum_{r=1}^{R} |S_r|^2,$$

where

$$S_r = \sum_{M<m\le 4M} \sum_{\substack{k\in\mathscr{K}_r \\ N<mk\le 2N}} b_m c_k f(mk).$$

By a second application of Cauchy's inequality we see that

$$|S_r|^2 \le \left(\sum_{M<m\le 4M} |b_m|^2 \right) \left(\sum_{M<m\le 4M} \left| \sum_{\substack{k\in\mathscr{K}_r \\ N<mk\le 2N}} c_k f(mk) \right|^2 \right).$$

Here the second factor above is

$$= \sum_{j\in\mathscr{K}_r} c_j \sum_{k\in\mathscr{K}_r} \overline{c_k} \sum_{\substack{M<m\le 4M \\ N<mj\le 2N \\ N<mk\le 2N}} f(mj)\overline{f(mk)}.$$

Since $|c_j\overline{c_k}| \le \frac{1}{2}|c_j|^2 + \frac{1}{2}|c_k|^2$, it follows that the above is

$$\le \sum_{k\in\mathscr{K}_r} |c_k|^2 \sum_{j\in\mathscr{K}_r} \left| \sum_{\substack{M<m\le 4M \\ N<mj\le 2N \\ N\le mk\le 2N}} f(mj)\overline{f(mk)} \right|.$$

Thus

$$|S|^2 \le \Delta^2 \left(\sum_{M<m\le 4M} |b_m|^2 \right) \left(\sum_{K<k\le 4K} |c_k|^2 \right)$$

for arbitrary b_m and c_k with

$$\Delta^2 = R \max_{1\le r\le R} \max_{k\in\mathcal{K}_r} \sum_{j\in\mathcal{K}_r} \Bigg| \sum_{\substack{M<m\le 4M\\ N<mj\le 2N\\ N\le mk\le 2N}} f(mj)\overline{f(mk)}\Bigg| . \tag{17.44}$$

When $R = 1$, this reduces to our former treatment. Let X denote the usual order of magnitude of the above sum over m when j and k range independently over the interval $(K, 4K]$. The variable j takes $\asymp K/R$ values above, which gives an overall order of magnitude KX, which is the same as when $R = 1$. But if the sum over m is smaller when j and k are constrained to lie in the same subinterval $\mathcal{K}_r$, then the estimate is improved. However, for each k there is a value of j, namely $j = k$, for which there is no cancellation in the sum over m. If $|f| \asymp 1$, then that contribution is $\asymp RM$. If the diagonal terms dominate our estimate of the contributions of the non-diagonal terms when $R = 1$, then taking $R > 1$ yields a weaker estimate. Thus we obtain an improvement over our original treatment if (i) our estimate of $\sum_{j\ne k} |\sum_m \cdots|$ is large compared with M, and (ii) our estimate of $|\sum_m \cdots|$ is better when $0 < |j-k| = o(K)$.

One of the foremost unsolved problems of prime number theory is to show that n^2+1 is prime for infinitely many integers n. In fact it is conjectured not just that there are infinitely many such n, but that the number of them with $n \le x$ is asymptotic to $C\,\mathrm{li}(x)$ as $x \to \infty$ where

$$C = \frac{1}{2}\prod_{p>2}\left(1 - \frac{\left(\frac{-1}{p}\right)}{p-1}\right). \tag{17.45}$$

While finding primes in sparse sequences is generally challenging, for sequences of the special form $\lfloor n^a \rfloor$ we have some success if a is not too large.

Theorem 17.2 (Piaetski-Shapiro 1953) *For a real number $a > 1$, let $\pi_a(x)$ denote the number of integers $n \le x$ such that $\lfloor n^a \rfloor$ is prime and put $\alpha = 1/a$. If $1 < a < 12/11$, then*

$$\pi_a(x) = \alpha \sum_{p\le x^a} p^{\alpha-1} + O\Big(x^{\frac{11a+14}{26}} \log x\Big).$$

By a quantitative form of the Prime Number Theorem and integration by parts we see that the main term above is

$$= \alpha \int_2^{x^a} \frac{u^{\alpha-1}}{\log u}\, du + O\big(x \exp\big(-c\sqrt{\log x}\big)\big).$$

By the change of variable $v = u^\alpha$ we see further that the main term above is $= \alpha\,\mathrm{li}(x) + O(1)$. Thus, in particular,

$$\pi_a(x) \sim \frac{x}{a \log x}$$

as $x \to \infty$, provided that $1 < a < 12/11$.

The prime number distribution model of Cramér asserts that a large integer n is prime with 'probability' $1/\log n$. This predicts that $\pi_a(x)$ should be approximately

$$\sum_{1<n\le x} \frac{1}{\log\lfloor n^a\rfloor} = \sum_{1<n\le x} \frac{1}{a\log n} + O\Bigg(\sum_{1<n\le x} \frac{1}{n^a(\log n)^2}\Bigg) = \alpha\,\mathrm{li}(x) + O(1).$$

Thus we interpret the Piatetski-Shapiro theorem as asserting that the sequence $\lfloor n^a\rfloor$ collects its fair share of primes, when $1 < a < 12/11$.

The bound 12/11 can be relaxed somewhat, but it is not clear by how much. We note that the sequence $\lfloor n^2\rfloor$ contains no prime. To prepare for the proof of Theorem 17.2 we first establish the basic estimate on which the proof will depend.

Lemma 17.3 *Let α be fixed, with $0 < \alpha < 1$, and suppose that $1 \le M \le M' \le 2M$. Then*

$$\sum_{M<m\le M'} e\big(cm^\alpha\big) \ll |c|^{1/2}M^{\alpha/2} + |c|^{-1/2}M^{1-\alpha/2}$$

uniformly for non-zero real numbers c.

Proof We may assume that $c < 0$, for if $c > 0$, then the sum is the complex conjugate of the value it would have if c was negative. In the van der Corput estimate of Theorem 16.7, take $f(x) = cx^\alpha$. Then $f''(x) = c\alpha(\alpha-1)x^{\alpha-2} \asymp -cM^{\alpha-2}$. The stated estimate is immediate. □

Proof of Theorem 17.2 Suppose that instead of counting integers $n \le x$ such that $\lfloor n^a\rfloor$ is prime, we count primes $p \le x^a$ such that $p = \lfloor n^a\rfloor$ for some n. If $n \le x$ and $\lfloor n^a\rfloor = p$, then $p \le n^a \le x^a$. Conversely, if $p \le x^a$ and $\lfloor n^a\rfloor = p$, then

$$n^a < p+1 \le x^a + 1 < x^a + x^{a-1} = x^a(1+1/x) < x^a(1+1/x)^a = (x+1)^a,$$

so that $n < x+1$. Thus when we sum over $p \le x^a$ such that $p = \lfloor n^a\rfloor$ for some n we obtain all the terms that arise when we sum over $n \le x$, plus at most one additional term. To say that there is an integer n such that $\lfloor n^a\rfloor = p$ is equivalent to saying that $p \le n^a < p+1$, which in turn is equivalent to $p^\alpha \le n < (p+1)^\alpha$ where $\alpha = 1/a$, which is to say that there is an integer in the interval $\left[p^\alpha, (p+1)^\alpha\right)$. This in turn is equivalent to saying that $\lfloor -p^\alpha \rfloor - \lfloor -(p+1)^\alpha \rfloor = 1$. Otherwise, this difference is 0. Thus

$$\pi_a(x) = \sum_{p \le x^a} (\lfloor -p^\alpha \rfloor - \lfloor -(p+1)^\alpha \rfloor) + O(1).$$

If the above sum is formed without taking integer parts, it becomes

$$\sum_{p \le x^a} (-p^\alpha + (p+1)^\alpha) = \sum_{p \le x^a} \alpha p^{\alpha-1} + O\left(\sum_{p \le x^a} p^{\alpha-2}\right).$$

Since $\alpha < 1$, the error term above is $O(\log\log x)$ uniformly in a.

Recall that $\{x\} = x - \lfloor x \rfloor$ denotes the fractional part of x. Thus to complete the proof it will suffice to show that

$$\sum_{p \le x^a} \{-p^\alpha\} - \{-(p+1)^\alpha\} \ll x^{\frac{11a+14}{26}} \log x \tag{17.46}$$

for $1 < a < 12/11$.

In §D.1 we defined the sawtooth function $s(x)$ to be $s(x) = \{x\}$ if x is not an integer, and $s(x) = 0$ if x is an integer, in Lemma D.1 we found that

$$s(x) = -\sum_{h=1}^{H} \frac{\sin 2\pi hx}{\pi k} + O\left(\min\left(1, \frac{1}{H\|x\|}\right)\right).$$

This same formula holds for $\{x\}$ since $\{x\}$ differs from $s(x)$ only when x is an integer, and the error term is $O(1)$ in that case. Moreover, in Theorem E.6 we defined a trigonometrical polynomial $g_H(x) = \sum_{h=-H}^{H} \widehat{g_H}(h) e(hx)$ such that

$$\min\left(1, \frac{1}{H\|x\}}\right) \ll g_H(x),$$

and such that

$$\widehat{g_H}(h) \ll \frac{1}{H} \log \frac{3H}{|h|+1}.$$

Thus if $N < N' \le 2N$ and H is a parameter to be chosen later, then

$$\begin{aligned} &\sum_{N<n\le N'} \Lambda(n)(\{-(n+1)^\alpha\} - \{-n^\alpha\}) \\ &= \sum_{N<n\le N'} \Lambda(n) \sum_{0<|h|\le H} \frac{e(-h(n+1)^\alpha) - e(-hn^\alpha)}{2\pi i h} \\ &+ O\left(\sum_{N<n\le N'} \Lambda(n) \left(\min\left(1, \frac{1}{H\|(n+1)^\alpha\|}\right) + \min\left(1, \frac{1}{H\|n^\alpha\|}\right)\right)\right). \end{aligned} \tag{17.47}$$

Here the error term is

$$\begin{aligned} &\ll (\log N) \sum_{N<n\le N'+1} g_H(n^\alpha) \\ &= (\log N) \sum_{h=-H}^{H} \widehat{g_H}(h) \sum_{N<n\le N'+1} e\big(hn^\alpha\big). \end{aligned} \tag{17.48}$$

If $h \neq 0$, then from Lemma 17.3 we see that

$$\sum_{N<n\le N'} e\big(hn^\alpha\big) \ll N^{\alpha/2}|h|^{1/2} + N^{1-\alpha/2}|h|^{-1/2}.$$

When $h = 0$ there is no cancellation in the sum over n, so expression (17.48) is

$$\begin{aligned} &\ll \frac{N(\log N)\log 3H}{H} + \frac{\log N}{H}\sum_{h=1}^{H}\big(N^{\alpha/2}h^{1/2} + N^{1-\alpha/2}h^{-1/2}\big)\log\frac{3H}{h} \\ &\ll (\log N)\big(NH^{-1}\log 3H + N^{\alpha/2}H^{1/2} + N^{1-\alpha/2}H^{-1/2}\big). \end{aligned} \tag{17.49}$$

When N is near x^a, this will require taking H to be somewhat larger than x^{a-1}. We postpone choosing H until further arguments are complete, so that we can choose H to minimize the sum of all error terms.

Concerning the main term (17.47) we note that

$$\frac{e\big(-h(n+1)^\alpha\big) - e\big(n^\alpha\big)}{2\pi i h} = -e\big(-hn^\alpha\big)\int_0^{(n+1)^\alpha - n^\alpha} e(-h\beta)\,d\beta.$$

Thus expression (17.47) is

$$= -\sum_{0<|h|\le H}\ \sum_{N<n\le N'} \Lambda(n)e\big(-hn^\alpha\big)\int_0^{(n+1)^\alpha-n^\alpha} e(-h\beta)\,d\beta.$$

Let $\delta(\nu) = (\nu+1)^\alpha - \nu^\alpha$. This is an increasing function of the real variable ν, so the inequality $\beta \le \delta(n)$ is equivalent to $n > \delta^{-1}(\beta)$. Hence the above is

$$= -\sum_{0<|h|\le H}\int_0^{(2N+1)^\alpha-(2N)^\alpha} e(-h\beta) \sum_{\substack{N<n\le N'\\ n>\delta^{-1}(\beta)}} \Lambda(n)e\big(-hn^\alpha\big)\,d\beta,$$

which by the triangle inequality is

$$\ll \sum_{0<|h|\le H}\int_0^{(2N+1)^\alpha-(2N)^\alpha} \Bigg|\sum_{\substack{N<n\le N'\\ n>\delta^{-1}(\beta)}} \Lambda(n)e\big(-hn^\alpha\big)\Bigg|\,d\beta.$$

Since $(2N+1)^\alpha - (2N)^\alpha \ll N^{\alpha-1}$, and h makes the same contribution as $-h$, the above is

$$\ll N^{\alpha-1} \sum_{0<|h|\le H} \max_{\substack{N1,N2\\ N\le N_1\le N_2\le 2N}} \left| \sum_{N_1<n\le N_2} \Lambda(n)e\big(-hn^\alpha\big)\right|. \tag{17.50}$$

To estimate the above sum over n we invoke the usual decomposition with $U = V = \lfloor N^{1/3} \rfloor$. Thus $S_1 = 0$, and

$$\begin{aligned} S_2 &= \sum_{t\le U^2} a(t) \sum_{N/t<r\le N'/t} e\big(h(rt)^\alpha\big) \\ &\ll \left|\sum_{t\le U} a(t) \sum_r \cdots\right| + \left|\sum_{U<t\le U^2} a(t) \sum_r \cdots\right| \\ &= S_2^{(1)} + S_2^{(2)}. \end{aligned} \tag{17.51}$$

By Lemma 17.3,

$$\begin{aligned} S_2^{(1)} &\ll \sum_{t\le U} (\log 2t)\Big(\big(ht^\alpha\big)^{1/2}(N/t)^{\alpha/2} + \big(ht^\alpha\big)^{-1/2}(N/t)^{1-\alpha/2}\Big) \\ &= \sum_{t\le U} \big(h^{1/2}N^{\alpha/2} + h^{-1/2}N^{1-\alpha/2}t^{-1}\big)\log 2t \\ &\ll h^{1/2}N^{\alpha/2}U\log N + h^{-1/2}N^{1-\alpha/2}(\log N)^2. \end{aligned}$$

Hence

$$N^{\alpha-1} \sum_{0<h\le H} |S_2^{(1)}(h)| \ll \big(H^{3/2}N^{3\alpha/2-2/3} + H^{1/2}N^{\alpha/2}\big)(\log N)^2.$$

Here the second term is majorized by the first when $\alpha \ge 2/3$, so with this restriction the above is

$$\ll H^{3/2}N^{3\alpha/2-2/3}(\log N)^2. \tag{17.52}$$

Also,

$$S_3 = \sum_{k\le U} \mu(k) \sum_{N/k<m\le N'/k} e\big(h(km)^\alpha\big)\log m,$$

which by (17.9) is

$$\ll (\log N) \sum_{k\le U} \max_{w\ge 1} \left| \sum_{\substack{N/k<m\le N'/k\\ m>w}} e\big(h(km)^\alpha\big)\right|.$$

By Lemma 17.3 this is

$$\begin{aligned}&\ll (\log N) \sum_{k \le U} \left(h^{1/2} N^{\alpha/2} + h^{-1/2} N^{1-\alpha/2} k^{-1}\right)\\ &\ll \left(h^{1/2} N^{\alpha/2+1/3} + h^{-1/2} N^{1-\alpha/2}\right)(\log N)^2.\end{aligned}$$

Hence

$$N^{\alpha-1} \sum_{0<h\le H} |S_3(h)| \ll \left(H^{3/2} N^{3\alpha/2-2/3} + H^{1/2} N^{\alpha/2}\right)(\log N)^2. \tag{17.53}$$

Here the second term is majorized by the first when $\alpha \ge 2/3$, so with this restriction the above is

$$\ll H^{3/2} N^{3\alpha/2} (\log N)^2. \tag{17.54}$$

We now consider Type II sums. From (17.44) we see that

$$\begin{aligned}&\left| \sum_{M<m\le 4M} \sum_{\substack{K<k\le 4K \\ N_1<mk\le N_2}} b_m c_k e\left(h(mk)^\alpha\right) \right|^2\\ &\qquad \le \Delta^2 \left(\sum_{M<m\le 4M} |b_m|^2 \right) \left(\sum_{K<k\le 4K} |c_k|^2 \right)\end{aligned} \tag{17.55}$$

with

$$\Delta^2 = R \max_{1\le r\le R} \max_{k\in\mathcal{K}_r} \sum_{j\in\mathcal{K}_r} \left| \sum_{\substack{M<m\le 4M \\ N_1<mj\le N_2 \\ N_1<mk\le N_2}} e\left(h(j^\alpha - k^\alpha)m^\alpha\right) \right|. \tag{17.56}$$

If $MK > 2N$ or $16MK < N$, then the bilinear form is empty, and $\Delta = 0$. Thus we may suppose that $MK \asymp N$. When $j = k$ in (17.56), the sum over m is $\ll M$. When $j \ne k$, by Lemma 17.3 the sum over m is

$$\ll h^{1/2} |j^\alpha - k^\alpha|^{1/2} M^{\alpha/2} + h^{-1/2} \left|j^\alpha - k^\alpha\right|^{-1/2} M^{1-\alpha/2}.$$

Since $\left|j^\alpha - k^\alpha\right| \asymp K^{\alpha-1}|j-k|$, the above is

$$\ll h^{1/2} K^{(\alpha-1)/2} |j-k|^{1/2} M^{\alpha/2} + h^{-1/2} K^{(1-\alpha)/2} |j-k|^{-1/2} M^{1-\alpha/2}.$$

Hence

$$\sum_{j\in\mathcal{K}_r}\left|\sum_{\substack{M<m\le 4M\\ N_1<mj\le N_2\\ N_1<mk\le N_2}} e\big(h(j^\alpha-k^\alpha)m^\alpha\big)\right|$$
$$\ll M + h^{1/2}K^{(\alpha-1)/2}(K/R)^{3/2}M^{\alpha/2} + K^{(1-\alpha)/2}\big(K/(hR)\big)^{1/2}M^{1-\alpha/2}$$
$$\ll M + h^{1/2}N^{\alpha/2}KR^{-3/2} + h^{-1/2}NR^{-1/2}.$$

Thus

$$\Delta^2 \ll RM + h^{1/2}N^{\alpha/2}KR^{-1/2} + h^{-1/2}N^{1-\alpha/2}R^{1/2}.$$

We now choose R so that $RM \sim h^{1/2}N^{\alpha/2}KR^{-1/2}$. That is, we set

$$R = \lfloor h^{1/3}N^{\alpha/3}M^{-2/3}K^{2/3}\rfloor,$$

which gives

$$\Delta \ll h^{1/6}N^{(1+\alpha)/6}K^{1/6} + h^{-1/6}N^{1/2-\alpha/4}h^{1/12}N^{\alpha/12}M^{-1/6}K^{1/6}.$$

Here $M^{-1/6} \asymp K^{1/6}N^{-1/6}$, so the above is

$$\ll h^{1/6}N^{(1+\alpha)/6}K^{1/6} + h^{-1/6}N^{1/3-\alpha/6}K^{1/3}.$$

We note that if $\alpha \ge 3/4$ and $K \ll N^{1/2}$, then the second term above is majorized by the first. Since $MK \asymp N$, if $K > N^{1/2}$ we simply interchange M and K. Thus for $N^{1/3} \ll K \ll N^{1/2}$ we have

$$\Delta \ll h^{1/6}N^{1/4+\alpha/6}.$$

From (17.14) it follows that

$$S_4(h) \ll H^{1/6}N^{3/4+\alpha/6}(\log N)^2.$$

Hence

$$N^{\alpha-1}\sum_{0<h\le H}|S_4(h)| \ll H^{7/6}N^{7\alpha/6-1/4}(\log N)^2.$$

Since $|a(t)| \le \log t$ in (17.51), it follows similarly that

$$N^{\alpha-1}\sum_{0<h\le H}|S_2^{(2)}| \ll H^{7/6}N^{7\alpha/6-1/4}(\log N)^2.$$

On combining these estimates with (17.52) and (17.53), it follows that the expression in (17.50) is

$$\ll \big(H^{7/6}N^{7\alpha/6-1/4} + H^{3/2}N^{3\alpha/2-2/3}\big)(\log N)^2.$$

We choose H so that the first expression inside the parentheses is approximately N/H. That is, we take $H = \lfloor N^{\frac{15}{26}-\frac{7}{13}\alpha}\rfloor$. The common order of magnitude is $N/H \sim N^{\frac{11}{26}+\frac{7}{13}\alpha}$. Other terms are smaller when $11/12 < \alpha < 1$. On combining our estimates we find that

$$\sum_{N<n\le N'} \Lambda(n)\big(\{-(n+1)^\alpha\} - \{-n^\alpha\}\big) \ll N^{\frac{11+14\alpha}{26}}(\log N)^2.$$

The desired estimate (17.46) follows from this by partial summation, so the proof is complete. □

17.3.1 Exercises

1. (a) In §13.2 we showed that the estimate $M(x) \ll_\varepsilon x^{1/2+\varepsilon}$ is a consequence of the Riemann Hypothesis (RH). (More precise conditional estimates were also derived.) Use those methods to show that if RH is true, then

$$\sum_{n\le x} \mu(n)n^{-it} \ll_\varepsilon x^{1/2+\varepsilon}\tau^\varepsilon.$$

   (b) Use integration by parts to show that if RH is true, and $\sigma > 1/2$ and $\varepsilon > 0$ are fixed, then

$$\sum_{d>y} \frac{\mu(d)}{y^s} \ll y^{1/2-\sigma+\varepsilon}\tau^\varepsilon.$$

2. (a) Show that if n is a positive integer, then

$$\sum_{d^2|n} \mu(d) = \begin{cases} 1 & \text{if } n \text{ is square-free,} \\ 0 & \text{otherwise.} \end{cases}$$

   (b) As usual, let $Q(x)$ denote the number of square-free integers not exceeding x. Show that

$$Q(x) = \sum_{\substack{d,m \\ d^2m\le x}} \mu(d).$$

   Let y be a parameter to be chosen later such that $1 < y < x^{1/2}$, and write the above as $\sum_{d\le y} + \sum_{y<d} = \Sigma_1 + \Sigma_2$.

   (c) Show that

$$\Sigma_1 = x\sum_{d\le y} \frac{\mu(d)}{d^2} - \frac{1}{2}M(y) - S(x,y),$$

   where

$$S(x,y) = \sum_{d\le y} \mu(d)B_1\big(\{x/d^2\}\big). \tag{17.57}$$

Here B_1 is the first Bernoulli polynomial, $B_1(z) = z - 1/2$, and $\{u\}$ denotes the fractional part of u, $\{u\} = u - \lfloor u \rfloor$. Thus $B_1(\{u\})$ is the same as the sawtooth function $s(u)$ except when u is an integer.

(d) Suppose that $c > 1$. Explain why

$$\Sigma_2 = \frac{1}{2\pi i}\int_{c-i\infty}^{c+i\infty} \zeta(s)\Bigg(\sum_{d>y}\frac{\mu(d)}{d^{2s}}\Bigg)\frac{x^s}{s}\,ds.$$

(e) Let $\mathscr{C}$ denote the rectilinear contour with vertices $1 + \frac{1}{\log x} - i\infty$, $1 + \frac{1}{\log x} - ix$, $1/2 + \varepsilon - ix$, $1/2 + \varepsilon + ix$, $1 + \frac{1}{\log x} + ix$, $1 + \frac{1}{\log x} + i\infty$. Explain why

$$\Sigma_2 = x\sum_{d>y}\frac{\mu(d)}{d^2} + \frac{1}{2\pi i}\int_{\mathscr{C}} \zeta(s)\Bigg(\sum_{d>y}\frac{\mu(d)}{d^{2s}}\Bigg)\frac{x^s}{s}\,ds.$$

Note that both these formulæ for Σ_2 hold unconditionally.

(f) Now assume RH, recall the result of 17.3.1.1(b), and that RH implies the Lindelöf Hypothesis (LH). Show that the term above involving the contour $\mathscr{C}$ is $\ll_\varepsilon x^{1/2+\varepsilon}y^{-1/2}$.

(g) By combining results, show that if RH holds, then

$$Q(x) = \frac{6}{\pi^2}x + O\big(y^{1/2+\varepsilon}\big) + O\big(x^{1/2+\varepsilon}y^{-1/2}\big) + O\big(S(x,y)\big).$$

Note that $y^{1/2+\varepsilon} \le x^{1/4+\varepsilon} \le x^{1/2+\varepsilon}y^{-1/2}$. Hence the first error term above is majorized by the second. It is trivial that $S(x,y) \ll y$. On taking $y = x^{1/3}$ it follows that the error term above is $\ll x^{1/3+\varepsilon}$. This was achieved in Exercise 13.2.1.16. In the exercises that follow we use the results of Exercises 17.1.1.7 and 17.1.1.8, together with (17.44) and the van der Corput method to derive a non-trivial estimate for $S(x,y)$ when y is a little larger than $x^{1/3}$.

3. By quoting an appropriate estimate established in §16.2, show that if $V > 0$ and $M \le M_1 < M_2 \le 4M$, then

$$\sum_{M_1<m\le M_2} e\big(V/m^2\big) \ll V^{1/2}M^{-1} + V^{-1/2}M^2.$$

4. (a) By Exercise 17.1.1.8, show that if N and V are positive real numbers with $N > V^2$, then

$$T := \sum_{N<n\le 2N}\mu(n)e\big(W/n^2\big) = T_1 + T_2,$$

where

$$T_1 = -\sum_{m\le V^2} b_m \sum_{N/m<k\le 2N/m} e\big(Wm^{-2}k^{-2}\big),$$
$$T_2 = -\sum_{\substack{m>V\ k>V \\ N<mk\le 2N}}\sum \mu(m)c_k e\big(Wm^{-2}k^{-2}\big),$$

and b_m is defined in (17.26) and c_k is defined in (17.27).

(b) Show that if $K < j < k \le 4K$, then

$$\left|W\left(\frac{1}{j^2}-\frac{1}{k^2}\right)\right| \asymp W\frac{|j-k|}{K^3}.$$

(c) Deduce that if $M < M_1 < M_2 \le 4M$, then

$$\sum_{M_1<m\le M_2} e\left(W\left(\frac{1}{(jm)^2}-\frac{1}{(km)^2}\right)\right)$$
$$\ll W^{1/2}K^{-3/2}M^{-1}|j-k|^{1/2} + W^{-1/2}K^{3/2}M^2|j-k|^{-1/2}.$$

(d) Deduce that if $k \in \mathscr{K}_r$, then

$$\left|\sum_{j\in\mathscr{K}_r}\sum_{M_1<m\le M_2} e\left(W\left(\frac{1}{(jm)^2}-\frac{1}{(km)^2}\right)\right)\right|$$
$$\ll M + W^{1/2}R^{-3/2}M^{-1} + W^{-1/2}K^2M^2R^{-1/2}.$$

(e) Let Δ be defined as in (17.44). Suppose that $W \ge M^4$. By taking $R = \lfloor W^{1/3}/M^{4/3}\rfloor$, show that if $\Delta = \Delta(W,M,K) = W^{1/6}M^{-1/6} + W^{-1/6}M^{2/3}K$, then

$$\sum_{\substack{M<m\le 4M\ K<k\le 4K \\ N<mk\le 2N}}\sum b_m c_k e\big(W/(mk)^2\big)$$
$$\ll \Delta\left(\sum_{M<m\le 4M}|b_m|^2\right)^{1/2}\left(\sum_{K<k\le 4K}|c_k|^2\right)^{1/2}.$$

(f) Show that if $K > M$, then $\Delta(W,M,K) > \Delta(W,K,M)$. Thus, whenever $K > M$, the above bilinear form inequality will be applied with the roles of K and M reversed.

(g) By setting some variables to 0, show that if $W \ge M^4$, $M < M' \le 4M$, and $K < K' \le 4K$, then

$$\sum_{\substack{M<m\le M' \\ N<mk\le 2N}} \sum_{K<k\le K'} b_m c_k e\big(W/(mk)^2\big)$$
$$\ll \Delta \Bigg(\sum_{M<m\le M'} |b_m|^2\Bigg)^{1/2} \Bigg(\sum_{K<k\le K'} |c_k|^2\Bigg)^{1/2}.$$

(h) Show that $m \le 2N/V$ for all m that occur in the definition of T_2. Deduce that as M runs from $N^{1/2}$ to N/V, the blocks $(M, 4M]$ cover all m in the interval $N^{1/2} \le m \le 2N/V$. Note further that if $V \ge N/W^{1/4}$, then $W \ge M^4$ for all these M.

(i) By considering dyadic blocks, show that if $V \ge NW^{-1/4}$, then

$$\sum_{\substack{m>V \\ N<mk\le 2N}} \sum_{k>V} b_m c_k e\big(Wm^{-2}k^{-2}\big)$$
$$\ll \big(N^{-1/12}W^{1/6} + N^{5/6}W^{-1/6}\big) \left(\sum_{\substack{m \\ N/2<mk\le 4N}} \sum_{k} |b_m c_k|^2\right).$$

(j) Take $V = NW^{-1/4}$, and show that

$$T_2 \ll \big(W^{1/6}N^{5/12} + W^{-1/6}N^{4/3}\big)(\log 2N)^2.$$

(k) Write

$$T_1 = \sum_{m\le V} + \sum_{V<m\le V^2} = T_1^{(1)} + T_1^{(2)}.$$

Show that if $N \le W^{3/8}$, then $V^2 \le N/V$. Treat $T_1^{(2)}$ as a Type II sum to show that

$$T_1^{(2)} \ll \big(W^{1/6}N^{5/12} + W^{-1/6}N^{4/3}\big)(\log 2N)^2.$$

(l) Use the bound from Exercise 17.3.1.3 above to show that

$$T_1^{(1)} \ll \sum_{m\le V} d(m)\big(W^{1/2}N^{-1} + N^2W^{-1/2}m^{-1}\big),$$

and show that the above is

$$\ll W^{1/2}N^{-1}V(\log 2V) + N^2W^{-1/2}(\log 2V)^2.$$

Show that if $W^{1/5} \le N \le W^{3/8}$, then the above is

$$\ll N^{5/12}W^{1/6}\log 2N.$$

(m) Conclude that if $W^{1/5} \le N \le W^{3/8}$, then

$$T \ll \big(N^{5/12}W^{1/6} + N^{4/3}W^{-1/6}\big)(\log 2N)^2.$$

5. (a) By quoting Lemma D.1, or otherwise, show that

$$B_1(\{\alpha\}) = -\sum_{0<|h|\le H} \frac{e(h\alpha)}{2\pi i h} + O\left(\min\left(1, \frac{1}{H\|\alpha\|}\right)\right).$$

(b) Deduce that $S := \sum_{N<n\le 2N} \mu(n)B_1\big(\{Xn^{-2}\}\big) = M + O(R)$ where

$$M = -\sum_{0<|h|\le H} \frac{1}{2\pi i h} \sum_{N<n\le 2N} \mu(n)e\big(hXn^{-2}\big),$$

$$R = \sum_{N<n\le 2N} \min\left(1, \frac{1}{H\|Xn^{-2}\|}\right).$$

(c) By using Theorem E.6, or otherwise, explain why

$$R \ll \frac{N}{H}\log 2H + \frac{1}{H}\sum_{h=1}^{H}\left(\log\frac{3H}{h}\right)\left|\sum_{N<n\le 2N} e\big(hX/n^2\big)\right|.$$

(d) Deduce that

$$R \ll NH^{-1}\log 2H + H^{1/2}X^{1/2}N^{-1} + H^{-1/2}X^{-1/2}N^2.$$

Explain why

$$M \ll \sum_{h=1}^{H}\frac{1}{h}\left|\sum_{N<n\le 2N} \mu(n)e\big(hXn^{-2}\big)\right|.$$

(e) Assuming that $(hX)^{1/5} \le N \le (hX)^{3/8}$ for $1 \le h \le H$, deduce that

$$M \ll \big(N^{1/5}X^{1/6}H^{1/6} + N^{4/3}X^{-1/6}\big)(\log 2N)^2.$$

(f) Assume the bound for M in (e) is valid. Show that, in general, neither of the terms $N^{1/5}X^{1/6}H^{1/6}(\log 2N)^2$ and $NH^{-1}\log 2H$ can be replaced by anything smaller. Deduce that no better bound than $N^{1/2}x^{1/7}(\log N)^2$ can be obtained for $S*(x, N)$ and that this is worse than trivial when $x > N^{7/2}$.

(g) Assuming the bound $S^*(x, N) \ll N^{1/2}x^{1/7}(\log x)^2$ and that $y > x^{5/14}$, deduce that the $S(x, y)$ of (17.54) satisfies $S(x, y) \ll x^{1/2}y^{-1/2} + y^{1/2}x^{1/7}(\log x)^2$ and that this is $\ll x^{\frac{9}{28}}$ when $y = x^{\frac{5}{14}}$.

(h) Show that if $1 \le h \le N^{1/2}x^{-1/7}$ and $x^{2/7} \le N \le x^{5/14}$, then $(hx)^{1/5} \le N \le (hx)^{3/8}$.

(i) Show that the other term in the upper bound for M, and the other two terms in the upper bound for R are smaller than $N^{1/7}X^{1/7}(\log N)^2$ for $X^{2/7} \le N \le X^{5/14}$.

(j) Conclude from the assumption of RH that

$$Q(x) = \frac{6}{\pi^2}x + O\big(x^{9/28+\varepsilon}\big).$$

17.4 Digit Sums of Primes

Let $n = \sum_i d_i 2^i$ be the binary expansion of n, so that each d_i is either 0 or 1. Then $s(n) = \sum_i d_i$ is the sum of the binary digits of n. Since $s(2n+1) = s(2n) + 1$, it follows that $\left|\sum_{0 \le n \le N}(-1)^{s(n)}\right| \le 1$ for all N. Our object now is to show that

$$\sum_{p \le x} (-1)^{s(p)} = o(\pi(x)) \tag{17.58}$$

as $x \to \infty$. We begin by establishing a simple estimate that makes our work shorter.

Lemma 17.4 *Let M and N be integers with $N \ge 2$. Then for each integer n there exists a weight $w(n)$ such that $w(n) \ge 1$ for $M+1 \le n \le M+N$, $w(n) \ge 0$ for all other n, and*

$$W(\alpha) = \sum_{n=-\infty}^{\infty} w(n)e(n\alpha)$$

has the properties that $\max_\alpha |W(\alpha)| = W(0) \ll N$ *and* $W(\alpha) = 0$ *if* $\|\alpha\| \ge 1/N$.

Suppose that $M+1 \le n \le M+N$. Then

$$\begin{aligned} 1 \le w(n) &= \int_0^1 W(\alpha)e(-n\alpha)\,d\alpha \\ &= \int_{-1/N}^{1/N} W(\alpha)e(-n\alpha)\,d\alpha \le \frac{2}{N}\max_\alpha |W(\alpha)|. \end{aligned}$$

Thus our bound for $\max|W(\alpha)|$ is optimal, apart from constants. If sharp constants were required, then we would appeal to Theorem E.5, but for our present purposes we have no need for such sophistication.

Proof We recall that if $f(x) = \max(0, 1-|x|)$, then $\widehat{f}(t) = \left(\frac{\sin \pi t}{\pi t}\right)^2$. If N is even, put $K = M + N/2$. If N is odd, put $K = M + (N+1)/2$. Thus in either case, K is an integer. After several changes of variable we deduce that if $g(x) = N \max(0, 1 - N|x + \alpha|)e(K(x + \alpha))$, then

$$\widehat{g}(t) = \left(\frac{\sin \pi(t - K)/N}{\pi(t - K)/N}\right)^2 e(t\alpha).$$

By the Poisson summation formula in the form of Theorem D.3, we find that $\sum_n g(n) = \sum_k \widehat{g}(k)$. Thus

$$\begin{aligned} W(\alpha) &= \frac{\pi^2}{4} \sum_{n=-\infty}^{\infty} \left(\frac{\sin \pi(n - K)/N}{\pi(n - K)/N}\right)^2 e(n\alpha) \\ &= \frac{\pi^2}{4} N \max(0, 1 - N\|\alpha\|)e(K\alpha) \end{aligned}$$

has the required properties. □

The function $(-1)^{s(n)}$ has a power series generating function

$$P(z) = \sum_{n=0}^{\infty} (-1)^{s(n)} z^n = \prod_{j=0}^{\infty} \left(1 - z^{2^j}\right) \tag{17.59}$$

for $|z| < 1$, but in our quest to prove (17.58) we find it easier to work on the unit circle with a truncated sum, so we set

$$T_J(\theta) = \sum_{0 \le n < 2^J} (-1)^{s(n)} e(n\theta) = \prod_{j=0}^{J-1} \left(1 - e(2^j\theta)\right). \tag{17.60}$$

We now derive a uniform upper bound for $|T_J(\theta)|$.

Lemma 17.5 *Let $T_J(\theta)$ be defined as in* (17.60). *Then*

$$\max_\theta |T_J(\theta)| \asymp \sqrt{3}^J.$$

Since $T_J(\theta)$ is a sum of 2^J unimodular terms, it is trivial that $|T_J(\theta)| \le 2^J$ for all θ. Put

$$\alpha = 1 - \frac{\log 3}{\log 4} = 0.20752. \tag{17.61}$$

Our lemma asserts that

$$\max_\theta |T_J(\theta)| \asymp 2^{(1-\alpha)J}. \tag{17.62}$$

Thus α measures the extent to which the maximum of $|T_J|$ is smaller than the trivial upper bound 2^J.

Proof From the identity $1 - e(\beta) = -e(\beta/2)(e(\beta/2) - e(-\beta/2))$, we see that

$$|T_J(\theta)| = \prod_{j=0}^{J-1} |2\sin \pi 2^j \theta|.$$

When $\theta = 1/3$, each factor on the right has the value $\sqrt{3}$; thus $|T_J(1/3)| = \sqrt{3}^J$, so $|T_J(\theta)|$ achieves the indicated size, and it remains to derive a uniform upper bound for $|T_J(\theta)|$.

Let $f(\theta) = \sin^2 \pi\theta \sin^3 2\pi\theta \sin 4\pi\theta$. We first show that

$$|f(\theta)| \le 27/64 \tag{17.63}$$

for all θ. By use of the double angle formulæ for sine and cosine we find that

$$f(\theta) = 32(\sin \pi\theta)^6(1 - \sin^2 \pi\theta)(1 - 2\sin^2 \pi\theta).$$

Let $p(u) = 32u^3(1-u)^2(1-2u)$. Thus $p(u) \ge 0$ for $0 \le u \le 1/2$, $p(u) \le 0$ for $1/2 \le u \le 1$, and it suffices to show that $|p(u)| \le 27/64$ for $0 \le u \le 1$. Now $p'(u) = -32u^2(u-1)(3u-1)(4u-3)$, so $p(u)$ is increasing for $0 \le u \le 1/3$, decreasing for $1/3 \le u \le 3/4$, and increasing for $3/4 \le u \le 1$. Hence $\max_{0\le u\le 1} p(u) = p(1/3) = 128/729 = 0.1756$ and $\min_{0\le u \le 1} p(u) = p(3/4) = -27/64 = -0.4219$, so (17.63) holds.

If $2K = J - 1$ or $J - 2$, then

$$\begin{aligned} |T_J(\theta)|^3 &= \prod_{j=0}^{J-1} |2\sin 2^j \pi\theta|^3 \\ &\ll |2\sin \pi\theta|^2 \left(\prod_{k=1}^{2K-1} |2\sin 2^k\pi\theta|^3\right) |2\sin 2^{2K}\pi\theta| \\ &= \prod_{k=0}^{K-1} |64 f(4^k\theta)| \le 27^K \ll 3^{3J/2}, \end{aligned}$$

so we have the desired upper bound. □

Since the functions $\log|1 - e(2^j\theta)|$ move on widely different periods, we expect them to be nearly independent, and so we expect that $\log|T_J(\theta)|$ should be distributed as if it were a sum of J independent random variables. As

$$\int_0^1 \log|1 - e(\theta)|\,d\theta = 0 \tag{17.64}$$

and

$$\int_0^1 \left(\log|1 - e(\theta)|\right)^2 d\theta = \frac{\pi^2}{12}, \tag{17.65}$$

we expect that

$$\exp\left(-C\sqrt{J}\right) \le |T_J(\theta)| \le \exp\left(C\sqrt{J}\right) \tag{17.66}$$

for most θ, if C is a large positive constant. On the other hand, by Parseval's identity it is trivial that

$$\int_0^1 |T_J(\theta)|^2\, d\theta = 2^J, \tag{17.67}$$

which is to say that $\|T_J\|_2 = 2^{J/2}$. This is much larger than the order of magnitude in (17.66), so we infer that the large value of the 2-norm is due to a small set of θ for which $|T_J(\theta)|$ is exceptionally large. If this is the case, then we would expect $\|T_J\|_1$ to be smaller than the root-mean-square, $\|T_J\|_2$. Lemma 17.6 helps us to show that this is the case.

Lemma 17.6 *Let* $g(\theta) = \sin \pi\theta \sin 2\pi\theta$, *and put*

$$h(\theta) = |g(\theta)| + |g(\theta + 1/4)| + |g(\theta + 1/2)| + |g(\theta + 3/4)|.$$

Then

$$\max_{\theta} h(\theta) = h(1/8) = \sqrt{2 + \sqrt{2}}.$$

Proof Clearly $h(\theta)$ has period $1/4$. Since $g(\theta)$ is even, it follows that $h(\theta)$ is also even. Hence we may restrict our attention to $0 \le \theta \le 1/8$. In this interval, $g(\theta) \ge 0$, $g(\theta + 1/4) \ge 0$, $g(\theta + 1/2) \le 0$, and $g(\theta + 3/4) \le 0$, so

$$\begin{aligned} h(\theta) &= (2 - \sqrt{2}) \sin^2 \pi\theta \cos \pi\theta + 2 \sin \pi\theta \cos^2 \pi\theta + \sqrt{2} \cos^3 \pi\theta \\ &= \frac{1}{2}(1 + \sqrt{2}) \cos \pi\theta + \frac{1}{2} \sin \pi\theta + \frac{1}{2}(\sqrt{2} - 1) \cos 3\pi\theta + \frac{1}{2} \sin 3\pi\theta. \end{aligned} \tag{17.68}$$

Consequently,

$$\begin{aligned} h'(\theta) &= -\frac{\pi}{2}(1 + \sqrt{2}) \sin \pi\theta + \frac{\pi}{2} \cos \pi\theta \\ &\quad - \frac{3\pi}{2}(\sqrt{2} - 1) \sin 3\pi\theta + \frac{3\pi}{2} \cos 3\pi\theta, \\ h''(\theta) &= -\frac{\pi^2}{2}(1 + \sqrt{2}) \cos \pi\theta - \frac{\pi^2}{2} \sin \pi\theta \\ &\quad - \frac{9\pi^2}{2}(\sqrt{2} - 1) \cos 3\pi\theta - \frac{9\pi^2}{2} \sin 3\pi\theta. \end{aligned}$$

In this last formula all terms are ≤ 0, so it is clear that $h''(\theta) \le 0$. But

$$\begin{aligned} h'(1/8) &= -\frac{\pi}{2}(1+\sqrt{2})\sqrt{\frac{1-1/\sqrt{2}}{2}} + \frac{\pi}{2}\sqrt{\frac{1+1/\sqrt{2}}{2}} \\ &\quad - \frac{3\pi}{2}(\sqrt{2}-1)\sqrt{\frac{1+1/\sqrt{2}}{2}} + \frac{3\pi}{2}\sqrt{\frac{1-1/\sqrt{2}}{2}} \\ &= \frac{\pi}{4}(2-\sqrt{2})\sqrt{2-\sqrt{2}} + \frac{\pi}{4}(4-3\sqrt{2})\sqrt{2+\sqrt{2}} \\ &= 0 \end{aligned}$$

because $\sqrt{2-\sqrt{2}} = (\sqrt{2}-1)\sqrt{2+\sqrt{2}}$ and $4-3\sqrt{2} = (1-\sqrt{2})(2-\sqrt{2})$. Hence the maximum is attained at $\theta = 1/8$, and from (17.68) we see that

$$h(1/8) = \frac{1}{2}(3-\sqrt{2})\sqrt{\frac{1+1/\sqrt{2}}{2}} + \frac{1}{2}\sqrt{2}\sqrt{\frac{1-1/\sqrt{2}}{2}} = \sqrt{2+\sqrt{2}}. \qquad \square$$

It is convenient to observe that

$$T_J(\theta) = T_{J-j}(\theta)T_j(2^{J-j}\theta). \tag{17.69}$$

Lemma 17.7 *Let $\beta = 0.057111674\ldots$ be determined by the relation $4^\beta = 2/\sqrt{2+\sqrt{2}}$. Then*

$$\sum_{a=1}^{2^J} |T_J(\theta + a/2^J)| \ll 2^{(3/2-\beta)J}$$

uniformly in θ.

By integrating this bound over $0 \le \theta \le 1/2^J$, it is immediate that

$$\int_0^1 |T_J(\theta)|\, d\theta \ll 2^{(1/2-\beta)J}. \tag{17.70}$$

Thus β reflects the margin by which we can say that $\|T_J\|_1$ is smaller than $\|T_J\|_2$.

Proof Let $S_J(\theta)$ denote the sum to be bounded. By taking $j = J-2$ in (17.69), we find that

$$S_J(\theta) = \sum_{a=1}^{2^J} |T_2(\theta + a/2^J)||T_{J-2}(4\theta + a/2^{J-2})|.$$

Here the second factor has period 2^{J-2} with respect to a, so the above

$$= 4\sum_{a=1}^{2^{J-2}} |T_{J-2}(4\theta + a/2^{J-2})| h(\theta + a/2^J)$$

in the notation of Lemma 17.6. Hence by that lemma it is immediate that

$$S_J(\theta) \le 4\sqrt{2+\sqrt{2}}S_{J-2}(4\theta).$$

We apply this $K = [J/2]$ times to see that

$$S_J(\theta) \le \left(4\sqrt{2+\sqrt{2}}\right)^K S_{J-2K}(2^{2K}\theta).$$

But $S_0(\theta) \ll 1$, $S_1(\theta) \ll 1$, and $2(2+\sqrt{2})^{1/4} = 2^{3/2-\beta}$, so we have the stated result. □

By applying (17.69) and then Lemma 17.7 with J replaced by $J - j$, we deduce that

$$\sum_{c=1}^{2^{J-j}} |T_J(\theta + c/2^{J-j})| \ll |T_j(2^{J-j}\theta)| 2^{(3/2-\beta)(J-j)}. \tag{17.71}$$

In Theorem F.2 it is shown that if θ is irrational, then the numbers $n\theta$ are uniformly distributed modulo 1; this is achieved by combining the simple exponential sum estimate of Lemma 16.4 with Weyl's criterion. In general, as we let n run from 1 to N, we expect that $n\theta$ will fall into a short interval I approximately the expected number of times. However, it can sometimes happen that a short interval is hit far more times than expected. We now show that this can only happen when θ has a rational approximation a/q that is exceptionally good, and with q unusually small.

Lemma 17.8 *Let θ be a given real number. Suppose that $\delta_1 \le \delta_2/12$, that $N \ge 3/\delta_2$, and that $n\theta \in I = [\phi - \delta_1, \phi + \delta_1]$ (mod 1) for at least $\delta_2 N$ of the integers $n \in [1, N]$. Then there is an integer q with $1 \le q \le 9/\delta_2$ such that*

$$\|q\theta\| \le \frac{3\delta_1}{\delta_2 N}. \tag{17.72}$$

By Dirichlet's theorem diophantine approximation we know that there is a $q \le N$ such that $\|q\theta\| \le 1/N$, but the q given by Lemma 17.8 gives a better approximation, and with a q that is quite small.

Proof Among the positive integers $q \le N$, let q be the one for which $\|q\theta\|$ is minimal. For $0 \le n \le N$, arrange the numbers $\{n\theta\}$ in increasing order, and consider the minimal gap between consecutive terms, say $\{n_1\theta\} \le \{n_2\theta\}$. Then $\|(n_1 - n_2)\theta\|$ is the length of this gap. But $0 < |n_1 - n_2| \le N$, and $\|q\theta\|$ is minimal, so we see that of all the gaps between the numbers $\{n\theta\}$, the gap between $\{q\theta\}$ and 0 (or 1) is minimal. With $n\theta \in I$ for at least $\delta_2 N$ values of N, we have $\ge \delta_2 N - 1$ gaps, each of length at least $\|q\theta\|$. Hence $\|q\theta\|(\delta_2 N - 1) \le 2\delta_1$. This implies (17.72) since $\delta_2 N \ge 3$.

We divide the interval $[1, N]$ into $\le N/q + 1$ intervals of length $\le q$. For a given n_0, we consider those n, $n_0 \le n < n_0 + q$ such that $n\theta \in I \pmod 1$. We put $\delta = \theta - a/q$, so that

$$n\theta = n_0\theta + (n - n_0)a/q + (n - n_0)\delta.$$

By (17.72) we know that $|\delta| \le 3\delta_1/(\delta_2 q N)$. Hence

$$|(n - n_0)\delta| \le \frac{3\delta_1}{\delta_2 N} \le \frac{1}{4N}$$

since $\delta_1 \le \delta_2/12$. Thus if $n\theta \in I$, then $(n - n_0)a/q \in J = [\phi - n_0\theta - \delta_1 - 1/(4N), \phi - n_0\theta + \delta + 1/(4N)]$. Since the numbers $(n - n_0)a/q$ are in arithmetic progression with common difference $1/q$, the number of n, $n_0 \le n < n_0 + q$, for which $(n - n_0)a/q \in J$ is $\le 1 + q(2\delta_1 + 1/(2N)) \le 2\delta_1 q + 3/2$ since $q \le N$. Consequently, the total number of $n \le N$ for which $n\theta \in I$ is

$$\le (N/q + 1)(2\delta_1 q + 3/2) = 2\delta_1 N + 3N/(2q) + 2\delta_1 q + 3/2.$$

Since $q \le N$, the first and third terms on the right-hand side sum to $\le 4\delta_1 N \le \delta_2 N/3$. The last term is $\le \delta_2 N/2$ since $N \ge 3/\delta_2$. Since the number of $n \le N$ for which $n\theta \in I$ is by hypothesis $\ge \delta_2 N$, we conclude that

$$\delta_2 N \le \frac{5}{6}\delta_2 N + \frac{3N}{2q},$$

which implies that $q \le 9/\delta_2$. □

With Lemmas 17.4–17.8 in hand, and most particularly with our non-trivial estimates of $\|T_J\|_\infty$ and of $\|T_J\|_1$, we are now in a position to apply Vaughan's identity with $f(n) = (-1)^{s(n)}$ to prove

Theorem 17.9 (Mauduit & Rivat 2010) *Let $s(n)$ denote the sum of the binary digits of n. Then*

$$\sum_{n \le N} (-1)^{s(n)} \Lambda(n) \ll N^{1-1/263}. \tag{17.73}$$

Proof For $N \ge 2$ we set

$$T(\theta) = \sum_{n \le N} (-1)^{s(n)} e(n\theta).$$

If we choose J so that $2^{J-1} < N \le 2^J$, then T is a truncation of the sum T_J, and hence by (E.22) and (E.24) we know that

$$\|T\|_\infty \ll N^{1-\alpha} \log N \tag{17.74}$$

and that

$$\|T\|_1 \ll N^{1/2-\beta} \log N, \tag{17.75}$$

where α and β are defined in (17.61) and Lemma 17.7. We take $f(n) = (-1)^{s(n)}$ in (17.6) in order to estimate $S = \sum_{n \le N} f(n)\Lambda(n)$. Our treatment of the Type I sums is very simple:

$$\sum_{\substack{n \le N \\ t|n}} f(n) = \frac{1}{t} \sum_{a=1}^{t} T(a/t).$$

By the triangle inequality and (17.74) it follows that

$$\sum_{t \le U} \left| \sum_{r \le N/t} f(rt) \right| \ll UN^{1-\alpha} \log N. \tag{17.76}$$

By replacing N in (17.74) by w and differencing, we see that

$$\max_{w} \left| \sum_{w \le n \le N} f(n)e(n\theta) \right| \ll N^{1-\alpha} \log N$$

uniformly in θ. Hence by the same reasoning,

$$\sum_{d \le V} \max_{w \ge 1} \left| \sum_{w \le h \le N/d} f(dh) \right| \ll VN^{1-\alpha} \log N.$$

Thus

$$S_3 \ll VN^{1-\alpha} (\log N)^2 \tag{17.77}$$

in the notation of (17.9). We write $S_2 = \sum_{t \le U} + \sum_{U < t \le UV} = S_I + S_{II}$; then

$$S_I \ll UN^{1-\alpha} (\log NUV)^2 \tag{17.78}$$

by (17.76). We treat S_{II} and S_4 as Type II sums, and for that we show that if $|b_m| \le 1$ for all m, $|c_k| \le 1$ for all k, and $M \le K$, then

$$\sum_{\substack{M < m \le 2M \\ K < k \le 2K \\ mk \le N}} b_m c_k f(mk) \ll K^{1+\varepsilon} M^{1-\beta/(3-4\beta)+\varepsilon} + K^{1-1/(10-8\beta)+\varepsilon} M^{1+(1-2\beta)/(10-8\beta)+\varepsilon}. \tag{17.79}$$

Here the second term is largest when $M \asymp K \asymp N^{1/2}$, at which point it is $\asymp N^{1-\beta/(10-8\beta)}$. Here $\beta/(10-8\beta) = 0.00598\ldots > 1/200$. The first term on the right in (17.79) becomes larger as M becomes smaller (with $K \asymp N/M$), but we take $U = V = N^{\alpha(3-4\beta)/(3-3\beta)}$, and note that then $NU^{-\beta/(3-4\beta)} = UN^{1-\alpha} = N^{1-\alpha\beta/(3-3\beta)}$. Here $\alpha\beta/(3-3\beta) = 0.0038104 > 1/263$. To treat a block with $M > K$ we simply reverse the roles of m and k. For S_4 we take

$$b_m = \mu(m), \qquad c_k = \frac{1}{\log N} \sum_{\substack{d|k \\ d>V}} \Lambda(d),$$

or vice versa if $M > K$. Conditions such as $m > U$ and $k > V$ can be met by stipulating that $b_m = 0$ if $m \le U$ and $c_k = 0$ if $k \le V$. To treat S_{II} we take

$$b_m = \begin{cases} b(m)/\log N & (m \ge U), \\ 0 & (m < U), \end{cases} \qquad c_k = \begin{cases} 1 & (k \le UV), \\ 0 & (k > UV), \end{cases}$$

or vice versa.

By Cauchy's inequality, the left-hand side of (17.79) is

$$\le M^{1/2} \left(\sum_{M<m\le 2M} \left| \sum_{\substack{K<k\le 2K \\ mk\le N}} c_k f(mk) \right|^2 \right)^{1/2}.$$

Thus to prove (17.79), it suffices to show that

$$\sum_{M<m\le 2M} \left| \sum_{\substack{K<k\le 2K \\ mk\le N}} c_k f(mk) \right|^2 \ll K^{2+\varepsilon} M^{1-2\beta/(3-4\beta)+\varepsilon} + K^{2-1/(5-4\beta)+\varepsilon} M^{1+(1-2\beta)/(5-4\beta)+\varepsilon}. \tag{17.80}$$

By van der Corput's lemma (Lemma 16.8) we see that

$$\begin{aligned} &\left| \sum_{\substack{K<k\le 2K \\ mk\le N}} c_k f(mk) \right|^2 \\ &\le \frac{K+H-1}{H} \sum_{\substack{K<k\le 2K \\ mk\le N}} |c_k f(mk)|^2 \\ &\quad + 2\,\mathrm{Re}\, \frac{K+H-1}{H} \sum_{h=1}^{H} (1-h/H) \sum_{\substack{K<k\le 2K-h \\ m(k+h)\le N}} c_{k+h}\overline{c_k} f(m(k+h)) f(mk). \end{aligned}$$

Here H is a parameter to be chosen later, subject to

$$H \le K. \tag{17.81}$$

The first term on the right-hand side above is $\ll K^2/H$. We sum the above over m to see that the left-hand side of (17.80) is

$$\ll \frac{K^2 M}{H} + \frac{K}{H} \sum_{h=1}^{H} \sum_{K<k\le 2K} \left| \sum_{\substack{M<m\le 2M \\ m(k+h)\le N}} f(m(k+h)) f(mk) \right|. \tag{17.82}$$

Let $n = \sum_j d_j 2^j$ be the binary expansion of n. We divide 2^J into n, so that $n = q2^J + r$. Then $r = \sum_{j<J} d_j 2^j$ and $s(n) = s(q) + s(r)$. Put $s_J(n) = \sum_{j<J} d_j =$

$s(r) = s(n) - s(q)$. Thus if $q2^J \le m, n < (q+1)2^J$, then $s(m) - s_J(m) = s(q) = s(n) - s_J(n)$. Put $f_J(n) = (-1)^{s_J(n)}$. Then $f(m(k+h))f(mk) = f_J(m(k+h))f_J(mk)$ unless there is a multiple of 2^J between mk and $m(k+h)$. We choose J so that 2^J is large compared with MH, but small compared with MK. Suppose that $mk < q2^J \le m(k+h)$. Then $\{mk/2^J\} \ge 1 - mh/2^J \ge 1 - 2MH/2^J$. Thus

$$\sum_{\substack{M<m\le 2M\\ m(k+h)\le N}} f(mk)f(m(k+h)) = \sum_{\substack{M<m\le 2M\\ m(k+h)\le N}} f_J(mk)f_J(m(k+h)) + O\Bigg(\sum_{\substack{M<m\le 2M\\ \{mk/2^J\}\ge 1-2MH/2^J}} 1\Bigg). \tag{17.83}$$

We group pairs m, k according to the value of mk to see that

$$\frac{K}{H}\sum_{h=1}^{H}\sum_{K<k\le 2K}\sum_{\substack{M<m\le 2M\\ \{mk/2^J\}\ge 1-2MH/2^J}} 1 \ll K^{1+\varepsilon}M^{\varepsilon}\sum_{\substack{n\le 4MK\\ \{n/2^J\}\ge 1-2MH/2^J}} 1$$

since $d(n) \ll (MK)^{\varepsilon}$. We divide the interval $(0, 4MK]$ into $\ll MK/2^J$ intervals of length 2^J. For n in an interval of length 2^J, the inequality $\{n/2^J\} \ge 1 - 4MH/2^J$ holds for $\ll MH$ values of n. Hence the above is

$$\ll (KM)^{2+\varepsilon}H/2^J. \tag{17.84}$$

The function f_J is periodic with period 2^J, and so has a finite Fourier transform,

$$\widehat{f_J}(a) = \frac{1}{2^J}\sum_{n=1}^{2^J} f_J(n)e(-an/2^J) = \frac{1}{2^J}T_J(-a/2^J),$$

so that

$$f_J(n) = \sum_{a=1}^{2^J}\widehat{f_J}(a)e(an/2^J).$$

Thus the first term on the right-hand side of (17.83) is

$$\sum_{a=1}^{2^J}\sum_{b=1}^{2^J}\widehat{f_J}(a)\widehat{f_J}(b)\sum_{\substack{M<m\le 2M\\ m(k+h)\le N}} e((am(k+h)+bmk)/2^J)$$
$$\ll \sum_{a=1}^{2^J}\sum_{b=1}^{2^J}|\widehat{f_J}(a)\widehat{f_J}(b)|\min(M, 1/\|(a(k+h)+bk)/2^J\|)$$

by (16.4). To (17.82) this contributes an amount

$$\ll \frac{K}{H}\sum_{h=1}^{H}\sum_{K<k\le 2K}\sum_{a=1}^{2^J}\sum_{b=1}^{2^J}|\widehat{f_J}(a)\widehat{f_J}(b)|\min(M, 1/\|(a(k+h)+bk)/2^J\|).$$

Our estimate for this depends on the power of 2 dividing $a+b$. Write $a+b = c2^j$ with c odd. We may assume that a and b are odd since $\widehat{f_J}(a) = 0$ if a is even. Thus $1 \le j \le J$, and the above is

$$\begin{aligned} &\ll \frac{K}{H} \sum_{a=1}^{2^J} \sum_{j=1}^{J} \sum_{\substack{c=1 \\ 2\nmid c}}^{2^{J-j}} |\widehat{f_J}(a)\widehat{f_J}(c2^j - a)| \\ &\qquad \times \sum_{h=1}^{H} \sum_{K<k\le 2K} \min(M, 1/\|ck/2^{J-j} + ah/2^J\|). \end{aligned} \tag{17.85}$$

Let $w_1(h)$ be weights that arise when Lemma 17.4 is applied to the interval $[1, H]$, and let $w_2(k)$ denote the weights when Lemma 17.4 is applied to the interval $[K, 2K]$. Then

$$\begin{aligned} &\sum_{h=1}^{H} \sum_{K<k\le 2K} \min(M, 1/\|ck/2^{J-j} + ah/2^J\|) \\ &\qquad \le \sum_{h=-\infty}^{\infty} \sum_{k=-\infty}^{\infty} w_1(h)w_2(k) \min(M, 1/\|ck/2^{J-j} + ah/2^J\|). \end{aligned}$$

Let $g(x)$ be defined as in Theorem E.6. Then the above is

$$\begin{aligned} &\ll \sum_{h=-\infty}^{\infty} \sum_{k=-\infty}^{\infty} w_1(h)w_2(k)g(ck/2^{J-j} + ah/2^J) \\ &= \sum_{m=-M}^{M} \widehat{g}(m) \sum_{h=-\infty}^{\infty} w_1(h)e(mah/2^J) \sum_{k=-\infty}^{\infty} w_2(k)e(mck/2^{J-j}) \\ &= \sum_{m=-M}^{M} \widehat{g}(m)W_1(ma/2^J)W_2(mc/2^{J-j}). \end{aligned}$$

By Lemma 17.4 and Theorem E.6, this is

$$\ll HK(\log M)\left(1 + \sum_{\substack{m=1 \\ \|ma/2^J\|<1/H \\ \|mc/2^{J-j}\|<1/K}}^{M} 1\right). \tag{17.86}$$

By (17.71) we see that

$$\sum_{c=1}^{2^{J-j}} |\widehat{f_J}(c2^j - a)| \ll |\widehat{f_j}(a)| 2^{(1/2-\beta)(J-j)}. \tag{17.87}$$

Hence

$$\sum_{a=1}^{J} |\widehat{f}_J(a)| \sum_{c=1}^{2^{J-j}} |\widehat{f}_J(c2^j - a)| \ll 2^{(1/2-\beta)(J-j)} \sum_{a=1}^{2^J} |\widehat{f}_J(a)\widehat{f}_j(a)|. \qquad (17.88)$$

By (17.69) we see that

$$\begin{aligned}\widehat{f}_J(a) = \frac{1}{2^J}T_J(a/2^J) &= \frac{1}{2^J}T_{J-j}(a/2^J)T_j(a/2^j) \\ &= \frac{1}{2^{J-j}}T_{J-j}(a/2^J)\widehat{f}_j(a). \end{aligned} \qquad (17.89)$$

Write $a = a_0 + a_1 2^j$. Then the right-hand side of (17.88) is

$$= 2^{(-1/2-\beta)(J-j)} \sum_{a_0=1}^{2^j} |\widehat{f}_j(a_0)|^2 \sum_{a_1=1}^{2^{J-j}} |T_{J-j}(a_0 + a_1/2^{J-j})|.$$

By Lemma 17.7 this is

$$\ll 2^{(1-2\beta)(J-j)} \sum_{a_0=1}^{j} |\widehat{f}_j(a_0)|^2.$$

Here the sum over a_0 is $= 1$ by Parseval's identity, so we conclude that

$$\sum_{a=1}^{J} |\widehat{f}_J(a)| \sum_{c=1}^{2^{J-j}} |\widehat{f}_J(c2^j - a)| \ll 2^{(1-2\beta)(J-j)}. \qquad (17.90)$$

Hence the term $HK \log M$ in (17.86), which reflects the mean value of $\min(M, 1/\|x\|)$, contributes to (17.85) an amount that is

$$\ll K^2 2^{(1-2\beta)J} \log M. \qquad (17.91)$$

It remains to estimate

$$K^2(\log M) \sum_{a=1}^{2^J} \sum_{j=1}^{J} \sum_{\substack{c=1 \\ 2\nmid c}}^{2^{J-j}} |\widehat{f}_J(a)\widehat{f}_J(c2^j - a)| \sum_{\substack{m=1 \\ \|ma/2^J\|<1/H \\ \|mc/2^{J-j}\|<1/K}}^{M} 1. \qquad (17.92)$$

The way that we proceed depends on the size of 2^{J-j}. Suppose first that $2^{J-j} \ge K$. Since $M \le K$, the numbers $m = 1, 2, \ldots, M$ comprise at most one complete system of residues modulo 2^{J-j}, and hence the number of them for which $\|mc/2^{J-j}\| \le 1/K$ is $\ll 2^{J-j}/K$ since c is odd. By (17.90), such a j contributes to (17.92) an amount

$$\ll K 2^{(2-2\beta)(J-j)} \log M,$$

and the sum over such j contributes

$$\ll K 2^{(2-2\beta)J} \log M. \tag{17.93}$$

Next, suppose that $M < 2^{J-j} < K$. For $1 \le m \le M < 2^{J-j}$ we have $m \not\equiv 0 \pmod{2^{J-j}}$, and hence $\|mc/2^{J-j}\| \ge 1/2^{J-j} > 1/K$. Thus in (17.90), the sum over m is empty when $M < 2^{J-j} < K$. Finally, suppose that $2^{J-j} \le M$. Since $K \ge M$, the inequality $\|mc/2^{J-j}\| < 1/K$ holds only when m is a multiple of 2^{J-j}. Write $m = r2^{J-j}$. Then we have to estimate

$$K^2(\log M) \sum_{a=1}^{2^J} \sum_{\substack{j \\ 1\le 2^{J-j} \le M}} \sum_{c=1}^{2^{J-j}} |\widehat{f}_J(a)\widehat{f}_J(c2^j - a)| \sum_{\substack{r=1 \\ \|ra/2^j\| < 1/H}}^{M/2^{J-j}} 1. \tag{17.94}$$

To the extent possible, we argue as before. By (17.87) the above is

$$\ll K^2(\log M) \sum_{a=1}^{2^J} \sum_{\substack{j \\ 1\le 2^{J-j} \le M}} 2^{(1/2-\beta)(J-j)} |\widehat{f}_J(a)\widehat{f}_j(a)| \sum_{\substack{r=1 \\ \|ra/2^j\| < 1/H}}^{M/2^{J-j}} 1.$$

We appeal to (17.89) and write $a = a_0 + a_1 2^j$ to see that the above is

$$= K^2(\log M) \sum_{\substack{j \\ 1\le 2^{J-j} \le M}} 2^{(-1/2-\beta)(J-j)} \sum_{a_0=1}^{2^j} |\widehat{f}_j(a_0)|^2$$
$$\times \sum_{\substack{r=1 \\ \|ra/2^j\| < 1/H}}^{M/2^{J-j}} \sum_{a_1=1}^{2^{J-j}} |T_{J-j}(a_0/2^J + a_1/2^{J-j})|.$$

We apply Lemma 17.7 to the sum over a_1, and thus see that the above is

$$\ll K^2(\log M) \sum_{\substack{j \\ 1\le 2^{J-j} \le M}} 2^{(1-2\beta)(J-j)} \sum_{a_0=1}^{2^j} |\widehat{f}_j(a_0)|^2 \sum_{\substack{r=1 \\ \|ra_0/2^j\| < 1/H}}^{M/2^{J-j}} 1. \tag{17.95}$$

In general, we would expect the sum over r to be about $M/(H2^{J-j})$ in size. Let B be chosen later, $B \le H$. The a_0 for which the sum over r is $\le M/(B2^{J-j})$ contribute an amount $\ll M/(B2^{J-j})$, by Parseval's identity. Now consider those a_0 for which the sum over r lies between $2^i M/(B2^{J-j})$ and $2^{i+1}M/(B2^{J-j})$. This is far more solutions than we expect, and by Lemma 17.8 it follows that there is a $q \ll B/2^i$ such that $\|qa_0/2^j\| \ll B2^{J-j}/(2^i HM)$. Let h denote an integer nearest $qa_0/2^j$. Then $1 \le h \le q$, and

$$\left|a_0 - \frac{2^j h}{q}\right| \ll \frac{B2^J}{2^i HMq},$$

so for each h there are $\ll 1 + B2^J/(2^i HMq) \ll B2^J/(2^i HMq)$ since $q2^i \ll B$ and we are assuming (17.81) so that $HM \le KM \ll N \ll 2^J$. On summing over h and over $q \ll B/2^i$, we find that the total number of such a_0 is $\ll B^2 2^J/(2^{2i} HM)$. Since $\widehat{f}_j(a_0) \ll 2^{-\alpha j}$ by (17.61), we find that the contribution of such a_0 is $\ll B2^{(1-2\alpha)j}/(2^i H)$. We sum over i, and combine our estimates to see that

$$\sum_{a_0=1}^{2^j} |\widehat{f}_j(a_0)|^2 \sum_{\substack{r=1 \\ \|ra_0/2^j\|<1/H}}^{M/2^{J-j}} 1 \ll \frac{M}{B2^{J-j}} + \frac{B2^{(1-2\alpha)j}}{H}.$$

To optimize this bound we take $B = M^{1/2} H^{1/2} 2^{-J/2+\alpha j}$, and thus see that

$$\sum_{a_0=1}^{2^j} |\widehat{f}_j(a)|^2 \sum_{\substack{r=1 \\ \|ra_0/2^j\|<1/H}}^{M/2^{J-j}} 1 \ll M^{1/2} H^{-1/2} 2^{-J/2+(1-\alpha)j}. \tag{17.96}$$

Hence the quantity (17.95) is

$$\ll K^2 M^{1/2} H^{-1/2} 2^{(1/2-\alpha)J} (\log M) \sum_{\substack{j \\ 1 \le 2^{J-j} \le M}} 2^{(\alpha-2\beta)(J-j)}.$$

But $\alpha - 2\beta > 0$, so the largest term occurs when $2^{J-j} \asymp M$, and hence the above is

$$\ll K^2 M^{1/2+\alpha-2\beta} H^{-1/2} 2^{(1/2-\alpha)J} \log M. \tag{17.97}$$

On combining this with (17.82), (17.84), (17.91), and (17.93), we conclude that the left-hand side of (17.80) is

$$\begin{aligned} &\ll K^2 M H^{-1} + (KM)^{2+\varepsilon} H 2^{-J} + K^2 2^{(1-2\beta)J} \log M \\ &\quad + K 2^{(2-2\beta)J} \log M + K^2 M^{1/2+\alpha-2\beta} H^{-1/2} 2^{(1/2-\alpha)J} \log M. \end{aligned}$$

Suppose that $2^J \asymp MHA$. Then (apart from the ε in the exponent), the first two terms are $\ll K^2 M(1/H + 1/A)$. If A and H are allowed to vary in such a way that AH is held constant, then the third and fourth terms above are fixed, and the sum of the first two terms is minimized by taking $A = H$. Accordingly, we take J so that $2^J \asymp MH^2$. Thus the above is

$$\begin{aligned} &\ll K^{2+\varepsilon} M^{1+\varepsilon} H^{-1} + K^2 M^{1-2\beta} H^{2-4\beta} \log M \\ &\quad + K M^{2-2\beta} H^{4-4\beta} \log M + K^2 H^{1/2-2\alpha} M^{1-2\beta} \log M. \end{aligned}$$

Here the last term is smaller than the second one, so may be ignored. If $K^{1-4\beta/3} \le M \le K$, then we take $H = K^{1/(5-4\beta)}/M^{(1-2\beta)/(5-4\beta)}$. Then the first and third terms are roughly equal and the second term is smaller. In this range, all terms are

$$\ll K^{1-1/(5-4\beta)+\varepsilon} M^{1+(1-2\beta)/(5-4\beta)+\varepsilon}.$$

For $2 \le M \le K^{1-4\beta/3}$, we take $H = M^{2\beta/(3-4\beta)}$. Then the first and second terms are nearly equal, and the third one is smaller. In this range, all terms are

$$\ll K^{2+\varepsilon} M^{1-2\beta/(3-4\beta)+\varepsilon}.$$

Thus we have (17.80), and the proof is complete. □

We note that (17.96) is worse than the trivial bound

$$\sum_{a_0=1}^{2^j} |\widehat{f}_j(a)|^2 \sum_{\substack{r=1 \\ \|ra_0/2^j\|<1/H}}^{M/2^{J-j}} 1 \ll M2^{-(J-j)} \sum_{a_0=1}^{2^j} |\widehat{f}_j(a)|^2 \ll M2^{-(J-j)}$$

when $2^{\alpha(J-j)} > M^\alpha H^{-1/2+2\alpha}$. Thus we could improve on (17.97), but this would not lead to a stronger conclusion because the bound in (17.97) makes a smaller contribution than the estimate (17.91).

17.4.1 Exercises

1. For $0 \le r \le 1$, let $f_r(\theta) = \log|1 - re(\theta)|$.
   (a) Show that if $0 \le r < 1$, then
   $$f_r(\theta) = -\sum_{n=1}^{\infty} \frac{r^n}{n} \cos 2\pi n\theta.$$
   (b) Show that if $\theta \notin \mathbb{Z}$, then $\sum_{n=1}^{\infty} (\cos 2\pi n\theta)/n$ converges.
   (c) By Abel's theorem (cf. §5.2), deduce that
   $$f_1(\theta) = -\sum_{n=1}^{\infty} \frac{\cos 2\pi n\theta}{n}$$
   when $\theta \notin \mathbb{Z}$.
   (d) Show that
   $$f_1(\theta) - f_r(\theta) \ll \min\left(\frac{1-r}{\|\theta\|}, \log\frac{1-r}{\|\theta\|}\right).$$
   (e) Deduce that $\|f_1 - f_r\|_1 \ll (1-r)\log(2/(1-r))$.
   (f) Show that if $0 \le r < 1$, then
   $$\widehat{f_r}(n) = \begin{cases} \dfrac{-r^{|n|}}{2|n|} & (n \ne 0), \\ 0 & (n = 0). \end{cases}$$

(g) By the inequality $|\widehat{f_r}(n) - \widehat{f_1}(n)| \le \|f_r - f_1\|_1$, deduce that

$$\widehat{f_1}(n) = \begin{cases} \dfrac{-1}{2n} & (n \ne 0), \\ 0 & (n = 0). \end{cases}$$

(h) Deduce (17.64).

(i) Deduce (17.65).

2. (a) Show that $|1 - e(\theta)| + |1 + e(\theta)| \le 2\sqrt{2}$ for all θ.

(b) Let $S_J(\theta)$ denote the sum in Lemma 17.7. Show that

$$S_J(\theta) \le 2\sqrt{2} S_{J-1}(2\theta).$$

3. For $1 \le n \le N$, let X_n denote independent random variables with $P(X_n = \pm 1) = 1/2$. For a generic point ω of our probability space, let $f_\omega(\theta) = \sum_{n=1}^N X_n e(n\theta)$ denote a random exponential polynomial.

(a) Show that

$$\int_0^1 |f_\omega(\theta)|^2 \, d\theta = N$$

for all ω.

(b) Show that

$$\int_0^1 |f_\omega(\theta)|^4 \, d\theta = \sum_{n=2}^{2N} \left(\sum_{\substack{1 \le m,k \le N \\ m+k=n}} X_m X_k \right)^2.$$

(c) Show that the number of pairs (m, k) with $1 \le m, k \le N$, and $m + k = n$ is $\max(0, N - |N + 1 - n|)$.

(d) Show that if n is odd, $2 \le n \le 2N$, then

$$E\left[\sum_{\substack{1 \le m_1,k_1,m_2,k_2 \le N \\ m_1+k_1=m_2+k_2=n}} X_{m_1} X_{k_1} X_{m_2} X_{k_2} \right] = 2(N - |N + 1 - n|).$$

(e) Show that if n is even, $2 \le n \le 2N$, then

$$E\left[\sum_{\substack{1 \le m_1,k_1,m_2,k_2 \le N \\ m_1+k_1=m_2+k_2=n}} X_{m_1} X_{k_1} X_{m_2} X_{k_2} \right] = 2(N - |N + 1 - n|) - 1.$$

(f) Show that

$$E\left[\int_0^1 |f_\omega(\theta)|^4 \, d\theta \right] = 2N^2 - N.$$

(g) Deduce that

$$P\left(\int_0^1 |f_\omega(\theta)|^4 \, d\theta > 4N^2\right) \le \frac{1}{2}.$$

(h) Show that

$$\int_0^1 |f|^2 \le \left(\int_0^1 |f|\right)^{2/3} \left(\int_0^1 |f|^4\right)^{1/3}$$

for all f.

(i) Show that

$$P\left(\int_0^1 |f_\omega(\theta)| \, d\theta \ge \frac{\sqrt{N}}{2}\right) \ge \frac{1}{2}.$$

With more work, it can be shown that $\int_0^1 |f_\omega(\theta)|^4 \, d\theta$ is usually near its expectation, with the result that the probability considered in (i) above tends rapidly to 1 as N tends to infinity. Also, it is unlikely that $\|f_\omega\|_\infty$ would be much larger than $\sqrt{N \log N}$. Hence in Lemma 17.7 and (17.62) we see that the coefficients $(-1)^{s(n)}$ produce behaviour that would be highly atypical for a random sequence.

4. Suppose that $0 < \delta_1 \le \delta_2/2$, that $N \ge 1/\delta_2$, and that $1 \le q \le 1/(2\delta_2)$. Choose a so that $(a, q) = 1$, put $\theta = a/q + \delta_1/(\delta_2 q N)$, and set $I = [0, 2\delta_1]$.

   (a) Show that $\|q\theta\| = \delta_1/(\delta_2 N)$.
   (b) Show that $n\theta \in I$ (mod 1) for at least $\delta_2 N$ values of n, $1 \le n \le N$.

5. (Mauduit, Montgomery, & Rivat 2018)

   (a) Explain why $|T_J(\theta)|^4 = |T_{J-2}(\theta)|^4 |T_2(2^{J-2}\theta)|^4$.
   (b) Explain why $|T_{J-1}(\theta)|^4 = |T_{J-2}(\theta)|^4 |T_1(2^{J-2}\theta)|^4$.
   (c) Write $|T_2(\alpha)|^4 - 2|T_1(\alpha)|^4 - 16 = \sum_{n=0}^{6} c_n \cos 2\pi n\alpha$. Show that $c_0 = c_1 = 0$.
   (d) Explain why $\int_0^1 |T_{J-2}(\theta)|^4 e(k\theta) \, d\theta = 0$ if $|k| \ge 2^{J-1} - 1$.
   (e) Put $u_J = \int_0^1 |T_J(\theta)|^4 \, d\theta$. Show that $u_0 = 1$, $u_1 = 6$, and that

   $$u_J = 2u_{J-1} + 16u_{J-2}$$

   for $J \ge 2$.
   (f) Show that

   $$u_J = \frac{17 + 5\sqrt{17}}{34}(1 + \sqrt{17})^J + \frac{17 - 5\sqrt{17}}{34}(1 - \sqrt{17})^J.$$

17.5 Notes

Section 17.1. The description of the various sums as being of Type I or Type II was introduced in Vaughan (1977b). The identity (17.6) connecting S with S_1, S_2, S_3, and S_4 was first displayed in Vaughan (1977a). The proof there was elementary, and the identity was discovered during an investigation of the properties of

$$\sum_{\substack{m|n \\ m\le U}} \mu(m).$$

Methods based on the identity (17.4) had already been used in Vaughan (1975), which had been noticed as an improvement of Gallagher's identity (Gallagher 1968)

$$-\frac{\zeta'}{\zeta} = -2\zeta' G + \zeta'\zeta G^2 - \zeta'\zeta\left(\frac{1}{\zeta} - G\right)^2.$$

This would now be considered a special case of Heath-Brown's identity (Heath-Brown 1982), which is discussed in Exercise 17.1.1.6. Montgomery then pointed out that (17.4) and (17.6) are simply different manifestations of the same underlying relationship.

The introduction of the relatively simple identity (17.6) led to a revived interest in a number of cognate problems that had otherwise been considered inaccessible. From among the many examples, we note the work of Heath-Brown & Patterson (1979) on the distribution of the arguments of Kummer sums (see also Heath-Brown 2000), and also the work of Green & Tao (2012) on the nature of the Möbius function.

Section 17.2. On the hypothesis of Theorem 17.1, Vinogradov (1937b) showed by a method based on the sieve of Eratosthenes that

$$S(\alpha) \ll \left(Nq^{-1/2} + N\exp(-\tfrac{1}{2}\sqrt{\log N}) + N^{1/2}q^{1/2}\right)(\log N)^{9/2},$$

see also Vinogradov (1947, 1954, Chapter 9, Theorem 1). Vinogradov later made a number of improvements to this and applied the technique to other situations. The ultimate result is Theorem 3 in Chapter 9 of Vinogradov (1954). For a general account of his work see the Royal Society obituary in Cassels and Vaughan (1985).

In response to a question of N. J. Fine, Besicovitch (1961) showed that there exist continuous 1-periodic real-valued non-constant even functions f such that $\sum_{a=1}^{q} f(a/q) = 0$ for all positive integers q. The construction of Bateman & Chowla (1963), found in Exercise 17.2.1.6, is simpler. The graph in Figure 17.1 is based on a rigorous computation of $\operatorname{Re} g(x)$ at 10,023 points together with linear interpolation.

Section 17.3. Concerning n^2+1, the quantitative conjecture with the constant C in (17.45) is Conjecture E in §5.42 of Hardy & Littlewood (1922).

Instead of breaking the interval $(K, 4K]$ into subintervals $\mathscr{K}_r$, we can restrict attention to terms near the diagonal by applying van der Corput differencing (Lemma 16.8), as we do in the proof of Theorem 17.9.

Theorem 17.2 is in Piaetski-Shapiro (1953). The connection with exponential sums and the van der Corput method has led to many refinements over the years. The best result currently is that the $\frac{12}{11}$ has been replaced by $\frac{243}{205}$ by Rivat & Wu (2001).

Section 17.4. For an integer $q > 1$ and a positive integer n, let $s_q(n)$ denote the sum of the digits in the base q expansion of n. Gel'fond (1967/1968) posed the problem of determining the distribution of $s_q(p)$ modulo m for arbitrary q and m greater than 2. This was settled by Mauduit & Rivat (2010), who showed that there is a $\theta_{q,m} < 1$ such that for all integers a,

$$\operatorname{card}\{p \le x : s_q(p) \equiv a \pmod{m}\} = \frac{d}{m}\pi(x; d, a) + O\big(x^{\theta_{q,m}}\big),$$

where $d = (m, q-1)$. Our discussion of the special case $q = m = 2$ follows an unpublished exposition of Green, and does not require some of the ideas needed to treat the general case. See also Drmota, Mauduit & Rivat (2020).

The result of Exercise 17.4.1.5(e) is a special case of the following result of Mauduit, Montgomery, & Rivat (2018): If k is a fixed positive integer, and

$$I_J = \int_0^1 |T_J(\alpha)|^{2k}\, d\alpha,$$

then the integrals I_J satisfy a linear recurrence of order k.

17.6 References

Bateman, P. T. & Chowla, S. (1963). Some special trigonometrical series related to the distribution of prime numbers, *J. London Math. Soc.* **38**, 372–374.

Besicovitch, A. S. (1961). Problems on continuity, *J. London Math. Soc.* **36**, 382–392.

Cassels, J. W. S. & Vaughan, R. C. (1985). Ivan Matveevich Vinogradov, *Biographical Memoirs of Fellows of the Royal Society* **31**, 613–631; *Bull. London Math. Soc.* **17**, 584–600.

Davenport, H. (1937a). On some infinite series involving arithmetical functions, *Quart. J. Math.* (2) **8**, 8–13; *Collected Works*, Vol. 4, London: Academic Press, 1977, pp. 1781–1786.

(1937b). On some infinite series involving arithmetical functions (II), *Quart. J. Math.* (2) **8**, 313–320; *Collected Works*, Vol. 4, London: Academic Press, 1977, pp. 1787–1794.

Drmota, M., Mauduit, C., & Rivat, J. (2020). Prime numbers in two bases, *Duke Math. J.* **169**, 1809–1876.

Gallagher, P. X. (1968). Bombieri's mean value theorem, *Mathematika*, **15**, 1–6.

Gel'fond, A. O. (1967/1968). Sur les nombres qui ont des propriétés additives et multiplicatives données, *Acta Arith.* **13**, 259–265.

Green, B. & Tao, T. (2012). The Möbius function is strongly orthogonal to nilsequences, *Ann. Math.* (2) **175**, 541–566.

Hardy, G. H. & Littlewood, J. E. (1922). Some problems of 'partitio numerorum'; III: On the expression of a number as a sum of primes, *Acta Math.* **44**, 1–70; *Collected Papers of G. H. Hardy*, Vol. I, Oxford: Clarendon Press, 1966, pp. 561–630.

Heath-Brown, D. R. (1982). Prime numbers in short intervals and a generalized Vaughan identity, *Canad. J. Math.* **34**, 1365–1377.

(2000). Kummer's conjecture for cubic Gauss sums, *Israel J. Math.*, **120**, 97–124.

Heath-Brown, D. R. & Patterson, S. J. (1979). The distribution of Kummer sums at prime arguments, *J. Reine Angew. Math.* **310**, 111–130.

Linnik, Y. V. (1960). All large numbers are sums of a prime and two squares (A problem of Hardy and Littlewood) I, *Mat. Sb. (N.S.)* **52** (94), 661–700.

(1961). All large numbers are sums of a prime and two squares (A problem of Hardy and Littlewood) II, *Mat. Sb. (N.S.)* **53** (95), 3–38.

Mauduit, C., Montgomery, H. L., & Rivat, J. (2018). Moments of a Thue–Morse generating function, *J. Anal. Math.* **135**, 713–724.

Mauduit, C. & Rivat, J. (2010). Sur un problème de Gelfond: la somme des chiffres des nombres premiers. [On a problem posed by Gel'fond: the sum of digits of primes] *Ann. Math.* **171**(3), 1591–1646.

Montgomery, H. L. & Vaughan, R. C. (1981). The distribution of squarefree numbers. In *Recent Progress in Analytic Number Theory* (Durham, 1979), Vol. 1, London: Academic Press, pp. 247–256.

Piatetski-Shapiro, I. I. (1953). On the distribution of prime numbers in sequences of the form $[f(n)]$, *Mat. Sbornik N.S.* **33** (75), 559–566.

Rivat, J. & Wu, Jie (2001). Prime numbers of the form $[n^c]$, *Glasg. Math. J.* **43**, 237–254,

Vaughan, R. C. (1975). Mean value theorems in prime number theory, *J. London Math. Soc. (2)* **10**, 153–162.

(1977a). Sommes trigonométriques sur les nombres premiers, *C. R. Acad. Sci. Paris A* **285**, 981–983.

(1977b). On the estimation of trigonometrical sums over primes, and related questions, *Institut Mittag-Leffler* **9**, 1–52.

Vinogradov, I. M. (1937a). Representation of an odd number as a sum of three primes, *Dokl. Akad. Nauk SSSR* **15**, 169–172.

(1937b). Some theorems concerning the theory of primes, *Mat. Sb.* **2** (44), 179–195.

(1947). *The Method of Trigonometrical Sums in the Theory of Numbers* [in Russian], Trav. Inst. Math. Stekloff 23, 109 pp.

(1954) *The Method of Trigonometrical Sums in the Theory of Numbers*, English translation and annotations by A. Davenport and K. F. Roth, London–New York: Interscience, 180 pp. Reprint, New York: Dover (2004).

18

Additive Prime Number Theory

18.1 Sums of Three Primes

We now address additive questions involving prime numbers, particularly the problem of expressing an integer as a sum of k primes:

$$n = p_1 + p_2 + \cdots + p_k. \tag{18.1}$$

The cases $k = 2$ and $k = 3$ of this were first enunciated by Goldbach in letters to Euler in 1742 and are know as the binary and ternary Goldbach problems, respectively. We employ the 'circle method' of Hardy–Littlewood, as later modified and improved by Vinogradov. For sums of three primes our method is successful. For sums of two primes our method fails, but we can nevertheless show that almost all even numbers can be expressed as a sum of two primes.

Let $r_k(n)$ denote the number of solutions of (18.1) in prime numbers p_i, and let $r_k(n, X)$ denote the corresponding number with no p_i exceeding X. Thus $r_k(n) = r_k(n, X)$ for $n \le X$, and the identity

$$\sum_n r_k(n, X)e(n\alpha) = \left(\sum_{p\le X} e(p\alpha)\right)^k$$

is an immediate consequence of writing the product on the right as a k-fold sum over $p_1, \ldots, p_k$, and then combining those terms for which $p_1 + \cdots + p_k = n$. Thus the generating function

$$S(\alpha) = \sum_{p\le X} e(p\alpha) \tag{18.2}$$

readily lends itself to the study of additive problems, and it is from its properties that we derive estimates for $r_k(n)$.

Since

$$\int_0^1 e(m\alpha)\, d\alpha = \begin{cases} 1 & \text{when } m = 0, \\ 0 & \text{otherwise,} \end{cases}$$

the functions $e(m\alpha)$ are orthonormal on the circle group $\mathbb{T} = \mathbb{R}/\mathbb{Z}$. Thus

$$r_k(n, X) = \int_0^1 S(\alpha)^k e(-n\alpha)\, d\alpha, \tag{18.3}$$

which is merely the formula for the nth Fourier coefficient of S^k.

The size of $S(\alpha)$ at an arbitrary point α depends on the extent to which α can be approximated by a rational number a/q with q relatively small. The primes are uniformly distributed among the reduced residue classes modulo q, but the reduced residue classes are not equally distributed, so we expect that the numbers $S(a/q)$ are sometimes large. Indeed,

$$S(a/q) = \sum_{h-1}^{q} e(ha/q)\pi(X; q, h),$$

which is approximately

$$\frac{1}{\varphi(q)} \operatorname{li}(X) \sum_{\substack{h=1 \\ (h,q)=1}}^{q} e(ha/q)$$

if $q \le (\log X)^A$. The inner sum above is Ramanujan's sum $c_q(a)$, and by Theorem 4.1 we know that $c_q(a) = \mu(q)$ when $(a, q) = 1$. Thus the above is

$$= \frac{\mu(q)}{\varphi(q)} \operatorname{li}(X)$$

if $(a, q) = 1$. By partial summation we find that $S(\alpha)$ has a peak of width comparable to $1/X$ at a/q when q is square-free. The principle of the Hardy–Littlewood method is to obtain an asymptotic formula for $r_k(n)$ by estimating the contributions to the integral (18.3) from those peaks when q is relatively small, and then to show that the remaining portions of $\mathbb{T}$ contribute *in toto* an amount of a smaller order of magnitude.

We now execute the approach outlined above in the ternary case.

Theorem 18.1 (Vinogradov 1937) *Let*

$$\mathfrak{S}_3(n) = \left(\prod_{p|n}\left(1 - \frac{1}{(p-1)^2}\right)\right)\prod_{p\nmid n}\left(1 + \frac{1}{(p-1)^3}\right)$$

and

$$\mathrm{ls}_3(n) = \sum_{\substack{m_1+m_2+m_3=n \\ m_i>1}} \prod_{i=1}^{3} \frac{1}{\log m_i}.$$

Then for any fixed positive number A,

$$r_3(n) = \mathfrak{S}_3(n)\,\mathrm{ls}_3(n) + O\big(n^2(\log n)^{-A}\big)$$

and

$$\mathrm{ls}_3(n) = \frac{1}{2}n^2(\log n)^{-3}\big(1 + O(1/\log n)\big).$$

The quantity $\mathfrak{S}_3$ is written above in the form of an Euler product, but we first encounter it below in expanded form, as an infinite series. In the parlance of Hardy–Littlewood, this is a *singular series*. Hence the use of the Fraktur letter S to denote it.

It is readily seen that $\mathfrak{S}_3(n) = 0$ for even n and that $\mathfrak{S}_3(n) \asymp 1$ for odd n. Consequently, all sufficiently large odd numbers can be expressed as a sum of three primes.

We begin with two lemmas. In the first of these, we find that $S(\alpha)$ is relatively small when α is not near a rational number with small denominator. The second relates to a sum that is useful in describing the peaks of $S(\alpha)$.

Lemma 18.2 *Suppose that $|\alpha - a/q| \le 1/q^2$, that $(a, q) = 1$, and that $S(\alpha)$ is defined as in* (18.2). *Then*

$$S(\alpha) \ll \big(Xq^{-1/2} + X^{4/5} + X^{1/2}q^{1/2}\big)(\log X)^{3/2}.$$

Proof Let $T(u) = \sum_{p\le u}(\log p)e(p\alpha)$. By Theorem 17.1 we see that

$$T(u) \ll \big(uq^{-1/2} + u^{4/5} + u^{1/2}q^{1/2}\big)(\log u)^{5/2}. \tag{18.4}$$

Then

$$S(\alpha) = \int_{2^-}^{X} (\log u)^{-1}\, dT(u) = \frac{T(X)}{\log X} + \int_2^X \frac{T(u)}{u(\log u)^2}\, du,$$

so the stated bound follows from (18.4). □

Lemma 18.3 *Let $U(\beta) = \sum_{1<m\le X} e(m\beta)/\log m$. Then*

$$U(\beta) \ll Y/\log Y$$

where $Y = \min\big(X, \|\beta\|^{-1}\big)$.

Proof When $\|\beta\| \le 1/X$, we argue that $|U(\beta)| \le U(0) \ll X/\log X$. When $\|\beta\| > 1/X$, we again bound the contribution of $m \le Y$ trivially. For the range $Y < m \le X$, we appeal to (16.4) and integrate by parts. □

Proof of Theorem 18.1 Let $P = (\log X)^B$ and $Q = X/P$ where B is to be selected later as a function of A. We now dissect $\mathbb{T}$ into appropriate arcs. For $q \le P$ and $(a, q) = 1$, let $\mathfrak{M}(q, a)$, called a *major arc*, denote the interval consisting of those α for which $|\alpha - a/q| \le 1/Q$. Further, let $\mathfrak{M}$ denote the union of these $\mathfrak{M}(q, a)$. If $\mathfrak{M}(q, a)$ and $\mathfrak{M}(q', a')$ are two major arcs with $a/q \ne a'/q'$, then

$$\left|\frac{a}{q} - \frac{a'}{q'}\right| \ge \frac{1}{qq'} \ge \frac{1}{P^2} > \frac{2}{Q},$$

so $\mathfrak{M}(q, a)$ and $\mathfrak{M}(q', a')$ are disjoint. We define $\mathfrak{m}$, the *minor arcs*, to be $\mathfrak{m} = \mathbb{T} \setminus \mathfrak{M}$.

From (18.3) we see that

$$r_3(n, X) = \int_{\mathfrak{m}} S(\alpha)^3 e(-n\alpha)\, d\alpha + \sum_{q \le P} \sum_{\substack{a=1 \\ (a,q)=1}}^{q} \int_{\mathfrak{M}(q,a)} S(\alpha)^3 e(-n\alpha)\, d\alpha.$$

We first estimate the integral over $\mathfrak{m}$. By Dirichlet's theorem, for any real number α and any $Q \ge 1$ there exist q and a with $q \le Q$, $(a, q) = 1$, and $|\alpha - a/q| \le 1/(qQ)$. If $q \le P$, then $\alpha \in \mathfrak{M}(q, a)$. Thus if $\alpha \in \mathfrak{m}$, then $P < q \le Q$, and hence

$$S(\alpha) \ll X(\log X)^{3/2-B/2}$$

by Lemma 18.2. By Parseval's identity and Chebyshev's estimate we have

$$\int_0^1 |S(\alpha)|^2\, d\alpha = \pi(X) \ll X/\log X.$$

Therefore

$$\int_{\mathfrak{m}} |S(\alpha)|^3\, d\alpha \le \big(\max_{\mathfrak{m}} |S(\alpha)|\big) \int_0^1 |S(\alpha)|^2\, d\alpha \ll X^2(\log X)^{1/2-B/2}.$$

Thus

$$\int_{\mathfrak{m}} S(\alpha)^3 e(-n\alpha)\, d\alpha \ll X^2(\log X)^{-A} \tag{18.5}$$

provided that

$$B \ge 2A + 1. \tag{18.6}$$

When $\alpha \in \mathfrak{M}(q, a)$, we can approximate $S(\alpha)$ *via* the Siegel–Walfisz theorem. Let

$$S(a/q, x) = \sum_{p \le x} e(pa/q).$$

The number of primes p with $(p,q) > 1$ is $\ll \log q$. Thus

$$S(a/q,x) = \sum_{\substack{h=1\\(h,q)=1}}^{q} \pi(x;q,h)e(ha/q) + O(\log q).$$

Let the *logarithmic sum* be

$$\operatorname{ls}(x) = \sum_{1<m\le x} (\log m)^{-1} = \operatorname{li}(x) + O(1).$$

By the Siegel–Walfisz theorem (Corollary 11.21),

$$\pi(x;q,h) = \frac{\operatorname{ls}(x)}{\varphi(q)} + O\big(X\exp(-c\sqrt{\log X})\big)$$

uniformly for $q \le P$ and $x \le X$. Hence

$$S(a/q,x) = \frac{\mu(q)}{\varphi(q)}\operatorname{ls}(x) + O\big(X\exp(-c\sqrt{\log X})\big). \tag{18.7}$$

Let $R(x) = S(a/q,x) - \operatorname{ls}(x)\mu(q)/\varphi(q)$, and set $\beta = \alpha - a/q$. Then

$$S(\alpha) = \int_1^X e(\beta x)\,dS(a/q,x) = \frac{\mu(q)}{\varphi(q)}\int_1^X e(\beta x)\,d\operatorname{ls}(x) + \int_1^X e(\beta x)\,dR(x).$$

Here the first integral on the right is $U(\beta)$, in the notation of Lemma 18.3. We estimate the final integral by integrating by parts and applying (18.7). Thus we find that

$$S(\alpha) = \frac{\mu(q)}{\varphi(q)}U(\beta) + O\big((1+|\beta|X)X\exp(-c\sqrt{\log X})\big).$$

For $\alpha \in \mathfrak{M}(q,a)$ we have $|\beta| \le 1/Q$, and for arbitrary complex numbers u and v we have $|u^3 - v^3| \le 3|u-v|\max\big(|u|^2,|v|^2\big)$. Therefore

$$S(\alpha)^3 = \frac{\mu(q)}{\varphi(q)^3}U(\beta)^3 + O\big(X^3\exp(-c\sqrt{\log X})\big).$$

Thus

$$\int_{\mathfrak{M}} S(\alpha)^3 e(-n\alpha)\,d\alpha = I(n)\sum_{q\le P} J(q) + O\big(|\mathfrak{M}|X^3\exp(-c\sqrt{\log X})\big), \tag{18.8}$$

where

$$I(n) = \int_{-1/Q}^{1/Q} U(\beta)^3 e(-n\beta)\,d\beta$$

and

$$J(q) = \frac{\mu(q)}{\varphi(q)^3} \sum_{\substack{a=1 \\ (a,q)=1}}^{q} e(-na/q).$$

Here $I(n)$ is what Hardy and Littlewood would have called the *singular integral*, and $\sum J(q)$ will turn out to be the singular series.

The measure of $\mathfrak{M}$ is $\le 2P^2/Q \ll X^{-1}(\log X)^{3B}$. Moreover,

$$\sum_{q>P} J(q) \ll \sum_{q>P} \varphi(q)^{-2} \ll P^{-1} = (\log X)^{-B} \quad \text{and} \quad \sum_{q=1}^{\infty} J(q) \ll 1.$$

By Lemma 18.3,

$$\int_{1/Q}^{1-1/Q} |U(\beta)|^3 \, d\beta \ll Q^2(\log Q)^{-3} \ll X^2(\log X)^{-2B-3}$$

and

$$\int_0^1 |U(\beta)|^3 \, d\beta \ll X^2(\log X)^{-3}.$$

Thus if (18.6) holds, then from (18.8) we deduce that

$$\int_{\mathfrak{M}} S(\alpha)^3 e(-n\alpha) \, d\alpha = \int_0^1 U(\beta)^3 e(-n\beta) \, d\beta \sum_{q=1}^{\infty} J(q) + O\left(\frac{X^2}{(\log X)^A}\right).$$

Clearly,

$$\int_0^1 U(\beta)^3 e(-n\beta) \, d\beta = \mathrm{ls}_3(n, X),$$

where

$$\mathrm{ls}_3(n, X) = \sum_{\substack{m_1+m_2+m_3=n \\ 1<m_i \le X}} \prod_{i=1}^{3} \frac{1}{\log m_i}.$$

Moreover, the sum in the definition of $J(q)$ is Ramanujan's sum $c_q(-n)$, which is a multiplicative function of q (as we recall from Theorem 4.1). Since $c_p(-n) = p - 1$ if $p|n$ and $c_p(-n) = -1$ otherwise, it follows that

$$\sum_{q=1}^{\infty} J(q) = \mathfrak{S}_3(n).$$

We take $B = 2A + 4$ so that (18.6) is satisfied, and conclude that

$$\int_{\mathfrak{M}} S(\alpha)^3 e(-n\alpha) \, d\alpha = \mathfrak{S}_3(n) \, \mathrm{ls}_3(n, X) + O\big(X^2(\log X)^{-A}\big). \tag{18.9}$$

This and (18.5) with $X = n$ give the first part of the theorem.

To establish the asymptotic formula for ls_3, we first observe that

$$\begin{aligned}\mathrm{ls}_3(n) &= \sum_{2\le m_1\le n-4} \frac{1}{\log m_1} \sum_{2\le m_2\le n-m_1-2} \frac{1}{(\log m_2)\log(n-m_1-m_2)} \\ &= \sum_{2\le m_1\le n-4} \frac{1}{\log m_1}\left(\int_2^{n-m_1-2} \frac{dx}{(\log x)\log(n-m_1-x)} + O\left(\frac{1}{\log n}\right)\right) \\ &= \int_2^{n-4}\int_2^{n-y-2} \frac{dx\,dy}{(\log x)(\log y)\log(n-x-y)} + O\left(\frac{n}{(\log n)^2}\right).\end{aligned} \tag{18.10}$$

To estimate this integral, we first observe that if j and k are fixed integers, then

$$\int_2^{X/2} \frac{dx}{(\log x)^j(\log(X-x))^k} = \frac{X}{2(\log X)^{j+k}} + O\left(\frac{X}{(\log X)^{j+k+1}}\right). \tag{18.11}$$

The point is that $(\log(X-x))^k = (\log X)^k(1+O(1/\log X))$ for $2 \le x \le X/2$, and then the remaining integral can be estimated by integrating by parts. Similarly,

$$\int_2^{X/2} \frac{x\,dx}{(\log x)^j(\log(X-x))^k} = \frac{X^2}{8(\log X)^{j+k}} + O\left(\frac{X^2}{(\log X)^{j+k+1}}\right). \tag{18.12}$$

From (18.11) we see that

$$\begin{aligned}\int_2^{X-2} \frac{dx}{(\log x)\log(X-x)} &= 2\int_2^{X/2} \frac{dx}{(\log x)\log(X-x)} \\ &= \frac{X}{(\log X)^2} + O\left(\frac{X}{(\log X)^3}\right).\end{aligned}$$

We take $X = n - y$, and insert this in (18.10) to see that

$$\mathrm{ls}_3(n) = \int_2^{n-4} \frac{n-y}{(\log y)(\log(n-y))^2}\,dy + O\left(\frac{n^2}{(\log n)^4}\right).$$

To estimate the contribution of the interval $2 \le y \le n/2$ we use both (18.11) and (18.12) with $j = 1$ and $k = 2$. To treat the interval $n/2 \le y \le n-4$ we replace y by $n - y$ and use (18.12) with $j = 2$ and $k = 1$. On assembling the various estimates we obtain the stated result. □

18.1.1 Exercises

1. Let $r_k(n)$ denote the number of representations of n as a sum of k primes.

   (a) Show that $r_k(n) = \sum_{p<n} r_{k-1}(n-p)$.

(b) Let

$$\mathfrak{S}_k(n) = \prod_{p|n}\left(1 + \frac{(-1)^k}{(p-1)^{k-1}}\right)\prod_{p\nmid n}\left(1 - \frac{(-1)^k}{(p-1)^k}\right) \tag{18.13}$$

and

$$\operatorname{ls}_k(n) = \sum_{\substack{m_1,m_2,\ldots,m_k \\ \sum m_i = n \\ m_i > 1}} \prod_{i=1}^{k} \frac{1}{\log m_i}. \tag{18.14}$$

For each fixed $k \ge 3$ and each fixed $A > 0$, show that

$$r_k(n) = \mathfrak{S}_k(n)\operatorname{ls}_k(n) + O\big(n^{k-1}(\log n)^{-A}\big).$$

(Do this by induction on k with the already completed case $k = 3$ as the basis of the induction. Do not use the circle method.)

2. Show that

$$\int_2^{X-2} \frac{dx}{(\log x)\log(X-x)} = \frac{X}{(\log X)^2} + \frac{2X}{(\log X)^3} + O\left(\frac{X}{(\log X)^4}\right).$$

3. (a) Use (16.4) to show that $\sum_{n\le X} c_q(n) \ll q\log q$ for $q > 1$.

(b) Show that $\sum_{n\le X} \mathfrak{S}_3(n) = X + O(1)$.

4. (Hooley 1998)

(a) Suppose that A is a fixed positive number and x is sufficiently large. Suppose further that a_1, a_2, a_3 are non-zero integers, not all of the same sign, which satisfy $|a_j| \le (\log x)^A$ and $(a_1, a_2, a_3) = 1$. Shew that the number $\Upsilon(x; \boldsymbol{a})$ of solutions of

$$a_1p_1 + a_2p_2 + a_3p_3 = 0$$

with $p_j \le x$ satisfies

$$\Upsilon(x;\boldsymbol{a}) \sim \mathfrak{S}(\boldsymbol{a})\Xi(x;\boldsymbol{a}),$$

where

$$\mathfrak{S}(\boldsymbol{a}) = \sum_{q=1}^{\infty}\sum_{\substack{a=1 \\ (a,q)=1}}^{q} \frac{c_q(a_1a)c_q(a_2a)c_q(a_3a)}{\varphi(q)^3}$$

and

$$\Xi(x;\boldsymbol{a}) = \sum_{\substack{m_1,m_2,m_3\le x \\ a_1m_1+a_2m_2+a_2m_3=0}} \frac{1}{(\log m_1)(\log m_2)(\log m_3)}.$$

(b) Let $T(x)$ denote the number of triples $p_1 < p_2 < p_3 \le x$ of primes in arithmetic progression. Prove that

$$T(x) \sim \frac{Cx^2}{2(\log x)^3},$$

where

$$C = \prod_{p>2} \left(\frac{p(p-2)}{(p-1)^2} \right).$$

18.2 Sums of Two Primes on Average

Our minor arc treatment fails in the binary case, but the major arc contributions suggest the conjecture that

$$r_2(n) \sim \mathfrak{S}_2(n)\,\mathrm{ls}_2(n)$$

as n tends to infinity through even values. Here $\mathfrak{S}_2(n)$ and $\mathrm{ls}_2(n)$ are defined as in (18.13) and (18.14), respectively. Although we are unable to prove the conjecture, we can prove that $r_2(n)$ is near $\mathfrak{S}_2(n)\,\mathrm{ls}_2(n)$ for most n. In order to display the flexibility of the circle method, we switch now to the von Mangoldt function $\Lambda(n)$ rather than count primes with weight 1.

Theorem 18.4 *Let*

$$\psi_2(n) = \sum_{m<n} \Lambda(m)\Lambda(n-m).$$

Then for any fixed $A > 0$,

$$\sum_{n \le X} \left(\psi_2(n) - \mathfrak{S}_2(n)n\right)^2 \ll X^3(\log X)^{-A}.$$

Corollary 18.5 *Let $E(X)$ denote the number of even natural numbers $n \le X$ such that n is not the sum of two primes. Then $E(X) \ll X/(\log X)^A$ for any fixed $A > 0$.*

Proof If n is even but not the sum of two primes, then $\psi_2(n) \ll n^{1/2}\log n$. Let $E_1(X)$ denote the number of even n, $X/2 < n \le X$, such that $\psi_2(n) < \mathfrak{S}_2(n)n/2$. We observe that $\mathfrak{S}_2(n) \gg 1$ uniformly for even n. Thus if n is counted by $E_1(X)$, then $|\psi_2(n) - \mathfrak{S}_2(n)n| \gg n \gg X$. By Theorem 18.4 it follows that $E_1(X) \ll X/(\log X)^A$. But $E(X) \ll 1 + \sum_r E_1(X/2^r)$, so we have the stated result. □

From (18.13) we see that

$$\mathfrak{S}_2(n) = \prod_p \left(1 + \frac{c_p(n)}{(p-1)^2}\right).$$

This product is absolutely convergent since $c_p(n) = -1$ for all but finitely many primes (namely the primes dividing n). Hence we may expand the product, and find that

$$\mathfrak{S}_2(n) = \sum_{q=1}^{\infty} \frac{\mu(q)^2}{\varphi(q)^2} c_q(n). \tag{18.15}$$

It is useful to be able to work with a truncation of this series. Thus, in preparation for the proof of Theorem 18.4, we establish a lemma.

Lemma 18.6 *Let*

$$\mathfrak{S}_2(n, P) = \sum_{q \le P} \frac{\mu(q)^2}{\varphi(q)^2} c_q(n).$$

Then

$$\sum_{n \le X} \left(\mathfrak{S}_2(n) - \mathfrak{S}_2(n, P)\right)^2 \ll P^{-2} X (\log X)^3$$

for $X \ge 2$.

Proof By (4.7) we see that

$$\mathfrak{S}_2(n) - \mathfrak{S}_2(n, P) = \sum_{q > P} \frac{\mu(q)^2}{\varphi(q)^2} c_q(n) \ll \sum_{q > P} \frac{1}{\varphi(q)^2} \sum_{d \mid (q,n)} d.$$

We write $q = dr$ and note that $\varphi(q) \ge \varphi(d)\varphi(r)$. Thus the above is

$$\ll \sum_{d \mid n} \frac{d}{\varphi(d)^2} \sum_{r > P/d} \frac{1}{\varphi(r)^2} \ll P^{-1} \sum_{d \mid n} \frac{d^2}{\varphi(d)^2}.$$

Put

$$f(n) = \left(\sum_{d \mid n} \frac{d^2}{\varphi(d)^2}\right)^2.$$

Then

$$\sum_{n \le X} \left(\mathfrak{S}_2(n) - \mathfrak{S}_2(n, P)\right)^2 \ll P^{-2} \sum_{n \le X} f(n),$$

so to complete the proof it suffices to show that

$$\sum_{n \le X} f(n) \ll X (\log X)^3. \tag{18.16}$$

But this follows from Corollary 2.15 since f is a non-negative multiplicative function, $f(p) = \big(1 + (p/(p-1))^2\big)^2 = 4 + O(1/p)$, and $f(p^k) \ll k^2$, so that $\sum_{p\le x} f(p)\log p \ll x$,

$$\prod_{p\le X}\left(1 + \frac{f(p)}{p} + \frac{f(p^2)}{p^2} + \cdots\right) \ll (\log X)^4,$$

and

$$\sum_{\substack{p^k\\ k\ge 2}} \frac{f(p^k)k\log p}{p^k} < \infty.$$

□

An alternative derivation of the estimate (18.16) that avoids the appeal to Corollary 2.15 is outlined in Exercise 18.2.1.4 below.

Proof of Theorem 18.4 The appropriate generating function is now

$$S(\alpha) = \sum_{n\le X} \Lambda(n)e(n\alpha).$$

Thus we define $\psi_2(n, X)$ by writing

$$S(\alpha)^2 = \sum_n \psi_2(n, X)e(n\alpha),$$

and we observe that $\psi_2(n, X) = \psi_2(n)$ for $n \le X$. In place of the auxiliary function U considered in §18.1, the appropriate function is

$$V(\beta) = \sum_{0\le n\le X} e(n\beta). \tag{18.17}$$

Let $w(n, X)$ denote the Fourier coefficients of $V(\beta)^2$, so that

$$V(\beta)^2 = \sum_n w(n, X)e(n\beta).$$

Thus $w(n, X) = n + 1$ for $n \le X$. We retain without modification the definitions of P, Q, and the major and minor arcs given in the proof of Theorem 18.1, although the dependence of B on A may be different. The main idea is to apply Parseval's identity, but before we do so we truncate $\mathfrak{S}_2$. By Lemma 18.6 we see that

$$\sum_{n\le X} (n+1)^2\big(\mathfrak{S}_2(n) - \mathfrak{S}_2(n, P)\big)^2 \ll X^3(\log X)^{3-2B}.$$

Since $\sum_{n\le X} \mathfrak{S}_2(n)^2 \ll X$, it suffices to show that

$$\sum_{0\le n\le X} \big(\psi_2(n) - \mathfrak{S}_2(n, P)(n+1)\big)^2 \ll X^3(\log X)^{-A} \tag{18.18}$$

if B is sufficiently large in terms of A. At this point, we require only that

$$B \geq (A+3)/2. \tag{18.19}$$

By Parseval's identity,

$$\sum_{n=0}^{\infty} \big(\psi_2(n,X) - \mathfrak{S}_2(n,P)w(n,X)\big)^2 = \int_0^1 \left|S(\alpha)^2 - T(\alpha)\right|^2 d\alpha, \tag{18.20}$$

where

$$T(\alpha) = \sum_{n=0}^{\infty} \mathfrak{S}_2(n,P)w(n,X)e(n\alpha) = \sum_{q \leq P} \frac{\mu(q)^2}{\varphi(q)^2} \sum_{\substack{a=1 \\ (a,q)=1}}^{q} V(\alpha - a/q)^2.$$

We first dispose of the minor arcs. By Cauchy's inequality,

$$\begin{aligned}|T(\alpha)|^2 &\ll \left(\sum_{q\leq P} \frac{1}{\varphi(q)}\right)\left(\sum_{q \leq P} \frac{1}{\varphi(q)^2} \sum_{\substack{a=1 \\ (a,q)=1}}^{q} |V(a/q)|^4\right) \\ &\ll (\log P) \sum_{q \leq P} \frac{1}{\varphi(q)^2} \sum_{\substack{a=1 \\ (a,q)=1}}^{q} \|\alpha - a/q\|^{-4}.\end{aligned}$$

For $q \leq P$ and $(a,q) = 1$,

$$\int_{\mathfrak{m}} \|\alpha - a/q\|^{-4}\, d\alpha \ll \int_{1/Q}^{\infty} \beta^{-4}\, d\beta \ll Q^3.$$

Hence

$$\int_{\mathfrak{m}} |T(\alpha)|^2\, d\alpha \ll Q^3 (\log P)^2 \ll X^3 (\log X)^{2-3B}.$$

From Theorem 17.1, as in the proof of Theorem 18.1, we find that

$$\max_{\mathfrak{m}} |S(\alpha)| \ll X(\log X)^{5/2 - B/2}.$$

Thus

$$\int_{\mathfrak{m}} |S(\alpha)|^4\, d\alpha \leq \big(\max_{\mathfrak{m}} |S(\alpha)|^2\big) \int_0^1 |S(\alpha)|^2\, d\alpha \ll X^3 (\log X)^{6-B}.$$

On combining these estimates we conclude that

$$\int_{\mathfrak{m}} \left|S(\alpha)^2 - T(\alpha)\right|^2\, d\alpha \ll X^3(\log X)^{6-B}. \tag{18.21}$$

For $\alpha \in \mathfrak{M}(q,a)$, let $\beta = \alpha - a/q$. Then

$$T(\alpha) = \frac{\mu(q)^2}{\varphi(q)^2}V(\beta)^2 + O\left(\sum_{r\le P}\frac{\mu(r)^2}{\varphi(r)^2}\sum_{\substack{b=1\\(b,r)=1\\b/r\ne a/q \bmod 1}}^{r}\|\alpha - b/r\|^{-2}\right)$$

by (16.4). For the b as in this last sum we have $\|a/q - b/r\| \ge 1/(qr)$. Hence

$$\sum_{\substack{b=1\\(b,r)=1\\b/r\ne a/q \bmod 1}}^{r}\|\alpha - b/r\|^{-2} \ll (qr)^2 + \sum_{m=1}^{r}(r/m)^2 \ll (qr)^2,$$

so that

$$T(\alpha) = \frac{\mu(q)^2}{\varphi(q)^2}V(\beta)^2 + O\big((\log X)^{3B}\big)$$

for $\alpha \in \mathfrak{M}(q,a)$. For such α, as in the proof of Theorem 18.1, we have

$$S(\alpha) = \frac{\mu(q)}{\varphi(q)}V(\beta) + O\big(X\exp\big(-c\sqrt{\log X}\big)\big), \tag{18.22}$$

whence

$$S(\alpha)^2 = \frac{\mu(q)^2}{\varphi(q)^2}V(\beta)^2 + O\big(X^2\exp\big(-c\sqrt{\log X}\big)\big).$$

By comparing our estimates for $S(\alpha)^2$ and $T(\alpha)$ we find that

$$\int_{\mathfrak{M}}\left|S(\alpha)^2 - T(\alpha)\right|^2\,d\alpha \ll |\mathfrak{M}|X^4\exp\big(-c\sqrt{\log X}\big) \ll X^3(\log X)^{-A}.$$

On combining this and (18.21) in (18.20), we deduce that

$$\sum_{n=0}^{\infty}\big(\psi_2(n,X) - \mathfrak{S}_2(n,P)w(n,X)\big)^2 \ll X^3(\log X)^{-A}$$

if $B \ge A+6$. Assuming that $A \ge 0$, we may take $B = A+6$, for then (18.19) is satisfied. Thus we have (18.18), and the proof is complete. □

18.2.1 Exercises

1. (Lavrik 1960) For positive integers k, let

$$T(X,k) = \sum_{k<n\le X}\Lambda(n)\Lambda(n-k).$$

Show that

$$\sum_{k\le X}\big(T(X,k) - \mathfrak{S}_2(k)(X-k)\big)^2 \ll X^3(\log X)^{-A}$$

for any fixed A.

2. Show that there exist infinitely many pairs a, b such that a, $a+b$, and $a+2b$ are all prime. Do this in two ways:
   (a) As a consequence of theorems already proved.
   (b) By using the circle method to derive an asymptotic formula for the number of solutions of the equation $2p = p_1 + p_2$.
3. Let $S(\alpha) = \sum_{n\le X} \Lambda(n)e(n\alpha)$.
   (a) Show that

$$\sum_{\substack{a=1\\(a,q)=1}}^{q} S(a/q+\beta) = \mu(q)S(\beta) + O\big(q(\log qX)^2\big).$$

   (b) Let $\mathfrak{M}(q,a) = [a/q - 1/X, a/q + 1/X]$. Show that

$$\sum_{\substack{a=1\\(a,q)=1}}^{q} \int_{\mathfrak{M}(q,a)} |S(\alpha)|\, d\alpha \gg 1$$

   provided that q is square-free and $q \le X/(\log X)^3$.
   (c) Show that

$$\int_0^1 |S(\alpha)|\, d\alpha \gg X^{1/2}.$$

4. Let $f(n)$ be defined as in the proof of Lemma 18.6.
   (a) Prove that

$$\sum_{n\le x} f(n) \ll \sum_{\substack{c\le X\\d\le X}} \frac{Xc^2d^2}{\varphi(c)^2\varphi(d)^2[c,d]}.$$

   (b) Show that

$$\begin{aligned}\sum_{\substack{c\le X\\d\le X}} \frac{c^2d^2}{\varphi(c)^2\varphi(d)^2[c,d]} &= \sum_{\substack{c\le X\\d\le X}} \frac{cd}{\varphi(c)^2\varphi(d)^2} \sum_{\substack{r|c\\r|d}} \varphi(r)\\ &= \sum_{r\le X}\Bigg(\sum_{\substack{d\le X\\r|d}} \frac{d}{\varphi(d)^2}\Bigg)^2.\end{aligned}$$

   (c) Write $d = rm$ and note that $\varphi(d) \ge \varphi(r)\varphi(m)$. Thus show that

$$\sum_{r\le X}\Bigg(\sum_{\substack{d\le X\\r|d}} \frac{d}{\varphi(d)^2}\Bigg)^2 \le \sum_{r\le X} \frac{r^2}{\phi(r)^3}\Bigg(\sum_{m\le X/r} \frac{m}{\varphi(m)^2}\Bigg)^2.$$

   (d) Deduce (18.16).

5. Let f be the multiplicative function for which $f(2)=0$, $f(p)=1/(p-2)$ for $p>2$, $f(p^r)=0$ for $r>1$, and put

$$C = 2\prod_{p>2}\left(1-\frac{1}{(p-1)^2}\right).$$

Then

$$C\sum_{d|n} f(d) = 2\prod_{\substack{p|n\\p>2}}\left(1+\frac{1}{p-1}\right)\prod_{\substack{p\nmid n\\p>2}}\left(1-\frac{1}{(p-1)^2}\right). \tag{18.23}$$

This is $\mathfrak{S}_2(n)$ if n is even.

(a) Show that

$$\sum_{n\le X}\mathfrak{S}_2(n) = C\sum_{m\le X/2}\sum_{d|2m} f(d).$$

(b) Deduce that

$$\sum_{n\le X}\mathfrak{S}_2(n) = \frac{1}{2}CX\sum_{d\le X/2} f(d)/d + O\left(\sum_{d\le X/2} f(d)\right).$$

(c) Show that

$$\sum_{d=1}^{\infty}\frac{f(d)}{d} = \frac{2}{C}.$$

(d) Conclude that

$$\sum_{n\le X}\mathfrak{S}_2(n) = X + O(\log X).$$

6. Recall that in Corollary 3.14 we established that if $x\ge 4$, then the number of $n\le x$ for which n and $n+r$ are both prime is $\ll \mathfrak{S}_2(r)x/(\log x)^2$ uniformly for even non-zero integers r. Deduce that

$$\sum_{n\le x}\left(\sum_{m=n+1}^{n+h}\Lambda(n+m)\right)^2 \ll hx\log x + h^2x.$$

7. With the notation used in the proof of Theorem 18.4 show that

$$\int_{\mathfrak{M}} |S(\alpha)|^2 d\alpha \sim X\log P$$

and

$$\int_{\mathfrak{m}} |S(\alpha)|^2 d\alpha \sim X\log Q.$$

In other words, in the twin prime problem, for example, there is essential cancellation deriving from the factor (2α) in the integrand.

18.3 Conditional Estimates

The theorems that we have established thus far can be greatly sharpened if we assume the Generalized Riemann Hypothesis (GRH).

Theorem 18.7 *Assume GRH. Then*

$$\sum_{1\le n\le X}\big(\psi_2(n)-\mathfrak{S}_2(n)n\big)^2 \ll X^{5/2}(\log X)^5.$$

Before proving this theorem, we first note two corollaries and establish three lemmas.

In the same way that we derived Corollary 18.5 from Theorem 18.4, we immediately have the following two corollaries.

Corollary 18.8 *Assume GRH. Let $E(X)$ denote the number of even integers $n \le X$ such that n is not the sum of two primes. Then $E(X) \ll X^{1/2}(\log X)^5$.*

Corollary 18.9 *Assume GRH. Let*

$$\psi_3(n) = \sum_{m_1+m_2+m_3=n} \Lambda(m_1)\Lambda(m_2)\Lambda(m_3).$$

Then

$$\psi_3(n) = \frac{1}{2}\mathfrak{S}_3(n)n^2 + O\big(n^{7/4}(\log n)^3\big),$$

where $\mathfrak{S}_3(n)$ is defined as in Theorem 18.1.

Proof For even n this is trivial. Hence we may assume that n is odd. We note that

$$\psi_3(n) = \sum_{m<n}\Lambda(m)\mathfrak{S}_2(n-m)(n-m) + \sum_{m<n}\Lambda(m)\Delta(n-m), \qquad (18.24)$$

where

$$\Delta(k) = \psi_2(k) - \mathfrak{S}_2(k)k.$$

By Cauchy's inequality and Theorem 18.7, the second sum in (18.24) is

$$\ll \Big(\sum_{m\le n}\Lambda(m)^2\Big)^{1/2}\big(n^{5/2}(\log n)^5\big)^{1/2} \ll n^{7/4}(\log n)^3.$$

Let C and $f(n)$ be defined as in Exercise 18.2.1.5. To estimate the first sum in (18.24) we use the formula (18.23), so that for odd n,

$$\sum_{m<n}\Lambda(m)\mathfrak{S}_2(n-m)(n-m) = C\sum_{d<n} f(d)\sum_{\substack{m<n\\ d\mid(n-m)}}\Lambda(m)(n-m) + O\big(n(\log n)^2\big).$$

Here the error term accounts for the contributions of those m that are powers of 2. If n is odd and m is a power of 2, then $\mathfrak{S}_2(n-m) = 0$, but the formula

(18.23) returns a value that is $\ll \log n$. If there is a prime p such that $p|n$ and $p|d$, then m must be a power of p. Thus the above is

$$= C \sum_{\substack{d<n \\ (d,n)=1}} f(d) \int_1^n \psi(x; d, n)\, dx + O\left(n(\log n)^2 \sum_{d<n} f(d)\right).$$

On GRH, if χ is a character (mod q), then

$$\begin{aligned}\int_1^X \psi(x,\chi)\, dx &= \frac{-1}{2\pi i}\int_{2-i\infty}^{2+i\infty} \frac{L'}{L}(s,\chi)\frac{X^{s+1}}{s(s+1)}\, ds \\ &= E_0(\chi)X^2/2 - \sum_\rho \frac{X^{\rho+1}}{\rho(\rho+1)} \\ &\quad - \frac{1}{2\pi i}\int_{1/4-i\infty}^{1/4+i\infty} \frac{L'}{L}(s,\chi)\frac{X^{s+1}}{s(s+1)}\, ds.\end{aligned}$$

By Theorem 10.17, the sum over ρ is $\ll X^{3/2}\log 2q$. On GRH, the formula of Lemma 11.1 is valid when $\operatorname{Re} s = 1/4$, so the integral on the right above is $\ll X^{5/4}\log 2q$. Thus if $(d,n) = 1$, then

$$\int_1^n \psi(x; d, n)\, dx = \frac{n^2}{2\varphi(d)} + O\big(n^{3/2}\log 2d\big).$$

Therefore

$$\begin{aligned}\sum_{m<n} \Lambda(m)\mathfrak{S}_2(n-m)(n-m) = \frac{1}{2}Cn^2 &\sum_{\substack{d<n \\ (d,n)=1}} \frac{f(d)}{\varphi(d)} \\ &+ O\left(n^{3/2}(\log n)\sum_{d<n} f(d)\right).\end{aligned}$$

From the estimates

$$\sum_{d<n} f(d) \ll \log n, \qquad \sum_{d>n} \frac{f(d)}{\varphi(d)} \ll n^{-1},$$

we obtain the stated result, upon observing that

$$C\sum_{\substack{d=1 \\ (d,n)=1}}^{\infty} \frac{f(d)}{\varphi(d)} = \mathfrak{S}_3(n). \qquad \square$$

We prove Theorem 18.7 by modifying our proof of Theorem 18.4. Corollary 18.9 could similarly be derived by modifying our proof of Theorem 18.1. Correspondingly, we could argue as above to derive Theorem 18.1 from Theorem 18.4, with the understanding that it would be necessary to derive a variant of

Theorem 18.4 that counts primes with weight 1, rather than integers with weight $\Lambda(n)$. To prepare for the proof of Theorem 18.7, we first establish several useful estimates.

Lemma 18.10 *Assume GRH. Let*

$$\psi(X,\chi,\beta) = \sum_{n\le X} \Lambda(n)\chi(n)e(n\beta),$$

and set $\psi'(X,\chi,\beta) = \psi(X,\chi,\beta) - E_0(\chi)V(\beta)$ *where* $V(\beta)$ *is defined in* (18.17). *Then for* $|\beta| \le 1$,

$$\psi'(X,\chi,\beta) = -\sum_{|\gamma|\le X^2} I(\beta,\gamma) + O\big((\log qX)^2\big),$$

where the numbers $1/2 + i\gamma$ *are the non-trivial zeros of* $L(s,\chi)$, *and*

$$I(\beta,\gamma) = \int_2^X e(\beta x)x^{-1/2+i\gamma}\,dx.$$

Proof By the explicit formula (12.14) with $T = X^2$, we see that

$$\begin{aligned}\psi'(X,\chi,\beta) &= \int_{2^-}^X e(\beta x)\,d\psi'(x,\chi)\\ &= -\sum_{|\gamma|\le X^2} I(\beta,\gamma)\\ &\quad + \int_{2^-}^X e(\beta x)\,dE_1(x,\chi) + \int_{2^-}^X e(\beta x)\,dE_2(x,X^2,\chi).\end{aligned}$$

By (12.15) we see that the integral with respect to E_1 is $\ll (\log qX)^2$. The integral with respect to E_2 is

$$= \Big[e(\beta x)E(x,X^2,\chi)\Big|_{2^-}^X - 2\pi i\beta\int_2^X e(\beta x)E_2(x,X^2,\chi)\,dx \ll (\log qX)^2$$

by (12.16) and (12.17). Thus we have the stated result. □

Lemma 18.11 *For real numbers* β *and* γ, *let* $I(\beta,\gamma)$ *be defined as in Lemma 18.10. If* $|\gamma| \le 1$, *then* $I(\beta,\gamma) \ll X^{1/2}$. *If* $1 \le |\gamma| \le 10|\beta|X$, *then* $I(\beta,\gamma) \ll |\beta|^{-1/2}$. *If* $|\gamma| \ge 1$ *and* $|\gamma| \ge 10|\beta|X$, *then* $I(\beta,\gamma) \ll X^{1/2}/|\gamma|$.

Proof The first estimate is trivial since by the triangle inequality, $|I(\beta,\gamma)| \le \int_2^X x^{-1/2}\,dx$. If $|\gamma| \ge 10|\beta|U$, then by Theorem 16.1 with $r(x) = x^{-1/2}$ and $\theta(x) = 2\pi\beta x + \gamma\log x$ we find that

$$\int_{U/2}^U e(\beta x)x^{-1/2+i\gamma}\,dx \ll \frac{U^{1/2}}{|\gamma|}. \tag{18.25}$$

If $|\gamma| \ge 10|\beta|X$, then we apply the above with $U = X2^{-r}$, and sum, to obtain the third estimate. Suppose now that $1 \le |\gamma| \le 10|\beta|X$. If β and γ have the same sign, then (by taking complex conjugates if necessary) we may assume that they are both positive, and in this case $\theta'(x) = 2\pi\beta + \gamma/x \ge \gamma/x$, and so (18.25) again holds. If β and γ have opposite signs, then put $x_0 = -\gamma/(2\pi\beta)$, and set $J_1 = [2, X] \cap [2, x_0/2]$, $J_2 = [2, X] \cap [x_0/2, 2x_0]$, and $J_3 = [2, X] \cap [2x_0, \infty)$. Thus $I(\beta,\gamma) = \int_{J_1} + \int_{J_2} + \int_{J_3} = I_1 + I_2 + I_3$, say. We cut J_1 into dyadic blocks and apply (18.25) to see that $I_1 \ll X/|\gamma|$. We apply Theorem 16.3 with $M \asymp |\beta|^{1/2}/|\gamma|^{1/2}$ and $\mu \asymp \beta^2/|\gamma|$ to see that $I_2 \ll |\beta|^{-1/2}$. If $U \ge 2x_0$, then by Theorem 16.1 we find that

$$\int_U^{2U} e(\beta x)x^{-1/2+i\gamma}\,dx \ll U^{-1/2}|\beta|^{-1}.$$

On summing over dyadic blocks, we deduce that $I_3 \ll |\beta|^{-1/2}|\gamma|^{-1/2}$. Thus we see that if $1 \le |\gamma| \le 10|\beta|X$, then $I(\beta,\gamma) \ll |\beta|^{-1/2}$, so the proof is complete. □

By Theorem 13.7 we know that the estimate $\psi'(x,\chi) \ll x^{1/2}(\log qx)^2$ is a consequence of GRH. By integrating by parts as in the proof of Theorem 18.1, we can deduce that

$$\psi'(X,\chi,\beta) \ll X^{1/2}(1 + |\beta|X)(\log qX)^2. \tag{18.26}$$

However, by utilizing the more detailed information provided by Lemma 18.10, and the estimates in Lemma 18.11 for the integrals $I(\beta,\gamma)$, we obtain a better estimate, as follows.

Lemma 18.12 *Assume GRH, and let* $\psi'(X,\chi,\beta)$ *be defined as in Lemma* 18.10. *Then*

$$\psi'(X,\chi,\beta) \ll \big(X^{1/2} + |\beta|^{1/2}X\big)(\log qX)^2. \tag{18.27}$$

When $|\beta|X \le 1$, the bounds (18.26) and (18.27) are comparable, but when $|\beta|X > 1$, the bound of (18.27) is smaller than that of (18.26) by a factor of $(|\beta|X)^{1/2}$. Despite this improvement over (18.26) we expect that more is true, and conjecture that

$$\psi'(X,\chi,\beta) \ll X^{1/2+\varepsilon}(\log q)^2. \tag{18.28}$$

Proof We may assume that $|\beta| \le 1$, for otherwise the estimate is trivial. By Lemma 18.11 we see that

$$\sum_{|\gamma|\le X^2} I(\beta,\gamma) \ll \sum_{|\gamma|\le 1} X^{1/2} + \sum_{1<|\gamma|\le 10|\beta|X} |\beta|^{-1/2} + \sum_{1<|\gamma|\le X^2} X^{1/2}|\gamma|^{-1}.$$

By Theorem 10.17 we know that the number of zeros $1/2 + i\gamma$ of $L(s, \chi)$ with $t < \gamma \le t + 1$ is $\ll \log q\tau$. Thus the right-hand side above is $\ll \big(X^{1/2} + |\beta|^{1/2}X\big)(\log qX)^2$, and the proof is complete. □

Proof of Theorem 18.7 Let $V(\beta)$, $w(n, X)$, $\mathfrak{S}_2(n, P)$, $\psi_2(n, X)$, $S(\alpha)$, and $T(\alpha)$ be defined as in the proof of Theorem 18.4, but we now take $P = Q = \lfloor X^{1/2} \rfloor$ and redefine the major arcs $\mathfrak{M}(q, a)$. Let $\mathcal{F}_Q$ denote the set of Farey fractions of order Q, which is to say the set of rational numbers a/q with $1 \le a \le q$, $(a, q) = 1$, and $q \le Q$. Let a'/q' and a''/q'' be the neighbours of $a/q \in \mathcal{F}_Q$ with $a'/q' < a/q < a''/q''$. Then we take $\mathfrak{M}(q, a) = \big[(a+a')/(q+q'), (a+a'')/(q+q'')\big)$. Since these intervals partition $\mathbb{T}$, we have no minor arcs.

By the method used to prove Theorem 18.4 we see that it suffices to show that

$$\int_0^1 \left|S(\alpha)^2 - T(\alpha)\right|^2 d\alpha \ll X^{5/2}(\log X)^5. \tag{18.29}$$

For $\alpha \in \mathfrak{M}(q, a)$, let

$$W(\alpha) = \frac{\mu(q)}{\varphi(q)} V(\alpha - a/q).$$

Thus $W(\alpha)^2$ is one of the terms comprising $T(\alpha)$, and

$$T(\alpha) - W(\alpha)^2 \ll \sum_{r \le Q} \frac{1}{\varphi(r)^2} \sum_{\substack{b=1 \\ (b,r)=1 \\ b/r \ne a/q \bmod 1}}^{r} \|\alpha - b/r\|^{-2}.$$

Suppose that $R/2 < r \le R \le Q$. Since $\alpha \notin \mathfrak{M}(r, b)$, we have $\|\alpha - b/r\| \ge 1/(r(r + r')) \ge 1/(2rQ) \ge 1/(2RQ)$ where $(b + b')/(r + r')$ is the endpoint of $\mathfrak{M}(r, b)$ lying between b/r and α. Also, if $b_1/r_1 \ne b_2/r_2$, then $\|b_1/r_1 - b_2/r_2\| \ge 1/(r_1r_2) \ge 1/R^2$. Therefore,

$$\sum_{R/2<r\le R} \sum_{\substack{b=1 \\ (b,r)=1 \\ b/r \ne a/q \bmod 1}}^{r} \|\alpha - b/r\|^{-2} \ll R^2Q^2 + \sum_{k=1}^{\infty} (k/R^2)^{-2} \ll R^2Q^2.$$

Thus

$$\sum_{R/2<r\le R} \frac{1}{\varphi(r)^2} \sum_{\substack{b=1 \\ (b,r)=1 \\ b/r \ne a/q \bmod 1}}^{r} \|\alpha - b/r\|^{-2} \ll Q^2(\log\log Q)^2,$$

whence

$$T(\alpha) - W(\alpha)^2 \ll X(\log X)^2.$$

Thus to prove (18.29) it suffices to show that

$$\int_0^1 |S(\alpha)^2 - W(\alpha)^2|^2\, d\alpha \ll X^{5/2}(\log X)^5. \tag{18.30}$$

To this end we first estimate $S(\alpha) - W(\alpha)$. Suppose that $\alpha \in \mathfrak{M}(q,a)$ and that $\beta = \alpha - a/q$. By the definition (9.3) of a Gauss sum and the basic orthogonality (4.15) of characters, we see that

$$\frac{1}{\varphi(q)} \sum_{\chi} \tau(\overline{\chi})\chi(b) = \begin{cases} e(b/q) & \text{if } (b,q) = 1, \\ 0 & \text{otherwise.} \end{cases} \tag{18.31}$$

We set $b = an$, multiply by $\Lambda(n)e(n\beta)$, and sum over $n \le X$ to see that

$$S(\alpha) = \frac{1}{\varphi(q)} \sum_{\chi} \tau(\overline{\chi})\chi(a)\psi(X,\chi,\beta) + O\Bigg(\sum_{\substack{n\le X\\(n,q)>1}} \Lambda(n)\Bigg),$$

where $\psi(X,\chi,\beta)$ is defined as in Lemma 18.10. Thus

$$S(\alpha) - \frac{\mu(q)}{\varphi(q)} V(\beta) = \frac{1}{\varphi(q)} \sum_{\chi} \tau(\overline{\chi})\chi(a)\psi'(X,\chi,\beta) + O\big((\log qX)^2\big). \tag{18.32}$$

By Lemma 18.12 this is

$$\ll q^{1/2}(X^{1/2} + |\beta|^{1/2}X)(\log X)^2 \ll (q^{1/2}X^{1/2} + Q^{-1/2}X)(\log X)^2$$

since $|\beta| \le 1/(qQ)$. Thus

$$S(\alpha) - W(\alpha) \ll X^{3/4}(\log X)^2 \tag{18.33}$$

uniformly in α.

By Parseval's identity,

$$\int_0^1 |S(\alpha)|^2\, d\alpha = \sum_{n\le X} \Lambda(n)^2 \ll X \log X,$$

while

$$\int_0^1 |W(\alpha)|^2\, d\alpha \ll \sum_{q\le Q} \varphi(q)^{-2} \sum_{\substack{a=1\\(a,q)=1}}^{q} \int_0^1 \min(X^2, \|\beta\|^{-2})\, d\beta$$

$$\ll X \sum_{q\le Q} \varphi(q)^{-1} \ll X \log X.$$

Thus from (18.33) we deduce that

$$\int_0^1 |S(\alpha)^2 - W(\alpha)^2|^2 \, d\alpha \ll X^{3/2}(\log X)^4 \int_0^1 |S(\alpha) + W(\alpha)|^2 \, d\alpha \ll X^{5/2}(\log X)^5.$$

Thus we have (18.30), and the proof is complete. □

The argument just completed may be expected to be inefficient in two respects. Some considerable cancellation should occur in the sum over χ in (18.32), and we also expect that the bound in Lemma 18.12 is weaker than the truth. Indeed, we expect that $S(\alpha) - W(\alpha) \ll X^{1/2+\varepsilon}$ for all α. It would then follow that

$$\sum_{n\le X} \big(\psi_2(n) - \mathfrak{S}_2(n)n\big)^2 \ll X^{2+\varepsilon}. \tag{18.34}$$

While we are unable to establish that the sum in (18.32) cancels uniformly, we can at least demonstrate the cancellation in mean square. By orthogonality,

$$\sum_{\substack{a=1\\(a,q)=1}}^{q} \left|S(a/q+\beta) - \frac{\mu(q)}{\varphi(q)}V(\beta)\right|^2 \ll q(\log xQ)^4 + \frac{1}{\varphi(q)}\sum_{\chi} |\tau(\overline{\chi})|^2 |\psi'(X,\chi,\beta)|^2. \tag{18.35}$$

By Lemma 18.12 this is

$$\ll (qX + q|\beta|X^2)(\log qX)^4. \tag{18.36}$$

For $|\beta| \le q^{-1}X^{-1/2}$, $q \le X^{1/2}$, this is uniformly $\ll X^{3/2}(\log X)^4$. By comparison, if were to estimate the left-hand side by applying (18.33) for each a, then the bound we would obtain would be much worse, namely $\ll X^2(\log X)^4$.

If $(a,q) = 1$, then $a/q + \beta \in \mathfrak{M}(q,a)$ precisely when

$$\frac{-1}{q(q+q')} \le \beta \le \frac{1}{q(q+q'')}.$$

Since the dimensions of this interval depend on a, we are not immediately able to apply (18.36). To circumvent this difficulty, we replace $\mathfrak{M}(q,a)$ by the slightly larger interval $\mathfrak{M}^*(q,a) = (a/q - 1/(qQ), a/q + 1/(qQ))$. Hence

$$\int_0^1 |W(\alpha)(S(\alpha) - W(\alpha))|^2 \, d\alpha \le \sum_{q\le Q} \varphi(q)^{-2} \sum_{\substack{a=1\\(a,q)=1}}^{q} \int_{-1/(qQ)}^{1/(qQ)} |V(\beta)|^2 \left|S(a/q+\beta) - \frac{\mu(q)}{\varphi(q)}V(\beta)\right|^2 d\beta.$$

By (18.36) this is

$$\ll \sum_{q \le Q} \varphi(q)^{-2} \int_0^1 \min\left(X^2, \beta^{-2}\right)\left(qX + q\beta X^2\right)(\log X)^4 \, d\beta$$
$$\ll X^2(\log X)^5 \sum_{q \le Q} q\varphi(q)^{-2} \ll X^2(\log X)^6. \tag{18.37}$$

Thus to sharpen Theorem 18.7 it suffices to improve our bound for

$$\int_0^1 |S(\alpha)(S(\alpha) - W(\alpha))|^2 \, d\alpha,$$

or equivalently, for $\int_0^1 |S(\alpha) - W(\alpha)|^4 \, d\alpha$. Such estimates remain to be established.

We next show that Lemma 18.12 can similarly be improved in mean square with respect to β.

Theorem 18.13 *Assume GRH. Let $\delta > 0$, and let χ be any character modulo q. Then*

$$\int_{-\delta}^{\delta} |\psi'(X, \chi, \beta)|^2 \, d\beta \ll \delta X (\log qX)^4.$$

By (18.35) we have an immediate corollary.

Corollary 18.14 *Assume GRH. Then*

$$\sum_{\substack{a=1 \\ (a,q)=1}}^{q} \int_{-\delta}^{\delta} \left| S(a/q + \beta) - \frac{\mu(q)}{\varphi(q)} V(\beta) \right|^2 \, d\beta \ll \delta q X (\log qX)^4.$$

Proof of Theorem 18.13 If $\delta X \le 1$, then it suffices to appeal to Lemma 18.12. Thus we assume that $\delta X \ge 1$. We may also assume that $\delta \le 1$ since it is trivial that

$$\int_0^1 |\psi'(X, \chi, \beta)|^2 \, d\beta \ll X \log X.$$

By Lemma 18.10 we see that

$$\int_{-\delta}^{\delta} |\psi'(X, \chi, \beta)|^2 \, d\beta \ll \int_{-\delta}^{\delta} \left| \sum_{|\gamma| \le X^2} I(\beta, \gamma) \right|^2 \, d\beta + \delta(\log qX)^4.$$

By Lemma 18.11 we know that $I(\beta, \gamma) \ll X^{1/2}|\gamma|^{-1}$ when $|\gamma| \ge 10\delta X$. Thus

$$\int_{-\delta}^{\delta} \left| \sum_{10\delta X < |\gamma| \le X^2} I(\beta, \gamma) \right|^2 \, d\beta \ll \delta X (\log qX)^4.$$

On the other hand,

$$\int_{-\delta}^{\delta} \Bigg| \sum_{|\gamma| \le 10\delta X} I(\beta, \gamma) \Bigg|^2 \, d\beta \le \int_{-\infty}^{\infty} \Bigg| \sum_{|\gamma| \le 10\delta X} \int_2^X e(\beta x) x^{-1/2+i\gamma} \, dx \Bigg|^2 \, d\beta,$$

which by Plancherel's formula is

$$= \int_2^X \Bigg| \sum_{|\gamma| \le 10\delta X} x^{i\gamma} \Bigg|^2 \frac{dx}{x}.$$

We make the change of variable $x = e^y$, and note that

$$\int_Y^{Y+1} \Bigg| \sum_{|\gamma| \le T} e^{i\gamma y} \Bigg|^2 \, dy \ll \sum_{\substack{|\gamma| \le T \\ |\gamma'| \le T}} \min(1, 1/|\gamma - \gamma'|).$$

For any given γ, the sum over γ' is $\ll (\log qT)^2$, as we see by using the bound of Theorem 10.17 in the same way that we did in the proof of Corollary 18.9. The number of γ is $\ll T \log qT$, so the above is $\ll T(\log qT)^3$. We take $T = 10\delta X$, and sum over $\ll \log X$ values of Y to see that

$$\int_{-\delta}^{\delta} \Bigg| \sum_{|\gamma| \le 10\delta X} I(\beta, \gamma) \Bigg|^2 \, d\beta \ll \delta X (\log qX)^4.$$

Thus the proof is complete. □

18.3.1 Exercises

1. Let $\mathfrak{M}(q, a)$ and $\mathfrak{M}^*(q, a)$ be defined as in the proof of (18.37).
   (a) Show that if a/q and a'/q' are neighbouring members of $\mathscr{F}_Q$, then $\mathfrak{M}^*(q, a)$ and $\mathfrak{M}^*(q', a')$ overlap.
   (b) Show that if a/q and a'/q' are neighbouring members of $\mathscr{F}_Q$, then $a/q \notin \mathfrak{M}^*(q', a')$.
   (c) Conclude that every α is in at least one, but not more than two of the arcs $\mathfrak{M}^*(q, a)$.
2. (a) By introducing appropriate weights before expanding and integrating, show that

$$\int_Y^{Y+1} \Bigg| \sum_{|\gamma| \le T} e^{i\gamma y} \Bigg|^2 \, dy \ll T(\log qT)^2.$$

   Here the γ are the imaginary parts of zeros of $L(s, \chi)$ and χ is a character modulo q.

(b) In the context of the proof of Theorem 18.13, show that

$$\int_{-\delta}^{\delta}\Big|\sum_{|\gamma|\le 10\delta X} I(\beta,\gamma)\Big|^2\, d\beta \ll \delta X(\log qX)^3.$$

18.4 A Lower Bound for the Error Term

We have estimated the mean square error in the binary Goldbach problem, and discussed the plausibility of sharper estimates such as (18.34). We now establish a bound in the opposite direction.

Theorem 18.15 *Suppose that* $1/2 < r < 1$, *and let* $R = 1/(1-r)$. *Then*

$$\sum_{n=1}^{\infty}\big(\psi_2(n) - \mathfrak{S}_2(n)n\big)^2 r^{2n} \gg R^2(\log R)^2.$$

Corollary 18.16 *As X tends to infinity,*

$$\sum_{n\le X}\big(\psi_2(n) - \mathfrak{S}_2(n)n\big)^2 = \Omega\big(X^2(\log X)^2\big)$$

and

$$\psi_2(n) - \mathfrak{S}_2(n)n = \Omega\big(n^{1/2}\log n\big).$$

Proof of Theorem 18.15 By Lemma 18.6 we see that

$$\sum_{N<n\le 2N}(n+1)^2\big(\mathfrak{S}_2(n) - \mathfrak{S}_2(n,Q)\big)^2 r^{2n} \ll Q^{-2}r^{2N}N^3(\log N)^3.$$

On setting $N = 2^k R$ and summing over k, we deduce that

$$\sum_{n=1}^{\infty}(n+1)^2\big(\mathfrak{S}_2(n) - \mathfrak{S}_2(n,Q)\big)^2 r^{2n} \ll Q^{-2}R^3(\log R)^3.$$

Since

$$\sum_{n=1}^{\infty}\mathfrak{S}_2(n)^2 r^{2n} \ll R,$$

it follows that

$$\sum_{n=1}^{\infty}\big(\mathfrak{S}_2(n)n - \mathfrak{S}_2(n,Q)(n+1)\big)^2 r^{2n} \ll Q^{-2}R^3(\log R)^3.$$

We take $Q = R^{\kappa}$ with $1/2 < \kappa < 1$. Thus it suffices to show that

$$\sum_{n=1}^{\infty}\big(\psi_2(n) - \mathfrak{S}_2(n,Q)(n+1)\big)^2 r^{2n} \gg R^2(\log R)^2.$$

By Parseval's identity the left-hand side is T_2 where

$$T_k = \int_0^1 \left| S(\alpha)^2 - \sum_{q \le Q} \frac{\mu(q)^2}{\varphi(q)^2} \sum_{\substack{a=1 \\ (a,q)=1}}^{q} (1 - re(\alpha - a/q))^{-2} \right|^k d\alpha$$

and

$$S(\alpha) = \sum_{n=1}^{\infty} \Lambda(n) r^n e(n\alpha).$$

By Cauchy's inequality, $T_2 \ge T_1^2$. But

$$\begin{aligned} T_1 \ge & \int_0^1 \left| \sum_{n=1}^{\infty} \Lambda(n) r^n e(n\alpha) \right|^2 d\alpha \\ & - \sum_{q \le Q} \frac{\mu(q)^2}{\varphi(q)^2} \sum_{\substack{a=1 \\ (a,q)=1}}^{q} \int_0^1 |1 - re(\alpha - a/q)|^{-2} \, d\alpha \\ = & \sum_{n=1}^{\infty} \Lambda(n)^2 r^{2n} - \sum_{q \le Q} \frac{\mu(q)^2}{\varphi(q)^2} (1 - r^2)^{-1}. \end{aligned}$$

Here the first sum is $\frac{1}{2} R \log R + O(R)$, and the sum over q is $\frac{1}{2} R \log Q + O(R)$, in view of Exercise 2.1.1.17. Thus the above is $\frac{1}{2} R \log R/Q \gg R \log R$, so the proof is complete. □

18.5 Prime k-Tuples

We begin by considering twin primes. The analysis and notation is similar to that in §18.2. In particular, we set

$$S(\alpha) = \sum_{n \le X} \Lambda(n) e(n\alpha) \quad \text{and} \quad V(\beta) = \sum_{n \le X} e(n\beta).$$

If h is a positive integer, then

$$\sum_{n \le X-h} \Lambda(n) \Lambda(n+h) = \int_0^1 |S(\alpha)|^2 e(h\alpha) \, d\alpha.$$

Let the major and minor arcs be defined as in the proofs of Theorems 18.1 and 18.4. From (18.22) we deduce that if $\alpha \in \mathfrak{M}(q, a)$, then

$$|S(\alpha)|^2 = \frac{\mu(q)^2}{\varphi(q)^2} |V(\beta)|^2 + O\big(X^2 \exp\big(-c\sqrt{\log X}\big)\big).$$

Hence

$$\int_{\mathfrak{M}} |S(\alpha)|^2 e(h\alpha)\, d\alpha = I(h) \sum_{q \le P} J(q) + O\big(|\mathfrak{M}| X^2 \exp\big(-c\sqrt{\log X}\big)\big),$$

where

$$\begin{aligned} I(h) &= \int_{-1/Q}^{1/Q} |V(\beta)|^2 e(h\beta)\, d\beta \\ &= \int_0^1 |V(\beta)|^2 e(h\beta)\, d\beta + O(Q) = X - h + O(Q) \end{aligned}$$

and

$$J(q) = \frac{\mu(q)^2}{\varphi(q)^2} c_q(h).$$

Here $c_q(h)$ is the Ramanujan sum, which we discussed in Theorem 4.1. In particular, it was shown that if q_1 and q_2 are relatively prime positive integers, then $c_{q_1q_2}(h) = c_{q_1}(h)c_{q_2}(h)$ for any integer h. Also, it was noted that

$$c_p(h) = \begin{cases} p-1 & \text{if } p|h, \\ -1 & \text{otherwise.} \end{cases}$$

From these properties it follows that if q is square-free and $h \ne 0$, then $|c_q(h)| \le |h|$. This is useful since it follows that the singular series $\mathfrak{S}_2(h) = \sum_q J(q)$ is absolutely convergent. Hence

$$\mathfrak{S}_2(h) = \prod_p \left(1 + \frac{c_p(h)}{(p-1)^2}\right) = \prod_{p|h} \left(1 + \frac{1}{p-1}\right) \prod_{p\nmid h} \left(1 - \frac{1}{(p-1)^2}\right).$$

Note that $\mathfrak{S}_2(h) = 0$ if h is odd. Based on this major arc treatment, we conjecture that if h is positive and even, then

$$\sum_{n \le X-h} \Lambda(n)\Lambda(n+h) \sim \mathfrak{S}_2(h) X \tag{18.38}$$

as $X \to \infty$. What we lack is a suitable treatment of the minor arcs. It would suffice to know that

$$\int_{\mathfrak{m}} |S(\alpha)|^2\, e(h\alpha) d\alpha = o(X).$$

This is not so much stronger than the estimate

$$\int_{\mathfrak{m}} |S(\alpha)|^2 d\alpha \sim X \log Q$$

as given by Exercise 18.2.1.7. We now turn to the main theme of this section, namely prime k-tuples with $k > 2$. Suppose that $h_1 < h_2 < \cdots < h_k$ are integers. Then the numbers $n+h_1, n+h_2, \ldots, n+h_k$ form a *prime k-tuple* if all the numbers $n+h_i$ are prime. We have already observed that if n is large, then n and $n+h$ cannot both be prime if h is odd. A similar phenomenon extends to prime k-tuples.

Definition 18.1 Let $\boldsymbol{h} = h_1, \ldots, h_k$ be a k-tuple of distinct non-negative integers and let $\nu_p(\boldsymbol{h})$ denote the number of different residue classes modulo p among the $h_1, \ldots, h_k$. If $\nu_p(\boldsymbol{h}) < p$ for every p, then $\boldsymbol{h}$ is called *admissible*.

If $\boldsymbol{h}$ is inadmissible, then there exists a prime p such that $\nu_p(\boldsymbol{h}) = p$, and hence for any n, the prime p divides at least one of the numbers $n+h_1, \ldots, n+h_k$. We conjecture that the necessary condition that $\boldsymbol{h}$ should be admissible is also sufficient to ensure the existence of infinitely many prime k-tuples with the spacing $\boldsymbol{h}$.

Conjecture 18.17 (The prime k tuple conjecture) *If $\boldsymbol{h}$ is admissible, then there are infinitely many positive integers n such that $n+h_1, n+h_2, \ldots, n+h_k$ are simultaneously prime.*

We note that a translation of an admissible k-tuple is again admissible since $\nu_p(\boldsymbol{h})$ is unchanged by translation. Also, if $\boldsymbol{h}$ is a k-tuple of integers, $\nu_p(\boldsymbol{h}) \le k$, and this is $< p$ if $p > k$. Thus to determine whether $\boldsymbol{h}$ is admissible it suffices to calculate $\nu_p(\boldsymbol{h})$ for $p \le k$. Also, if the members of $\boldsymbol{h}$ lie in an interval of length N, then $\nu_p(\boldsymbol{h}) = k$ for all $p > N$. Useful admissible k-tuples are provided by Theorem 18.18.

Theorem 18.18 *Suppose that $k \ge 2$, and that the primes $p_1, p_2, \ldots, p_k$ satisfy*

$$k < p_1 < p_2 < \cdots < p_k.$$

Then the k-tuple $\boldsymbol{h} = (p_1, p_2, \ldots, p_k)$ is admissible. If these p_j are the least distinct primes $> k$, then $p_k - p_1 < k \log k + k \log\log k + O(k)$.

Proof If $p > k$, then $\nu_p(\boldsymbol{h}) \le k < p$. If $p < k$, then $p_j \not\equiv 0 \pmod p$ for $1 \le j \le k$, and so $\nu_p(\boldsymbol{h}) \le p-1 < p$. Let P_n denote then nth prime. From a quantitative version of the Prime Number Theorem it follows that

$$P_n = n \log n + n \log\log n + O(n).$$

In Exercise 6.2.1.5 a more precise estimate for P_n was proposed, but the weaker estimate above is sufficient to give the desired estimate. □

In §7.3 we introduced the functions

$$\rho(y) = \limsup_{x\to\infty}(\pi(x+y) - \pi(x)), \qquad \overline{\rho}(N) = \max_M \sum_{\substack{n=M+1 \\ p|n \implies p>N}}^{M+N} 1.$$

It is clear that $\rho(N) \le \overline{\rho}(N)$. If $k = \overline{\rho}(N)$, then the n counted in the above sum form an admissible k-tuple, so the k-tuple conjecture implies that $\rho(N) = \overline{\rho}(N)$ for all positive N. Also, in Theorem 7.16 we showed that there is a positive constant C such that $\overline{\rho}(N) \ge \pi(N) + CN(\log N)^{-2}$ for all sufficiently large N. In the reverse direction, in Theorem 3.3 we showed that $\overline{\rho}(N) \le 2\pi(N) + O\big(N(\log N)^{-2}\big)$ for all $N \ge 2$.

We have already failed to prove what we want when $k = 2$, and the situation is of course no better for larger k, but we can still make some useful observations and formulate a quantitative conjecture, similar to the one in (18.38). We work now with the k-tuple $\boldsymbol{h} = (0, h_1, \ldots, h_{k-1})$ where $0 < h_1 < \cdots < h_k$. Set

$$R(X, \boldsymbol{h}) = \sum_{n_0 \le X - h_{k-1}} \Lambda(n_0)\Lambda(n_0 + h_1) \cdots \Lambda(n_0 + h_{k-1}).$$

With $\boldsymbol{\alpha} = (\alpha_1, \alpha_2, \ldots, \alpha_{k-1})$ we find that

$$\begin{aligned}
&= \int_{\mathbb{T}^{k-1}} S(\alpha_1 + \cdots + \alpha_{k-1}) \prod_{j=1}^{k-1} \big(S(-\alpha_j)e(h_j\alpha_j)\big)\, d\boldsymbol{\alpha} \\
&= \sum_{n_0 \le X} \Lambda(n_0) \cdots \sum_{n_{k-1} \le X} \Lambda(n_{k-1}) \prod_{j=1}^{k-1} \int_{\mathbb{T}} e((n_0 - n_j + h_j)\alpha_j)\, d\alpha_j \\
&= R(X, \boldsymbol{h}).
\end{aligned}$$

In this new setting, the analogue of a major arc is a small $(k-1)$-dimensional block. To identify the blocks that we should attend to, we appeal to Dirichlet's theorem on Diophantine approximation (Lemma 15.10), which asserts that for any $\boldsymbol{\alpha} \in \mathbb{T}^{k-1}$ and any integer $Q \ge 1$, there exists in integer q, $1 \le q \le Q^{k-1}$, such that $\|q\alpha_j\| \le 1/Q$ for $1 \le j \le k-1$. Let a_j be the integer nearest $q\alpha_j$. Then

$$\left|\alpha_j - \frac{a_j}{q}\right| \le \frac{1}{qQ}.$$

Let $d = (a_1, a_2, \ldots, a_{k-1}, q)$. By replacing each a_j by a_j/d and q by q/d, we may suppose that $(a_1, a_2, \ldots, a_{k-1}, q) = 1$. The largest contributions are made by the smallest values of q. Let $P = N^\delta$. We restrict our attention to $q \le P$ and $|\beta_j| \le 1/P$ where $\beta_j = \alpha_j - a_j/q$. The approximation

$$S(\alpha) \sim \frac{\mu(q)}{\varphi(q)} V(\alpha - a/q)$$

applies only when $|\alpha - a/q|$ is small and $(a, q) = 1$. In our current situation, it may be that $(a_j, q) > 1$, but we note that

$$\frac{c_q(a)}{\varphi(q)} = \frac{\mu\left(\frac{q}{(a,q)}\right)}{\varphi\left(\frac{q}{(a,q)}\right)},$$

so

$$S(\alpha_j) \sim \frac{c_q(a_j)}{\varphi(q)} V(\beta_j)$$

for all j. Writing $\boldsymbol{a} = (a_1, \ldots, a_{k-1})$ with $1 \le a_j \le q$ and $b = a_1 + \cdots + a_{k-1}$, put

$$f(q; \boldsymbol{h}) = \sum_{\boldsymbol{a}}{}^{*} \frac{c_q(b)}{\varphi(q)^k} \prod_{j=1}^{k-1} \big(c_q(-a_j) e(a_j h_j / q)\big), \tag{18.39}$$

where $\sum^*$ runs over $\boldsymbol{a}$ subject to $(a_1, \ldots, a_{k-1}, q) = 1$ and $b = a_1 + \cdots + a_{k-1}$. Set

$$\mathfrak{S}(\boldsymbol{h}; P) = \sum_{q \le P} f(q; \boldsymbol{h}),$$

and put

$$J(P) = \int_{|\beta_j| \le 1/P} V(\beta_1 + \cdots + \beta_{k-1}) \prod_{j=1}^{k-1} \big(V(-\beta_j) e(h_j \beta_j)\big)\, d\boldsymbol{\beta}.$$

Then we expect that $R(X, \boldsymbol{h}) \sim J(P)\mathfrak{S}(\boldsymbol{h}; P)$. It is not hard to show that $J(P)$ is within $O\big(P^{-1+\varepsilon}\big)$ of

$$\int_{\mathbb{T}^{k-1}} V(\beta_1 + \cdots + \beta_{k-1}) \prod_{j=1}^{k-1} \big(V(-\beta_j) e(h_j \beta_j)\big)\, d\boldsymbol{\beta}$$
$$= \sum_{\substack{n_0, \ldots, n_{k-1} \\ 0 < n_j \le X}} \prod_{j=1}^{k-1} \int_{\mathbb{T}} e((n_0 + h_j - n_j)\beta_j)\, d\beta_j = X - h_{k-1} + O(1).$$

In order to assess the size of $f(q; \boldsymbol{h})$ it is helpful to observe that this quantity is a multiplicative function of q. Moreover, if $r > 1$ and $(a, p) = 1$, then $c_{p^r}(a) = 0$, so $f(p^r, \boldsymbol{h}) = 0$ since $(a_j, p) = 1$ for at least one of the a_j. Thus $f(q, \boldsymbol{h})$ is supported on square-free integers. This is not such a surprise since $S(\alpha)$ has its peaks at Farey points with square-free denominators. Now suppose that $q = p$. In the sum over $\boldsymbol{a}$, the only term that must be avoided is $\boldsymbol{a} = (p, \ldots, p)$. That single term, if it were included, would contribute exactly 1. So we sum over the a_j without restriction, and then subtract 1. We expand the sum $c_p(b)$ to see that

$$\sum_{\substack{a_1,\dots,a_{k-1}\\ 1\le a_j\le p}} c_p(b)\prod_{j=1}^{k-1}\left(c_p(-a_j)e(a_jh_j/p)\right)$$
$$=\sum_{n_0=1}^{p-1}\prod_{j=1}^{k-1}\left(\sum_{a_j=1}^{p} c_p(-a_j)e(a_j(n_0+h_j)/p)\right).$$

Here $c_p(-a_j) = p-1$ if $a_j = p$, and $c_p(a_j) = -1$ if $0 < a_j < p$, so the sum over a_j is

$$p-1-\sum_{a_j=1}^{p-1} e(a_j(n_0+h_j)/p) = p-\sum_{a_j=1}^{p} e(a_j(n_0+h_j)/p)$$
$$=\begin{cases} p & \text{if } h_j \not\equiv -n_0 \pmod p,\\ 0 & \text{if } h_j \equiv -n_0 \pmod p.\end{cases}$$

The product of these sums is therefore p^{k-1} if there is no j, $1 \le j \le k-1$, such that $h_j \equiv -n_0 \pmod p$. Since n_0 runs through all $p-1$ non-zero residue classes (mod p), this first alternative arises exactly $p-1-\nu'_p(\boldsymbol{h})$ times, where $\nu'_p(\boldsymbol{h})$ is the number of non-zero residue classes (mod p) found among the h_j with $1 \le j \le k-1$. Since $\boldsymbol{h} = (0, h_1, \dots, h_{k-1})$, $\nu_p(\boldsymbol{h}) = 1 + \nu'_p(\boldsymbol{h})$. In case $h_j \equiv -n_0 \pmod p$ for one or more values of j, the product is 0, so our expression is $p^{k-1}(p-\nu_p(\boldsymbol{h}))$ and

$$f(p;\boldsymbol{h}) = \frac{(p-\nu_p(\boldsymbol{h}))p^{k-1}}{(p-1)^k} - 1. \tag{18.40}$$

Let

$$D = \prod_{i<j}(h_j - h_i).$$

If $p \nmid D$, then $\nu_p(\boldsymbol{h}) = k$, and so $f(p;\boldsymbol{h}) \ll p^{-2}$ for such p. Hence $\mathfrak{S}(\boldsymbol{h};P)$ converges absolutely to $\mathfrak{S}(\boldsymbol{h})$ as $P \to \infty$ where

$$\mathfrak{S}(\boldsymbol{h}) = \sum_{q=1}^{\infty} f(q;\boldsymbol{h}) = \prod_p (1+f(p;\boldsymbol{h})) = \prod_p \left(1-\frac{\nu_p(\boldsymbol{h})}{p}\right)\left(1-\frac{1}{p}\right)^{-k} \tag{18.41}$$

and

$$\mathfrak{S}(\boldsymbol{h}) \ll_k (\log\log(3D))^k \ll_k (\log\log(3\max_j |h_j|))^k. \tag{18.42}$$

Thus when the h_j are distinct, if $\boldsymbol{h}$ is inadmissible, then $\mathfrak{S}(\boldsymbol{h}) = 0$. If $\boldsymbol{h}$ is admissible, then $\nu_p(\boldsymbol{h}) \le \min(k, p-1)$, and so $1 - \nu_p(\boldsymbol{h})/p \ge 1/p$ when $p \le k$ and is $\ge 1 - k/p$ when $p > k$. Thus there is a positive constant $C(k)$ such that, when the h_j are distinct, $\boldsymbol{h}$ is admissible if and only if

$$\mathfrak{S}(\boldsymbol{h}) > C(k). \tag{18.43}$$

As an extension of the quantitative twin prime conjecture, we have the quantitative prime k-tuple conjecture.

Conjecture 18.19 *Suppose that $\boldsymbol{h} = (h_1, h_2, \ldots, h_k)$ is an admissible k-tuple of distinct integers. Then*

$$R(X; \boldsymbol{h}) \sim X\mathfrak{S}(\boldsymbol{h})$$

as $X \to \infty$.

As with twin primes, the barrier to proving this conjecture is our lack of suitable bounds for the size of the integrand outside the regions that we have identified as major 'arcs'. It is generally believed that there are no secondary main terms, and that the error term is $\ll X^{1/2+\varepsilon}$. We note that the quantity X on the right-hand side above reflects the size of the singular integral, which in turn is the density of solutions of our system in real variables. Also, the factor $1 + f(p; \boldsymbol{h})$ of $\mathfrak{S}(\boldsymbol{h})$ is the density of p-adic solutions of our system. Thus the right-hand side is the product of local densities, extended over all valuations of the rational field. While we seem at present to be very far from proving the prime k-tuple conjecture, we accept it as a guide to our thinking as to how primes are distributed in short intervals.

In Exercise 18.2.1.5 we noted that the mean value of singular series $\mathfrak{S}_2(n)$ is asymptotically 1. We now extend this to prime k-tuples.

Theorem 18.20 (Gallagher 1976) *Let $k \ge 2$ be fixed, and let H run through positive integers. Further, let $\mathcal{H}$ denote the set of k-tuples $\mathbf{h}$ of distinct integers $h_1, \ldots, h_k$ with $1 \le h_j \le H$, and let $\mathcal{A}$ be the subset of those $\mathbf{h}$ that are also admissible. Then*

$$\sum_{\boldsymbol{h} \in \mathcal{A}} \mathfrak{S}(\boldsymbol{h}) = H^k + O\big(H^{k-1+\varepsilon}\big).$$

Proof We first show that the case $k = 2$ is an easy consequence of the result of Exercise 18.2.1.5, whose notation we adopt. Thus

$$\sum_{\substack{0<h_1,h_2\le H \\ h_1 \ne h_2}} \mathfrak{S}(h_1, h_2) = 2\sum_{h_2=2}^{H}\sum_{h_1=1}^{h_2-1} \mathfrak{S}_2(h_2 - h_1),$$

which by Exercise 18.2.1.5(d) is

$$= 2\sum_{h_2=2}^{H}\big(h_2 + O(\log h_2)\big) = H^2 + O(H\log H).$$

From now on we assume that $k \ge 3$. Since $\mathfrak{S}(\boldsymbol{h}) = 0$ if $\boldsymbol{h}$ is inadmissible, it suffices to prove the conclusion with $\mathscr{A}$ replaced by $\mathscr{H}$. We argue from the original definition (18.39) of $f(q,\boldsymbol{h})$. We note that $f(1,\boldsymbol{h}) = 1$ for all $\boldsymbol{h}$, which gives the main term $H(H-1)\cdots(H-k+1) = H^k + O\big(H^{k-1}\big)$. It remains to bound the contributions of $q > 1$. From (18.40) we see that if $\nu_p(\boldsymbol{h}) = k$, then

$$|f(p;\boldsymbol{h})| \le \frac{C_k}{p^2}$$

and otherwise

$$|f(p;\boldsymbol{h})| \le \frac{C_k}{p},$$

where C_k is a suitable positive number. Let $D = \prod_{1\le i<j\le k}|h_j - h_i|$, so that $D \le H^{k(k-1)/2}$. Then

$$|f(q;\boldsymbol{h})| \le q^{-2}C_k^{\omega(q)}(D,q) \ll_\varepsilon q^{\varepsilon-2}(D,q).$$

For convenience we introduce the parameter $Q \ge 1$ which is at our disposal. Then

$$\sum_{q>Q}|f(q;\boldsymbol{h})| \ll \sum_{r|D} r \sum_{\substack{q>Q\\(D,q)=r}} q^{\varepsilon-2} \ll \sum_{r|D} r^{\varepsilon-1}\sum_{t>Q/r} t^{\varepsilon-2} \ll Q^{\varepsilon-1}d(D).$$

Hence

$$\sum_{q>Q}|f(q;\boldsymbol{h})| \ll Q^{\varepsilon-1}H^{\varepsilon}. \tag{18.44}$$

For convenience we write

$$g(q;\boldsymbol{h}) = \varphi(q)^k f(q;\boldsymbol{h}). \tag{18.45}$$

Crudely, from (18.39) we have

$$|g(q\,;\boldsymbol{h})| \le g^*(q)$$

for any $\boldsymbol{h}$ where

$$g^*(q) = \sum_{\substack{\boldsymbol{a}\\(\boldsymbol{a},q)=1}} |c_q(a_1+\cdots+a_{k-1})c_q(-a_1)\cdots c_q(-a_{k-1})|.$$

This is also a multiplicative function of q (with its support on the squarefree numbers). Consider the k numbers $-a_1-\cdots-a_{k-1}, a_1,\ldots,a_{k-1}$. When

$(\boldsymbol{a}, p) = 1$ at least two of these numbers are not multiples of p. Moreover, in $g^*(p)$ the terms with exactly j of the $a_2, \dots, a_k, a_2 + \cdots + a_k$ divisible by p contribute $(p-1)^j$, and since the $a_1, \dots, a_{k-1}, a_1 + \cdots + a_{k-1}$ are linearly dependent the number of such terms is at most $\binom{k}{j}(p-1)^{k-1-j}$. Hence $g^*(p) \le 2^k(p-1)^{k-1}$ and $g^*(q)\varphi(q)^{-k} \ll q^{\varepsilon-1}$. Therefore

$$\sum_{\boldsymbol{h}\in[1,H]^k\setminus\mathscr{H}} \sum_{1<q\le Q} f(q; \boldsymbol{h}) \ll H^{k-1} \sum_{1<q\le Q} q^{\varepsilon-1} \ll H^{k-1}Q^{\varepsilon}. \tag{18.46}$$

Returning to (18.39), when $q > 1$ at least two of $a_1, \dots, a_{k-1}, -a_1 - \cdots - a_{k-1}$ are non-zero modulo q. If there are at least two such a_i, then we pick two and call them b_1, b_2. The remaining a_i can be listed in the form $b_3, \dots, b_{k-1}$ so that $-a_1 - \cdots - a_{k-1} = -b_1 - b_2 - \cdots - b_{k-1}$. If only one of the a_i is non-zero modulo q, then call it b_1 and take $b_2 = -a_1 - \cdots - a_{k-1}$. In that case any one of the other a_i can be rewritten in the form in the form $-b_1 - b_2 - s \pmod q$ where s is the sum of the remaining a_t. Thus

$$\begin{aligned}\sum_{\boldsymbol{h}\in[1,H]^k} g(q; \boldsymbol{h}) \ll H^{k-2} &\sum_{b_1=1}^{q-1} \frac{|c_q(b_1)|}{\|b_1/q\|} \sum_{b_2=1}^{q-1} \frac{|c_q(b_2)|}{\|b_2/q\|} \\ &\times \sum_{\boldsymbol{b}\in[1,q]^{k-3}} |c_q(b_3)\cdots c_q(b_{k-1})c_q(b_1 + \cdots + b_{k-1})|,\end{aligned}$$

where $\boldsymbol{b} = b_3, \dots, b_{k-1}$ and where the summand over $\boldsymbol{b}$ is taken to be $|c_q(b_1 + b_2)|$ when $k = 3$. In general this multiple sum does not exceed

$$\varphi(q)\Big(\sum_{b=1}^{q} |c_q(b)|\Big)^{k-3}.$$

Since $|c_q(b)| \le (q, b)$ the sum here is at most

$$\sum_{r|q} r\varphi(q/r) \le d(q)q.$$

Similarly

$$\sum_{b=1}^{q-1} \frac{|c_q(b)|}{\|b/q\|} \le \sum_{r|q} r \sum_{a=1}^{q/r-1} \|a/(q/r)\|^{-1} \ll d(q)q\log q.$$

Therefore

$$\sum_{\boldsymbol{h}\in[1,H]^k} \sum_{1<q\le Q} f(q,\ \boldsymbol{h}) \ll H^{k-2}Q^{1+\varepsilon}.$$

Hence, by (18.44) and (18.46) the choice $Q = H$ secures the theorem. □

18.5.1 Exercises

1. (a) We observed that if the k-tuple $\boldsymbol{h}$ is inadmissible, then the numbers $n+h_1, n+h_2, \ldots, n+h_k$ are simultaneously prime for at most finitely many non-negative n. Show that in fact the number of such n is $\le k$.

   (b) Suppose that p is a prime for which $\nu_p(\boldsymbol{h}) = p$ and that the h_i are all non-negative. Show that the numbers $n+h_1, n+h_2, \ldots, n+h_k$ are not all prime if $n > p$.

2. Suppose that $k \ge 2$ and that $1 < q_1 < q_2 < \cdots < q_k$. Suppose further that none of the q_j is divisible by a prime $p \le k$. Show that $q_1, \ldots, q_k$ forms an admissible set.
3. Let $h_j = (2j-1)^2$ for $j = 1, \ldots, k$. Prove that $\boldsymbol{h}$ is an admissible set.
4. Given a set $\boldsymbol{h}$ of distinct non-negative integers $h_1, \ldots, h_k$, for each prime p let $\nu_p(\boldsymbol{h})$ denote the number of different residue classes modulo p^2 amongst the $h_1, \ldots, h_k$. If $\nu_p(\boldsymbol{h}) < p^2$, then we call $\boldsymbol{h}$ *sf–admissible*. Let $S(x; \boldsymbol{h})$ denote the number of $n \le x$ such that $n+h_1, \ldots, n+h_k$ are simultaneously square-free.

   (a) Let $f(n)$ denote the characteristic function of the square-free numbers. Prove that

   $$S(x;\boldsymbol{h}) = \sum_{n\le x} f(n+h_1)\cdots f(n+h_k)$$

   and

   $$f(n) = \sum_{d^2|n} \mu(d).$$

   (b) Suppose that $0 < \delta < 1/(3k)$ and let $y = x^\delta$ and

   $$f(n;y) = \sum_{\substack{d\le y\\ d^2|n}} \mu(d).$$

   Prove that for $j = 1, \ldots, k$,

   $$S(x;\boldsymbol{h}) = T_j(x;y) + O(x^{1+\varepsilon}y^{-1}),$$

   where

   $$T_j(x;y) = \sum_{n\le x} f(n+h_1;y)\cdots f(n+h_j;y) f(n+h_{j+1})\cdots f(n+h_k).$$

   (c) Given a k-tuple of positive integers $\boldsymbol{d} = d_1, \ldots, d_k$, let $d = d_1\cdots d_k$ and given another k-tuple $\boldsymbol{r}$ we use $\boldsymbol{d}|\boldsymbol{r}$ to mean $d_j|r_j$ for $j = 1, \ldots, k$ and $\boldsymbol{d}^2$ to mean $d_1^2, \ldots, d_k^2$. Write $n+\boldsymbol{h}$ for the k-tuple $n+h_1, \ldots, n+h_k$.

Let $\rho(\boldsymbol{d})$ denote the number of solutions of $\boldsymbol{d}^2|n+\boldsymbol{h}$ in n modulo d^2. Prove that $\rho(\boldsymbol{d}) \le d^2$ and that

$$T_k(x;y) = x \sum_{d_1 \le y,\ldots,d_k \le y} \frac{\mu(d_1)\cdots\mu(d_k)}{d^2}\rho(\boldsymbol{d}) + O(y^{3k}).$$

(d) Let $\rho^*(\boldsymbol{d})$ denote the number of solutions of $\boldsymbol{d}^2|n+\boldsymbol{h}$ in n modulo $\mathrm{lcm}[d_1,\ldots,d_k]^2$. Prove that $\rho(\boldsymbol{d}) = d^2\mathrm{lcm}[d_1,\ldots,d_k]^{-2}\rho^*(\boldsymbol{d})$ and that $\rho^*(\boldsymbol{d}) \le 1$.

(e) Prove that

$$\sum_{\max(d_j)>y} \frac{\mu(d_1)\cdots\mu(d_k)}{d^2}\rho(\boldsymbol{d}) \ll \sum_{\max(d_j)>y} \frac{\mu(d_1)^2\cdots\mu(d_k)^2}{[d_1,\ldots,d_k]^2}$$
$$\ll \sum_{m>y} \frac{2^{k\omega(m)}}{m^2} \ll y^{\varepsilon-1}$$

and deduce that

$$T_k(x,y) = x\sum_{m=1}^{\infty} \frac{g(m)}{m^2} + O\big(xy^{\varepsilon-1}\big),$$

where

$$g(m) = \sum_{\substack{\boldsymbol{d} \\ [d_1,\ldots,d_k]=m}} \mu(d_1)\cdots\mu(d^k)\rho^*(\boldsymbol{d}).$$

(f) Prove that $\rho(\boldsymbol{d})$ is multiplicative, i.e., given $\boldsymbol{d}, \boldsymbol{e}$, define

$$\boldsymbol{de} = d_1e_1,\ldots,d_ke_k$$

and deduce that if $(d,e)=1$, then $\rho(\boldsymbol{de}) = \rho(\boldsymbol{d})\rho(\boldsymbol{e})$.

(g) Prove that $g(m)$ is multiplicative and has its support on the square-free numbers.

(h) Deduce that

$$\sum_{m=1}^{\infty} \frac{g(m)}{m^2} = \prod_p \left(1 + g(p)p^{-2}\right).$$

(i) Prove that $1 + g(p)p^{-2} = 1 - \nu_p(\boldsymbol{h})p^{-2}$.

(j) (Pillai, 1936) Prove that $S(x;h) = x\prod_p \left(1 - \frac{\nu_p(\boldsymbol{h})}{p^2}\right) + O(x^{1-\delta})$ and that if $\boldsymbol{h}$ is sf-admissible, then there are infinitely many n such that $n + h_j$ are simultaneously square-free for $j = 1,\ldots,k$.

5. Find the minimal diameter of 20-tuples which are sf-admissible, i.e., $\max h_j - h_i$ is minimal.

18.6 The Distribution of Primes in Short Intervals

For a k-tuple $\boldsymbol{h} = (h_1, \ldots, h_k)$ of distinct integers let $\pi(M; \boldsymbol{h})$ denote the number of integers m, $1 \le m \le M$, for which $m + h_1, m + h_2, \ldots, m + h_k$ are all prime. We now use Conjecture 18.19 to derive conjectures concerning moments of the number of primes in short intervals. It is clear that if H and n are positive integers, then

$$\sum_{m \le M} (\pi(m+h) - \pi(m))^n = \sum_{m \le M} \sum_{m < p_1, \ldots, p_n \le m+H} 1.$$

Let k denote the number of distinct primes among the $p_i, i = 1, 2, \ldots, n$. Think of indices i_1 and i_2 as being 'related' if $p_{i_1} = p_{i_2}$. Thus the p_i partition the set $\{1, 2, \ldots, n\}$ into k non-empty subsets $\mathcal{S}_1, \ldots, \mathcal{S}_k$ of related indices. For $i \in \mathcal{S}_j$ the prime p_i depends only on j; call it $p(j)$. Suppose further that the labelling of the subsets has been chosen so that $p(1) < p(2) < \cdots < p(k)$. Put $h_j = p(j) - m$. Then $p(1), \ldots, p(k)$ is a k-tuple of primes counted by $\pi(M; \boldsymbol{h})$, and this k-tuple has the property that $0 < h_1 < \cdots < h_k \le H$. Let $\left\{ {n \atop k} \right\}$ denote the number of ways of partitioning $\{1, 2, \ldots, n\}$ into k unordered non-empty subsets. This is a *Stirling number of the second kind* and is the number of ways of choosing the subsets $\mathcal{S}_j$, before they are given names with subscripts. There are $k!$ ways to order them, so the right-hand side above is

$$= \sum_{k=1}^{n} \left\{ {n \atop k} \right\} k! \sum_{0 < h_1 < \cdots < h_k \le H} \pi(M; \boldsymbol{h}) = \sum_{k=1}^{n} \left\{ {n \atop k} \right\} \sum_{\substack{0 < h_1, \ldots, h_k \le H \\ h_j \text{ distinct}}} \pi(M; \boldsymbol{h}). \tag{18.47}$$

Suppose that the relation

$$\pi(M; \boldsymbol{h}) \sim \mathfrak{S}(\boldsymbol{h}) \frac{M}{(\log M)^k}$$

holds uniformly for all admissible k-tuples $\boldsymbol{h}$ with $0 < h_1 < h_2 < \cdots < h_k \le H$, $k \le n$, and $H \le C \log M$ where C is an arbitrarily large constant. Suppose that $0 < \lambda \le C$, and that $H = \lfloor \lambda \log M \rfloor$. Then the expression (18.47) is

$$\sim \sum_{k=1}^{n} \left\{ {n \atop k} \right\} \frac{M}{(\log M)^k} \sum_{\substack{0 < h_1, \ldots, h_k \le H \\ h_j \text{ distinct}}} \mathfrak{S}(\boldsymbol{h}),$$

which by Theorem 18.20 is $\sim m_n(\lambda) M$ as $M \to \infty$ where

$$m_n(\lambda) = \sum_{k=1}^{n} \left\{ {n \atop k} \right\} \lambda^k. \tag{18.48}$$

This suggests a subsidiary conjecture.

Conjecture 18.21 *Let $m_n(\lambda)$ be defined as in* (18.48), *and let C be an arbitrary positive number. Then for any given n,*

$$\lim_{M\to\infty}\frac{1}{M}\sum_{m=1}^{M}(\pi(m+\lambda\log m)-\pi(m))^n = m_n(\lambda)$$

uniformly for $0 < \lambda \le C$.

It is very significant that the moments $m_n(\lambda)$ that arise here are precisely the moments of a Poisson random variable X with parameter λ (see Exercise 18.6.1.4). Such a variable takes non-negative integer values, with the probabilities

$$P(X = r) = e^{-\lambda}\frac{\lambda^r}{r!}$$

for $r = 0, 1, \ldots$. It can happen that two different distributions have the same moments. However, if the moment generating function is entire, it follows that the two distributions must in fact be the same. In Exercise 18.6.1.5 we establish the (well-known) fact that the moment generating function of a Poisson random variable is entire. Thus the distribution of $\pi(m+\lambda\log m) - \pi(m)$ should be close to Poisson. This suggests a further conjecture.

Conjecture 18.22 *Let $P_r(M,\lambda)$ be the number of $m \le M$ for which the interval $(m, m+\lambda\log m]$ contains exactly r primes. Then*

$$P_r(M,\lambda) \sim e^{-\lambda}\frac{\lambda^r}{r!}M$$

as $M \to \infty$, provided that $|r-\lambda| \ll \sqrt{1+\lambda}$.

It may be the case that the constraint on r can be gradually relaxed as $M \to \infty$, but the question of how quickly depends more on arithmetic than on probability theory. The case $r = 0$ is of course of great interest, and $P(X = 0) = e^{-\lambda}$ for a Poisson variable X, but this is at the extreme end of the distribution when λ is large, and the incidence of very long gaps between primes is expected to be a more complicated issue.

Concerning the Stirling numbers of the second kind, it is customary to set $\left\{{0 \atop 0}\right\} = 1$, and $\left\{{n \atop 0}\right\} = 0$ for $n > 0$. Given a partitioning of $\{1, \ldots, n-1\}$ into $k-1$ parts, we can derive a partitioning of $\{1, \ldots, n\}$ into k parts by introducing the new part $\{n\}$. Alternatively, given a partitioning of $\{1, \ldots, n-1\}$ into k parts, we can derive a partitioning of $\{1, \ldots, n\}$ into k parts by adjoining n to any of the given k subsets. Thus

$$\left\{{n \atop k}\right\} = \left\{{n-1 \atop k-1}\right\} + k\left\{{n-1 \atop k}\right\}. \tag{18.49}$$

This Pascal-like recurrence gives rise to a triangular array of numbers (see Table 18.1). Stirling numbers of the first kind, which may be denoted by $\left[{n \atop k}\right]$, count the number of permutations of $\{1, \ldots, n\}$ with exactly k cycles in their cycle decomposition. Rather obviously, $\left[{n \atop k}\right] = \left[{n-1 \atop k-1}\right] + (n-1)\left[{n-1 \atop k}\right]$. The 'factorial power' is defined to be $x^{\underline{n}} = x(x-1)\cdots(x-n+1)$ with $x^{\underline{0}} = 1$. Just as $1, x, x^2, x^3, \ldots$ form a basis for polynomials, so also do $x^{\underline{0}}, x^{\underline{1}}, x^{\underline{3}}, \ldots$. In Exercise 18.6.1.2 below we use Stirling numbers of the second kind to express an ordinary power as a linear combination of factorial powers. Stirling numbers of the first kind are similarly useful in writing a factorial power as a linear combination of ordinary powers.

$n \setminus k$	0	1	2	3	4	5	6	7	8	9	10
0	1										
1	0	1									
2	0	1	1								
3	0	1	3	1							
4	0	1	7	6	1						
5	0	1	15	20	10	1					
6	0	1	31	90	65	15	1				
7	0	1	63	301	350	140	21	1			
8	0	1	127	966	1701	1050	266	28	1		
9	0	1	255	3025	7770	6951	2646	642	36	1	
10	0	1	511	9330	34105	42525	22827	7140	930	45	1

Table 18.1 Stirling numbers of the second kind.

18.6.1 Exercises

1. (a) By inclusion–exclusion, or otherwise, show that the number of surjective maps from a set of n elements to a set of k elements is

$$\sum_{j=0}^{k} (-1)^j \binom{k}{j} (k-j)^n.$$

   (b) Show that the number of surjective maps from a set of n elements to a set of k elements is $\left\{{n \atop k}\right\} k!$.

   (c) Conclude that $\left\{{n \atop k}\right\} = \frac{1}{k!} \sum_{j=0}^{k} (-1)^j \binom{k}{j} (k-j)^n$.

   (d) Explain why $\sum_{j=0}^{k} (-1)^j \binom{k}{j} (k-j) = 0$ for $k > 1$.

   (e) Show that $\left\{{p \atop k}\right\} \equiv 0 \pmod{p}$ for $1 < k < p$.

2. Use the recurrence (18.49) to give a proof by induction that

$$\sum_{k=0}^{n} \left\{{n \atop k}\right\} x^{\underline{k}} = x^n. \tag{18.50}$$

   Hint: Note that $x \cdot x^{\underline{k}} = x^{\underline{k+1}} + k x^{\underline{k}}$.

3. Suppose that a and n are integers with $q > n \ge 0$, and consider the q^n n-tuples $(a_1, \ldots, a_n)$ with $1 \le a_i \le q$. The object of this exercise is to establish a relationship between Stirling numbers of the second kind and ordinary powers.

   (a) Consider n-tuples as above, but with the restriction that the coordinates take on exactly k different values. Show that the number of such n-tuples is $\left\{{n \atop k}\right\} q^{\underline{k}}$.
   (b) Deduce that $\sum_{k=0}^{n} \left\{{n \atop k}\right\} q^{\underline{k}} = q^n$.
   (c) Argue that since each side in part (b) is a polynomial in q, and since these two polynomials are equal at infinitely many arguments, they must be identically equal.

4. If X is a Poisson random variable with parameter λ, then its nth moment is

$$E\left[X^n\right] = e^{-\lambda} \sum_{r=0}^{\infty} r^n \frac{\lambda^r}{r!}. \tag{18.51}$$

   By taking $x = r$ in (18.50), or otherwise, show that $E\left[X^n\right] = m_n(\lambda)$ where $m_n(\lambda)$ defined in (18.48).

5. If X is a random variable, then by definition, its *moment generating function* is $\sum_{n=0}^{\infty} E\left[X^n\right] \frac{z^n}{n!}$. Use (18.51) to show that

$$\sum_{n=0}^{\infty} m_n(\lambda) \frac{z^n}{n!} = e^{-\lambda} \exp\left(\lambda e^z\right).$$

6. Show that

$$\sum_{k=1}^{n} k \left\{ {n \atop k} \right\} x^{\underline{k}} = x^{n+1} - x(x-1)^n.$$

7. Let $X_1, \ldots, X_k$ be independent identically distributed random variables each with the distribution $P(X_j = a) = 1/p$ for $a = 1, 2, \ldots, p$. Let $\nu(\boldsymbol{X})$, a dependent random variable, denote card$\{X_1, \ldots, X_k\}$. This random variable takes values from 1 to k.

   (a) Show that

$$P(\nu(\boldsymbol{X}) = r) = \left\{ {k \atop r} \right\} p^{\underline{r}} p^{-k}$$

   for $r = 1, 2, \ldots, k$.
   (b) Deduce that

$$E\left[1 - \frac{\nu(\boldsymbol{X})}{p}\right] = \left(1 - \frac{1}{p}\right)^k.$$

(c) Conclude that

$$E\left[\left(1-\frac{\nu(X)}{p}\right)\left(1-\frac{1}{p}\right)^{-k}\right]=1.$$

18.7 Notes

Section 18.1. There is a copy of Goldbach's letter at `https://en.wikipedia.org/wiki/Christian_Goldbach#/media/File:Letter_Goldbach-Euler.jpg`

Hardy and Littlewood developed their eponymous method in a series of eight papers with the the generic title 'On some problems of partitio numerorum'. Only two are directly connected with the Goldbach problems, and the first of those is only the third in the sequence (the seventh was never published although a number of people had copies). They produced at least three other publications connected with the development of the circle method, including the published version of Hardy's inaugural lecture at Oxford, the published version of his lecture to the Copenhagen Mathematical Society, and an overview of the method as applied to the Goldbach problems in a paper in the *Proceedings of the Cambridge Philosophical Society*. There is also a seminal paper of Hardy and Ramanujan on the partition function and sums of squares which should be considered a prolegomenon to the method. See Hardy (1966, Volume 1, Section 2), especially parts (b), (c), and (d).

Vinogradov also made significant contributions to the development of the method, for which see Vinogradov (1985, Section 1).

To get a view of the overall scope of the modern versions of the method, see Davenport (2005) and Vaughan (1997). Hardy & Littlewood (1922) determined the asymptotic number of representations of a large odd number as a sum of three primes, assuming GRH. Vinogradov (1937) gave the first unconditional proof.

Section 18.2. After the publication of Vinogradov in 1937, van der Corput (1937), Chudakov (1938), and Estermann (1938) independently established Theorem 18.4, and with it the estimate for $E(X)$ found in Corollary 18.5. This stood as the best-known estimate for many years, but Vaughan (1972) showed that $E(X) \ll X\exp\left(-c\sqrt{\log X}\right)$. Then Montgomery & Vaughan (1975) followed a suggestion of Gallagher to show that there is an effectively computable constant $\delta > 0$ such that $E(X) \ll X^{1-\delta}$ for all large X. Chen & Liu (1989) showed that one can take $\delta = 0.05$, and admissible values of δ were established in small increments by (Li 1999, 2000), and Lu (2010). Recently, Pintz (2023) announced his intent to publish a proof that $E(X) \ll X^{3/4}$.

Section 18.3. Corollary 18.9 is a special case of Theorem A in §11.3 of Hardy & Littlewood (1922), and Lemmas 18.10–18.12 are substantially the same of those found in Hardy & Littlewood (1922).

Section 18.4. The result here is due to Montgomery & Vaughan (1973).

Section 18.5. Hardy & Littlewood (1922), pp. 54–62, gave a conditional determination of the asymptotic number of prime k-tuples. Lavrik (1961) showed that the proposed formulæ are correct in mean square.

Section 18.6. This section is based on Gallagher (1976), in which the proof of Theorem 18.20 is based on the product formula (18.41) for the singular series. Many systems of notation for the Stirling numbers have been used, with none of them dominant. We have followed the example of Graham, Knuth, & Patashnik (1989), who also provide a large collection of interesting identities.

18.8 References

Chen J.-R. & Liu J. M. (1989). The exceptional set of Goldbach numbers III, *Chinese Quart. J. Math.* **4**, no. 1, 1–15.

van der Chudakov, N. G. (1938). On the density of the set of even numbers which are not representable as a sum of two odd primes, *Izcv. Akad. Nauk SSSR* **2**, 25–40.

Corput, J. G. (1937). Sur l'hypothèse de Goldbach pour presque tous les nombres pairs, *Acta Arith.* **2**, 266–290.

Davenport, H. (2005). *Analytic Methods for Diophantine Equations and Diophantine Inequalities*, Cambridge: Cambridge University Press.

Estermann, T. (1938). On Goldbach's problem: Proof that almost all even positive integers are sums of two primes, *Proc. London Math. Soc.* (2) **44**, 307–314.

Gallagher, P. X. (1976). On the distribution of primes in short intervals, *Mathematika* **23**, 4–9; Corrigendum 28 (1981), 86.

Graham, R. L., Knuth, D. E., & Patashnik, O. (1989). *Concrete Mathematics*, Reading: Addison-Wesley.

Hardy, G. H. (1966). *Collected Papers of G. H. Hardy*, Vol. 1, Oxford: Clarendon Press.

Hardy, G. H. & Littlewood, J. E. (1922). Some problems of 'partitio numerorum'; III: On the expression of a number as a sum of primes, *Acta Mathematica* **44**, 1–70.

Hooley, C. (1998). On the Barban–Davenport–Halberstam theorem VIII, *J. Reine Angew. Math.* **499**, 1–46.

Lavrik, A. F. (1960). On the twin prime hypothesis of the theory of primes by the method of I. M. Vinogradov, *Dokl. Akad. Nauk SSSR* **132**, 1013–1015; *Soviet Math. Dokl.* **1**, 700–702.

(1961). On the theory of distribution of primes based on I. M. Vinogradov's method of trigonometric sums, *Trudy Mat. Inst. Steklov.* **64**, 90–125.

Li, H. (1999). The exceptional set of Goldbach numbers I, *Quart. J. Math.* Ser. (2) **50**, 471–482.

(2000). The exceptional set of Goldbach numbers II, *Acta Arith.* **92**, 71–88,

Lu, W. C. (2010). Exceptional set of Goldbach number, *J. Number Theory* **130**, 2359–2392.

Montgomery, H. L. & Vaughan, R. C. (1973). Error terms in additive prime number theory, *Quart. J. Math. Oxford Ser.* **24**, 207–216.

(1975). The exceptional set in Goldbach's problem, *Acta Arith.* **27**, 353–370.

Pillai, S. S. (1936). On sets of square-free integers, *J. Indian Math. Soc. N.S.* **2**, 116–118.

Pintz, J. (2023). A new explicit formula in the additive theory of primes with applications I. The explicit formula for the Goldbach problem and the generalized twin prime problem, *Acta Arith.* **210**, 53–94.

Vaughan, R. C. (1972). On Goldbach's problem, *Acta Arith.* **22**, 21–48.

(1997). *The Hardy–Littlewood Method*, Second edition, Cambridge Tracts in Math. 125, Cambridge: Cambridge University Press.

Vinogradov, I. M. (1937). Some theorems concerning the theory of primes, *Mat. Sb.* **2** (44), 179–195.

(1985). *Selected Works*. Berlin: Springer Verlag.

19

The Large Sieve

19.1 Trigonometric Polynomials

The large sieve takes various forms, as a mean square upper bound for a trigonometric polynomial at well-spaced points, as a mean square upper bound for the distribution of a set of integers into arithmetic progressions, and as a mean square upper bound for character sums. We take the trigonometric form to be fundamental and derive the other versions from it.

Let

$$T(x) = \sum_{n=M+1}^{M+N} c_n e(nx) \tag{19.1}$$

be a trigonometric polynomial. Suppose that $\delta > 0$, and that the points x_r are well spaced (mod 1) in the sense that

$$\|x_r - x_s\| \geq \delta \tag{19.2}$$

whenever $r \neq s$. We seek an inequality of the form

$$\sum_{r=1}^{R} |T(x_r)|^2 \leq \Delta \sum_{n=M+1}^{M+N} |c_n|^2, \tag{19.3}$$

which is to hold for all possible choices of c_n. Our objective is to determine how Δ must depend on N and δ. When $R = 1$ it is easy to establish an inequality of this form since by Cauchy's inequality,

$$|T(x_1)|^2 \leq N \sum_{n=M+1}^{M+N} |c_n|^2. \tag{19.4}$$

This is best possible, for if $c_n = e(-nx_1)$ for all n, then $T(x_1) = N$. Thus if (19.3) holds for all c_n, then $\Delta \ge N$. We also observe that

$$\int_0^1 \sum_{r=1}^{R} |T(x + r/R)|^2 \, dx = R \int_0^1 |T(x)|^2 \, dx = R \sum_{n=M+1}^{M+N} |c_n|^2.$$

Hence there is an x for which

$$\sum_{r=1}^{R} |T(x + r/R)|^2 \ge R \sum_{n=M+1}^{M+N} |c_n|^2.$$

For any given $\delta > 0$ we can choose $R = \lfloor 1/\delta \rfloor$, and then the points $x + r/R$ satisfy (19.2). Thus if Δ satisfies (19.3), then $\Delta \ge R \ge 1/\delta - 1$.

We now show that (19.3) holds with a value of Δ not much larger than necessitated by the above considerations. Our first result in this direction is somewhat inferior, but the approach is very direct, and generalizes usefully to other situations. For each r, let $\mathfrak{M}_r = (x_r - \delta/2, x_r + \delta/2)$ be a short interval centred at x_r. We note that if the x_r satisfy (19.2), then the intervals $\mathfrak{M}_r$ are disjoint (mod 1). The idea is that $|T(x_r)|^2$ is approximately the average of $|T(x)|^2$ over $\mathfrak{M}_r$ unless $T'(x)$ is very large, in which case the integral of $|T'(x)|^2$ over $\mathfrak{M}_r$ is large. To put this intuitive principle on a sound footing we prove the following.

Lemma 19.1 (Sobolev) *Suppose that $a < b$ and that f is a continuous complex-valued function with a piecewise continuous and bounded first derivative on the interval $[a, b]$. Then*

$$\left| f\left(\frac{a+b}{2}\right) \right| \le \frac{1}{b-a} \int_a^b |f(x)| \, dx + \frac{1}{2} \int_a^b |f'(x)| \, dx \tag{19.5}$$

and

$$|f(x)| \le \frac{1}{b-a} \int_a^b |f(u)| \, du + \int_a^b |f'(u)| \, du \tag{19.6}$$

for any $x \in [a, b]$.

Proof Suppose that $a \le x \le b$. By integration by parts we see that

$$\begin{aligned} \int_x^b f(u) \, du &= \Big[f(u)(u-b) \Big|_x^b - \int_x^b f'(u)(u-b) \, du \\ &= (b-x) f(x) - \int_x^b f'(u)(u-b) \, du, \end{aligned}$$

and similarly that

$$\int_a^x f(u)\,du = \Big[f(u)(u-a)\Big|_a^x - \int_a^x f'(u)(u-a)\,du$$
$$= (x-a)f(x) - \int_a^x f'(u)(u-a)\,du.$$

On adding these two identities we deduce that

$$(b-a)f(x) = \int_a^b f(u)\,du + \int_a^x f'(u)(u-a)\,du + \int_x^b f'(u)(u-b)\,dx.$$

Hence, by the triangle inequality,

$$(b-a)|f(x)| \le \int_a^b |f(u)|\,du + (x-a)\int_a^x |f'(u)|\,du + (b-x)\int_x^b |f'(u)|\,du.$$

Now $x-a \le b-a$ and $b-x \le b-a$, so we have (19.6). If $x = (a+b)/2$, then $x-a = b-x = (a+b)/2$, which gives (19.5). □

Lemma 19.2 (Gallagher 1967) *Let $g(x)$ be a continuous function with period 1, with a piecewise continuous and bounded first derivative. Suppose that $\delta > 0$, and that $x_1, x_2, \ldots, x_R$ are well spaced modulo 1 in the sense that (19.2) holds. Then*

$$\sum_{r=1}^R |g(x_r)|^2 \le \frac{1}{\delta}\int_0^1 |g(x)|^2\,dx + \left(\int_0^1 |g(x)|^2\,dx\right)^{1/2}\left(\int_0^1 |g'(x)|^2\,x\right)^{1/2}.$$

Proof Let $\mathfrak{M}_r = (x_r - \delta/2, x_r + \delta/2)$ for $1 \le r \le R$. By the Sobolev lemma with $f(x) = g(x)^2$ and $(a,b) = \mathfrak{M}_r$ we find that

$$|g(x_r)|^2 \le \frac{1}{\delta}\int_{\mathfrak{M}_r} |g(x)|^2\,dx + \int_{\mathfrak{M}_r} |g(x)g'(x)|\,dx.$$

The arcs $\mathfrak{M}_r$ are pairwise disjoint modulo 1, so

$$\sum_{r=1}^R |g(x_r)|^2 \le \frac{1}{\delta}\int_0^1 |g(x)|^2\,dx + \int_0^1 |g(x)g'(x)|\,dx.$$

To complete the proof we apply the Cauchy–Schwarz inequality to the last term. □

Suppose that $U(x)$ is a trigonometric polynomial of the special form

$$U(x) = \sum_{k=-K}^K b_k e(kx).$$

By Gallagher's lemma,

$$\sum_{r=1}^{R} |U(x_r)|^2 \le \frac{1}{\delta} \sum_{k=-K}^{K} |b_k|^2 + \left(\sum_{k=-K}^{K} |b_k|^2 \right)^{1/2} \left(\sum_{k=-K}^{K} |2\pi i k b_k|^2 \right)^{1/2}.$$

Since $|2\pi i k| \le 2\pi K$ for $-K \le k \le K$, it follows that

$$\sum_{r=1}^{R} \left| \sum_{k=-K}^{K} b_k e(kx_r) \right|^2 \le \left(\frac{1}{\delta} + 2\pi K \right) \sum_{k=-K}^{K} |b_k|^2.$$

This is a special case of (19.3). To obtain the general case, let $K = \lfloor N/2 \rfloor$, put $L = K + M + 1$, and set $U(x) = T(x)e(-Lx)$. Then $U(x)$ is of the required shape, $|U(x)| = |T(x)|$, and $2K \le N$, so we have proved a useful inequality.

Theorem 19.3 *Suppose that M and N are integers, $N \ge 1$, and that $T(x)$ is a trigonometric polynomial as given in* (19.1). *Suppose that $\delta > 0$, and that the points x_r are well spaced in the sense that* (19.2) *holds. Then*

$$\sum_{r=1}^{R} |T(x_r)|^2 \le \left(\frac{1}{\delta} + \pi N \right) \sum_{n=M+1}^{M+N} |c_n|^2.$$

For the purposes of estimating character sums, Theorem 19.3 is perfectly satisfactory, but when dealing with arithmetic progressions the coefficient of N on the right-hand side becomes important. To optimize this dependence we adopt a different line of attack. The quantity to be estimated is a bilinear form in the coefficients c_n. Often, when presented with the problem of estimating a bilinear form we simply expand, take the outer summation inside, and estimate the resulting innermost sum:

$$\sum_{r=1}^{R} \left| \sum_{n=M+1}^{M+N} c_n e(nx_r) \right|^2 = \sum_{m=M+1}^{M+N} \sum_{n=M+1}^{M+N} c_m \overline{c_n} \sum_{r=1}^{R} e((m-n)x_r).$$

Unfortunately, we have little control over the inner sum on the right, so this approach, in its most direct form, leads nowhere. However, every bilinear form inequality has a dual, and we have the option of passing to the dual before performing the above manipulations. More precisely, by Theorem G.1 we see that the inequality (19.3) holds for all choices of the c_n if and only if

$$\sum_{n=M+1}^{M+N} \left| \sum_{r=1}^{R} y_r e(nx_r) \right|^2 \le \Delta \sum_{r=1}^{R} |y_r|^2 \tag{19.7}$$

for all y_r. On expanding and taking the sum over n inside we find that the left-hand side above is

$$= \sum_{r=1}^{R} \sum_{s=1}^{R} y_r \overline{y_s} \sum_{n=M+1}^{M+N} e(n(x_r - x_s)).$$

By applying (16.4) to estimate the innermost sum we could demonstrate that $\Delta \le N + O(\delta^{-1} \log \delta^{-1})$, which is good for N but inferior for δ. The extra logarithm results from the inverse first power decay of the exponential sum, which, in turn, is attributable to the jump discontinuity of the characteristic function $\chi_{\mathcal{J}}(x)$ of the interval $\mathcal{J} = [M+1, M+N]$. To obtain an exponential sum that decays faster, we introduce a smooth weighting factor.

Theorem 19.4 *Suppose that M and N are integers, $N \ge 1$, and that $T(x)$ is a trigonometric polynomial as given by* (19.1). *Suppose that $0 < \delta \le 1$, and that the points x_r are well spaced in the sense that* (19.2) *holds. Then*

$$\sum_{r=1}^{R} |T(x_r)|^2 \le \left(N + \frac{1}{\delta} - 1\right) \sum_{n=M+1}^{M+N} |c_n|^2.$$

Proof If $R = 1$, then we have the stated result by Cauchy's inequality, as in (19.4). If $R \ge 2$, then $\delta \le 1/2$. We proceed to (19.7), but before expanding we introduce a weighting factor $w(n)$. If $\chi_{\mathcal{J}}(n) \le w(n)$ for all integers n, then the left-hand side of (19.7) is

$$\le \sum_{n\in\mathbb{Z}} w(n) \left| \sum_{r=1}^{R} y_r e(nx_r) \right|^2.$$

Suppose that $\sum_{n\in\mathbb{Z}} w(n) < \infty$, and put $W(x) = \sum_{n\in\mathbb{Z}} w(n)e(nx)$. Thus $W(x)$ is a continuous function with period 1 whose Fourier coefficients are the $w(n)$. On expanding the above we see that it is

$$= \sum_{r=1}^{R} \sum_{s=1}^{R} y_r \overline{y_s} \sum_{n\in\mathbb{Z}} w(n) e(n(x_r - x_s)) = \sum_{r=1}^{R} \sum_{s=1}^{R} y_r \overline{y_s} W(x_r - x_s).$$

Let A be a positive parameter, and set

$$w(n) = \begin{cases} 0 & (n \le M + 1 - A), \\ \frac{1}{A}(n - M - 1 + A) & (M + 1 - A \le n \le M + 1), \\ 1 & (M + 1 \le n \le M + N), \\ \frac{1}{A}(N + A - n) & (N \le n \le N + A), \\ 0 & (n \ge N + A), \end{cases}$$

then $W(x)$ decays like an inverse square, and by choosing A carefully with $A \asymp 1/\delta$ we can show that $\Delta \le N + 2/\delta$. However, by employing the more sophisticated weighting given in Theorem E.5 we find that we can actually ensure that $W(x) = 0$ for $\|x\| \ge \delta$, so that the bilinear form above consists only of diagonal terms. Moreover, with this choice of the $w(n)$ we find that $W(0) = N - 1 + 1/\delta$, so the proof is complete. □

In most arithmetic applications of the large sieve, the x_r are simply taken to be the Farey fractions of order Q, as in Corollary 19.5.

Corollary 19.5 *Let M and N be integers, $N \ge 1$, and suppose that $T(x)$ is a trigonometric polynomial of the form* (19.1). *Then for any positive integer Q,*

$$\sum_{q=1}^{Q} \sum_{\substack{a=1 \\ (a,q)=1}}^{q} |T\,(a/(q))\,|^2 \le (N + Q^2) \sum_{n=M+1}^{M+N} |c_n|^2.$$

Proof By Cauchy's inequality, $|T(1)|^2 \le N \sum_n |c_n|^2$, which suffices. For $Q \ge 2$, the numbers $1/2$ and $1/1$ are among the Farey fractions of order Q, with the result that two adjacent Farey fractions a/q and a'/q' differ by at most $1/2$. Thus

$$\left\| \frac{a}{q} - \frac{a'}{q'} \right\| = \left| \frac{a}{q} - \frac{a'}{q'} \right| = \frac{|aq' - a'q|}{qq'} \ge \frac{1}{qq'}.$$

Thus we may take $\delta = 1/Q^2$. □

19.1.1 Exercises

1. Let $T(x)$ be defined as in (19.1).

   (a) Show that

   $$\int_0^1 \sum_{q=1}^{Q} \sum_{\substack{a=1 \\ (a,q)=1}}^{q} |T(x + a/q)|^2 \, dx = \left(\sum_{q \le Q} \varphi(q) \right) \sum_{n=M+1}^{M+N} |c_n|^2.$$

   (b) Deduce that there is an x such that

   $$\sum_{q=1}^{Q} \sum_{\substack{a=1 \\ (a,q)=1}}^{q} |T(x + a/q)|^2 \, dx \gtrsim \frac{3}{\pi^2} Q^2 \sum_{n=M+1}^{M+N} |c_n|^2.$$

2. Suppose that $f(x)$ is a complex-valued function with a continuous first derivative, and that $f(x) \to 0$ as $x \to \pm\infty$. Show that

   $$|f(x)| \le \frac{1}{2} \int_{-\infty}^{\infty} |f'(u)| \, du$$

   for all real x. (Thus $\|f\|_\infty \le \frac{1}{2} \|f'\|_1$.)

3. Suppose that $a > 0$, and that $f(x)$ has a continuous first derivative for $-a \le x \le a$. Show that if $-a \le x \le a$, then

$$\left| f(x) - \frac{1}{2a}\int_{-a}^{a} f(u)\,du \right| \le \int_{-a}^{a} |f'(u)|\,du.$$

4. Suppose that $f'(x)$ is continuous for $0 \le x \le 1$.

   (a) Show that if $0 \le x \le 1$, then

$$f(x) = \int_0^1 f(u)\,du + \int_0^x f'(u)u\,du + \int_x^1 f'(u)(u-1)\,du.$$

   (b) Deduce that

$$\max_{0\le x\le 1} |f(x)| \le \int_0^1 |f(u)|\,du + \int_0^1 |f'(u)|\,du.$$

5. Suppose that $f(x, y)$ has continuous derivatives through the second order on $[0, 1]^2$.

   (a) Show that if $0 \le x \le 1$ and $0 \le y \le 1$, then

$$|f(x,y)| \le \int_0^1\int_0^1 |f(u,v)| + |f_1(u,v)| + |f_2(u,v)| + |f_{12}(u,v)|\,du\,dv.$$

   (b) Show that

$$|f(1/2, 1/2)| \le \int_0^1\int_0^1 |f(u,v)| + \frac{1}{2}|f_1(u,v)| + \frac{1}{2}|f_2(u,v)| + \frac{1}{4}|f_{12}(u,v)|\,du\,dv.$$

6. (a) Suppose that $\int_0^\infty f(x)\,dx$ is a convergent improper Riemann integral. Show that if $f'(x) \to 0$ as $x \to \infty$, then $f(x) \to 0$ as $x \to \infty$.

   (b) Show that if $g(x) \to 0$ as $x \to \infty$ and $g''(x) \to 0$ as $x \to \infty$, then $g'(x) \to 0$ as $x \to \infty$.

7. Let $x_1, x_2, \ldots, x_R$ be points in $\mathbb{T}$. For $\delta > 0$, let $N_\delta(x)$ denote the number r for which $\|x_r - x\| < \delta$.

   (a) Show that

$$\sum_{\substack{1\le r\le R\\ \|x_r - x\|\le\delta/2}} \frac{1}{N_\delta(x_r)} \le 1$$

   for all $x \in \mathbb{T}$.

(b) Show that if M and N are integers, $N \ge 1$, and $T(x)$ is given by (19.1), then

$$\sum_{r=1}^{R} \frac{|T(x_r)|^2}{N_\delta(x_r)} \le \left(\frac{1}{\delta} + \pi N\right) \sum_{n=M+1}^{M+N} |c_n|^2$$

for all $\delta > 0$.

(c) Show that the result of part (b) includes Theorem 19.3.

8. Let μ be a non-negative measure on $\mathbb{T}$.

(a) Show that if $T(x)$ is given as in (19.1), and if $\delta > 0$, then

$$\int_{\mathbb{T}} |T(x)|^2 \, d\mu(x) \le \left(\max_{x\in\mathbb{T}} \mu((x-\delta/2, x+\delta/2))\right) \left(\frac{1}{\delta} + \pi N\right) \sum_{n=M+1}^{M+N} |c_n|^2.$$

(b) Derive Theorem 19.3 from the result of part (a).

9. (P. J. Cohen, oral communication 1977) Suppose that M and N are integers, $N \ge 1$, and that $T(x)$ is a trigonometric polynomial as given in (19.1). Suppose that $\delta > 0$ and that there are constants A, B and a real-valued function $f(N, \delta)$ such that

$$N^{-1} \sup_{\delta} f(N, \delta) \to 0 \text{ as } N \to \infty.$$

Suppose further that whenever the points x_r are well spaced in the sense that (19.2) holds and the c_n are arbitrary complex numbers, we have

$$\sum_{r=1}^{R} |T(x_r)|^2 \le (AN + B\delta^{-1} + f(N, \delta)) \sum_{n=M+1}^{M+N} |c_n|^2.$$

Let H be a positive integer, and define

$$x_{rh} = \frac{x_r + h}{H}, \qquad 1 \le r \le R,\ 0 \le h < H,$$

$$b_n = \begin{cases} c_{n/H} & \text{when } H|n, \\ 0 & \text{when } H\nmid n, \end{cases}$$

$$T^*(x) = \sum_{n=HN+H}^{HM+HN} b_n e(nx).$$

(a) Prove that $\min \|x_{rh} - x_{sj}\| \ge \delta/H$ where the minimum is taken over pairs r, h and s, j with $r, h \ne s, j$.

(b) Prove that

$$\sum_{r=1}^{R}\sum_{h=0}^{H-1}|T^*(x)|^2 \le \Delta \sum_{n=M+1}^{M+N}|c_n|^2,$$

where

$$\Delta = A(HN - H + 1) + \frac{BH}{\delta} + f\left(HN - H + 1, \frac{\delta}{H}\right).$$

(c) Prove that

$$\sum_{r=1}^{R}\sum_{h=0}^{H-1}|T^*(x)|^2 = H\sum_{r=1}^{R}|T(x_r)|^2.$$

(d) Prove that

$$\sum_{r=1}^{R}|T(x_r)|^2 \le (A(N-1) + B\delta^{-1})\sum_{n=M+1}^{M+N}|c_n|^2.$$

10. Suppose that $\delta > 0$ and that the points x_r satisfy (19.2). Show that for any y_r there is a number θ, $-1 \le \theta \le 1$, such that

$$\sum_{n=M+1}^{M+N}\left|\sum_{r=1}^{R} y_r e(nx_r)\right|^2 = \left(N - 1 + \frac{\theta}{\delta}\right)\sum_{r=1}^{R}|y_r|^2.$$

11. (Bombieri & Davenport, 1969, §5) Take

$$T(x) = \sum_{k=1}^{K} e(kRx)$$

and set $x_r = r/R$ for $r = 1, 2, \ldots, R$.

(a) Show that when this particular trigonometric polynomial is expressed in the notation of (19.1), the parameter $N = KR - R + 1$.

(b) Compute all quantities in (19.3) in terms of K and R, and show that equality is achieved.

(c) Show that $(N-1)\delta$ is an integer.

12. (Montgomery 1978) Let M and $N \ge 1$ be integers, and suppose that $T(x)$ is given by (19.1). For given positive integers Q, X, let $\Delta = \Delta(N, Q, X)$ be the optimal constant in the inequality

$$\sum_{q\in\mathcal{Q}}\sum_{\substack{a=1\\(a,q)=1}}^{q}|T(a/q)|^2 \le \Delta\sum_{n=M+1}^{M+N}|c_n|^2,$$

where the c_n are arbitrary and $\mathcal{Q}$ is a set of X positive integers not exceeding Q. Show that $\Delta(N, Q, X) \asymp \min(N + Q^2, X(N + Q))$.

13. (Burgess 1971) Let $N_\delta(x)$ be defined as in Exercise 19.1.1.7, but take the x_r to be the points a/q with $(a,q)=1$ and $q \in \mathcal{Q}$.

    (a) Show that if $\delta = (QX)^{-1}$, then

$$\sum_{q\in\mathcal{Q}} \sum_{\substack{a=1\\(a,q)=1}}^{q} N_\delta(a/q) \ll QX.$$

    (b) By using Cauchy's inequality and applying part (a) and Exercise 19.1.1.7, show that

$$\sum_{q\in\mathcal{Q}} \sum_{\substack{a=1\\(a,q)=1}}^{q} |T(a/q)| \ll \Bigl(QX(N+QX)\sum_{n=M+1}^{M+N} |c_n|^2\Bigr)^{1/2}.$$

14. We have discussed a bilinear form with a coefficient matrix of the form $[e(nx_r)]$ where the n are consecutive integers and the x_r are well spaced modulo 1. We now consider a more general bilinear form with a coefficient matrix of the form $[e(\lambda_m\mu_n)]$ where the λ_m and μ_n are both well-spaced sequences. Specifically, suppose that $-L/2 \le \lambda_m \le L/2$ for all m, and $|\lambda_m - \lambda_{m'}| \ge \eta > 0$ for $m \ne m'$, while $0 \le \mu_n \le M$ for all n, and $|\mu_n - \mu_{n'}| \ge \delta > 0$ for $n \ne n'$. Our objective is to find a number $\Delta = \Delta(L,\eta,M,\delta)$ such that

$$\sum_n \Bigl|\sum_m x_m e(\lambda_m\mu_n)\Bigr|^2 \le \Delta^2 \sum_m |x_m|^2 \qquad (19.8)$$

for all choices of the variables x_m.

    (a) Let $S(\mu) = \sum_m x_m e(\lambda_m\mu)$. Show that

$$\sum_n |S(\mu_n)|^2 \le \frac{1}{\delta}\int_{-\frac12\delta}^{M+\frac12\delta} |S(\mu)|^2\,d\mu + \int_{-\frac12\delta}^{M+\frac12\delta} |S(\mu)S'(\mu)|\,d\mu.$$

    (b) Let $S_+(\mu)$ be Selberg's majorant function as provided by Theorem E.3 and chosen so that it majorizes the characteristic function of the interval $[a,b]$, its Fourier transform has support in $(-\eta,\eta)$, and $S_+(0) = b - a + 1/\eta$. Show that if $T(\mu) = \sum_m c_m e(\lambda_m\mu)$, then

$$\int_a^b |T(\mu)|^2\,d\mu \le \int_{-\infty}^{\infty} S_+(\mu)|T(\mu)|^2\,d\mu = (b-a+1/\eta)\sum_m |c_m|^2.$$

(c) Deduce that

$$\int_{-\frac{1}{2}\delta}^{M+\frac{1}{2}\delta} |S(\mu)|^2\, d\mu \le (M+\delta+1/\eta)\sum_m |x_m|^2,$$

$$\int_{-\frac{1}{2}\delta}^{M+\frac{1}{2}\delta} |S'(\mu)|^2\, d\mu \le \pi^2 L^2 (M+\delta+1/\eta)\sum_m |x_m|^2.$$

(d) Deduce that (19.8) holds with $\Delta^2 = (\pi L + 1/\delta)(M+\delta+1/\eta)$.

(e) Show that (19.8) holds with Δ given by part (d) when the intervals $[-L/2, L/2]$, $[0, M]$ are replaced by $[A, A+L]$, $[B, B+M]$ for any A and B.

(f) Show that the number of m is $\le 1 + L/\eta$. Deduce that $|S(\mu_1)|^2 \le (1 + L/\eta)\sum_m |x_m|^2$. Show that if there are two or more values of n, then $\delta \le M$, in which case $\Delta^2 \le (\pi L + 1/\delta)(2M + 1/\eta)$.

Selberg (1991, pp. 221–224) used a different method to show that one can take $\Delta^2 = (L+1/\delta)(M+1/\eta) + 1 + \min(\delta L, \eta M)$, and speculated that the inequality will still hold without the last term ($\min(\cdots)$). Preissmann (1985) had shown earlier that the inequality is in general false when $\Delta^2 = (L+1/\delta)(M+1/\eta)$.

19.2 Mean Square Distribution in Arithmetic Progressions

Suppose that we have a sequence of numbers c_n for $M+1 \le n \le M+N$. We now consider how these numbers are distributed when n falls in various arithmetic progressions. Let

$$Z(q,h) = \sum_{\substack{n=M+1\\ n\equiv h\,(q)}}^{M+N} c_n. \tag{19.9}$$

If $T(x)$ is given as in (19.1), then

$$T(a/q) = \sum_{h=1}^{q} Z(q,h)e(ah/q),$$

and hence by the orthogonality of the additive characters (mod q) (or, in other words, Parseval's identity for the finite Fourier transform, as used in §4.1) it follows that

$$\sum_{a=1}^{q} |T(a/q)|^2 = q\sum_{h=1}^{q} |Z(q,h)|^2. \tag{19.10}$$

Let

$$Z = Z(1,0) = T(0) = \sum_{n=M+1}^{M+N} c_n. \tag{19.11}$$

Thus the average of the $Z(q,h)$ is Z/q. It is natural to consider the mean square difference of the $Z(q,h)$ from its mean (called the 'variance' in probability theory). We now express this variance in terms of T.

Lemma 19.6 *Let T, $Z(q,h)$, and Z be defined as in* (19.1), (19.9), *and* (19.11), *respectively. Then*

$$q\sum_{h=1}^{q} |Z(q,h) - Z/q|^2 = \sum_{a=1}^{q-1} |T(a/q)|^2$$

for arbitrary complex numbers c_n.

Proof On expanding, we see that the left-hand side above is

$$= q\sum_{h=1}^{q} |Z(q,h)|^2 - 2\operatorname{Re}\overline{Z}\sum_{h=1}^{q} Z(q,h) + |Z|^2.$$

Here the second sum is Z, so the above is

$$= q\sum_{h=1}^{q} |Z(q,h)|^2 \ - \ |Z|^2.$$

The stated identity now follows by appealing to (19.10) and (19.11). □

In Lemma 19.6 we have restricted a to non-zero residue classes modulo q, but not to reduced residue classes, as would be required in order to appeal to Corollary 19.5. However, for a prime modulus the reduced residues and non-zero residues coincide, so we have the following theorem.

Theorem 19.7 *Let $\mathcal{N} \subseteq [M+1, M+N]$ be a set of Z integers. Let $Z(q,h)$ denote the number of $n \in \mathcal{N}$ such that $n \equiv h \pmod q$. Then for any positive integer Q,*

$$\sum_{p\le Q} p\sum_{h=1}^{p}(Z(p,h) - Z/p)^2 \le (N + Q^2)Z.$$

Proof Take $c_n = 1$ if $n \in \mathcal{N}$, and $c_n = 0$ otherwise. In Lemma 19.6, replace q by p, sum over $p \le Q$, and then apply Corollary 19.5. □

By Theorem 19.7 we see that if $Z > N^{1/2+\varepsilon}$, then most of the numbers $Z(p,h)$ are near their mean, Z/p, for $p \le N^{1/2}$. In particular, we note the following consequence.

Corollary 19.8 *Let $\mathcal{N} \subseteq [M+1, M+N]$ be a set of Z integers. Choose τ, $0 < \tau \le 1$, and let $\mathcal{P}$ denote the set of primes $p \le Q$ such that $Z(p,h) = 0$ for at least τp residue classes $h \pmod p$. Put $P = \operatorname{card}(\mathcal{P})$. Then*

$$Z \le \frac{N + Q^2}{\tau P}.$$

Here we finally see how the large sieve got its name: We are estimating how many integers remain in an interval after a large amount of sifting has been done. We find, not surprisingly, that Z is small if P is large, and vice versa.

Proof If $p \in \mathcal{P}$, then the inner sum in Theorem 19.7 is $(Z/p)^2$ for at least τp values of h. Hence the prime p contributes at least τZ^2 to the left-hand side, so we see that

$$\tau P Z^2 \le (N + Q^2) Z.$$

If $Z = 0$, then there is nothing to prove. Otherwise $Z > 0$, and we may cancel Z from both sides to obtain the stated inequality. □

To exemplify the sorts of arithmetic applications that these tools might find, we apply Corollary 19.8 to show that the least quadratic non-residue of a prime $p > 2$ is not often very large. For an odd prime p, let $n_2(p)$ denote the least positive quadratic non-residue. The distribution of this quantity is quite easy to determine: We first observe that $n_2(p)$ is a prime number, for if $n_2(p) = ab$, then $\left(\frac{ab}{p}\right) = -1$, and hence $\left(\frac{a}{p}\right) = -1$ or $\left(\frac{b}{p}\right) = -1$. We note by quadratic reciprocity that $n_2(p) = 2$ if $p \equiv \pm 3 \pmod 8$, which by the Prime Number Theorem for arithmetic progressions is the case for asymptotically $1/2$ of the primes p. Also, $n_2(p) = 3$ if $\left(\frac{2}{p}\right) = 1$ and $\left(\frac{3}{p}\right) = -1$. That is, either $p \equiv 1 \pmod 8$ and $\left(\frac{p}{3}\right) = -1$ or if $p \equiv 7 \pmod 8$ and $\left(\frac{p}{3}\right) = 1$. Hence $n_2(p) = 3$ for asymptotically $1/4$ of the primes. Let $p_1 < p_2 < p_3 < \cdots$ be the prime numbers listed in increasing order. Then by continuing in this way we see that $n_2(p) = p_k$ for asymptotically 2^{-k} of the primes. Using the Siegel–Walfisz theorem (Corollary 11.19) to the modulus $q = 4\prod_{j=1}^{k} p_j < e^{(1+\varepsilon)k \log k}$, we can state this quantitatively:

$$\operatorname{card}\{p \le x : n_2(p) = p_k\} = \frac{\operatorname{li} x}{2^k} + O\big(x \exp(-c\sqrt{\log x})\big) \tag{19.12}$$

for $p_k \ll \log\log x$. For somewhat larger k we can appeal to the Brun–Titchmarsh inequality (Theorem 3.9). Thus we see that

$$\operatorname{card}\{p \le x : n_2(p) \ge p_k\} \ll \frac{x}{2^k \log x} \tag{19.13}$$

uniformly for $p_k \le \frac{1}{2}\log x$. The presumption that this bound might hold for still larger p_k suggests the conjecture that

$$n_2(p) \le (1 + o(1))(\log p)\log\log p \tag{19.14}$$

for large primes. This is stronger than the bound we derived from the Generalized Riemann Hypothesis (cf. Theorem 13.11) and much stronger than the conjecture

$$n_2(p) \ll p^{\varepsilon} \tag{19.15}$$

usually attributed to Vinogradov. To bound the frequency with which $n_2(p)$ might be larger than asserted in (19.14), we employ Corollary 19.8.

Theorem 19.9 *Let a be fixed, $a > 2$. The number of primes p, $2 < p \le x$, for which $n_2(p) > (\log p)^a$ is $< x^{2/a+o(1)}$.*

Proof We apply Corollary 19.8 with $N = Q^2$, $\mathcal{N} = \{n \in [1, N] : p|n \Rightarrow p < (\log Q/2)^a\}$, and $\mathcal{P} = \{p \in (Q/2, Q] : n_2(p) > (\log p)^a\}$. Thus if $n \in \mathcal{N}$ and $p \in \mathcal{P}$, then all the prime factors of n are so small that they are quadratic residues (mod p), and hence $\left(\frac{n}{p}\right) = 1$. Hence $Z(p, h) = 0$ for at least $(p + 1)/2$ residue classes h (mod p), and so we may take $\tau = 1/2$. By Corollary 7.9 we know that $Z = N^{1-1/a+o(1)}$. Consequently $P \ll N^{1/a+o(1)} = Q^{2/a+o(1)}$. To complete the proof it suffices to set $Q = 2^{-j}x$ and sum over $j = 0, 1, 2, \ldots$. □

Although (19.15) has not yet been proved for all p, we can use the above method to show that any possible exceptions are exceedingly rare.

Theorem 19.10 (Linnik 1942) *If $\delta > 0$, then the number of primes $p \le x$ for which $n_2(p) > p^{\delta}$ is $\ll_{\delta} \log\log x$.*

Proof Let $\mathcal{P}$ be the set of primes p, $Q^{1/2} < p \le Q$, for which $n_2(p) > p^{\delta}$. We show that $P \ll_{\delta} 1$; then the stated result follows on summing over $Q = x^{1/2^j}$. Put $N = Q^2$, and let $\mathcal{N}$ be the set of integers n, $1 \le n \le N$, composed entirely of prime numbers not exceeding $N^{\delta/4}$. As in the preceding proof we may take $\tau = 1/2$ in Corollary 19.8. By Dickman's theorem (Theorem 7.2) we know that $Z \gg_{\delta} N$. Hence by Corollary 19.8 we see that $P \ll_{\delta} 1$, and the proof is complete. □

Since we have determined the distribution function of the $n_2(p)$, and have also shown that large values of $n_2(p)$ are rare, we can deduce that a moment of the $n_2(p)$ tends to the moment of the distribution function.

Theorem 19.11 (Erdős 1961) *Suppose that $\delta \ge 0$ is chosen so that $n_2(p) < p^{\delta+\varepsilon}$ for all $p > p_0(\varepsilon)$. Let γ be a fixed real number such that $\gamma < 1/\delta$. Then*

$$\sum_{2<p\le x} n_2(p)^{\gamma} \sim c(\gamma)\pi(x) \tag{19.16}$$

as x tends to infinity, where

$$c(\gamma) = \sum_{k=1}^{\infty} p_k^{\gamma}/2^k$$

and $2 = p_1 < p_2 < \cdots$ are the primes in increasing order.

From (19.16) it is easy to deduce that $n_2(p) \ll p^{1/\gamma}$, so it is to be expected that we can prove (19.16) only under the assumption that $\gamma < 1/\delta$. By our remarks following Theorem 9.27 we may take $\delta = 1/(4\sqrt{e})$. Thus it follows that (19.16) holds for all $\gamma < 4\sqrt{e} = 6.59\ldots$.

Proof By (19.12) we see that the primes $p \le x$ for which $n_2(p) \le \log\log x$ contribute to the left-hand side of (19.16) an amount that is asymptotic to the right-hand side of (19.16). Thus it remains to show that those p for which $n_2(p)$ is larger make a smaller contribution. Suppose that $\log\log x < p_k \le \frac{1}{2}\log x$. By (19.13) we see that the number of $p \le x$ for which $n_2(p) = p_k$ is $\ll 2^{-k}\pi(x)$. On summing this over the appropriate range of k we obtain a contribution that is $o(\pi(x))$. Next suppose that $\frac{1}{2}\log x < p_k \le (\log x)^C$ where C is to be determined later. By (19.13) we see that the number of $p \le x$ for which $n_2(p)$ falls in this range is $\ll x\exp(-c(\log x)/\log\log x)$. The maximum contribution made by such a prime to the sum in (19.16) is $(\log x)^{C\gamma}$. Since the product of these last two quantities is $o(\pi(x))$ this suffices. Finally, consider primes p for which $n_2(p) > (\log x)^C$. By Theorem 19.9 the number of such primes is $< x^{2/C+\varepsilon}$. The maximum contribution made by such a prime is $< x^{\gamma(\delta+\varepsilon)}$. Hence the total contribution by all such primes is $< x^{\gamma(\delta+\varepsilon)+2/C+\varepsilon}$. Now $\gamma\delta < 1$, so we may choose $\varepsilon > 0$ so small that $\gamma(\delta+\varepsilon) \le 1 - 3\varepsilon$. If we take $C = 2/\varepsilon$, then the contribution in question is $< x^{1-\varepsilon} = o(\pi(x))$, so the proof is complete. □

Suppose that we try to use Theorem 19.7 as a small sieve. For example, suppose that $\mathcal{N} = \{p : N^{1/2} < p \le N\}$ and that $Q = N^{1/2}$. Then $Z(p, 0) = 0$ for all $p \le Q$, and hence we obtain the estimate

$$Z \ll \frac{N}{\sum_{p \le Q} 1/p} \ll \frac{N}{\log\log N},$$

which is vastly inferior to the bounds we obtained by Selberg's sieve (see Theorem 21.1) (cf. Theorem 3.3). Of course the $\log\log N$ arises because the sum is restricted to primes. If we were able to sum over all $q \le Q$, then we might expect to get a bound $O(N/\log N)$, which is comparable to our prior estimates. We now show that this can be done.

Lemma 19.12 (Montgomery 1968) *For $M + 1 \le n \le M + N$ let the numbers c_n be given. For each prime p let $\mathcal{D}(p)$ be the collection of those residue classes*

$d \pmod p$ for which $c_n = 0$ whenever $n \equiv d \pmod p$. Let $\delta(p) = \operatorname{card}\mathscr{D}(p)$, and let $\mathscr{R}(p)$ be the complementary set of $p - \delta(p)$ residue classes $\pmod p$. Finally, let $T(x)$, $Z(q,h)$, and Z be defined as in (19.1), (19.9), and (19.11), respectively. If q is square-free, then

$$\sum_{\substack{a=1\\(a,q)=1}}^{q} |T(a/q)|^2 \ge |Z|^2 \prod_{p|q} \frac{\delta(p)}{p-\delta(p)}. \tag{19.17}$$

We think of the residue classes $\mathscr{D}(p)$ as being *deleted*, so that $\mathscr{R}(p)$ is the set of residue classes that *remain*. We note that if we replace the c_n by $c_n e(n\beta)$, then the numbers $\delta(p)$ are unchanged, so that not only do we have (19.17), but more generally,

$$\sum_{\substack{a=1\\(a,q)=1}}^{q} |T(a/q+\beta)|^2 \ge |T(\beta)|^2 \prod_{p|q} \frac{\delta(p)}{p-\delta(p)} \tag{19.18}$$

for any real number β.

Proof We proceed by induction on the number of primes dividing q. The assertion is trivial when $q = 1$. Suppose that q is prime, say $q = p$. By Lemma 19.6 we know that

$$\sum_{a=1}^{p-1} |T(a/p)|^2 = p\sum_{h=1}^{p} \left|Z(p,h) - \frac{Z}{p}\right|^2. \tag{19.19}$$

Clearly,

$$p\sum_{h\in\mathscr{D}(p)} \left|Z(p,h) - \frac{Z}{p}\right|^2 = p\sum_{h\in\mathscr{D}(p)} \left|\frac{Z}{p}\right|^2 = |Z|^2\frac{\delta(p)}{p}. \tag{19.20}$$

On the other hand,

$$\sum_{h\in\mathscr{R}(p)} \left(Z(p,h) - \frac{Z}{p}\right) = Z - (p-\delta(p))\frac{Z}{p} = \frac{\delta(p)}{p}Z,$$

so by Cauchy's inequality,

$$\frac{\delta(p)^2}{p^2}|Z|^2 = \left|\sum_{h\in\mathscr{R}(p)} \left(Z(p,h) - \frac{Z}{p}\right)\right|^2 \le (p-\delta(p))\sum_{h\in\mathscr{R}(p)} \left|Z(p,h) - \frac{Z}{p}\right|^2.$$

Thus

$$p\sum_{h\in\mathscr{R}(p)} \left|Z(p,h) - \frac{Z}{p}\right|^2 \ge \frac{\delta(p)^2}{p(p-\delta(p))}|Z|^2.$$

On combining this with (19.20), we find that

$$p\sum_{h=1}^{p}\left|Z(p,h)-\frac{Z}{p}\right|^2 \ge |T(0)|^2\frac{\delta(p)}{p-\delta(p)},$$

which is (19.17) when q is prime.

Now suppose that q is the product of two or more primes, so that we may write $q = q_1q_2$ with $(q_1, q_2) = 1$, $q_1 > 1$, $q_2 > 1$. Since q_1 and q_2 each has fewer prime factors than q, by the inductive hypothesis we know that the inequality (19.17) (and hence also (19.18)) hold for q_1 and for q_2. By the Chinese remainder theorem we see that

$$\sum_{\substack{a=1\\(a,q)=1}}^{q}|T(a/q)|^2 = \sum_{\substack{a_1=1\\(a_1,q_1)=1}}^{q_1}\ \sum_{\substack{a_2=1\\(a_2,q_2)=1}}^{q_2}|T(a_1/q_1+a_2/q_2)|^2.$$

By taking $\beta = a_1/q_1$ in (19.18) we see that the above is

$$\ge \sum_{\substack{a_1=1\\(a_1,q_1)=1}}^{q_1}|T(a_1/q_1)|^2\prod_{p|q_2}\frac{\delta(p)}{p-\delta(p)}.$$

By (19.17) this is

$$\ge |T(0)|^2\prod_{p|q_1}\frac{\delta(p)}{p-\delta(p)}\prod_{p|q_2}\frac{\delta(p)}{p-\delta(p)} = |T(0)|^2\prod_{p|q}\frac{\delta(p)}{p-\delta(p)},$$

so the induction is complete. □

Theorem 19.13 *Let $\mathcal{N}$ be a set of Z integers in the interval $M+1 \le n \le M+N$. For each prime p let $\delta(p)$ denote the number of residue classes* (mod p) *not represented by any member $n \in \mathcal{N}$. Then for any integer $Q \ge 1$,*

$$Z \le \frac{N+Q^2}{L},$$

where

$$L = \sum_{q\le Q}\mu^2(q)\prod_{p|q}\frac{\delta(p)}{p-\delta(p)}.$$

Precisely the same estimate can be obtained by Selberg's sieve (see Theorem 21.1), if Theorem E.5 is used to eliminate the non-diagonal terms (see Exercise 19.2.1.5 below).

Proof By Lemma 19.12 it is clear that

$$Z^2\mu^2(q)\prod_{p|q}\frac{\delta(p)}{p-\delta(p)} \le \sum_{\substack{a=1\\(a,q)=1}}^{q} |T(a/q)|^2. \tag{19.21}$$

We sum this over $q \le Q$ and apply Corollary 19.5 to see that

$$Z^2 L \le (N+Q^2)Z.$$

If $Z = 0$, then there is nothing to prove. If $Z > 0$, then we cancel Z from both sides to obtain the stated inequality. □

We now give a second proof of Lemma 19.12 by exhibiting an explicit expression for the difference between the two sides of the inequality (19.17).

Theorem 19.14 (Huxley 1972) *For $M+1 \le n \le M+N$ let the numbers c_n be given. For each prime p let $\mathscr{D}(p)$ be the collection of those residue classes d* (mod p) *for which $c_n = 0$ whenever $n \equiv d$* (mod p)*. Let $\delta(p) = \operatorname{card}\mathscr{D}(p)$, and let $\mathscr{R}(p)$ be the complementary set of $p - \delta(p)$ residue classes* (mod p)*. For general q let $\mathscr{R}(q)$ be the collection of those residue classes r* (mod q) *such that $r \in \mathscr{R}(p)$ for all $p|q$. Put $r(q) = \operatorname{card}\mathscr{R}(q)$. Finally, let $T(x)$, $Z(q,h)$, and Z be defined as in* (19.1), (19.9), *and* (19.11), *respectively. If q is square-free, then*

$$\sum_{\substack{a=1\\(a,q)=1}}^{q} |T(a/q)|^2 = |Z|^2\prod_{p|q}\frac{\delta(p)}{p-\delta(p)} + \sum_{\substack{a=1\\(a,q)=1}}^{q}\left|\sum_{r\in\mathscr{R}(q)} e(ar/q)\big(Z(q,r)-Z/r(q)\big)\right|^2. \tag{19.22}$$

Proof We first show that if $k \in \mathscr{R}(q)$, then

$$\sum_{h\in\mathscr{R}(q)} c_q(h-k) = \prod_{p|q}\delta(p), \tag{19.23}$$

where $c_q(n)$ is Ramanujan's sum (cf. §4.1). We recall from Theorem 4.1 that $q_q(n) = \sum_{d|(q,n)} d\mu(q/d)$. Thus the left-hand side above is

$$= \sum_{h\in\mathscr{R}(q)}\sum_{\substack{d|q\\d|(h-k)}} d\mu(q/d) = \sum_{d|q} d\mu(q/d)\sum_{\substack{h\in\mathscr{R}(q)\\h\equiv k\ (d)}} 1.$$

In the inner sum, h (mod p) is fixed if $p|d$, but is free to take on any value in $\mathscr{R}(p)$ if $p \nmid d$. Thus there are $\prod_{p|q/d}(p-\delta(p))$ such h, and hence the expression above is

$$= \sum_{d|q} d\mu(q/d) \prod_{p|q/d} (p - \delta(p)) = \prod_{p|q} (p - (p - \delta(p))),$$

and so we have (19.23).

To establish (19.22) we expand the second term on the right-hand side, and find that it is

$$\sum_{\substack{a=1\\(a,q)=1}}^{q} \Big| \sum_{h\in\mathcal{R}(q)} Z(q,h)e(ah/q) \Big|^2$$
$$- 2\operatorname{Re} \frac{\overline{Z}}{r(q)} \sum_{\substack{a=1\\(a,q)=1}}^{q} \sum_{h\in\mathcal{R}(q)} Z(q,h) \sum_{k\in\mathcal{R}(q)} e(a(h-k)/q)$$
$$+ \frac{|Z|^2}{r(q)^2} \sum_{\substack{a=1\\(a,q)=1}}^{q} \Big| \sum_{h\in\mathcal{R}(q)} e(ah/q) \Big|^2$$
$$= T_1 - 2\operatorname{Re} T_2 + T_3,$$

say. Clearly T_1 is equal to the left-hand side of (19.22). By taking the sum over a inside and applying (19.23), we see that

$$T_2 = \frac{\overline{Z}}{r(q)} \sum_{h\in\mathcal{R}(q)} Z(q,h) \prod_{p|q} \delta(p) = \overline{Z}\Big(\prod_{p|q} \frac{\delta(p)}{p-\delta(p)}\Big) \sum_{h\in\mathcal{R}(q)} Z(q,h)$$
$$= |Z|^2 \prod_{p|q} \frac{\delta(p)}{p-\delta(p)}.$$

Finally, we see that

$$T_3 = \frac{|Z|^2}{r(q)^2} \sum_{k\in\mathcal{R}(q)} \sum_{h\in\mathcal{R}(q)} c_q(h-k),$$

which by (19.23) is

$$= \frac{|Z|^2}{r(q)^2} \sum_{k\in\mathcal{R}(q)} \prod_{p|q} \delta(p).$$

The number of terms in the sum is $r(q)$, so

$$T_3 = |Z|^2 \prod_{p|q} \frac{\delta(p)}{p-\delta(p)}.$$

On combining these observations we obtain (19.22) and the proof is complete. □

19.2.1 Exercises

1. Let $\mathcal{Q}$ be a set of pairwise coprime positive integers not exceeding Q, suppose that $T(x)$ is given as in (19.1), and that $Z(q,h)$ is defined by (19.9).

   (a) Show that
$$\sum_{q\in\mathcal{Q}}\sum_{a=1}^{q-1}|T(a/q)|^2 \le (N+Q^2)\sum_{n=M+1}^{M+N}|c_n|^2.$$

   (b) Show that
$$\sum_{q\in\mathcal{Q}} q\sum_{h=1}^{q}|Z(q,h)-Z/q|^2 \le (N+Q^2)\sum_{n=M+!}^{M+N}|c_n|^2.$$

   (c) Show that this includes Theorem 19.7.

2. Let $\mathcal{R}(p)$ and $\delta(p)$ be defined as in Theorem 19.14. Show that if q is square-free, then
$$\sum_{h=1}^{q}\left(\prod_{\substack{p|q\\ h\in\mathcal{R}(p)}}\frac{\delta(p)}{p-\delta(p)}\right)^2 = q\prod_{p|q}\frac{\delta(p)}{p-\delta(p)}.$$

3. (Montgomery 1968) Let $T(x)$ be defined as in (19.1), and $Z(q,h)$ be defined as in (19.9). Put $f(a)=T(a/q)$ if $(a,q)=1$, $f(a)=0$ otherwise. Let $\widehat{f}(h)=\frac{1}{q}\sum_{a=1}^{q}f(a)e(-ah/q)$ be the discrete Fourier transform of f.

   (a) Show that
$$\widehat{f}(h)=\frac{1}{q}\sum_{d|q}\mu(q/d)dZ(d,h).$$

   (b) Deduce that
$$\sum_{\substack{a=1\\(a,q)=1}}^{q}|T(a/q)|^2=\frac{1}{q}\sum_{h=1}^{q}\left|\sum_{d|q}\mu(q/d)dZ(d,h)\right|^2.$$

4. Let $w_\pm(n)$ be defined as in Theorem E.5. Show that if $q\le 1/\delta$, then
$$\sum_{n\equiv a\,(q)}w_\pm(n)=W_\pm(0)/q$$
for all $a \pmod q$.

5. Let w_+ and W_+ be as in Theorem E.5, let $f(x)$ be a polynomial with integral coefficients, and let $\delta(p)$ denote the number of solutions of the congruence $f(x) \equiv 0 \pmod{p}$. Suppose that λ_d is real and subject to the conditions $\lambda_1 = 1$, $\lambda_d = 0$ for $d > z$. Assume that P is a positive square-free integer.

   (a) Explain why

$$\sum_{\substack{n=M+1 \\ (f(n),P)=1}}^{M+N} 1 \le \sum_{\substack{n \\ (f(n),P)=1}} w_+(n) \le \sum_n w_+(n) \Biggl(\sum_{\substack{d|f(n) \\ d|P}} \lambda_d \Biggr)^2. \tag{19.24}$$

   (b) Show that if q is square-free, then

$$\sum_{\substack{n \\ q|f(n)}} w_+(n) = W_+(0)\frac{\delta(q)}{q}$$

   if $q \le 1/\delta$, where $\delta(q) = \prod_{p|q} \delta(p)$.

   (c) Explain why it may be assumed that $\delta(p) > 0$ for all $p|P$.

   (d) Set $\delta = z^{-2}$, and show that the right-hand side of (19.24) is

$$= (N - 1 + z^2) \sum_{\substack{d_i|P \\ d_i \le z \\ i=1,2}} \frac{\lambda_{d_1}\delta(d_1)}{d_1} \frac{\lambda_{d_2}\delta(d_2)}{d_2} \frac{(d_1, d_2)}{\delta((d_1, d_2))}. \tag{19.25}$$

   (e) Put $g(q) = \prod_{p|q}(p - \delta(p))/\delta(p)$. Show that if q is square-free, then

$$\sum_{d|q} g(d) = \frac{q}{\delta(q)}.$$

   (f) Show that the right-hand side of (19.25) is

$$= (N - 1 + z^2) \sum_{\substack{q|P \\ q \le z}} g(q) y_q^2, \tag{19.26}$$

   where

$$y_q = \sum_{\substack{d \\ q|d|P \\ d \le z}} \frac{\lambda_d \delta(d)}{d}. \tag{19.27}$$

(g) Show that if the y_q are given as in part (f), then

$$\lambda_d = \frac{d}{\delta(d)} \sum_{\substack{q \\ d|q|P \\ q\le z}} \mu(q/d) y_q. \tag{19.28}$$

(h) Show that the y_q are real, that $y_q = 0$ if $q > z$, and that

$$\sum_{\substack{q|P \\ q\le z}} \mu(q) y_q = 1.$$

(i) Show that configurations of y_q with the properties described in the preceding part are in one-to-one correspondence with admissible choices of the λ_d.

(j) Show that the sum in (19.26) is

$$= \sum_{\substack{q|P \\ q\le z}} g(q)\left(y_q - \frac{\mu(q)}{g(q)L}\right)^2 + \frac{1}{L},$$

where

$$L = \sum_{\substack{q|P \\ q\le z}} \frac{\mu^2(q)}{g(q)}.$$

(k) Show that $y_q = \mu(q)/(g(q)L)$ is an admissible choice of the y_q, and hence deduce that

$$\operatorname{card}\{n \in [M+1, M+n] : (f(n), P) = 1\} \le \frac{N - 1 + z^2}{L}.$$

(l) Suppose that P is a positive integer, and that for each prime $p|P$ a set $\mathscr{D}(p)$ of $\delta(p)$ residue classes is given. Show that there is a polynomial $f(x)$ such that if $p|P$, then $f(x) \equiv 0 \pmod{p}$ if and only if $x \in \mathscr{D}(p)$.

An old conjecture, which perhaps dates to Gauss, is that if a is a given integer, then there exist infinitely many primes for which a is a primitive root, unless $a = -1, 0, 1$ or a is a perfect square. Suppose now that a meets these requirements, and let $N_a(x)$ denote the number of primes not exceeding x for which a is a primitive root. *Artin's conjecture* is the statement that

$$N_a(x) \sim A(a)\frac{x}{\log x} \qquad (x \to \infty).$$

Artin overlooked some considerations, with the result that his proposed formula for the constant $A(a)$ was incorrect. The definition was amended by Heilbronn. Hooley (1967) showed that the (adjusted) Artin conjecture is true, provided that the Riemann Hypothesis for the Dedekind zeta functions of a certain family of Galois number fields is true. From the next exercise we find that any possible exceptions to Artin's Conjecture are quite rare.

6. (Gallagher 1967)

   (a) Let p be an odd prime. Prove that the number of primitive roots modulo p is $\varphi(p-1)$.

   (b) Use the Siegel–Walfisz theorem and the Brun–Titchmarsh inequality to show that

   $$\sum_{p \le X} \frac{\varphi(p-1)}{p-1} = c \operatorname{li} x + O\big(X(\log X)^{-A}\big)$$

   for $X \ge 2$, where $c = \prod_p \left(1 - \frac{1}{p(p-1)}\right)$.

   (c) In Theorem 19.7, let $\mathcal{N}$ be the set of those integers n, $1 \le n \le N$, such that n is not a primitive root (mod p) for any prime $p \le \sqrt{N}$. Set $Q = \lfloor \sqrt{N} \rfloor$. Explain why

   $$\sum_{h=1}^{p} (Z(p,h) - Z/p)^2 \ge Z^2 \frac{\varphi(p-1)}{p}$$

   for all $p \le Q$.

   (d) Conclude that card $\mathcal{N} \ll N^{1/2} \log N$. Note that $\mathcal{N}$ includes squares, so card $\mathcal{N} \gg N^{1/2}$. Vaughan (1973) derived a better bound for card $\mathcal{N}$ by arguing instead from Theorem 19.13.

7. Let p be a prime with $(p, 10) = 1$.

   (a) Let h be the order of 10 modulo p. Show that the decimal expansion of $1/p$ is periodic with least period h.

   (b) Deduce that the decimal expansion of $1/p$ has least period $p-1$ if and only if 10 is a primitive root of p. (The first such primes are 7, 17, 19, 23, 29, 47, ...)

8. Suppose that p and q are primes, with $p = 4q+1$. Show that 2 is a primitive root of p. (To show that there are infinitely many such (p, q) pairs would be similar to proving the twin prime conjecture. One would conjecture that there are infinitely many such pairs, the first few being (13,3), (29,7), (53,13), (149,37).)

9. (Vaughan 1973) Erdős (1947) conjectured that 7, 15, 21, 45, 75, and 105 are the only values of n for which $n - 2^k$ is prime for all positive integers

k for which this expression is positive. Let $E(N)$ be the number of such n not exceeding N. Prove that there is a positive constant c such that

$$E(N) \ll N \exp\left(-\frac{c(\log N)\log\log N}{\log\log N}\right).$$

10. Suppose that $k \geq 2$ and that $h_1, \ldots, h_k$ are k distinct non-negative admissible integers in the sense of Definition 18.1. Define $\nu_p(\boldsymbol{h})$ to be the number of different residue classes modulo p amongst the $\boldsymbol{h}$ and, when $N \in \mathbb{N}$, $R(N; \boldsymbol{h})$ to be the number of $n \leq N$ such that the $n + h_j$ are simultaneously prime.

    (a) Suppose that $Q \geq 1$. Prove that

    $$R(N; \boldsymbol{h}) \leq \frac{N + Q^2}{L(Q)} + O_k(Q),$$

    where

    $$L(Q) = \sum_{q \leq Q} \mu(q)^2 \prod_{P|q} \frac{\nu_p(\boldsymbol{h})}{p - \nu_p(\boldsymbol{h})}.$$

    (b) Suppose that $Q \geq 3$. Prove that

    $$L(Q) = \frac{(\log Q)^k}{k!\mathfrak{S}(\boldsymbol{h})} + O_{\boldsymbol{h}}\big((\log Q)^{k-1}\big),$$

    where $\mathfrak{S}(\boldsymbol{h})$ is given by (18.41).

    (c) Suppose that $N \geq 3$. Prove that

    $$R(N; \boldsymbol{h}) \leq 2^k k!\mathfrak{S}(\boldsymbol{h}) \frac{N}{(\log N)^k} + O_{\boldsymbol{h}}\left(\frac{N \log\log N}{(\log N)^{k-1}}\right).$$

11. (The 'Larger Sieve' of Gallagher 1971) Suppose that $Q \geq 1$ and $N \geq 1$ and M are integers, and $\{c_n\}$ is a sequence of non-negative real numbers such that $c_n > 0$ only when $M + 1 \leq n \leq M + N$. Define $Z(q, h) = \sum_{n \equiv h \,(\mathrm{mod}\, q)} c_n$, $Z = Z(1, 0)$, and let $\mathcal{A}_q$ be a set of residue classes h modulo q such that $Z(q, h) = 0$ when $h \notin \mathcal{A}_q$ and let $r(q) = \operatorname{card} \mathcal{A}_q$.

    (a) Suppose that $r(q) \neq 0$. Prove that

    $$\sum_{h \in \mathcal{A}_q} \left(Z(h, q) - \frac{Z}{r(q)}\right)^2 = \sum_{h \in \mathcal{A}_q} Z(q, h)^2 - \frac{Z^2}{r(q)}.$$

    (b) Let $\mathcal{Q}$ be a finite set of prime powers such that for $q \in \mathcal{Q}$, $r(q) \neq 0$. Group pairs n_1, n_2 of members of $[M + 1, M + N]$ according to their common difference, and hence show that

    $$\sum_{q \in \mathcal{Q}} \Lambda(q) \sum_{h \in \mathcal{A}_q} Z(h, q)^2 = \sum_{n_1, n_2} c_{n_1} c_{n_2} \sum_{\substack{q \in \mathcal{Q} \\ q | n_2 - n_1}} \Lambda(q).$$

(c) Prove that

$$\sum_{\substack{n_1,n_2\\ n_1\ne n_2}} c_{n_1}c_{n_2} \sum_{\substack{q\in\mathcal{Q}\\ q\mid n_2-n_1}} \Lambda(q) \le \left(Z^2 - \sum_n c_n^2\right)\log N.$$

(d) Deduce that

$$\sum_{q\in\mathcal{Q}} \Lambda(q) \sum_{h\in\mathcal{A}_q} Z(q,h)^2 \le \left(Z^2 - \sum_n c_n^2\right)\log N + \sum_n c_n^2 \sum_{q\in\mathcal{Q}} \Lambda(q),$$

and so

$$0 \le \left(Z^2 - \sum_n c_n^2\right)\log N + \sum_n c_n^2 \sum_{q\in\mathcal{Q}} \Lambda(q) - Z^2 \sum_{q\in\mathcal{Q}} \frac{\Lambda(q)}{r(q)}.$$

(e) Conclude that

$$Z^2 \le \frac{\sum_{q\in\mathcal{Q}} \Lambda(q) - \log N}{\sum_{q\in\mathcal{Q}} \Lambda(q)/r(q) - \log N} \sum_{n=M+1}^{M+N} c_n^2$$

provided that the denominator is positive, and that if $c_n = 0$ or 1 for every n, then

$$Z \le \frac{\sum_{q\in\mathcal{Q}} \Lambda(q) - \log N}{\sum_{q\in\mathcal{Q}} \Lambda(q)/r(q) - \log N}.$$

12. Let $\mathcal{N}$ denote the set of those integers n, $1 \le n \le N$, such that n is a quadratic or zero residue modulo p for every $p \le \sqrt{N}$. Show that card $\mathcal{N} \ll \sqrt{N}$. (This is best possible since squares $\le N$ are members of $\mathcal{N}$.)
13. (Kowalski & Vaughan 2025, §3, Vaughan 2014) Prove that if $n \in \mathbb{N}$, then the number $R(n)$ of solutions of $x^3 + y^2 = n$ in positive integers x and y satisfies $R(n) \ll n^{1/6}$.

19.3 Character Sums

Let

$$S(\chi) = \sum_{n=M+1}^{M+N} c_n\chi(n). \tag{19.29}$$

We reduce the question of the mean square size of $S(\chi)$ for primitive characters to the mean square size of the corresponding trigonometric polynomial.

Lemma 19.15 *Let $S(\chi)$ be defined as in (19.29), and $T(x)$ be defined as in (19.1). Then*

$$\frac{q}{\varphi(q)} \sum_{\chi}{}^{\star} |S(\chi)|^2 \le \sum_{\substack{a=1 \\ (a,q)=1}}^{q} |T(a/q)|^2, \tag{19.30}$$

where $\sum_{\chi}^{\star}$ denotes a sum over all primitive characters (mod q).

Proof We recall (cf. Theorem 9.7) that if χ is a primitive character (mod q), then $\chi(n)$ can be expressed in a simple way in terms of the additive characters $e(an/q)$, namely

$$\tau(\overline{\chi})\chi(n) = \sum_{a=1}^{q} \overline{\chi}(a) e(an/q) \tag{19.31}$$

for all n. On multiplying by c_n and summing, we see that

$$\tau(\overline{\chi}) S(\chi) = \sum_{a=1}^{q} \overline{\chi}(a) T(a/q).$$

From Theorem 9.7 we know that $|\tau(\chi)| = \sqrt{q}$ for all primitive χ, so on taking the modulus-squared and summing over primitive χ it follows that

$$q \sum_{\chi}{}^{\star} |S(\chi)|^2 = \sum_{\chi}{}^{\star} \left| \sum_{a=1}^{q} \overline{\chi}(a) T(a/q) \right|^2 .$$

On the right-hand side we drop the condition that χ be primitive, and invoke the orthogonality of characters (as expressed in (4.14)) to see that the above is

$$\le \sum_{\chi} \left| \sum_{a=1}^{q} \overline{\chi}(a) T(a/q) \right|^2 = \varphi(q) \sum_{\substack{a=1 \\ (a,q)=1}}^{q} |T(a/q)|^2.$$

This gives the stated result. □

On combining Lemma 19.15 with Corollary 19.5 we obtain the Dirichlet character version of the large sieve inequality.

Theorem 19.16 *Let M and N be integers with $N \ge 1$, and let $S(\chi)$ be defined as in (19.29). Then for any integer $Q \ge 1$,*

$$\sum_{q \le Q} \frac{q}{\varphi(q)} \sum_{\chi}{}^{\star} |S(\chi)|^2 \le (N + Q^2) \sum_{n=M+1}^{M+N} |c_n|^2$$

for arbitrary complex numbers c_n.

19.3.1 Exercises

1. Some parts of this exercise may be familiar. Let $s(n)$ denote the 'square-free part' of n, which is to say that $s(n)$ is the largest square-free divisor of n.
   (a) Show that if n is square-free, then
   $$\frac{1}{\varphi(n)} = \frac{1}{n}\prod_{p\mid n}\left(1+\frac{1}{p}+\frac{1}{p^2}+\cdots\right) = \sum_{\substack{m>0\\ s(m)=n}}\frac{1}{m}.$$
   (b) Deduce that
   $$\sum_{n\le x}\frac{\mu(n)^2}{\varphi(n)} = \sum_{\substack{m>0\\ s(m)\le x}}\frac{1}{m}.$$
   (c) Show that the above is
   $$\ge \sum_{m\le x}\frac{1}{m}.$$
   (d) By considering the sum above to be a Riemann sum, show that
   $$\sum_{n\le x}\frac{\mu(n)^2}{\varphi(n)} \ge \log x.$$
   (e) (van Lint & Richert 1965) Show that if q is a positive integer, then
   $$\sum_{n\le x}\frac{\mu(n)^2}{\varphi(n)} \le \left(\sum_{d\mid q}\frac{\mu(d)^2}{\varphi(d)}\right)\left(\sum_{\substack{m\le x\\ (m,q)=1}}\frac{\mu(m)^2}{\varphi(m)}\right).$$
   (f) Conclude that
   $$\sum_{\substack{m\le x\\ (m,q)=1}}\frac{\mu(m)^2}{\varphi(m)} \ge \frac{\varphi(q)}{q}\log x.$$
2. (Bombieri & Davenport 1968) Recall from Theorem 9.5 that (19.31) holds for all χ modulo q, if $(n,q)=1$.
   (a) Show that if $c_n = 0$ whenever $(n,q) > 1$, then
   $$\sum_{\chi}|\tau(\chi)|^2|S(\chi)|^2 = \varphi(q)\sum_{a=1}^{q}{}^{\star}\,|T(a/q)|^2.$$
   (b) Suppose that $c_n = 0$ whenever n has a prime factor $\le Q$. Show that
   $$\sum_{q\le Q}\frac{1}{\varphi(q)}\sum_{\chi}|\tau(\chi)|^2|S(\chi)|^2 \le (N+Q^2)\sum_{n=M+1}^{M+N}|c_n|^2.$$

(c) Suppose that $c_n = 0$ whenever $(n, q) > 1$, and that the character $\chi \pmod q$ is induced by the primitive character $\chi^\star \pmod d$. Show that $S(\chi) = S(\chi^\star)$.

(d) Recall from Theorem 9.10 that

$$\tau(\chi) = \begin{cases} \tau(\chi^\star)\mu(q/d)\chi^\star(q/d) & \text{if } (q/d, d) = 1, \\ 0 & \text{otherwise.} \end{cases}$$

Also, recall from Theorem 9.7 that $|\tau(\chi)| = \sqrt{q}$ if χ is a primitive character modulo q. Show that if the c_n are as in (b), then

$$\sum_{q\le Q} \frac{q}{\varphi(q)} \Big({\sum_{\chi}}^{\star} |S(\chi)|^2 \Big) \Bigg(\sum_{\substack{k\le Q/q \\ (k,q)=1}} \frac{\mu(k)^2}{\varphi(k)} \Bigg) \le (N + Q^2) \sum_{n=M+1}^{M+N} |c_n|^2.$$

(e) Show that if the c_n are as in (b), then

$$\sum_{q\le Q} (\log Q/q) {\sum_{\chi}}^{\star} |S(\chi)|^2 \le (N + Q^2) \sum_{n=M+1}^{M+N} |c_n|^2.$$

(f) Let $\mathcal{N}$ be the set of those integers $n \in [M+1, M+N]$ such that $(n, q) = 1$ for all $q \le Q$. Put $Z = \operatorname{card} \mathcal{N}$. Show that

$$Z^2 \log Q + \sum_{1<q\le Q} (\log Q/q) {\sum_{\chi}}^{\star} \Big| \sum_{n\in\mathcal{N}} \chi(n) \Big|^2 \le (N + Q^2) Z.$$

(g) Now suppose that $M = 0$, that $Q = N^{1/2}/\log N$, that $c_p = \log p$ for $N^{1/2} < p \le N$, and that $c_n = 0$ otherwise. Then $\sum_{n=1}^{N} |c_n|^2 = N \log N + O(N)$, and the first term on the left-hand side is $\sim \frac{1}{2} N^2 \log N$. If there exists an exceptional real character χ_1 with conductor $q_1 < N^\varepsilon$, then this character also contributes an amount $\sim \frac{1}{2} N^2 \log N$. The consequence is that the combined contribution of all other primitive characters is $o\big(N^2 \log N\big)$.

3. (Erdős & Shapiro 1957) Let χ be a primitive character modulo q.

(a) Show that

$$\sum_{m=1}^{q} \chi(m + n_1)\overline{\chi}(m + n_2) = c_q(n_1 - n_2)$$

where c_q is Ramanujan's sum, as defined in (4.5). (Suggestion: Write $\overline{\chi}$ in terms of additive characters, as in Corollary 9.8.)

(b) Deduce that for arbitrary numbers b_n,

$$\sum_{m=1}^{q}\left|\sum_{n=1}^{q} b_n\chi(m+n)\right|^2 = \sum_{\substack{a=1\\(a,q)=1}}^{q}\left|\sum_{n=1}^{q} b_n e(an/q)\right|^2.$$

(c) Explain why the right-hand side above is

$$\le q\sum_{n=1}^{q}|b_n|^2. \tag{19.32}$$

(d) Show that

$$\left|\sum_{m=1}^{q}\sum_{n=1}^{q} a_m b_n\chi(m+n)\right| \le \sqrt{q}\left(\sum_{m=1}^{q}|a_m|^2\right)^{1/2}\left(\sum_{n=1}^{q}|b_n|^2\right)^{1/2}$$

for arbitrary numbers a_m and b_n.

(e) Show that equality holds in (19.32) if $b_n = e(cn/q)$ with $(c,q) = 1$.

(f) Show that equality holds in (19.32) if $b_n = \chi(n)$.

(g) Compare the results here with those of Exercise G.2.1.17.

4. (Norton 1972) Let χ be a primitive character modulo q.

(a) Show that if $1 \le h \le q$, then

$$\sum_{n=1}^{q}\left|\sum_{m=1}^{h}\chi(m+n)\right|^2 = \sum_{\substack{a=1\\(a,q)=1}}^{q}\left|\sum_{n=1}^{h} e(an/q)\right|^2.$$

(b) Deduce that

$$\sum_{n=1}^{q}\left|\sum_{m=1}^{h}\chi(m+n)\right|^2 = qh - h^2 - \sum_{\substack{1<a<q\\(a,q)>1}}\left(\frac{\sin \pi ah/q}{\sin \pi a/q}\right)^2.$$

(Norton conjectured that the left-hand side above is $< qh$ for all non-principal χ; this was proved by Burgess 1975.)

(c) Show that if χ is non-principal (mod p) and $1 \le h \le p$, then

$$\sum_{n=1}^{p}\left|\sum_{m=1}^{h}\chi(m+n)\right|^2 = ph - h^2.$$

5. Suppose that $q > 1$ is an integer, that $(b,q) = 1$, and that b has order h modulo q. Show that

$$\prod_{\chi}(1-\chi(b)z) = (1-z^h)^{\varphi(q)/h}$$

for all z. (Hint: Recall Exercise 4.2.1.4(c).)

6. Suppose that χ is a character modulo q, and that h is the order of χ.

   (a) Show that for each integer a, the number of residue classes modulo q for which $\chi(n) = e(a/h)$ is exactly $\varphi(q)/h$.

   (b) Show that

$$\prod_{\substack{n=1\\(n,q)=1}}^{q} (1 - \chi(n)z) = \left(1 - z^h\right)^{\varphi(q)/h}.$$

19.4 Maximal Variants

We begin with a somewhat inferior bound, but one that suffices in many applications.

Theorem 19.17 (Uchiyama 1972) *For given real or complex numbers c_n, let*

$$T^*(x) = \max_{1\le n\le N}\left|\sum_{m=M+1}^{M+n} c_m e(mx)\right|.$$

Suppose that $x_1, x_2, \ldots, x_K$ are well spaced to the extent that $\|x_j - x_k\| \ge \delta$ for $j \ne k$. Then

$$\sum_{k=1}^{K} T^*(x_k)^2 \ll \left(N\log 2N + \delta^{-1}(\log 2N)^2\right)\sum_{n=M+1}^{M+N} |c_n|^2.$$

Proof From (E.29) we see that

$$T^*(x_k)^2 \le R\sum_{r=1}^{R}\sum_{s=0}^{2^{r-1}-1}\left|\sum_{\frac{Ns}{2^{r-1}}<n\le\frac{Ns}{2^{r-1}}+\frac{N}{2^r}} c_n e(nx_k)\right|^2,$$

where $R = \lceil(\log N)/(\log 2)\rceil$. Thus

$$\sum_{k=1}^{K} T(x_k)^2 \le R\sum_{r=1}^{R}\sum_{s=0}^{2^{r-1}-1}\sum_{k=1}^{K}\left|\sum_{M+\frac{Ns}{2^{r-1}}<n\le M+\frac{Ns}{2^{r-1}}+\frac{N}{2^r}} c_n e(nx_k)\right|^2, \qquad (19.33)$$

which by Theorem 19.4 is

$$
\begin{aligned}
&\le R\sum_{r=1}^{R}\sum_{s=0}^{2^{r-1}-1}\left(\frac{N}{2^r}+\delta^{-1}\right)\sum_{M+\frac{Ns}{2^{r-1}}<n\le M+\frac{Ns}{2^{r-1}}+\frac{N}{2^r}}|c_n|^2\\
&\le \sum_{r=1}^{R}\left(\frac{N}{2^r}+\delta^{-1}\right)\sum_{n=M+1}^{M+N}|c_n|^2\\
&\ll R\big(N\log 2N+\delta^{-1}(\log 2N)^2\big)\sum_{n=M+1}^{M+N}|c_n|^2. \qquad \square
\end{aligned}
$$

For $f\in L^1(\mathbb{T})$, let $\widehat{f}(n)=\int_0^1 f(x)e(-nx)\,dx$ denote its Fourier coefficients, and set

$$
s_N(f;x)=\sum_{n=-N}^{N}\widehat{f}(n)e(nx), \qquad s^\star(f,x)=\sup_{N\ge 1}|s_N(f;x)|.
$$

Kolmogorov (1926) exhibited an $f\in L^1(\mathbb{T})$ for which the sequence $s_N(f,\ x)$ diverges for all x, but Carleson (1966) showed that $s_N(f;x)\to f(x)$ as $N\to\infty$ for almost all x, provided that $f\in L^2(\mathbb{T})$. Hunt (1968) extended this to $f\in L^p(\mathbb{T})$ for all $p>1$, and established a quantitative inequality: $\|s^\star(f)\|_p\ll_p\|f\|_p$. The case $p=2$ is particularly useful for us: There is an absolute constant C_{H} ('Hunt's constant') such that

$$
\int_0^1\max_{1\le K\le N}\left|\sum_{n=1}^{K}a_ne(nx)\right|^2dx\le C_{\mathrm{H}}\sum_{n=1}^{N}|a_n|^2 \tag{19.34}
$$

for arbitrary complex numbers a_n. We now use this bound to derive a more precise maximal variant of the large sieve.

Theorem 19.18 *Let $x_1,x_2,\ldots,x_R$ be points of $\mathbb{T}$ that satisfy* (19.2), *let C_{H} be defined as in* (19.34), *and let a_n be arbitrary complex numbers, for $M+1\le n\le M+N$. Then*

$$
\sum_{r=1}^{R}\max_{1\le K\le N}\left|\sum_{n=M+1}^{M+K}a_ne(nx_r)\right|^2\le C_{\mathrm{H}}(\delta^{-1}+\pi N)\sum_{n=M+1}^{M+N}|a_n|^2.
$$

Proof Let

$$
S(x)=\sum_{n=M+1}^{M+N}a_ne(nx).
$$

If we replace $S(x)$ by $e(-Lx)S(x)$, then each partial sum is multiplied by the unimodular factor $e(-Lx)$. Thus the size of the largest partial sum, and its length, are unchanged. Hence through a suitable choice of L we may assume that the interval $[M+1, M+N]$ is a subset of $[-N/2, N/2]$. Also, by replacing a_{n+L} by a_n we can suppose that the support of the a_n lies in $[-N/2, N/2]$. Let $K(x)$ be chosen so that

$$\left|\sum_{n=M+1}^{M+K(x)} a_n e(nx)\right| = \max_{1\le K\le N}\left|\sum_{n=M+1}^{M+K} a_n e(nx)\right|,$$

and let $S^\star(x)$ denote this common value. Here $K(x)$ is piecewise constant, with at most finitely many jump discontinuities. The function $S^\star(x)$ is continuous, and differentiable except possibly at the discontinuities of $K(x)$. By Lemma 19.2 it follows that

$$\begin{aligned}\sum_{r=1}^{R} |S^\star(x_r)|^2 \le \frac{1}{\delta}\int_0^1 |S^\star(x)|^2\,dx \\ + \left(\int_0^1 |S^\star(x)|^2\,dx\right)^{1/2}\left(\int_0^1 \left|\frac{d}{dx}S^\star(x)\right|^2\,dx\right)^{1/2}.\end{aligned}$$

From (19.34) we see that

$$\int_0^1 |S^\star(x)|^2\,dx \le C_{\mathrm{H}} \sum_{n=M+1}^{M+N} |a_n|^2.$$

If $f(x)$ is a complex-valued differentiable function of the real variable x, then $\left|\frac{d}{dx}|f(x)|\right| \le |f'(x)|$. Hence

$$\begin{aligned}\left|\frac{d}{dx}S^\star(x)\right| \le \left|\frac{d}{dx}\sum_{n=M+1}^{M+K(x)} a_n e(nx)\right| = \left|\sum_{n=M+1}^{M+K(x)} 2\pi i n a_n e(nx)\right| \\ \le 2\pi \max_{1\le K\le N}\left|\sum_{n=M+1}^{M+K} n a_n e(nx)\right|.\end{aligned}$$

From (19.34) we deduce that

$$\int_0^1 \left|\frac{d}{dx}S^\star(x)\right|^2\,dx \le 4\pi^2 C_{\mathrm{H}} \sum_{n=M+1}^{M+N} |na_n|^2 \le \pi^2 N^2 C_{\mathrm{H}} \sum_{n=M+1}^{M+N} |a_n|^2$$

since $|n| \le N/2$ when $M+1 \le n \le M+N$. These bounds combine to give the stated result. □

Theorem 19.19 *If M, N, and Q are positive integers, then*

$$\sum_{q\le Q}\frac{q}{\varphi(q)}\sum_{\chi}{}^{\star}\sup_{y}\left|\sum_{\substack{1\le m\le M\\1\le n\le N\\mn\le y}}a_m b_n\chi(mn)\right|$$
$$\ll\left(M+Q^2\right)^{1/2}\left(N+Q^2\right)^{1/2}\left(\sum_{m=1}^{M}|a_m|^2\right)^{1/2}\left(\sum_{n=1}^{N}|b_n|^2\right)^{1/2}\log 2MN \quad (19.35)$$

for arbitrary complex numbers a_m and b_n.

Proof By Cauchy's inequality and the large sieve in the form of Theorem 19.16, we see that

$$\sum_{q\le Q}\frac{q}{\varphi(q)}\sum_{\chi}{}^{\star}\left|\sum_{m=1}^{M}\sum_{n=1}^{N}a_m b_n\chi(mn)\right|$$
$$\le\left(\sum_{q\le Q}\frac{q}{\varphi(q)}\sum_{\chi}{}^{\star}\left|\sum_{m=1}^{M}a_m\chi(m)\right|^2\right)^{1/2}\left(\sum_{q\le Q}\frac{q}{\varphi(q)}\sum_{\chi}{}^{\star}\left|\sum_{n=1}^{N}b_n\chi(n)\right|^2\right)^{1/2}$$
$$\ll\left(M+Q^2\right)^{1/2}\left(N+Q^2\right)^{1/2}\left(\sum_{m=1}^{M}|a_m|^2\right)^{1/2}\left(\sum_{n=1}^{N}|b_n|^2\right)^{1/2}. \quad (19.36)$$

In order to truncate this to $mn \le y$, we use a device discussed in Appendix E.4.1. Specifically, by (E.27) we find that

$$\sup_{y}\left|\sum_{\substack{1\le m\le M\\1\le n\le N\\mn\le y}}a_m b_n\chi(mn)\right|$$
$$\ll\int_{-T}^{T}\left|\sum_{\substack{1\le m\le M\\1\le n\le N}}a_m b_n\chi(mn)(mn)^{-it}\right|\min(\log MN, 1/|t|)\,dt$$
$$+\frac{MN}{T}\sum_{\substack{1\le m\le M\\1\le n\le N}}|a_m b_n|.$$

By Cauchy's inequality, the last term is

$$\ll\frac{M^{3/2}N^{3/2}}{T}\left(\sum_{m=1}^{M}|a_m|^2\right)^{1/2}\left(\sum_{n=1}^{N}|b_n|^2\right)^{1/2}.$$

In order that this term should not be troublesome, we take $T=(MN)^{3/2}$. Since

$$\int_{-T}^{T}\min(\log MN, 1/|t|)\,dt\ll\log(T\log 2MN),$$

the desired result follows from (19.36). □

19.4.1 Exercises

1. (Uchiyama 1972) Show that for arbitrary integers $M, N > 1$, $Q > 1$, and complex numbers a_n,

$$\sum_{q\le Q}\frac{q}{\varphi(q)}\sum_{\chi}{}^{\star}\max_{1\le\nu\le N}\left|\sum_{n=M+1}^{M+\nu}a_n\chi(n)\right|^2 \ll (Q^2(\log N)^2+N\log N)\sum_{n=M+1}^{M+N}|a_n|^2.$$

2. Let C_{H} be defined as in (19.34). Show that

$$\sum_{q\le Q}\frac{q}{\varphi(q)}\max_{1\le K\le N}\sum_{\chi}{}^{\star}\left|\sum_{n=M+1}^{M+K}a_n\chi(n)\right|^2 \le C_{\mathrm{H}}\big(\pi N+Q^2\big)\sum_{n=M+1}^{M+N}|a_n|^2$$

for arbitrary complex numbers a_n, $M+1\le n\le M+N$.

3. Show that

$$\sum_{q\le Q}\frac{q}{\varphi(q)}\sum_{\chi}{}^{\star}\max_{1\le K\le N}\left|\sum_{n=M+1}^{M+K}a_n\chi(n)\right|^2 \ll \big(N+Q^2\big)(\log N)^2\sum_{n=M+1}^{M+N}|a_n|^2$$

for arbitrary complex numbers a_n, $M+1\le n\le M+N$.

19.5 Notes

Section 19.1. The large sieve was introduced by Linnik (1941), with the aim of treating the least quadratic non-residue in Linnik (1942) already in mind. Rényi (1947, 1948) used the large sieve to show that every large even number $2n$ can be written in the form

$$2n = p + P_k, \tag{19.37}$$

where P_k denotes a number that is a product of at most k prime numbers. Rényi developed the large sieve in several papers, culminating in Rényi (1959), where the result, stronger than those in prior formulations, is equivalent to the assertion that the bound (G.9) implies the bound (G.7) in Theorem G.12. When the vectors $\boldsymbol{\phi}_r$ in that theorem are taken to have coordinates $\chi_r(n)$ for $M < n \le M+N$, and χ_r runs over all primitive characters with conductor

$q \le Q$, it follows by the Pólya–Vinogradov inequality (Theorem 9.18) that $\Delta^2 \ll N + Q^3 \log Q$. Consequently,

$$\sum_{q \le Q} \sum_{\chi}{}^{\star} \left| \sum_{n=M+1}^{M+N} c_n \chi_r(n) \right|^2 \ll (N + Q^3 \log Q) \sum_{n=M+1}^{M+N} |c_n|^2 \tag{19.38}$$

for arbitrary complex numbers c_n. Barban (1961, 1963) used the above estimate to show that most L functions do not have a zero at small height and real part near 1, and from those estimates deduced a bound of the form

$$\sum_{q \le x^{a-\varepsilon}} \mu(q)^2 \max_{\substack{h \,(\mathrm{mod}\, q)\\ (h,q)=1}} \left| \pi(x; q, h) - \frac{\operatorname{li} x}{\varphi(q)} \right| \ll x(\log x)^{-A} \tag{19.39}$$

with $a = 1/3$. By introducing an appeal to the sixth moment estimate of Linnik (1960), he improved this to $a = 3/8$. To obtain the representation (19.37), we fix a large even number $2n$, allow p to range over all primes $\le 2n$, and sift the numbers $2n - p$. To do this, we need to know, for small d, the number of p such that $d|(2n - p)$; that is, $\pi(2n; d, 2n)$. When $(d, 2n) = 1$, this number should be close to $(\operatorname{li} 2n)/\varphi(d)$ for most d. The small sieve only requires that the approximation here is good on average, as in the above bound. Thus Levin (1963a,b) showed that Rényi's theorem (19.37) holds for $k = 4$ if the above holds for some $a \ge 0.3058$ and for $k = 3$ if the above holds for some $a \ge 0.401$. Hence Barban's result implies that (19.37) holds with $k = 4$. Pan (1963) independently achieved $a = 3/8$ and $k = 4$. Wang (1960) earlier showed that GRH implies that one can take $k = 3$.

Barban (1966) wrote a detailed survey of the large sieve and its applications, as it existed up to 1964. However, Roth (1965) revolutionized the subject with the brilliant idea of taking the vectors $\boldsymbol{\phi}_r$ to have coordinates $e(nx_r)$ for $M < n \le M + N$. Thus $\boldsymbol{\phi}_r$ and $\boldsymbol{\phi}_s$ are nearly orthogonal if $\|x_r - x_s\|$ is large compared with $1/N$. These vectors are much closer to being orthogonal than were Rényi's. Arithmetic applications follow by taking the x_r to be Farey points of order Q, in which case the x_r differ by $1/Q^2$. Thus where Rényi had $Q^3 \log Q$, Roth had $Q^2 \log Q$. Bombieri (1965), working independently of Roth, reduced Rényi's $Q^3 \log Q$ to Q^2, and derived improved zero-density estimates for L-functions, which yielded (19.39) with $a = 1/2$. Previously, this was only known as a consequence of GRH. Conjecture 20.4 (known as the Elliott–Halberstam Hypothesis) asserts that (19.39) holds for $a = 1$, but this is not known to hold for any $a > 1/2$, even under the assumption of GRH.

Let f be a measurable function defined over $\mathbb{R}^n$. In seminal work, Sobolev (1938) initiated the study of bounds for norms of f in terms of norms of partial derivatives of f. Leoni (2017) and Saloff-Coste (2002) have provided introductions to this subject. Gallagher (1967) used Lemma 19.2 to prove Theorem 19.3. The idea that Lemma 19.1 can be used to derive a discrete mean upper bound at well-spaced points from a continuous mean value is invaluable, and has been used in many other situations. Theorem 19.4 in the slightly weaker form with $N + \frac{1}{\delta} - 1$ replaced by $N + \frac{1}{\delta}$ is due to Montgomery & Vaughan (1973, 1974). Paul Cohen (unpublished; see Exercise 19.1.1.9) observed that from the large sieve with this larger factor one can deduce the smaller one. The proof we give is due to Selberg (see Montgomery (1978)). The factor $N - 1 + 1/\delta$ is quite sharp when $N\delta$ is large, and indeed equality can be achieved when $(N-1)\delta$ is an integer, as we see in Exercise 19.1.1.11. Bombieri & Davenport (1969) showed that when $N\delta \le 1/4$, the large sieve (19.3) holds with $\Delta = \frac{1}{\delta}\big(1 + 270(N\delta)^3\big)$. Bombieri and Selberg (see Bombieri (1971) and Chapter 1 of Montgomery (1971)) were among the first to consider the use of bilinear forms (as we discuss in §G.1 and §G.2) in the context of the large sieve. Soon after, Matthews (1972a, 1972b, 1973), Kobayashi (1971, 1973), and Elliott (1971) also treated the large sieve in this way.

Section 19.2. Theorems of the general kind (Theorem 19.7 and Corollary 19.8) were first obtained by Linnik (1941) and developed by Rényi (1948, 1949a,b, 1950). Theorem 19.10 is in Linnik (1942). Erdős (1961) established the case $\gamma = 1$ of Theorem 19.11 and the argument displayed here is a simple generalization. Theorem 19.13 is due to Montgomery (1968), although the special case $\delta(p) = 1$ was obtained first by Bombieri & Davenport (1968). Montgomery established the critical inequality (19.17) by a judicious application of Cauchy's inequality coupled with Möbius inversion. Our simple proof of (19.17) is due to Gallagher. Bombieri (1974) gave an overview of the large sieve with a range of interesting applications.

Ramaré (2007, 2009, 2010) has studied the large sieve in great detail, while Baier (2006), Baier & Bansal (2020), Baier & Zhao (2005, 2008, 2012, 2019), Wolke (1971/1972), and Zhao (2004a,b, 2005, 2007) have considered sparse sets of moduli. Huxley (1968, 1970, 1971) extended the large sieve to algebraic number fields, and Hlawka (1970) established a version of the large sieve for $\mathbb{T}^n$. Kowalski (2006, 2008) has extended the large sieve in a number of directions, including to arithmetic geometry and to discrete groups. As it stands, the large sieve is only an upper bound, but Conrey, Iwaniec, & Soundararajan (2011, 2013, 2019) have constructed a more elaborate asymptotic large sieve.

The larger sieve of Gallagher (1971), established in Exercise 19.2.1.11 and subsequently applied, has had many applications and has been especially useful in establishing the density of square-free values of polynomials. See, for example, Hooley (1976, 2009) and Kowalski & Vaughan (2005).

Artin's conjecture, described in the preamble to Exercise 19.2.1.6, was communicated by Emil Artin to Helmut Hasse on September 27, 1927, according to the latter's diary. For more detail concerning the correction to the constant, see the introduction to Hooley (1967).

Section 19.3. Lemma 19.15 is due to Gallagher (1967).

Section 19.4. Theorem 19.17 is due to Uchiyama (1972). Hunt's theorem, even when restricted to $p = 2$, remains challenging to prove. For an accessible exposition of this, see Lacey (2004).

Maximal variants that flow from the Hardy–Littlewood maximal theorem are available without appealing to the Carleson–Hunt theorem, as follows: Let f be a bounded measureable function with period 1 and let the maximal function $M_f(x)$ be defined as in (E.28). Montgomery (1982) showed that if $\Delta(f,\delta)$ has the property that

$$\sum_{r=1}^{R} |f(x_r)|^2 \le \Delta(f,\delta)$$

whenever the x_r are well spaced as in (19.2), then

$$\sum_{r=1}^{R} M_f(x_r)|^2 \le 200\Delta(f,\delta).$$

Thus if T is defined as in (19.1), then by Theorem 19.4 it follows that

$$\sum_{r=1}^{R} M_T(x_r)|^2 \le 200\big(N+\delta^{-1}-1\big)\sum_{n=M+1}^{M+N} |c_n|^2. \tag{19.40}$$

The estimate (19.36) and Theorem 19.19 are Lemmas 1 and 2 of Vaughan (1980). For further discussion of the material considered in this section see Montgomery & Vaaler (1989).

The large sieve continues to be of central interest in analytic number theory. Some examples of recent work in the area are Baier & Bansal (2020), Boca & Radziwiłł (2020), Gao & Zhao (2022), and Kinnunen, Lehrbäck & Vähäkangas (2021).

19.6 References

Baier, S. (2006). On the large sieve with sparse sets of moduli, *J. Ramanujan Math. Soc.* **21**, 279–295.

Baier, S. & Bansal, A. (2020). Large sieve with sparse sets of moduli for $\mathbb{Z}[i]$, *Acta Arith.* **196**, 17–34.

Baier, S. & Zhao, Liangyi (2005). Large sieve inequality with characters for powerful moduli, *Int. J. Number Theory* **1**, 265–279.

(2008). An improvement for the large sieve for square moduli, *J. Number Theory* **128**, 154–174.

(2012). Large sieve inequalities for quartic character sums, *Q. J. Math.* **63**, 891–917.

(2019). A lower bound for the large sieve with square moduli, *Bull. Aust. Math. Soc.* **100**, 225–229.

Barban, M. B. (1961). New applications of the 'great sieve' of Ju. V. Linnik (Russian), *Akad. Nauk Uzbek. SSR Trudy Inst. Mat.* **22**, 1–20; A remark on the author's paper 'New applications of the "large sieve" of Ju. V. Linnik', *Theory Probability Math. Statist.*, Tashkent: Izdat. 'Nauka' Uzbek. SSR (1964), pp. 130–133.

(1963). The 'density' of the zeros of Dirichlet L-series and the problem of the sum of primes and 'near primes', *Mat. Sb. (N.S.)* **61** (103), 418–425.
(1966). The 'large sieve method' and its application to number theory (Russian), *Uspehi Akad. Nauk SSSR* **21**, 51–102; *Russian Math. Surveys* **21**, 49–103.
Boca, F. P. & Radziwiłł, M. (2020). Limiting distribution of eigenvalues in the large sieve matrix, *J. Eur. Math. Soc.* **22**, 2287–2329.
Bombieri, E. (1965). On the large sieve, *Mathematika* **12**, 201–225.
(1971). A notc on the large sieve, *Acta Arithmetica* **18**, 401–404.
(1974). *Le Grande Crible dans la Théorie Analytique des Nombres* (Première Édition), Astérisque **18**, Paris: Société Mathématique de France, i+87 pp.
Bombieri, E. & Davenport, H. (1968). On the large sieve method. In *Abhandlungen aus Zahlentheorie und Analysis: zur Erinnerung an Edmund Landau, 1877–1938*, Berlin: Deutch Verlag Wiss., pp. 11–22.
(1969). Some inequalities involving trigonometrical polynomials, *Ann. Scuola Norm. Sup. Pisa* (3) **23**, 223–241.
Burgess, D. A. (1971). The average of the least primitive root modulo p^2, *Acta Arith.* **18**, 263–271.
(1975). On a conjecture of Norton, *Acta Arith.* **27**, 265–267.
Carleson, L. (1966). On convergence and growth of partial sums of Fourier series, *Acta Math.* **116**, 135–157.
Conrey, J. B., Iwaniec, H., & Soundararajan, K. (2011). Asymptotic large sieve, *arXiv* **1105.1176** [math.NT], 26 pp.
(2013). Critical zeros of Dirichlet L-functions, *J. Reine Angew. Math.* **681**, 175–198.
(2019). The mean square of the product of a Dirichlet L-function and a Dirichlet polynomial, *Funct. Approx. Comment. Math.* **61**, 147–177.
Elliott, P. D. T. A. (1971). On inequalities of large sieve type, *Acta Arith.* **18**, 405–422.
Erdős, P. (1947). On the integers of the form $2^r + p$ and some related problems, *Summa Brasil. Math.* **2**, 113–123.
(1961). Remarks on number theory I (in Hungarian), *Mat. Lapok* **12**, 10–17.
Erdős, P. & Shapiro, H. N. (1957). On the least primitive root of a prime, *Pacific J. Math.* **7**, 861–865.
Gallagher, P. X. (1967). The large sieve, *Mathematika* **14**, 14–20.
(1971). A larger sieve, *Acta Arith.* **18**, 77–81.
Gao, P. & Zhao, L. (2022). The large sieve with power moduli in imaginary quadratic number fields (English summary), *Int. J. Number Theory* **18**(8), 1713–1733.
Hlawka, E. (1970). Bemerkungen zum großen Sieb von Linnik, *Österreich. Akad. Wiss. Math.-Natur. Kl. S.-B.* II **178**, 13–18.
Hooley, C. (1967). On Artin's conjecture, *J. Reine Angew. Math.* **225**, 209–220.
(1976). *Applications of Sieve Methods to the Theory of Numbers*, Cambridge Tracts in Math. 70, Cambridge: Cambridge University Press.
(2009). On the power-free values of polynomials in two variables: II, *J. Number Theory* **129**, 1443–1455.
Hunt, R. A. (1968). On the convergence of Fourier series. In *Orthogonal Expansions and Their Continuous Analogues* (Edwardsville, IL 1967), Carbondale: Southern Illinois University Press, pp. 235–255.
Huxley, M. N. (1968). The large sieve inequality for algebraic number fields, *Mathematika* **15**, 178–187.
(1970). The large sieve inequality for algebraic number fields, II. Means of moments of Hecke zeta-functions, *Proc. London Math. Soc.* (3) **21**, 108–128.
(1971). The large sieve inequality for algebraic number fields, III. Zero-density results, *J. London Math. Soc.* (2) **3**, 233–240.

(1972). Irregularity in sifted sequences, *J. Number Theory* **4**, 437–454.

Kinnunen, J., Lehrbäck, J.,& Vähäkangas, A. (2021). *Maximal Function Methods for Sobolev Spaces*, Math. Surveys Vol. 257, Providence: Amer. Math. Soc., xii+ 338 pp.

Kobayashi, I. (1971). Remarks on the large sieve method, *Seminar on Modern Methods in Number Theory*, Paper No. 21. Institute of Statistical Mathematics, Tokyo.

(1973). A note on the Selberg sieve and the large sieve, *Proc. Japan Acad.* **49**, 1–5.

Kolmogorov, A. N. (1926). Une série de Fourier–Lebesgue divergente partout, *Comptes Rendus Acad. Sci. Paris*, **183**, 1327–1328.

Kowalski, E. (2006). The large sieve, monodromy and zeta functions of curves, *J. Reine Angew. Math.* **601**, 29–69.

(2008). *The Large Sieve and Its Applications*, Cambridge Tracts in Math. 175, Cambridge: Cambridge University Press, xxi+293 pp.

Kowalski, J. M. & Vaughan, R. C. (2025). Squarefree density of polynomials II, *Math. Ann.* **391**(4), 6037–6056.

Lacey, M. T. (2004). Carleson's theorem: proof, complements, variations, *Publ. Mat.* **48**, 251–307.

Leoni, G. (2017). *A First Course in Sobolev Spaces*, Second edition, Grad. Stud. Math. 181, Providence: Amer. Math. Soc., 734 pp.

Levin, B. V. (1963a). *The Sieve Method and Its Applications*, Doctoral Dissertation, Moscow: Moscow State University.

(1963b). Distribution of 'near primes' in polynomial sequences, *Mat. Sb. (N.S.)* **61** (103), 389–407.

Linnik, Yu. V. (1941). The large sieve, *Doklady Akad. Nauk URSS (N.S.)* **30**, 292–294.

(1942). A remark on the least quadratic non-residue, *Doklady Akad. Nauk URSS (N.S.)* **36**, 119–120.

(1960). An asymptotic formula in the Hardy–Littlewood additive problem, *Izv. Akad. Nauk SSSR, Ser. Mat.* **24**. 629–706; *Soviet Math. Dokl.* **1**, 927–928.

van Lint, J. H. & Richert, H.-E. (1965). On primes in arithmetic progressions, *Acta Arith.* **11**, 209–216,

Matthews, K. R. (1972a). On an inequality of Davenport and Halberstam, *J. London Math. Soc.* **4**, 638–642.

(1972b). On a bilinear form associated with the large sieve, *J. London Math. Soc.* **5**, 567–570.

(1973). Hermitian forms and the large and small sieves, *J. Number Theory* **5**, 16–23.

Montgomery, H. L. (1968). A note on the large sieve, *J. London Math. Soc.* **43**, 93–98.

(1971). *Topics in Multiplicative Number Theory*, Lecture Notes in Math. 227, Berlin: Springer-Verlag, x+178 pp.

(1978). The analytic principle of the large sieve, *Bull. Amer. Math. Soc.* **84**, 547–567.

(1982). Maximal variants of the large sieve, *J. Fac. Sci. Univ. Tokyo 1A* **28**, 805–812.

Montgomery, H. L. & Vaaler, J. D. (1989). Maximal variants of basic inequalities. In *Proc. Congress on Number Theory* (Zarauz, 1984), Bilbao: Univ. País Vasco, pp. 181–197.

Montgomery, H. L. & Vaughan, R. C. (1973). The large sieve, *Mathematika* **20**, 119–134.

(1974). Hilbert's inequality, *J. London Math. Soc.* (2) **8** 73–82.

Norton, K. K. (1972). On character sums and power residues, *Trans. Amer. Math. Soc.* **167**, 203–226; Erratum, **174**, 507.

Pan, Cheng Dong (1963). On the representation of even numbers as the sum of a prime and a product of not more than 4 primes, *Sci. Sinica* **12**, 455–473.

Preissmann, E. (1985). On the norm of the matrix $(e(x_r y_s))$, *Linear and Multilinear Algebra* **17**, 39–40.

Ramaré, O. (2007). Eigenvalues in the large sieve inequality, *Funct. Approx. Comment. Math.* **37**, 399–427.

(2009). *Arithmetical Aspects of the Large Sieve Inequality*, Harish-Chandra Research Institute Lecture Notes, 1, New Delhi: Hindustan Book Agency, x+201 pp.

(2010). Eigenvalues in the large sieve inequality, II, *J. Théor. Nombres Bordeaux* **22**, 181–196.

Rényi, A. (1947). On the representation of an even number as the sum of a single prime and single almost-prime number (Russian), *Doklady Akad. Nauk SSSR. Ser. Mat. (N.S.)* **56**, 455–458.

(1948). On the representation of an even number as the sum of a single prime and single almost-prime number (Russian), *Izvestiya Akad. Nauk SSSR Ser. Mat.* **12**, 57–78; reprinted in *Twelve Papers on Number Theory and Function Theory*, Amer. Math. Soc. Transl. (2) **19** (1962), 299–321; `http://dx.doi.org/10.1090/trans2/019/12`.

(1949a). Probability methods in number theory, *Publ. Math. Collectae Budapest* **1**(21) 9 pp.

(1949b). Un nouveau théorème cocnernant les fonctions indépendantes et ses application à la théorie des nombres, *J. Math. Pures* (9) **28**, 137–149.

(1950). On the large sieve of Ju. V. Linnik, *Compositio Math.* **8**, 68–75.

(1959). New version of the probabilistic generalization of the large sieve (Russian summary), *Acta Math. Acad. Sci. Hungar.* **10**, 217–226.

Roth, K. F. (1965). On the large sieves of Linnik and Rényi, *Mathematika* **12**, 1–9.

Saloff-Coste, L. (2002). *Aspects of Sobolev-Type Inequalities*, London Math. Soc. Lecture Notes 289, Cambridge: Cambridge University Press.

Selberg, A. (1991). *Collected Papers*, Vol. II, Berlin: Springer.

Sobolev, S. L. (1938). On a theorem of functional analysis, *Mat. Sb. (N.S.)* **4**, 471–497; in *Eleven Papers on Analysis*, Amer. Math. Soc. Transl. Ser. 2, **34** (1963), pp. 39–68, Providence: Amer. Math. Soc. (1963), `http://dx.doi.org/10.1090/trans2/034/02`.

Uchiyama, S. (1972). The maximal large sieve, *Hokkaido Math. J.* **1**, 117–126.

Vaughan, R. C. (1973). Some applications of Montgomery's sieve, *J. Number Theory* **5**, 64–79.

(1980). An elementary method in prime number theory, *Acta Arith.* **37**, 111–115.

(2014). Integer points on elliptic curves, *Rocky Mountain J. Math.* **44**, 1377–1382.

Wang, Y. (1960). On the representation of large integer as a sum of a prime and an almost prime, *Acta Math. Sinica* **10**, 168–181; *Chinese Math.* **1**, 181–195.

Wolke, D. (1971/19722). On the large sieve with primes, *Acta Math. Acad. Sci. Hungar.* **22**, 239–247.

Zhao, L. (2004a). Large sieve inequalities for special characters to prime square moduli, *Funct. Approx. Comment. Math.* **32**, 99–106.

(2004b). Large sieve inequality for characters to square moduli, *Acta Arith.* **112**, 297–308.

(2005). An improvement of a large sieve inequality in high dimensions, *Mathematika* **52**, 93–100.

(2007). Large sieve inequality with quadratic amplitudes, *Monatsh. Math.* **151**, 165–173.

20

Primes in Arithmetic Progressions: III

20.1 Averages of $|\psi(x,\chi)|$

Our best unconditional bounds for $\psi(x,\chi)$, such as Theorem 11.16, are not very good owing to our rather limited knowledge of the zero-free region of $L(s,\chi)$. It is possible to extend the range for the modulus q of χ at the expense of the error term or by including zeros close to 1. See Corollary 11.20 or the chapter on Vinogradov's mean value theorem in the forthcoming Volume III. If we assume GRH, then we have a much better estimate (cf. Theorem 13.7). However, in some situations, a good bound for an average of $|\psi(x,\chi)|$ is all that is required, and such bounds can be obtained unconditionally by combining our methods of Chapter 17 with the large sieve.

As in §19.3, we let $\sum_\chi^\star$ denote a sum over all primitive characters modulo q. With this notation, we have the following theorem.

Theorem 20.1 *For arbitrary $Q \ge 1$ and $x \ge 2$,*

$$\sum_{q\le Q} \frac{q}{\varphi(q)} \sideset{}{^\star}\sum_{\chi} \sup_{y\le x} |\psi(y,\chi)| \ll \left(x + x^{5/6}Q + x^{1/2}Q^2\right)(\log x)^3. \tag{20.1}$$

The term $q = 1$ contributes an amount $\sim x$, but otherwise we believe that $|\psi(y,\chi)|$ is usually of the size $y^{1/2}(\log q)^{1/2}$. Thus we expect that the left-hand side above is $\ll x + Q^2x^{1/2}(\log Q)^{1/2}$. On the assumption of GRH it would follow that it is $\ll x + Q^2x^{1/2}(\log Qx)^2$.

Proof If $Q^2 > x$, then we obtain (20.1) from (19.35) by taking $M = 1$, $a_1 = 1$, $N = [x]$, and $b_n = \Lambda(n)$. Suppose that $Q \le x^{1/2}$. By Vaughan's identity

(17.6) with $f(n) = \chi(n)$ we find that $\psi(y,\chi) = S_1 + S_2 + S_3 + S_4$ where by (17.6)–(17.10) and (17.12),

$$S_1(y,\chi) = \sum_{n\le U} \Lambda(n)\chi(n), \tag{20.2}$$

$$S_2(y,\chi) \ll (\log UV) \sum_{t\le UV} \Big|\sum_{r\le y/t} \chi(rt)\Big|, \tag{20.3}$$

$$S_3(y,\chi) \ll (\log y) \sum_{k\le V} \sup_w \Big|\sum_{w\le m\le y/k} \chi(m)\Big|, \tag{20.4}$$

$$S_4(y,\chi) = \sum_{U<m\le y/V} b(m) \sum_{V<k\le y/m} \mu(k)\chi(mk), \tag{20.5}$$

and $b(m)$ satisfies $b(m) \ll \log m$. Thus by (19.35),

$$\sum_{q\le Q} \frac{q}{\varphi(q)} \sum_\chi{}^{\star} \sup_{y\le x} \Bigg|\sum_{\substack{M<m\le 2M\\ U<m\le y/V}} b(m) \sum_{V<k\le y/m} \mu(k)\chi(mk)\Bigg|$$
$$\ll \big(x + QxM^{-1/2} + Qx^{1/2}M^{1/2} + Q^2x^{1/2}\big)(\log x)^2.$$

On summing this over $M = 2^\ell$ for $U/2 \le M = 2^\ell \le x/V$, we deduce that

$$\sum_{q\le Q} \frac{q}{\varphi(q)} \sum_\chi{}^{\star} \sup_{y\le x} |S_4(y,\chi)|$$
$$\ll \big(x + QxU^{-1/2} + QxV^{-1/2} + Q^2x^{1/2}\big)(\log x)^3. \tag{20.6}$$

We write

$$S_2(y,\chi) = \sum_{t\le UV} = \sum_{t\le U} + \sum_{U<t\le UV} = S_2'(y,\chi) + S_2''(y,\chi), \tag{20.7}$$

and treat S_2'' in the same way that we treated S_4. Thus

$$\sum_{q\le Q} \frac{q}{\varphi(q)} \sum_\chi{}^{\star} \sup_{y\le x} |S_2''(y,\chi)|$$
$$\ll \big(x + QxU^{-1/2} + Qx^{1/2}U^{1/2}V^{1/2} + Q^2x^{1/2}\big)(\log x)^3. \tag{20.8}$$

For $q = 1$, $S_2'(y,\chi) \ll y(\log U)^2$. For $q > 1$ we apply the Pólya–Vinogradov inequality (Theorem 9.18) to see that

$$S_2'(y,\chi) \ll q^{1/2}U(\log qU)^2.$$

Hence

$$\sum_{q\le Q} \frac{q}{\varphi(q)} \sum_\chi{}^{\star} \sup_{y\le x} |S_2'(y,\chi)| \ll \big(x + Q^{5/2}U\big)(\log Ux)^2. \tag{20.9}$$

We treat S_3 in the same way that we treated S_2', and hence find that

$$\sum_{q\le Q}\frac{q}{\varphi(q)}\sum_{\chi}{}^{\star}\sup_{y\le x}|S_3(y,\chi)| \ll \left(x+Q^{5/2}V\right)(\log Vx)^2. \tag{20.10}$$

Finally, it is trivial that

$$\sum_{q\le Q}\frac{q}{\varphi(q)}\sum_{\chi}{}^{\star}\sup_{y\le x}|S_1(y,\chi)| \ll Q^2U. \tag{20.11}$$

On combining (20.6)–(20.11), we conclude that

$$\begin{aligned}\sum_{q\le Q}\frac{q}{\varphi(q)}\sum_{\chi}{}^{\star}&\sup_{y\le x}|\psi(y,\chi)|\\ &\ll \big(x+QxU^{-1/2}+QxV^{-1/2}+Q^2x^{1/2}\\ &\quad +U^{1/2}V^{1/2}Qx^{1/2}+Q^{5/2}U+Q^{5/2}V\big)(\log xUV)^3.\end{aligned}$$

By allowing U and V to vary with UV held constant, we see that $U=V$ is optimal. For $x^{1/3}\le Q\le x^{1/2}$, we obtain the stated bound by taking $U=V=x^{2/3}/Q$. For $1\le Q\le x^{1/3}$, we obtain the stated bound by taking $U=V=x^{1/3}$. □

20.1.1 Exercises

1. Let $\pi(x,\chi)$, $\pi(x;q,a)$, $\vartheta(x,\chi)$, and $\vartheta(x;q,a)$ be defined as in (11.20) and (11.21).
   (a) Show that $|\psi(x,\chi)-\vartheta(x,\chi)|\le\psi(x)-\vartheta(x)\ll x^{1/2}$.
   (b) Show that
   $$\sum_{q\le Q}\frac{q}{\varphi(q)}\sum_{\chi}{}^{\star}\sup_{y\le x}|\vartheta(y,\chi)| \ll \left(x+x^{5/6}Q+x^{1/2}Q^2\right)(\log x)^3.$$
   (c) Show that
   $$\pi(x,\chi)=\frac{\psi(x,\chi)}{\log x}+\int_2^x\frac{\psi(u,\chi)}{u(\log u)^2}\,du.$$
   (d) Show that
   $$\pi(x,\chi)\ll\frac{1}{\log x}\sup_{x^{1/2}\le y\le x}|\psi(y,\chi)|+x^{1/2}.$$
   (e) Show that
   $$\sup_{y\le x}|\pi(y,\chi)|\ll\frac{1}{\log x}\sup_{y\le x}|\psi(y,\chi)|+x^{1/2}.$$

(f) Conclude that if $x \ge 2$, then

$$\sum_{q \le Q} \frac{q}{\varphi(q)} \sum_{\chi}{}^{\star} \sup_{y \le x} |\pi(y,\chi)| \ll \left(x + x^{5/6}Q + x^{1/2}Q^2\right)(\log x)^3.$$

(g) Show that

$$\sum_{\chi} \left| \sum_{n=M+1}^{M+N} c_n \chi(n) \right|^2 = \varphi(q) \sum_{\substack{h=1 \\ (h,q)=1}}^{q} \left| \sum_{\substack{n=M+1 \\ n \equiv h \ (\mathrm{mod}\ q)}}^{M+N} c_n \right|^2,$$

where $\sum_\chi$ denotes a sum over all characters modulo q.

(h) Show that

$$\sum_{\chi} \left| \sum_{n=M+1}^{M+N} c_n \chi(n) \right|^2 \le (N+q) \sum_{n=M+1}^{M+N} |c_n|^2.$$

(i) Show that

$$\sum_{\chi} \left| \sum_{m=1}^{M} \sum_{n=1}^{N} a_m b_n \chi(mn) \right| \le (M+q)^{1/2}(N+q)^{1/2} \times \left(\sum_{m=1}^{M} |a_m|^2 \right)^{1/2} \left(\sum_{n=1}^{N} |b_n|^2 \right)^{1/2}.$$

(j) Show that

$$\sum_{\chi} \sup_{y} \left| \sum_{\substack{1 \le m \le M \\ 1 \le n \le N \\ mn \le y}} a_m b_n \chi(mn) \right| \ll (M+q)^{1/2}(N+q)^{1/2} \left(\sum_{m=1}^{M} |a_m|^2 \right)^{1/2} \left(\sum_{n=1}^{N} |b_n|^2 \right)^{1/2} \log 2MN.$$

(k) Show that if $q \ge x$, then

$$\sum_{\chi} \sup_{y \le x} |\psi(y,\chi)| \ll q x^{1/2} (\log 2x)^{3/2}.$$

2. (a) Show that

$$\sum_{\chi} \sup_{y \le x} \left| \sum_{\substack{M < m \le 2M \\ U < m \le y/V}} b(m) \sum_{V < k \le y/m} \mu(k)\chi(mk) \right| \ll \left(x + q^{1/2}xM^{-1/2} + q^{1/2}x^{1/2}M^{1/2} + qx^{1/2}\right)(\log x)^2.$$

(b) Deduce that

$$\sum_{\chi} \sup_{y \le x} |S_4(y,\chi)| \ll \big(x + q^{1/2}xU^{-1/2} + q^{1/2}xV^{-1/2} + qx^{1/2}\big)(\log x)^3.$$

(c) Let S_2' and S_2'' be defined as in (20.7). Show that

$$\sum_{\chi} \sup_{y \le x} |S_2''(y,\chi)| \ll \big(x + q^{1/2}xU^{-1/2} + q^{1/2}x^{1/2}U^{1/2}V^{1/2} + qx^{1/2}\big)(\log x)^3.$$

(d) Show that if $1 < q \le x$, then

$$\sum_{\chi} \sup_{y \le x} |S_2'(y,\chi)| \ll q^{3/2}U(\log xU)^2.$$

(e) Show that if $1 < q \le x$, then

$$\sum_{\chi} \sup_{y \le x} |S_3(y,\chi)| \ll q^{3/2}V(\log x)^2.$$

(f) Conclude that if $x \ge 2$ and $q > 1$, then

$$\sum_{\chi} \sup_{y \le x} |\psi(y,\chi)| \ll \big(x + q^{1/6}x^{2/3} + qx^{1/2}\big)(\log 2x)^3.$$

20.2 The Bombieri–Vinogradov Theorem

For $(a,q) = 1$, let

$$E(x;q,a) = \psi(x;q,a) - \frac{x}{\varphi(q)}, \tag{20.12}$$

put

$$E(x,q) = \sup_{\substack{a \\ (a,q)=1}} |E(x;q,a)|, \tag{20.13}$$

and set

$$E^*(x,q) = \sup_{y \le x} E(y,q). \tag{20.14}$$

We show that $E^*(x,q)$ is considerably smaller than $x/\varphi(q)$ for most $q \le x^{1/2}(\log x)^{-A}$.

Theorem 20.2 (The Bombieri–Vinogradov Theorem) *Let A be a fixed positive number. Then*

$$\sum_{q \le Q} E^*(x,q) \ll x^{1/2} Q (\log x)^3 \tag{20.15}$$

for $x^{1/2}(\log x)^{-A} \le Q \le x^{1/2}$.

The implicit constant in (20.15) is non-effective since our proof will involve an appeal to the Siegel–Walfisz theorem (Corollary 11.19).

Let $\mathcal{Q}$ be the set of those $q \le Q$ for which $E^*(x,q) > x/(\varphi(q)(\log x)^B)$. Since $\varphi(q) \le q$, we deduce that the number of members of $\mathcal{Q}$ is

$$\ll Q^2 x^{-1/2} (\log x)^{B+3}.$$

This is small compared with Q if $x^{1/2}(\log x)^{-A} \le Q$ and

$$Q = o\big(x^{1/2}(\log x)^{-B-3}\big).$$

We recall (11.22), which is to say that

$$\psi(x;q,a) = \frac{1}{\varphi(q)} \sum_{\chi} \overline{\chi}(a) \psi(x,\chi).$$

If we assume GRH, then we have a good estimate for $\psi(x,\chi)$, namely (by Theorem 13.7)

$$\psi(x,\chi) = E_0(\chi)x + O\big(x^{1/2}(\log x)(\log qx)\big),$$

where

$$E_0(\chi) = \begin{cases} 1 & (\chi = \chi_0), \\ 0 & (\text{otherwise}). \end{cases}$$

Put $\psi'(x,\chi) = \psi(x,\chi) - E_0(\chi)x$. Then

$$\psi(x;q,a) - x/\varphi(q) = \frac{1}{\varphi(q)} \sum_{\chi} \overline{\chi}(a) \psi'(x,\chi), \tag{20.16}$$

and so

$$E(x,q) \le \frac{1}{\varphi(q)} \sum_{\chi} |\psi'(x,\chi)| \tag{20.17}$$

by the triangle inequality. Thus on GRH,

$$E(x,q) \ll x^{1/2}(\log x)^2, \tag{20.18}$$

as was already noted in Corollary 13.8. In view of the Brun–Titchmarsh inequality (Theorem 3.9) we know that $E(x,q) \ll x/\varphi(q)$ for $q \le x^{1-\delta}$. Thus the estimate (20.18) – despite being a consequence of GRH – is worse than trivial when $q > x^{1/2}/\log x$. Here GRH gives a weak result (when q is large) because we have eliminated any possible cancellation that presumably occurs in the sum over χ in (20.16). Indeed, by Corollary 13.10 we know that on GRH the root-mean-square size of $E(x; q, a)$ is $\ll x^{1/2}\varphi(q)^{-1/2}(\log x)^2$ when $q \le x$, and we expect that $E(x, q)$ is not much larger.

Conjecture 20.3 *If $(a, q) = 1$ and $q \le x$, then*

$$\psi(x; q, a) = \frac{x}{\varphi(q)} + O\big(x^{1/2+\varepsilon}/q^{1/2}\big).$$

For many purposes, it would be enough to know this on average:

Conjecture 20.4 (The Elliott–Halberstam Hypothesis) *Let A and ε be fixed positive numbers. In the notation of* (20.14),

$$\sum_{q \le Q} E^*(x, q) \ll x(\log x)^{-A}$$

provided that $Q \le x^{1-\varepsilon}$.

Proof of Theorem 20.2 From (20.17) we see that

$$E^*(x, q) \le \frac{1}{\varphi(q)} \sum_{\chi} \sup_{y \le x} |\psi'(y, \chi)|.$$

Suppose that $d|q$ and that the character $\chi \pmod q$ is induced by the primitive character $\chi^\star \pmod d$. Then

$$\begin{aligned} \psi'(y, \chi^\star) - \psi'(y, \chi) &= \sum_{p|q} \sum_{\substack{k \\ p^k \le y}} \chi^\star(p)^k \log p \\ &\ll \sum_{p|q} \log y = \omega(q) \log y \ll (\log qy)^2. \end{aligned} \tag{20.19}$$

Hence

$$\sum_{q \le Q} E^*(x, q) \ll \sum_{d \le Q} \sum_{\chi^\star} \Big(\sup_{y \le x} |\psi'(y, \chi^\star)| + O\big((\log Qx)^2\big)\Big) \sum_{\substack{q \le Q \\ d|q}} \frac{1}{\varphi(q)}.$$

Write $q = dm$. Since $\varphi(dm) \ge \varphi(d)\varphi(m)$, it follows that

$$\sum_{m \le Q/d} \frac{1}{\varphi(dm)} \le \frac{1}{\varphi(d)} \sum_{m \le Q/d} \frac{1}{\varphi(m)}.$$

Now

$$\sum_{m\le y}\frac{1}{\varphi(m)} \le \sum_{\substack{m\\ p|m\Rightarrow p\le y}}\frac{1}{\varphi(m)} = \prod_{p\le y}\left(1+\frac{1}{p-1}+\frac{1}{p(p-1)}+\cdots\right)$$
$$= \prod_{p\le y}\left(1+\frac{p}{(p-1)^2}\right) = \prod_{p\le y}\left(1-\frac{1}{p}\right)^{-1}\left(1+\frac{1}{p(p-1)}\right)$$
$$\ll \log 2y$$

by Mertens' formula (Theorem 2.7(e)). (Alternatively, we could appeal to (2.32) with $\kappa = 1$, and then integrate by parts. The asymptotic formula of Exercise 2.1.1.13(d) would be overkill at this point.) Hence

$$\sum_{\substack{q\le Q\\ d|q}}\frac{1}{\varphi(q)} \le \frac{1}{\varphi(d)}\sum_{m\le Q/d}\frac{1}{\varphi(m)} \ll \frac{1}{\varphi(d)}\log\frac{2Q}{d} \tag{20.20}$$

for $d \le Q$, so

$$\sum_{q\le Q} E^*(x,q) \ll Q(\log Qx)^2 + \sum_{d\le Q}\frac{\log 2Q/d}{\varphi(d)}\sum_{\chi}{}^{\star}\sup_{y\le x}|\psi'(y,\chi)|. \tag{20.21}$$

Put $Q_1 = (\log x)^{A+1}$, and suppose that $Q_1 \le U \le Q$. By Theorem 20.1 we see that

$$\sum_{U<q\le 2U}\frac{\log 2Q/q}{\varphi(q)}\sum_{\chi}{}^{\star}\sup_{y\le x}|\psi'(y,\chi)|$$
$$\ll \frac{\log 4Q/U}{U}\sum_{U<q\le 2U}\frac{q}{\varphi(q)}\sum_{\chi}{}^{\star}\sup_{y\le x}|\psi(y,\chi)|$$
$$\ll \left(x/U + x^{5/6} + x^{1/2}U\right)(\log x)^3\log 4Q/U.$$

On summing over $U = 2^k Q_1$, we deduce that

$$\sum_{Q_1<q\le Q}\frac{\log 2Q/q}{\varphi(q)}\sum_{\chi}{}^{\star}\sup_{y\le x}|\psi'(y,\chi)|$$
$$\ll xQ_1^{-1}(\log x)^4 + x^{5/6}(\log x)^5 + x^{1/2}Q(\log x)^3$$
$$\ll x^{1/2}Q(\log x)^3 \tag{20.22}$$

since $Q \ge x^{1/2}(\log x)^{-A}$. Suppose that χ is a primitive character modulo q with $q \le Q_1$. By the Siegel–Walfisz theorem (Corollary 11.19) we know that $\sup_{y\le x}|\psi'(y,\chi)| \ll x\exp\left(-c_1\sqrt{\log x}\right)$. Hence

$$\sum_{q\le Q_1}\frac{\log 2Q/q}{\varphi(q)}\sum_{\chi}{}^{\star}\sup_{y\le x}|\psi'(y,\chi)| \ll x\exp\left(-c_2\sqrt{\log x}\right) \ll x^{1/2}Q(\log x)^3$$

since $Q \ge x^{1/2}(\log x)^{-A}$. We combine this with (20.22) in (20.21) to obtain the desired bound. □

After we proved the Siegel–Walfisz Theorem for $\psi(x;q,a)$ in §11.3, in Corollary 11.20 we derived analogues for $\vartheta(x;q,a)$ and $\pi(x;q,a)$. We follow the approach used there to deduce concomitant analogues.

Corollary 20.5 *For $(a,q)=1$ let*

$$E_\vartheta(x;q,a) = \vartheta(x;q,a) - \frac{x}{\varphi(q)}, \qquad E_\pi(x;q,a) = \pi(x;q,a) - \frac{\mathrm{li}(x)}{\varphi(q)},$$

$$E_\vartheta(x,q) = \sup_{\substack{a\\(a,q)=1}} |E_\vartheta(x;q,a)|, \qquad E_\pi(x,q) = \sup_{\substack{a\\(a,q)=1}} |E_\pi(x;q,a)|,$$

$$E_\vartheta^*(x,q) = \sup_{y\le x} E_\vartheta(y,q), \qquad E_\pi^*(x,q) = \sup_{y\le x} E_\pi(y,q),$$

and let $A>0$ be fixed. Then

$$\sum_{q\le Q} E_\vartheta^*(x,q) \ll x^{1/2}Q(\log x)^3 \tag{20.23}$$

and

$$\sum_{q\le Q} E_\pi^*(x,q) \ll x^{1/2}Q(\log x)^2 \tag{20.24}$$

provided that $x^{1/2}(\log x)^{-A} \le Q \le x^{1/2}$.

Proof We first observe that

$$\psi(y;q,a) - \vartheta(y;q,a) \le \psi(y) - \vartheta(y) \ll y^{1/2}.$$

Hence

$$|\vartheta(y;q,a) - y/\varphi(q)| \ll E(y;q,a) + y^{1/2}.$$

Thus

$$E_\vartheta^*(x,q) \ll E^*(x,q) + x^{1/2},$$

so (20.23) follows from Theorem 20.2. As for $\pi(y;q,a)$, we write

$$\pi(y;q,a) = \int_{2^-}^{y} \frac{1}{\log u}\, d\vartheta(u;q,a) = \frac{\mathrm{li}(y)}{\varphi(q)} + \int_{2^-}^{y} \frac{1}{\log u}\, d(\vartheta(u;q,a) - u/\varphi(q)).$$

By partial integration this last integral is

$$= \frac{\vartheta(u;q,a) - u/\varphi(q)}{\log u}\Big|_{2^-}^{y} - \int_2^y \frac{\vartheta(u;q,a) - u/\varphi(q)}{u(\log u)^2}\, du.$$

For $2 \le u \le \sqrt{x}$ we use the trivial bound $\vartheta(u; q, a) \ll u(\log u)/q$, and for $\sqrt{x} \le u \le y$ we use the inequality $|E_\vartheta(u; q, a)| \le E^*_\vartheta(y, q)$. Thus

$$E^*_\pi(x; q, a) \ll x^{1/2}/q + E^*_\vartheta(x; q)/\log x.$$

Hence (20.24) follows from (20.23), and the proof is complete. □

The following variant of the Bombieri–Vinogradov theorem is convenient in some applications.

Corollary 20.6 *Let $A > 0$ be fixed. Then*

$$\sum_{q \le Q} qE^*(x, q)^2 \ll x^{3/2}Q(\log x)^4, \tag{20.25}$$

$$\sum_{q \le Q} qE^*_\vartheta(x, q)^2 \ll x^{3/2}Q(\log x)^4, \tag{20.26}$$

and

$$\sum_{q \le Q} qE^*_\pi(x, q)^2 \ll x^{3/2}Q(\log x)^2 \tag{20.27}$$

provided that $x^{1/2}(\log x)^{-A} \le Q \le x^{1/2}$.

Proof If $q \le x$, then there are $\ll x/q$ integers $n \le x$ such that $n \equiv a \pmod{q}$. Thus it is trivial that $\psi(x; q, a) \ll x(\log x)/q$. (The Brun–Titchmarsh inequality gives a better bound.) Hence $qE^*(x, q)^2 \ll E^*(x, q)x \log x$, and so (20.25) follows from Theorem 20.2. Similarly, (20.26) follows from (20.23). For $\pi(x; q, a)$ the trivial bound is $\pi(x; q, a) \ll x/q$, so $qE^*_\pi(x, q)^2 \ll E^*_\pi(x, q)x$, and thus (20.27) follows from (20.24). □

20.3 Applications of the Bombieri–Vinogradov Theorem

The twin prime problem is to show that there are infinitely many prime numbers p such that $p + 2$ is also prime. One way of attacking this problem would be to sieve the numbers $p + 2$, and try to estimate the number of survivors. However, in order for a sieve to be applicable, we must know approximately how many multiples of d are in the set $\{p + 2 : p \le x\}$. That is, we need to know that $\pi(x; d, -2)$ is approximately $\mathrm{li}(x)/\varphi(d)$ for most odd d up to a certain size. The Bombieri–Vinogradov theorem gives us precisely the sort of information we need for sifting up to $x^{1/2}(\log x)^{-A}$. By Selberg's lambda squared method we can show that the number of primes $p \le x$ for which $p + 2$ is prime is

$$\le (4 + o(1))cx/(\log x)^2,$$

where

$$c = 2\prod_{p>2}\left(1-\frac{1}{(p-1)^2}\right). \tag{20.28}$$

(The details are outlined in Exercises 20.3.1.6–8.) This is a factor of 2 better than the bound in Theorem 3.10, but is still a factor 4 larger than the conjectured truth. The Bombieri–Vinogradov theorem is also useful when a lower bound sieve is applied to the twin prime problem. This will be explored in Chapter 21.

Theorem 20.7 *The number of representations of a positive integer n as a sum of a prime and the product of two positive integers is*

$$\sum_{p<n} d(n-p) = \frac{n\zeta(2)\zeta(3)}{\zeta(6)}\prod_{p\mid n}\left(1-\frac{p}{p^2-p+1}\right) + O\left(\frac{n\log\log 3n}{\log n}\right).$$

Here the main term is $\gg n/\log\log 3n$, so the main term is definitely of a larger order of magnitude than the error term.

Proof Let $\mathscr{P}$ denote the set of primes, and put $Q = n^{1/2}/(\log n)^A$. Then by the method of the hyperbola, where we initially consider the two regions $d \le n/Q, e \le n/d$ and $e \le Q, d \le n/e$ minus the over-counted region $d \le n/Q, e \le Q$, and then we separate from the first region the subregion $Q < d \le n/Q, e \le n/d$,

$$\begin{aligned}\sum_{p<n} d(n-p) = \sum_{\substack{d,e\\ de\le n\\ n-de\in\mathscr{P}}} 1 &= \sum_{d\le Q}\pi(n;d,n) + \sum_{Q<d\le n/Q}\pi(n;d,n)\\ &\quad + \sum_{e\le Q}\pi(n;e,n) - \sum_{\substack{d\le n/Q\\ e\le Q\\ n-de\in\mathscr{P}}} 1\\ &= \Sigma_1+\Sigma_2+\Sigma_3-\Sigma_4,\end{aligned} \tag{20.29}$$

say. If $(d,n) > 1$, then $\pi(n;d,n) \le 1$. Thus

$$\Sigma_1 = \sum_{\substack{d\le Q\\ (d,n)=1}}\pi(n;d,n) + O(Q) = \mathrm{li}(n)\sum_{\substack{d\le Q\\ (d,n)=1}}\frac{1}{\varphi(d)} + \sum_{\substack{d\le Q\\ (d,n)=1}} E_\pi(n;d,n) + O(Q).$$

From Exercise 2.1.1.16(c) we see that

$$\begin{aligned}\sum_{\substack{n\le Q\\ (n,q)=1}}\frac{1}{\varphi(n)} &= \frac{\zeta(2)\zeta(3)}{\zeta(6)}\prod_{p\mid q}\left(1-\frac{p}{p^2-p+1}\right)\\ &\quad\times\left(\log Q + C_0 + \sum_{p\mid q}\frac{\log p}{p-1} - \sum_{p\nmid q}\frac{\log p}{p^2-p+1}\right)\\ &\quad + O\big(2^{\omega(q)}(\log Q)/Q\big).\end{aligned}$$

By considering ‘record-breaking’ q as in the proof of Theorem 2.9 we see that

$$\sum_{p|q} \frac{\log p}{p-1} \ll \log\log 3q$$

uniformly for $q \geq 1$. Thus by Corollary 20.5 we deduce that

$$\begin{aligned}\Sigma_1 = {} & \frac{\zeta(2)\zeta(3)}{\zeta(6)} \prod_{p|n} \left(1 - \frac{p}{p^2 - p + 1}\right) n \frac{\log Q}{\log n} \\ & + O\left(\frac{n \log\log n}{\log n}\right) + O\big(n(\log n)^{-A+2}\big).\end{aligned} \tag{20.30}$$

By the Brun–Titchmarsh inequality (Theorem 3.9),

$$\Sigma_2 \ll \frac{n}{\log n} \sum_{Q<d\leq n/Q} \frac{1}{\varphi(d)} \ll \frac{n \log\log n}{\log n}. \tag{20.31}$$

Clearly $\Sigma_3 = \Sigma_1$. We note that

$$\Sigma_4 = \sum_{e\leq Q} \sum_{\substack{n-ne/Q\leq p\leq n \\ p\equiv n \ (\mathrm{mod}\ e)}} 1.$$

Thus by the Brun–Titchmarsh inequality,

$$\Sigma_4 \ll \frac{n}{Q \log n} \sum_{e\leq Q} \frac{e}{\varphi(e)} \ll \frac{n}{\log n}. \tag{20.32}$$

We take $A = 3$, and note that $\log Q = \frac{1}{2} \log n + O(\log\log n)$. Thus the stated result follows on combining (20.30)–(20.32) in (20.29). □

Let

$$E = \liminf_{n\to\infty} \frac{p_{n+1} - p_n}{\log p_n}. \tag{20.33}$$

By the Prime Number Theorem, $E \leq 1$. In an unpublished manuscript, Hardy & Littlewood showed (assuming GRH) that $E \leq 2/3$; see the chapter notes (§20.5). Indeed it was the requirements of this problem which motivated Bombieri's work on Theorem 20.2. In the arguments that follow, we use the Bombieri–Vinogradov theorem to show unconditionally that $E \leq 1/2$. In Chapter 22 we show not only that $E = 0$ but also that $\liminf_{n\to\infty} p_{n+1} - p_n$ is bounded.

Let

$$S(\alpha) = \sum_{p\leq N} (\log p) e(p\alpha), \qquad T(\alpha) = \sum_{h=1}^{H} e(h\alpha).$$

Then

$$|T(\alpha)|^2 = \sum_{h=-H}^{H} (H - |h|)e(h\alpha),$$

so

$$\begin{aligned}\int_0^1 |S(\alpha)T(\alpha)|^2\, d\alpha &= \sum_{h=-H}^{H} (H - |h|) \int_0^1 |S(\alpha)|^2 e(h\alpha)\, d\alpha \\ &= H \sum_{p\le N} (\log p)^2 + 2\sum_{h=1}^{H} (H - h)R(N; h),\end{aligned}$$

where

$$R(N; h) = \sum_{\substack{p,p'\le N \\ p-p'=h}} (\log p)(\log p').$$

Since $\sum_{p\le N} (\log p)^2 = N \log N + O(N)$ by the Prime Number Theorem, this gives the following lemma.

Lemma 20.8 *Let $S(\alpha)$, $T(\alpha)$, and $R(N; h)$ be defined as above. Then*

$$\int_0^1 |S(\alpha)T(\alpha)|^2\, d\alpha = HN \log N + 2\sum_{h=1}^{H} (H - h)R(N; h) + O(HN).$$

Our objective is to derive a lower bound for the integral in Lemma 20.8 that is sufficiently large to prove that $R(N; h) > 0$ for at least one h. To this end, we apply the large sieve, which in the form of Corollary 19.5 gives

$$\sum_{q\le Q} \sum_{\substack{a=1 \\ (a,q)=1}}^{q} |S(a/q)T(a/q)|^2 \le \left(N + H + Q^2\right) \int_0^1 |S(\alpha)T(\alpha)|^2\, d\alpha. \quad (20.34)$$

We anticipate that $S(a/q)$ is often near $\frac{\mu(q)}{\varphi(q)}N$. Thus the next lemma provides an asymptotic evaluation of the main term that we expect will emerge.

Lemma 20.9 *Let $T(\alpha)$ be defined as above. If $2 \le H \le Q$, then*

$$\sum_{q\le Q} \frac{\mu(q)^2}{\varphi(q)^2} \sum_{\substack{a=1 \\ (a,q)=1}}^{q} |T(a/q)|^2 = H \log Q + H^2 + O(H \log H).$$

Proof The left-hand side above is

$$\sum_{q\le Q}\frac{\mu(q)^2}{\varphi(q)^2}\sum_{\substack{a=1\\(a,q)=1}}^{q}\sum_{h=-H}^{H}(H-|h|)e(ha/q)$$
$$=\sum_{q\le Q}\frac{\mu(q)^2}{\varphi(q)^2}\sum_{h=-H}^{H}(H-|h|)c_q(h)$$
$$=H\sum_{q\le Q}\frac{\mu(q)^2}{\varphi(q)}+2\sum_{h=1}^{H}(H-h)\sum_{q\le Q}\frac{\mu(q)^2}{\varphi(q)^2}c_q(h).$$

As we observed already in (18.15), if $h\neq 0$, then

$$\sum_{q=1}^{\infty}\frac{\mu(q)^2}{\varphi(q)^2}c_q(h)=\prod_p\left(1+\frac{c_p(h)}{(p-1)^2}\right)$$
$$=\prod_{p|h}\left(1+\frac{1}{p-1}\right)\prod_{p\nmid h}\left(1-\frac{1}{(p-1)^2}\right)=\mathfrak{S}_2(h)$$

is the singular series for twin primes. As for the tail in this series, we note that if $h\neq 0$, then

$$\sum_{q>Q}\frac{\mu(q)^2}{\varphi(q)^2}c_q(h)=\sum_{q>Q}\frac{\mu(q)^2}{\varphi(q)^2}\sum_{\substack{d|q\\d|h}}d\mu(q/d)\ll\sum_{d|h}d\sum_{\substack{q>Q\\d|q}}\frac{\mu(q)^2}{\varphi(q)^2}$$
$$\ll\sum_{d|h}\frac{d}{\varphi(d)^2}\sum_{m>Q/d}\frac{1}{\varphi(m)^2}\ll Q^{-1}\sum_{d|h}\frac{d^2}{\varphi(d)^2}$$
$$\ll Q^{-1}d(h)(h/\varphi(h))^2.$$

Now multiply both sides of the above by $H-h$ and sum over h, to see that

$$\sum_{h=1}^{H}(H-h)\sum_{q>Q}\frac{\mu(q)^2}{\varphi(q)^2}c_q(h)\ll Q^{-1}H^2\log 2H.$$

Thus

$$\sum_{q\le Q}\frac{\mu(q)^2}{\varphi(q)^2}\sum_{\substack{a=1\\(a,q)=1}}^{q}|T(a/q)|^2=H\sum_{q\le Q}\frac{\mu(q)^2}{\varphi(q)}+2\sum_{h=1}^{H}(H-h)\mathfrak{S}_2(h)$$
$$+O\big(Q^{-1}H^2\log 2H\big).$$

In Exercise 2.1.1.17 it was shown that $\sum_{q\le Q}\mu(q)^2/\varphi(q)=\log Q+O(1)$. (Actually, a more precise estimate was proved, with lower order terms.) In Exercise

3.4.1.3(a), and again in Exercise 18.2.1.5, it was shown that $\sum_{h=1}^{H} \mathfrak{S}_2(h) = H + O(\log H)$. Thus

$$2\sum_{h=1}^{H}(H-h)\mathfrak{S}_2(h) = 2\sum_{h=1}^{H-1}\sum_{m=1}^{h}\mathfrak{S}_2(m)$$
$$= 2\sum_{h=1}^{H-1}\big(h + O(\log 2h)\big) = H^2 + O(H\log H).$$

Thus the proof is complete. □

We now derive a lower bound for $|S(a/q)|^2$ in terms of the distribution of primes into residue classes modulo q.

Lemma 20.10 *Let $E_\vartheta(N;q,a)$ be defined as in Corollary 20.5, and put*

$$U(a/q) = 2\frac{\mu(q)}{\varphi(q)}N\,\mathrm{Re}\sum_{\substack{m=1\\(m,q)=1}}^{q} E_\vartheta(N;q,m)e(am/q). \tag{20.35}$$

If $(a,q) = 1$ and $q \le N$, then

$$|S(a/q)|^2 \ge \frac{\mu(q)^2}{\varphi(q)^2}N^2 + U(a/q) + O(N\log N).$$

Proof We write

$$S(a/q) = \sum_{p\le N}(\log p)e(pa/q) = \sum_{m=1}^{q} e(am/q)\vartheta(N;q,m).$$

If $(m,q) > 1$, then $\vartheta(N;q,m) = 0$ unless m is a prime dividing q, in which case $\vartheta(N;q,m) \le \log m$. Since $\sum_{p|q}\log p \le \log q$, it follows that the above is

$$= \sum_{\substack{m=1\\(m,q)=1}}^{q} e(am/q)\vartheta(N;q,m) + O(\log q).$$

We write $\vartheta(N;q,m) = N/\varphi(q) + E_\vartheta(N;q,m)$ to see that the above is

$$= \frac{\mu(q)}{\varphi(q)}N + \sum_{\substack{m=1\\(m,q)=1}}^{q} E_\vartheta(N;q,m)e(am/q) + O(\log q).$$

Here we have used the fact that $c_q(m) = \mu(q)$ if $(m,q) = 1$. If A and B are complex numbers, then $|A|^2 = |A+B|^2 + O\big(|AB| + |B|^2\big)$. Take $A = S(a/q)$ and note that $S(a/q) \ll N$. Take B to be the error term above. Hence

$$|S(a/q)|^2 = \left|\frac{\mu(q)}{\varphi(q)}N + \sum_{\substack{m=1\\(m,q)=1}}^{q} E_\vartheta(N;q,m)e(am/q)\right|^2 + O(N\log N).$$

The modulus-squared on the right-hand side is

$$\frac{\mu(q)^2}{\varphi(q)^2}N^2 + U(a/q) + \left|\sum_{\substack{m=1\\(m,q)=1}}^{q} E_\vartheta(N;q,m)e(am/q)\right|^2.$$

Here the last term is non-negative, so we have the stated lower bound. □

The following simple estimate will be useful.

Lemma 20.11 *For positive integers q and H,*

$$\sum_{m=1}^{q}\left|\sum_{\substack{a=1\\(a,q)=1}}^{q} |T(a/q)|^2 e(am/q)\right| \le qd(q)H^2.$$

Proof We note that

$$\begin{aligned}\sum_{\substack{a=1\\(a,q)=1}}^{q} |T(a/q)|^2 e(am/q) &= \sum_{\substack{a=1\\(a,q)=1}}^{q}\sum_{h=-H}^{H}(H-|h|)e(a(h+m)/q)\\ &= \sum_{h=-H}^{H}(H-|h|)c_q(m+h)\\ &= \sum_{h=-H}^{H}(H-|h|)\sum_{\substack{d\mid q\\ d\mid(m+h)}} d\mu(q/d).\end{aligned}$$

Hence

$$\left|\sum_{\substack{a=1\\(a,q)=1}}^{q} |T(a/q)|^2 e(am/q)\right| \le \sum_{h=-H}^{H}(H-|h|)\sum_{\substack{d\mid q\\ d\mid(m+h)}} d.$$

The sum over m of this upper bound is

$$\sum_{h=-H}^{H}(H-|h|)\sum_{d\mid q}\sum_{\substack{m=1\\ m\equiv -h \pmod d}}^{q} d = qd(q)H^2.$$ □

Lemma 20.12 *If* $N^{1/2}(\log N)^{-A} \le Q \le N^{1/2}$, *then*

$$\sum_{q\le Q} \sum_{\substack{a=1\\(a,q)=1}}^{q} |T(a/q)|^2 U(a/q) \ll H^2 N^{7/4} Q^{1/2} (\log N)^4.$$

Proof From the definition of $U(a/q)$ we see that the expression to be bounded is exactly

$$2N \sum_{q\le Q} \frac{\mu(q)}{\varphi(q)} \sum_{\substack{m=1\\(m,q)=1}}^{q} E_\vartheta(N;q,m) \operatorname{Re} \sum_{\substack{a=1\\(a,q)=1}}^{q} |T(a/q)|^2 e(am/q).$$

By the triangle inequality, the above is

$$\ll N \sum_{q\le Q} \frac{\mu(q)^2}{\varphi(q)} E_\vartheta(N;q) \sum_{m=1}^{q} \left| \sum_{\substack{a=1\\(a,q)=1}}^{q} |T(a/q)|^2 e(am/q) \right|.$$

By Lemma 20.11, this is

$$\ll H^2 N \sum_{q\le Q} \frac{\mu(q)^2 d(q) q}{\varphi(q)} E_\vartheta(N;q).$$

By Cauchy's inequality, this is

$$\le H^2 N \left(\sum_{q\le Q} \frac{\mu(q)^2 d(q)^2 q}{\varphi(q)^2} \right)^{1/2} \left(\sum_{q\le Q} q E_\vartheta(N;q)^2 \right)^{1/2}.$$

The first sum over q is

$$\le \prod_{p\le Q} \left(1 + \frac{4p}{(p-1)^2}\right) \asymp \prod_{p\le Q} \left(1 - \frac{1}{p}\right)^{-4} \ll (\log Q)^4.$$

By the Bombieri–Vinogradov theorem in the form of Corollary 20.6, the second sum above over q is $\ll N^{3/2} Q (\log N)^4$. These estimates give the stated bound. □

Theorem 20.13 (Bombieri & Davenport 1966) *Let E be defined as in* (20.33). *Then* $E \le 1/2$.

Proof We take $Q \sim N^{1/2}(\log N)^{-10}$. From Lemma 20.10 we deduce that

$$\sum_{q\le Q}\sum_{\substack{a=1\\(a,q)=1}}^{q}|S(a/q)T(a/q)|^2 \ge N^2\sum_{q\le Q}\frac{\mu(q)^2}{\varphi(q)^2}\sum_{\substack{a=1\\(a,q)=1}}^{q}|T(a/q)|^2$$
$$+\sum_{q\le Q}\sum_{\substack{a=1\\(a,q)=1}}^{q}|T(a/q)|^2U(a/q)$$
$$+O\left(N(\log N)\sum_{q\le Q}\sum_{\substack{a=1\\(a,q)=1}}^{q}|T(a/q)|^2\right).$$

Suppose that $H \asymp \log N$. By Lemma 20.9 we know that the first term on the right above is $= N^2\big(\frac{1}{2}H\log N + H^2\big) + O\big(N^2(\log N)(\log\log N)\big)$. By Lemma 20.12, the second term above contributes $\ll N^2\log N$. The final error term above we estimate trivially: $|T(\alpha)| \le H$ for all α, and the double sum has $\le Q^2$ summands. Thus this final error term is $\ll N^2(\log N)^{-17}$. Hence, by (20.34),

$$\int_0^1 |S(\alpha)T(\alpha)|^2 d\alpha \ge \frac{1}{H}N\log N + NH^2 + O((N\log N)\log\log N).$$

Therefore by Lemma 20.8,

$$\frac{1}{2}HN + 2\sum_{h=1}^{H}(H-h)R(N;h) \ge H^2N + O\big(N(\log N)(\log\log N)\big). \quad (20.36)$$

Set $H \sim c\log N$ with $c > 1/2$. Then the sum over h must be positive, and indeed

$$\sum_{h=1}^{H}(H-h)R(N;h) \gg_c N(\log N)^2.$$

Thus $p_{n+1} - p_n \le c\log N$ for many primes $p_n \le N$. □

20.3.1 Exercises

In Exercise 2.1.1.17, a crude version of an estimate of Ward (1927) for $\sum_{n\le x}\mu(n)^2/\varphi(n)$ was proposed, without indicating the method of approach. In Exercise 20.3.1.1 we sketch an elegant treatment.

Let $Q(x)$ denote the number of square-free integers not exceeding x; an elementary estimate for this was established in Theorem 2.2. In §6.2 it was noted that the analytic method used to prove the Prime Number Theorem can also be used to show that $M(x) \ll x\exp\big(-c\sqrt{\log x}\big)$. From this we argued by elementary methods that

$$Q(x) = \frac{6}{\pi^2}x + O\big(x^{1/2}\exp\big(-c\sqrt{\log x}\big)\big)$$

in Exercises 6.2.1.8 and 6.2.1.19. In Exercise 20.3.1.3 we sketch a correspondingly improved estimate for Ward's important sum. Of course we know that RH implies the better estimate for $Q(x)$ found in Exercise 17.3.1.5(j), which would yield (assuming RH) a smaller error term.

1. (a) Explain why

$$\sum_{n\le y}\frac{\mu(n)^2 n}{\varphi(n)} \le \sum_{n\le y}\sum_{m|n}\frac{\mu(m)^2}{\varphi(m)} \le \sum_{m\le y}\frac{\mu(m)^2 y}{\varphi(m)m} \ll y.$$

By integrating by parts, or otherwise, show that

$$\sum_{n>y}\frac{\mu(n)^2}{n\varphi(n)} \ll y^{-1}.$$

(b) Let f be the multiplicative function defined by the relations $f(p) = -f(p^2) = \frac{1}{p(p-1)}$, $f(p^k) = 0$ for $k > 0$. Let $g(n) = 1/n$ for all n. Show that

$$\frac{\mu(n)^2}{\varphi(n)} = \sum_{m|n} f(m)g(n/m).$$

(c) Show that

$$\sum_{n\le x}\frac{\mu(n)^2}{\varphi(n)} = \sum_{m\le x} f(m)\left(\log\frac{x}{m} + C_0\right) + O\left(\sum_{m\le x}|f(m)|m/x\right).$$

(d) Show that if m is cube-free, then m is uniquely of the form $m = d_1 d_2^2$ where d_1 and d_2 are square-free and $(d_1, d_2) = 1$. Show also that

$$f(m) = \frac{\mu(d_2)}{d_1 d_2 \varphi(d_1 d_2)}$$

for such d_i.

(e) Show that

$$\sum_{m\le x}|f(m)|m \ll \sum_{d_1\le x}\frac{\mu(d_1)^2}{\varphi(d_1)}\sum_{d_2\le (x/d_1)^{1/2}}\frac{\mu(d_2)^2 d_2}{\varphi(d_2)} \ll x^{1/2}.$$

(f) Show that

$$\sum_{m>x}|f(m)| \ll \sum_{d_1 d_2^2 > x}\frac{\mu(d_1)^2\mu(d_2)^2}{d_1\varphi(d_1)d_2\varphi(d_2)}$$

$$\ll \sum_{d_1\le x}\frac{\mu(d_1)^2(x/d_1)^{-1/2}}{d_1\varphi(d_1)} + \sum_{d_1>x}\frac{\mu(d_1)^2}{d_1\varphi(d_1)} \ll x^{-1/2}.$$

(g) Deduce that

$$\sum_{m>x} |f(m)| \log \frac{m}{x} = \int_x^\infty \sum_{m>y} |f(m)| \frac{dy}{y} \ll x^{-1/2}.$$

(h) Show that

$$\sum_{\substack{m=1\\(m,q)=1}}^\infty f(m) = 1$$

for all positive integers q.

(i) Deduce that

$$\sum_{n\le x} \frac{\mu(n)^2}{\varphi(n)} = \log x + C_0 - \sum_{m=1}^\infty f(m)\log m + O\big(x^{-1/2}\big).$$

(j) Show that

$$\sum_{m=1}^\infty f(m)\log m = \sum_{d=1}^\infty \Lambda(d) \sum_{m=1}^\infty f(md).$$

(k) Show that

$$\sum_{m=1}^\infty f(pm) = 0, \qquad \sum_{m=1}^\infty f(p^2m) = \frac{-1}{p(p-1)}.$$

(l) Conclude that

$$\sum_{n\le x} \frac{\mu(n)^2}{\varphi(n)} = \log x + C_0 + \sum_p \frac{1}{p(p-1)} + O\big(x^{-1/2}\big).$$

2. Let $R(x)$ be defined by the relation $Q(x) = \frac{6}{\pi^2}x + R(x)$. In Exercises 6.2.1.8 and 6.2.1.19 it was shown that $R(x) \ll x^{1/2}\exp\big(-c\sqrt{\log x}\big)$. Deduce that there is a constant D such that

$$\sum_{n\le x} \frac{\mu(n)^2}{n} = \frac{6}{\pi^2}\log x + D + O\big(x^{-1/2}\exp\big(-c\sqrt{\log x}\big)\big).$$

3. (a) Show that if $\sigma > 1$, then

$$\sum_{n=1}^\infty \frac{\mu(n)^2}{\varphi(n)n^{s-1}} = \frac{\zeta(s)}{\zeta(2s)} \prod_p \left(1 + \frac{1}{(p-1)(p^s+1)}\right) = \frac{\zeta(s)}{\zeta(2s)} F(s),$$

say.

(b) Show that $F(s) = \sum_n f(n)n^{-s}$ where

$$f(n) = \lambda(n) \prod_{p|n} \frac{1}{1-p}.$$

(c) Deduce that

$$\frac{\mu(n)^2}{\varphi(n)} = \sum_{dm=n} \frac{f(d)}{d} \frac{\mu(m)^2}{m}.$$

(d) Let

$$G(s) = \prod_p \left(1 + \frac{1}{(p-1)(p^s-1)}\right).$$

Note that this product is absolutely convergent for $\sigma > 0$. Show that $G(s) = \sum_n g(n)n^{-s}$ where

$$g(n) = \prod_{p|n} \frac{1}{p-1} = |f(n)|.$$

(e) Show that

$$\sum_{n\le x} \frac{\mu(n)^2}{\varphi(n)} = \sum_{d\le x} \frac{f(d)}{d} \sum_{m\le x/d} \frac{\mu(m)^2}{m}.$$

(f) Deduce that

$$\sum_{n\le x} \frac{\mu(n)^2}{\varphi(n)} = \sum_{d\le x} \frac{f(d)}{d}\left(\frac{6}{\pi^2}\log x/d + D\right) + O\Big(x^{-1/2}\exp\big(-\tfrac{c}{2}\sqrt{\log x}\big) \sum_{d\le x^{3/4}} \frac{g(d)}{d^{1/2}}\Big) + O\Big(x^{-1/2} \sum_{x^{3/4}<d<x} \frac{g(d)}{d^{1/2}}\Big). \quad (20.37)$$

(g) Show that $\sum_{n\le x} g(n) \le G(\varepsilon)x^{\varepsilon}$. Thus $\sum_{n\le x} g(n) \ll_\varepsilon x^{\varepsilon}$. (This can also be established by appealing to Theorem 1.3.)

(h) Let $\alpha > 0$ be fixed. Show that

$$\sum_{d>y} \frac{g(d)}{d^\alpha} \ll_\varepsilon y^{-\alpha+\varepsilon}.$$

(i) Deduce that the second error term in (20.37) is $\ll_\varepsilon x^{-7/8+\varepsilon}$.

(j) (Ward 1927) Note that $F(1) = \zeta(2)$. Conclude that

$$\sum_{n\le x} \frac{\mu(n)^2}{\varphi(n)} = \log x + E + O\big(x^{-1/2}\exp\big(-\tfrac{c}{2}\sqrt{\log x}\big)\big)$$

for some constant E (which is determined in Exercise 20.3.1.1).

4. (a) Suppose that a is a positive integer and

$$f(p) = \begin{cases} \sum_{t|(p+1)/a} |\mu(t)| d\left(\frac{p+1}{at}\right) & \text{when } p \equiv -1 \pmod{a}, \\ 0 & \text{otherwise.} \end{cases}$$

Prove that there is a positive number $C(a)$ such that for $X \geq X_0(a)$ we have

$$\sum_{\substack{p \leq X \\ p \equiv -1 \pmod{a}}} f(p) > C(a) X (\log X)^2.$$

(b) (Vaughan 1970) Let $E_a(N)$ denote the number of natural numbers n not exceeding N such that

$$\frac{a}{n} = \frac{1}{x} + \frac{1}{y} + \frac{1}{z}$$

is insoluble in integers. Prove that there is a positive number $C(a)$ such that

$$E_a(N) \ll N \exp\big(-C(a)(\log N)^{2/3}\big).$$

5. Let $\tau(n)$ denote the number of square-free divisors of n,

$$\tau(n) = \sum_{m|n} \mu(m)^2.$$

Prove that

$$\sum_{p \leq x} \tau(p+1) = x + O\left(\frac{x \log\log x}{\log x}\right).$$

6. In this exercise, combined with the next two exercises, we establish an improved upper bound for the number of twin primes. Let $f(n)$ and $g(n)$ be multiplicative functions defined as follows:

$$f(n) = \prod_{p^\alpha \| n} \frac{1}{(p-1)^\alpha}, \qquad g(n) = \prod_{p^\alpha \| n} \frac{p^{\alpha-1}}{(p-1)^\alpha}.$$

(a) Show that $nf(n) = \sum_{d|n} g(d)$.

(b) Show that

$$\sum_{\substack{n \leq z \\ 2 \nmid n}} f(n) = \sum_{\substack{d \leq z \\ 2 \nmid d}} \frac{g(d)}{d} \sum_{\substack{m \leq z/d \\ 2 \nmid m}} \frac{1}{m}.$$

(c) Show that

$$\sum_{\substack{m \leq w \\ 2 \nmid m}} \frac{1}{m} = \frac{1}{2} \log w + C_1 + O(1/w)$$

where $C_1 = (C_0 + \log 2)/2$.

(d) Show that

$$\sum_{\substack{d=1\\2\nmid d}}^{\infty} \frac{g(d)}{d} = \frac{2}{c},$$

where c is defined as in (20.28).

(e) Show that

$$\sum_{\substack{n\le z\\2\nmid n}} f(n) = \frac{\log z}{c} + C_2 + O((\log z)/z),$$

where

$$C_2 = \frac{C_0 + \log 2}{c} - \frac{1}{2}\sum_{\substack{d=1\\2\nmid d}}^{\infty} \frac{g(d)\log d}{d}.$$

7. Let

$$\varphi_2(n) = n\prod_{p|n}\left(1 - \frac{2}{p}\right).$$

(a) Show that

$$\sum_{\substack{n\le z\\2\nmid n}} \frac{\mu(n)^2}{\varphi_2(n)} = \sum_{\substack{n\le z\\2\nmid n}} \frac{\mu(n)^2}{\varphi(n)}\prod_{p|n}\left(1 + \frac{1}{p-1} + \frac{1}{(p-1)^2} + \cdots\right).$$

Let $f(n)$ be defined as in Exercise 20.3.1.6. Explain why the right-hand side above is

$$\ge \sum_{\substack{n\le z\\2\nmid n}} f(n).$$

(b) Conclude that

$$\sum_{\substack{n\le z\\2\nmid n}} \frac{\mu(n)^2}{\varphi_2(n)} \ge \frac{\log z}{c} + O(1),$$

where c is defined as in (20.28).

8. Put $P = \prod_{2<p\le z} p$, and let Λ_d be real numbers such that $\Lambda_1 = 1$ and $\Lambda_d = 0$ for $d > z$.

(a) Explain why the number of primes $p \le x$ for which $p + 2$ is prime does not exceed

$$\pi(z) + \sum_{p\le x}\left(\sum_{d|(p+2)} \Lambda_d\right)^2.$$

(b) Show that the multiple sum in part (a)

$$= \sum_{\substack{d|P \\ e|P}} \Lambda_d \Lambda_e \pi(x; [d,e], -2)$$

$$= \operatorname{li}(x) \sum_{\substack{d|P \\ e|P}} \frac{\Lambda_d \Lambda_e}{\varphi([d,e])} + \sum_{\substack{d|P \\ e|P}} \Lambda_d \Lambda_e E_\pi(x; [d,e], -2). \qquad (20.38)$$

(c) Show that if f is a multiplicative function, then $f((d,e))f([d,e]) = f(d)f(e)$.

(d) Let $\varphi_2(n)$ be defined as in Exercise 20.3.1.7. Show that if n is square-free, then

$$\varphi(n) = \sum_{d|n} \varphi_2(d).$$

(e) Show that the first sum in (20.38) is $\sum_{\delta|P} \varphi_2(\delta) y_\delta^2$ where

$$y_\delta = \sum_{\substack{d|P \\ \delta|d}} \frac{\Lambda_d}{\varphi(d)}.$$

(f) Show that if $\Lambda_d = 0$ for $d > z$, then $y_\delta = 0$ for $\delta > z$.

(g) Show that

$$\Lambda_d = \varphi(d) \sum_{\substack{\delta|P \\ d|\delta}} \mu(\delta/d) y_\delta.$$

(h) Show that if $y_\delta = 0$ for $\delta > z$, then $\Lambda_d = 0$ for $d > z$.

(i) Explain why $\sum_{\delta|P} \mu(\delta) y_\delta = 1$.

(j) Put

$$L = \sum_{\substack{\delta \le z \\ 2\nmid\delta}} \frac{\mu(\delta)^2}{\varphi_2(\delta)}.$$

Show that

$$\sum_{\substack{\delta|P \\ \delta\le z}} \varphi_2(\delta) y_\delta^2 = \frac{1}{L} + \sum_{\substack{\delta|P \\ \delta\le z}} \varphi_2(\delta) \big(y_\delta - \mu(\delta)/(L\varphi_2(\delta))\big)^2.$$

(k) Take $y_\delta = \mu(\delta)/(L\varphi_2(\delta))$ for $\delta|P$, $\delta \le z$. Show that the first term in (20.38) is

$$\le \frac{c \operatorname{li}(x)}{\log z} + O\big(x/((\log x)(\log z)^2)\big).$$

(l) Show that

$$\Lambda_d = \frac{\mu(d)\varphi(d)}{L\varphi_2(d)} \sum_{\substack{r\le z/d\\(r,2d)=1}} \frac{\mu(r)^2}{\varphi_2(r)}.$$

(m) Explain why

$$\frac{\varphi(d)}{\varphi_2(d)} \sum_{\substack{r\le z/d\\(r,2d)=1}} \frac{\mu(r)^2}{\varphi_2(r)} \le L,$$

and hence deduce that $|\Lambda_d| \le 1$ for all d.

(n) Show that if $q|P$, then

$$\sum_{\substack{d,e\\ [d,e]=q}} |\Lambda_d\Lambda_e| \le 3^{\omega(q)}.$$

(o) Show that the second term in (20.38) has absolute value not exceeding

$$\sum_{\substack{q\le z^2\\ 2\nmid q}} \mu(q)^2 3^{\omega(q)} E_\pi(x,q).$$

(p) Show that

$$\sum_{\substack{q\le z^2\\ 2\nmid q}} \frac{\mu(q)^2 9^{\omega(q)}}{q} \le \prod_{2<p\le z^2}\left(1+\frac{9}{p}\right) \ll (\log z)^9.$$

(q) Deduce by (20.27) that the second term in (20.38) is

$$\ll x^{3/4} z (\log xz)^6.$$

(r) Take $z = x^{1/4}(\log x)^{-9}$. Conclude that the number of primes $p \le x$ for which $p+2$ is prime does not exceed

$$\frac{4cx}{(\log x)^2}\left(1+O\left(\frac{\log\log x}{\log x}\right)\right),$$

where c is defined as in (20.28). This bound is smaller by a factor of 2 than the bound we obtained in §3.4.

20.4 Mean Square Distribution

We begin with an upper bound for the mean square error in the Prime Number Theorem for arithmetic progressions, which we then use to derive an asymptotic estimate for the same quantity.

Theorem 20.14 *Let A be fixed. If $x/(\log x)^A \le Q \le x$, then*

$$\sum_{q\le Q} \sum_{\substack{a=1\\(a,q)=1}}^{q} (\psi(x;q,a) - x/\varphi(q))^2 \ll Qx\log x. \tag{20.39}$$

Proof We start by recalling the identity (20.16). From the orthogonality property of Dirichlet characters (as in (4.12) or Exercise 4.2.1.3), it follows that

$$\sum_{\substack{a=1\\(a,q)=1}}^{q} (\psi(x;q,a) - x/\varphi(q))^2 = \frac{1}{\varphi(q)} \sum_{\chi} |\psi'(x,\chi)|^2.$$

If $\chi^\star$ is the primitive character that induces χ, then $\psi'(x,\chi)$ differs little from $\psi'(x,\chi^\star)$, as we see from (20.19). Hence the left-hand side of (20.39) is

$$\ll \sum_{q\le Q} \frac{1}{\varphi(q)} \sum_{\chi} (|\psi'(x,\chi^\star)|^2 + (\log qx)^4)$$

$$\ll \sum_{d\le Q} \Bigg(\sideset{}{^\star}\sum_{\chi \,(\mathrm{mod}\, d)} |\psi'(x,\chi)|^2 \Bigg) \Bigg(\sum_{\substack{q\le Q\\ d\mid q}} \frac{1}{\varphi(q)} \Bigg) + Q(\log Qx)^4.$$

From the estimate (20.20) we see that it suffices to show that

$$\sum_{q\le Q} \frac{\log \frac{2Q}{q}}{\varphi(q)} \sideset{}{^\star}\sum_{\chi} |\psi'(x,\chi)|^2 \ll Qx\log x. \tag{20.40}$$

By the Siegel–Walfisz theorem in the form of Corollary 11.18 we know that

$$\psi'(x,\chi) \ll x\exp\big(-c\sqrt{\log x}\big)$$

for $q \le (\log x)^{A+2}$. The contribution of such q is therefore

$$\ll x^2(\log x)^{A+3}\exp\big(-c\sqrt{\log x}\big) \ll x^2(\log x)^{-A} \ll Qx.$$

Consider now a range $Q_1 < q \le 2Q_1$ with $1 < Q_1 \le Q$. Then $\psi'(x,\chi) = \psi(x,\chi)$, and the contribution is

$$\ll \frac{\log \frac{2Q}{Q_1}}{Q_1} \sum_{Q_1<q\le 2Q_1} \frac{q}{\varphi(q)} \sideset{}{^\star}\sum_{\chi} |\psi(x,\chi)|^2.$$

By the large sieve (Theorem 19.16) this is

$$\ll \frac{\log \frac{2Q}{Q_1}}{Q_1}(x+Q_1^2)\sum_{n\le x}\Lambda(n)^2 \ll (x^2Q_1^{-1}+xQ_1)(\log x)\log\frac{2Q}{Q_1}.$$

We cover the interval $(\log x)^{A+2} \le q \le Q$ with ranges of the above sort, and sum, to obtain (20.40). Thus the proof is complete. □

For many applications the estimate of Theorem 20.13 is sufficient, but it is interesting to note that with a little more work we can obtain not just an upper bound but an asymptotic estimate. To prepare for the main argument we first establish a lemma.

Lemma 20.15 *There exist absolute constants a and b such that*

$$\sum_{n\le y}\frac{(1-n/y)^2}{\varphi(n)} = \frac{\zeta(2)\zeta(3)}{\zeta(6)}\log y + a + \frac{\log y}{y} + \frac{b}{y} + O_\varepsilon\big(y^{-3/2+\varepsilon}\big) \quad (20.41)$$

for $y \ge 1$.

Proof By manipulating Euler products we see that

$$\begin{aligned}\sum_{n=1}^{\infty}\frac{1}{\varphi(n)n^s} &= \zeta(s+1)\prod_p\left(1+\frac{1}{(p-1)p^{s+1}}\right)\\ &= \zeta(s+1)\zeta(s+2)\prod_p\left(1+\frac{1}{(p-1)p^{s+2}}-\frac{1}{(p-1)p^{2s+3}}\right)\\ &= \zeta(s+1)\zeta(s+2)F(s),\end{aligned}$$

say. By taking $k = 2$ in (5.19), we see that in (20.41) the left-hand side is

$$= \frac{2}{2\pi i}\int_{\sigma_0-i\infty}^{\sigma_0+i\infty}\zeta(s+1)\zeta(s+2)F(s)\frac{y^s}{s(s+1)(s+2)}\,ds,$$

where $\sigma_0 > 0$. The Euler product $F(s)$ is absolutely convergent for $\sigma > -3/2$, and is uniformly bounded for $\sigma \ge -3/2+\delta$. We let σ_1 be slightly larger than $-3/2$, and apply Cauchy's theorem with a path from $\sigma_0 - iT$ to $\sigma_0 + iT$ to $\sigma_1 + iT$ to $\sigma_1 - iT$ to $\sigma_0 - iT$. By Corollaries 1.17 and 10.5 we see that $\zeta(s+1)\zeta(s+2) \ll \tau^{3/2}$ on this contour. Thus the integral from $\sigma_1 + iT$ to $\sigma_1 - iT$ is $\ll y^{\sigma_1}$. Within the contour the integrand has double poles at $s = 0$ and at $s = -1$. The residue at $s = 0$ is

$$\zeta(2)F(0)\left(C_0 + \frac{\zeta'}{\zeta}(2) + \frac{F'}{F}(0) - \frac{3}{2}\right).$$

This gives the first two main terms since $G(0) = \zeta(3)/\zeta(6)$. At $s = -1$, the residue is

$$-2\zeta(0)F(-1)y^{-1}\left(\frac{\zeta'}{\zeta}(0) + C_0 + \frac{F'}{F}(-1) + \log y\right).$$

We recall (10.11), which asserts that $\zeta(0) = -1/2$. Since $F(-1) = 1$, we have the remaining main terms. □

Theorem 20.16 *Let $A > 0$ be fixed. If $x/(\log x)^A \le Q \le x$, then*

$$\sum_{q\le Q}\sum_{\substack{a=1\\(a,q)=1}}^{q}(\psi(x;q,a) - x/\varphi(q))^2 = Qx\log Q + O(Qx). \tag{20.42}$$

Proof Let $Q_1 = x(\log x)^{-A-1}$. By Theorem 20.14, the contribution of $q \le Q_1$ to the above is $\ll x^2(\log x)^{-A} \ll Qx$. Thus we may restrict our attention to the range $Q_1 \le q \le Q$. The inner sum on the left-hand side is

$$= \sum_{\substack{a=1\\(a,q)=1}}^{q}\psi(x;q,a)^2 - 2\frac{x}{\varphi(q)}\sum_{\substack{a=1\\(a,q)=1}}^{q}\psi(x;q,a) + \frac{x^2}{\varphi(q)}. \tag{20.43}$$

Here the second sum is

$$\begin{aligned}&= \sum_{\substack{n\le x\\(n,q)=1}}\Lambda(n) = \psi(x) - \sum_{p\mid q}\left[\frac{\log x}{\log p}\right]\log p\\ &= x + O\big((\log qx)^2\big) + O\big(x\exp\big(-c\sqrt{\log x}\big)\big).\end{aligned}$$

The first sum in (20.43) is

$$\sum_{\substack{m,n\le x\\ m\equiv n\ (q)\\ (mn,q)=1}}\Lambda(m)\Lambda(n).$$

If the condition $(mn,q) = 1$ is omitted, then the value of the above is changed by not more than

$$\sum_{p\mid q}\left[\frac{\log x}{\log p}\right]^2(\log p)^2 \ll (\log qx)^3.$$

In Exercise 2.1.1.16(c) it was established that

$$\sum_{q\le x}\frac{1}{\varphi(q)} = \frac{\zeta(2)\zeta(3)}{\zeta(6)}\left(\log x + C_0 - \sum_p\frac{\log p}{p^2-p+1}\right) + O\Big(\frac{\log x}{x}\Big).$$

Hence

$$\sum_{Q_1<q\le Q}\frac{1}{\varphi(q)}=\frac{\zeta(2)\zeta(3)}{\zeta(6)}\log\frac{Q}{Q_1}+O\Big(\frac{\log Q_1}{Q_1}\Big).$$

Thus we deduce that

$$\begin{aligned}\sum_{Q_1<q\le Q}\ &\sum_{\substack{a=1\\(a,q)=1}}^{q}(\psi(x;q,a)-x/\varphi(q))^2\\ &=\sum_{Q_1<q\le Q}\ \sum_{\substack{m,n\le x\\ m\equiv n\ (q)}}\Lambda(m)\Lambda(n)-\frac{\zeta(2)\zeta(3)}{\zeta(6)}x^2\log\frac{Q}{Q_1}+O(Qx).\end{aligned}$$

The terms with $m=n$ contribute an amount

$$(Q-Q_1+O(1))\sum_{n\le x}\Lambda(n)^2=Qx\log x+O(Qx).$$

Hence to obtain the stated result it suffices to show that

$$\begin{aligned}\sum_{Q_1<q\le Q}\ &\sum_{\substack{m<n\le x\\ m\equiv n\ (q)}}\Lambda(m)\Lambda(n)\\ &=\frac{\zeta(2)\zeta(3)}{2\zeta(6)}x^2\log\frac{Q}{Q_1}-\frac{1}{2}Qx\log\frac{x}{Q}+O(Qx).\end{aligned}\tag{20.44}$$

To this end we show that

$$\begin{aligned}\sum_{y<q\le x}\ &\sum_{\substack{m<n\le x\\ m\equiv n\ (q)}}\Lambda(m)\Lambda(n)\\ &=\frac{\zeta(2)\zeta(3)}{2\zeta(6)}x^2\log\frac{x}{y}+\frac{a}{2}x^2+\frac{1}{2}xy\log\frac{x}{y}+O(xy)\end{aligned}\tag{20.45}$$

for $x(\log x)^{-A-1}\le y\le x$, where a is the constant in Lemma 20.15. This suffices, for on taking $y=Q_1$ and $y=Q$, and differencing, we obtain (20.44).

The left-hand side of (20.45) is

$$\begin{aligned}\sum_{y<q\le x}\ \sum_{0<k\le x/q}\ &\sum_{0<m\le x-kq}\Lambda(m)\Lambda(m+kq)\\ &=\sum_{0<k\le x/y}\ \sum_{y<q\le x/k}\ \sum_{0<m\le x-kq}\Lambda(m)\Lambda(m+kq)\\ &=\sum_{0<k\le x/y}\ \sum_{0<m\le x-ky}\Lambda(m)\sum_{y<q\le(x-m)/k}\Lambda(m+kq)\\ &=\sum_{0<k\le x/y}\ \sum_{0<m\le x-ky}\Lambda(m)(\psi(x;k,m)-\psi(m+ky;k,m)).\end{aligned}$$

If m is a prime power and $(m,k)>1$, then $m=p^r$, say, where $p|k$, and the prime powers congruent to m modulo k are powers of the same prime p. Thus the pairs $m,\ k$ for which $(m,k)>1$ contribute to the above an amount

$$\ll \sum_{k\le x/y}\sum_{p|k}\left[\frac{\log x}{\log p}\right](\log p)^2 \ll \sum_{k\le x/y}(\log kx)^3 \ll (\log x)^{A+4}.$$

On the other hand, by the Siegel–Walfisz theorem (Corollary 11.19), the pairs $k,\ m$ for which $(k,m)=1$ contribute the amount

$$\sum_{0<k\le x/y}\frac{1}{\varphi(k)}\sum_{\substack{0<m\le x-ky\\(m,k)=1}}\Lambda(m)(x-m-ky)$$
$$+O\left(\sum_{0<k\le x/y}\sum_{m\le x-ky}\Lambda(m)x\exp\left(-c\sqrt{\log x}\right)\right).$$

The error term here is $\ll(\log x)^{A+1}x^2\exp\big(-c\sqrt{\log x}\big)\ll x^2(\log x)^{-A}$, so can be ignored. In the main term, if the condition that $(m,k)=1$ is dropped, then the expression is altered by an amount that is

$$\ll x\sum_{0<k\le x/y}\sum_{p|k}\left[\frac{\log x}{\log p}\right]\log p \ll x(\log x)^3 \ll x^2(\log x)^{-A}.$$

By the Prime Number Theorem we know that

$$\sum_{m\le z}\Lambda(m)(z-m)=\frac{1}{2}z^2+O\big(z^2\exp\big(-c\sqrt{\log z}\big)\big).$$

On taking $z=x-ky$, we see that the remaining main term is

$$\frac{1}{2}\sum_{0<k\le x/y}\frac{(x-ky)^2}{\varphi(k)}+O\left(x^2\exp\left(-c\sqrt{\log x}\right)\sum_{0<k\le x/y}\frac{1}{\varphi(k)}\right).$$

By Lemma 20.15 this is

$$=\frac{\zeta(2)\zeta(3)}{2\zeta(6)}x^2\log\frac{x}{y}+\frac{a}{2}x^2+\frac{1}{2}xy\log\frac{x}{y}+O(xy).$$

Thus we have (20.45), and the proof is complete. □

20.4.1 Exercises

1. Suppose that $q \ge x$. Explain why

$$\sum_{\substack{a=1\\(a,q)=1}}^{q}\left(\psi(x;q,a)-\frac{x}{\varphi(q)}\right)^2 \ll \frac{x^2}{\varphi(q)}+\sum_{n\le x}\Lambda(n)^2$$

$$\ll \frac{x^2}{q}\log q + x\log x \ll x\log x.$$

2. The object of Exercise 4.2.1.2 was to show that

$$\sum_{\chi}\left|\sum_{n=1}^{q}c_n\chi(n)\right|^2 = \varphi(q)\sum_{\chi}|c_\chi|^2,$$

where the c_n are arbitrary and χ runs over all Dirichlet characters modulo q in the sum on the left. By a suitable application of this, or otherwise, show that

$$\sum_{\substack{a=1\\(a,q)=1}}^{q}\left(\psi(x;q,a)-\frac{x}{\varphi(q)}\right)^2 = \frac{1}{\varphi(q)}\sum_{\chi\ne\chi_0}|\psi(x,\chi)|^2+\frac{(\psi(x;\chi_0)-x)^2}{\varphi(q)}.$$

20.5 Notes

Section 20.1. Let $N(\alpha, T)$ denote the number of zeros of the Riemann zeta function in the rectangle $\alpha \le \sigma \le 1$, $0 < t \le T$. An estimate for $N(\alpha, T)$ is known as a *zero-density theorem*, although the estimate is not actually a density. To the extent that it can be shown that the zeta function does not have many zeros with large real part, various consequences can be derived concerning the distribution of prime numbers. For a Dirichlet character χ, let $N(\alpha, T; \chi)$ denote the number of zeros of $L(s, \chi)$ in the rectangle $\alpha \le \sigma \le 1$, $|t| \le T$. Bombieri (1965) used his form of the large sieve to derive upper bounds for quantities roughly of the form $\sum_{q\le Q}\sum_{\chi}^{\star} N(\alpha, T, \chi)$. The bounds obtained were then used to estimate sums of $|\psi(x, \chi)|$, and those bounds were in turn used to derive the Bombieri–Vinogradov theorem. Bombieri's derivation of zero-density estimates involved much work; Gallagher (1968) was the first to obtain corresponding bounds for sums of $|\psi(x, \chi)|$ without considering zero densities, although his arguments still involved inverse Mellin transforms and contour integrals. Vaughan (1975) simplified Gallagher's arguments somewhat, but it was in Vaughan (1977, 1980) that he introduced his decomposition (17.5) of $\Lambda(n)$, which allows us to derive estimates for sums of $|\psi(x,\chi)|$ from the large sieve in an entirely elementary way.

Section 20.2. Bombieri (1965) based his argument on a deep application of his version of the large sieve to give suitable estimates for the distribution of zeros of Dirichlet L-functions, while the work of A. I. Vinogradov (1965), which was entirely independent, did not involve the large sieve, was much more complicated, and led to slightly weaker estimates.

In some situations we do not need an estimate for each individual $E(x;q,a)$; rather a bound for a sum of such quantities suffices. Following Wang (1962), we say that the primes are distributed with level α if

$$\sum_{q\le x^{\alpha-\varepsilon}} \max_{(a,q)=1} |E(x;q,a)| = O\big(x(\log x)^{-A}\big) \tag{20.46}$$

for arbitrarily large fixed $A>0$. Barban (1963) and Pan (1963, 1964) claimed proofs that $\alpha = 3/8$ could be achieved, but before their complicated work could be evaluated, Bombieri and Vinogradov achieved $\alpha = 1/2$, which is exactly what follows from GRH.

The assertion (20.46) with $\alpha = 1$ is the Elliott–Halberstam conjecture (Conjecture 20.4; see Elliott & Halberstam (1970)).

The question arises as to the extent to which one can increase the range for q in the Bombieri–Vinogradov theorem when one relaxes conditions such as taking the maximum over a or the absolute value of

$$\psi(x;q,a) - \frac{x}{\varphi(q)}.$$

In this context there is a series of papers, Fouvry (1982, 1984), Fouvry & Iwaniec (1980, 1983), Bombieri, Friedlander, & Iwaniec (1986, 1987, 1989, 2019), and more recently Assing, Blomer, & Li (2021) in which the main innovation is the introduction of estimates for incomplete Kloosterman sums. In the last of these papers it is shown *inter alia* that

$$\sum_{\substack{q\le Q\\(q,a_1a_2)=1\\q\equiv c_0 \bmod c}} \left(\sum_{\substack{n\le x\\ a_2n\equiv a_1 \bmod q\\ n\equiv d_0 \bmod d}} \Lambda(n) - \frac{1}{\varphi(qd)} \sum_{\substack{n\le x\\(n,qd)=1}} \Lambda(n) \right) \ll_{C,A} x(\log x)^{-A} \tag{20.47}$$

provided that $Q \le x^{1/2+\delta}$ for some small positive δ and c, d, c_0d_0, a_1, a_2 satisfy various conditions, including

$$0<|a_1|\le x^{1+\delta},\ 0<|a_2|\le x^{\delta},\ c,d\le(\log x)^C,\ (d_0,d)=(c_0,c)=1.$$

In a different direction as a significant part of his work on bounded gaps between primes, Zhang (2014) showed that, for $Q \le x^{1/2+\delta}$,

$$\sum_{q \le \mathcal{A}(Q,R)} \sum_{c \in \mathcal{C}(q)} \Bigg| \sum_{\substack{x \le n < 2x \\ n \equiv c \bmod q}} \Lambda(n) - \frac{1}{\varphi(q)} \sum_{\substack{x \le n < 2x \\ (n,q)=1}} \Lambda(n) \Bigg| \ll_A (\log x)^{-A},$$

where $\mathcal{A}(Q, R)$ is the set of R-factorable numbers q not exceeding Q, i.e., the q with no prime factor exceeding R, and where $\mathcal{C}(q)$ is a set of solutions of a special polynomial congruence modulo q.

Section 20.3. The constant 4 has been improved slightly by complicated combinatorial arguments based on systems of weights. As far as we have been able to ascertain, the best bound so far is in Wu (2004) where the 4 is replaced by 3.91045. Wu gives a review of earlier work on the subject. For other sieve applications of the Bombieri–Vinogradov Theorem see §21.5 and §22.2.

Titchmarsh (1930) showed that

$$\sum_{a<p\le n} d(p-a) = \frac{\zeta(2)\zeta(3)}{\zeta(6)} \prod_{p|a} \Big(1 - \frac{p}{p^2-p+1}\Big) n + O\Big(\frac{n \log\log 3n}{\log n}\Big).$$

for fixed non-zero integer a as $n \to \infty$, assuming that GRH holds. This has become known as the Titchmarsh divisor problem. A proof independent of this hypothesis was given by Linnik (1961), with a slightly worse form of the error term; see also Bredihin (1963). Rodriguez (1965) gave a proof based on Bombieri's theorem. The cognate problem considered in Theorem 2.5 was treated by Walfisz (1953) assuming GRH, but note the correction supplied by the reviewer Bateman in *Mathematical Reviews*. Our proof of Theorem 2.5 follows the pattern set by Rodriguez. The various extensions of the Bombieri–Vinogradov theorem with a fixed and using incomplete Kloosterman sums can be used to improve the error term in the cognate problem. Let E be defined as in (20.33). In an unpublished manuscript *Partitio Numerorum* VII (ca 1926), Hardy & Littlewood showed (assuming GRH) that $E \le 2/3$. Erdős (1940) gave the first unconditional proof that $E < 1$. Let $\pi(x,k)$ denote the number of primes $p \le x$ such that $p+k$ is prime. Erdős showed that if $\pi(x,k) < (c+\varepsilon)\mathfrak{S}_2(k)x/(\log x)^2$ for all k and all large x, then $E \le 1-1/(2c)$. For a detailed derivation of this result, see Exercise 3.4.1.3. Ricci (1954) observed that Selberg's method gives $c = 8$, and hence that $E \le 15/16$. Bombieri (1965) showed that one can take $c = 4$, which gives $E \le 7/8$. Rankin (1940) refined the Hardy–Littlewood argument to obtain $E \le 3/5$ on GRH, and Rankin (1950) showed that $E \le (42/43)(3/5) = 0.5860\ldots$ on GRH by combining his method with that of Erdős. It might seem strange that these authors obtained weaker results from GRH than Bombieri & Davenport (1966) achieved unconditionally. The explanation is that in the last line of the proof of our Lemma 20.10, we discarded a non-negative quantity. It seems that Hardy, Littlewood, and Rankin estimated the size of that term, without recognizing that this is unnecessary. Bombieri & Davenport (1966) combined their results with Erdős's method to show that $E \le (2+\sqrt{3})/8 = 0.466506\ldots$. To see how this is done, see Exercises 21.1.1.1–3. More refined kernels $T(\alpha)$ were introduced by Pil'tjaĭ (1972) and Huxley (1973, 1977) to obtain small further improvements,

$$E \le 0.4571\ldots, \qquad E \le 0.4463\ldots, \qquad E \le 0.4425\ldots,$$

respectively. Maier (1988) contributed a larger improvement by adapting his matrix method (see Volume III) to the situation so as to take advantage of oscillations in the primes over short intervals. This led to the known bound being reduced by a factor of e^{-C_0} where C_0 is Euler's constant and gives $E \le 0.2484\ldots$. This work was completely overtaken by that described in Chapter 22.

Section 20.4. Barban (1963) showed that

$$\sum_{q \le Q} \sum_{\substack{a=1 \\ (a,q)=1}}^{q} (\psi(x;q,a) - x/\varphi(q))^2 \ll x^2(\log x)^{-A}$$

provided that $Q \le x(/\log x)^{-B}$ where $B = B(A)$. Davenport & Halberstam (1966) showed that one may take $B = A + 5$. Then Gallagher (1967) showed that one may take $B = A + 1$ and Montgomery (1970, 1971) showed that if $x(\log x)^{-A} \le Q \le x$, then the above is $= Qx\log x + O\big(Qx\log(2x/Q)\big)$. Hooley (1975) (see also Hooley 1974) then introduced his inversion method which we use to prove Theorem 2.16. This can be seen in the switch from q to k in the treatment of (20.45). Using it, Hooley established that for Q in this same interval the above is $= Qx\log Q - cQx + O\big(Q^{5/4}x^{3/4}\big)$ where $c = C_0 + \log(2\pi) + \sum_p \frac{\log p}{p(p-1)}$. Hooley then followed this over a period of 40 years with a long sequence of papers with the same title investigating various aspects and generalizations of this result. Harper & Soundararajan (2017) and de la Bretèche & Fiorilli (2023) have given lower bounds for the expession displayed above when $x^{1/2} < Q \le x$.

Hooley (1974, 2002) conjectured that

$$\sum_{\substack{a=1 \\ (a,q)=1}}^{q} \left(\psi(x;q,a) - \frac{\psi(x,\chi_0)}{\varphi(q)}\right)^2 \sim x\log q$$

for q in some range (depending on x). *Hooley's Conjecture* is not known to hold in any range, but Fiorilli (2015) has conjectured that it holds for $(\log\log x)^{1+\delta} \le q \le x$ for any fixed $\delta > 0$. Fiorilli & Martin (2023) have shown that the expression above can be much larger than $x \log q$ when $q \asymp \log\log x$. For a survey of the many different generalizations of the Montgomery–Hooley asymptotic formula see Vaughan (2025).

More is known on the Generalized Riemann Hypothesis. The best that is known is due to Goldston & Vaughan (1996). That and Montgomery (1970) are based on applications of the Hardy–Littlewood methods which, whilst more complicated than the Hooley inversion method, sometimes gives more insight and suggests possible improvements. See for example Vaughan (2001, 2003a,b) and the corresponding question concerning the distribution of square-free numbers in arithmetic progression (Vaughan 2005) and more general sequences (Vaughan 1998a,b). This is a very active area with many associated aspects. See the survey article of Vaughan (2025), and references therein.

20.6 References

Assing, E., Blomer, V., & Li, Junxin (2021). Uniform Titchmarsh divisor problems, *Adv. Math.* **393**, Paper No. 108076, 51 pp.

Barban, M. B. (1963). The 'density' of the zeros of Dirichlet L-series and the problem of the sum of primes and 'near primes' (Russian), *Mat. Sb. (N.S.)* **61** (103), 418–425.

Bombieri, E. (1965). On the large sieve, *Mathematika* **12**, 201–225.

Bombieri, E. & Davenport, H. (1966). Small differences between prime numbers, *Proc. Roy. Soc. Ser. A* **293**, 1–18.

Bombieri, E., Friedlander, J. B., & Iwaniec, H. (1986). Primes in arithmetic progressions to large moduli, *Acta Math.* **156**, 203–251.

(1987). Primes in arithmetic progressions to large moduli. II, *Math. Ann.* **277**, 361–393.

(1989). Primes in arithmetic progressions to large moduli. III, *J. Amer. Math. Soc.* **2** no. 2, 215–224.

(2019). Some corrections to an old paper, arXiv:1903.01371, `https://doi.org/10.48550/arXiv.1903.01371`

Bredihin, B. M. (1963). Applications of the dispersion method in binary additive problems.(in Russian), *Dokl. Akad. Nauk SSSR* **149**, 9–11.

de la Bretèche, R. & Fiorilli D. (2023). Moments of moments of primes in arithmetic progressions, *Proc. London Math. Soc.* **127**, 165–220.

Davenport, H. & Halberstam, H. (1966). Primes in arithmetic progressions, *Michigan Math. J.* **13**, 485–489; Corrigendum, *ibid* **14** (1967), 229–232.

Elliott, P. D. T. A. & Halberstam, H. (1970). A conjecture in prime number theory, *Symposia Mathematica*, Vol. IV (INDAM, Rome, 1968/69), pp. 59–72, London: Academic Press, 1970.

Erdős, P. (1940). The difference of consecutive primes, *Duke Math. J.* **6**, 438–441.

Fiorilli, D. (2015). The distribution of the variance of primes in arithmetic progressions, *Int. Math. Res. Not.* **12**, 4421–4448.

Fiorilli, D. & Martin, G. (2023). Disproving Hooley's conjecture, *J. Eur. Math. Soc.* **25**, 4791–4812

Fouvry, É. (1982). Répartition des suites dans les progressions arithmétiques, *Acta Arith.* **41**, 359–382.

(1984). Autour du théorème de Bombieri–Vinogradov, *Acta Math.* **152**, 219–244.

Fouvry, É. & Iwaniec, H. (1980). On a theorem of Bombieri–Vinogradov type, *Mathematika* **27**, 135–172.

(1983). Primes in arithmetic progressions, *Acta Arith.* **42**, no. 2, 197–218.

Gallagher, P. X. (1967). The large sieve, *Mathematika* **14**, 14–20.

(1968). Bombieri's mean value theorem, *Mathematika* **15**, 1–6.

Goldston, D. A. & Vaughan, R. C. (1996). On the Montgomery–Hooley asymptotic formula. In *Proceedings of the Symposium on Sieve methods, Exponential Sums, and Their Applications in Number Theory* (Cardiff 1995), Cambridge: Cambridge University Press, pp. 117–142.

Harper, A. J. & Soundararajan, K. (2017). Lower bounds for the variance of sequences in arithmetic progressions: Primes and divisor functions, *Q. J. Math.* **68**, 97–123.

(1974). The distribution of sequences in arithmetic progressios. In *Proc. International Congress Math.* (Vancouver, BC, 1974), Vol. 1, pp. 357–364, Montreal: Canad. Math. Congress, 1975.

(1975). On the Barban–Davenport–Halberstam theorem. I. Collection of articles dedicated to Helmut Hasse on his seventy-fifth birthday, III., *J. Reine Angew. Math.* **274/275**, 206–223.

(2002). On theorems of Barban–Davenport–Halberstam type. In *Number Theory for the Millennium, II* (Urbana, IL, 2000), pp. 195–228, Natick: A K Peters, 2002.

Huxley, M. N. (1973). Small differences between consecutive primes, *Mathematika* **20**, 229–232.

(1977). Small differences between consecutive primes, II, *Mathematika* **24**, 142–152.

Linnik, J. V. (1961). The dispersion method in binary additive problems/ (Russian), Leningrad: Izdat. Leningrad. Univ.; English translation, Providence: Amer. Math. Soc. (1963).

Maier, H. (1988). Small differences between prime numbers, *Michigan Math. J.* **35**, 323–344.

Montgomery, H. L. (1970). Primes in arihmetic progressions, *Michigan Math. J.* **17**, 33–39.

(1971). *Topics in Multiplicative Number Theory*, Springer Lecture Notes 227, Berlin: Springer, ix+178 pp.

Pan, Cheng Dong (1963). A note on the large sieve method and its applications, *Acta Math. Sinica* **13**, 262–268.

(1964). A new application of the Ju. V. Linnik large sieve method, *Acta Math. Sinica* **14**, 597–606; translated as *Chinese Math.–Acta* **5** (1964), 642–652.

Pil'tjaĭ, G. Z. (1972). On the size of the difference between consecutive primes, *Izdat. Saratov. Univ., Saratov*, 73–79.

Rankin, R. A. (1940). The difference between consecutive prime numbers. II, *Proc. Cambridge Philos. Soc.* **36**, 255–266.

(1950). The difference between consecutive prime numbers. IV, *Proc. Amer. Math. Soc.* **1**, 143–150.

Ricci, G. (1954). Sull'andamento della differenza di numeri primi consecutivi, *Riv. Mat. Univ. Parma* **5**, 3–54.

Rodriquez, G. (1965). Sul problema dei divisori di Titchmarsh, *Bollettino Unione Matematica Italiana* (3) **20**, 358–366.

Titchmarsh, E. C. (1930). A divisor problem, *Rend. Circ. Mat. Palermo* (2) **54**, 414–429; Correction: *ibid* **57** (1933), 478–479.

Vaughan, R. C. (1970). On a problem of Erdős, Straus and Schinzel, *Mathematika* **17**, 193–198.

(1975). Mean value theorems in prime number theory, *J. London Math Soc.* (2) **10**, 153–162.

(1977). Sommes trigonométriques sur les nombres premiers. (French) *C. R. Acad. Sci. Paris Sér. A-B* **285**, no. 16, A981–A983.

(1980). An elementary method in prime number theory, *Acta Arith.* **37**, 111–115.

(1998a). On a variance associated with the distribution of general sequences in arithmetic progressions I, *Phil. Trans. Royal Soc. Lond. A* **356**, 781–791.

(1998b). On a variance associated with the distribution of general sequences in arithmetic progressions II, *Phil. Trans. Royal Soc. Lond. A* **356**, 793–809.

(2001). On a variance associated with the distribution of primes in arithmetic progressions, *Proc. London Math. Soc.* **82**, 533–553.

(2003a). Moments for primes in arithmetic progressions, I, *Duke Math. J.* **120**, 371–383.

(2003b). Moments for primes in arithmetic progressions, II, *Duke Math. J.*, **120**, 385–403.

(2005). A variance for k-free numbers in arithmetic progressions, *Proc. London Math. Soc.* (3) **91**, 573–597 .

(2025). The generalized Montgomery–Hooley formula; A survey. In *Essays in Analytic Number Theory: In Honor of Helmut Maier's 70th Birthday*. Edited by John Friedlander, Carl Pomerance, Michael Th. Rassias, Springer Verlag, x, 350, to appear. ISBN978-3-032-00868-8, 978-3-032-00871-8, 978-3-032-00869-5.

Vinogradov, A. I. (1965). The density hypothesis for Dirichlet L-functions, *Isv. Akad. Nauk SSSR Ser. Mat.* **29**, 903–934; Corrigendum, (1966) **30**, 719–720.

Walfisz, A. Z. (1953). On the theory of prime numbers (Russian), *Soobshch. Akad. Nauk Gruzin. SSR* **14**, 77–83. MR0062158.

Wang, Yuan (1962). On the representation of large integer as a sum of a prime and an almost prime, *Sci. Sinica* **11**, 1033–1054, Appendix.

Ward, D. R. (1927). Some series involving Euler's function, *J. London Math. Soc.* **2**, 210–214. `https://doi.org/10.1112/jlms/s1-2.4.210`

Wu, J. (2004). Chen's double sieve, Goldbach's conjecture, and the twin prime problem, *Acta Arith.* **114**, 215–273.

Zhang, Yitang (2014). Bounded gaps between primes, *Ann. Math.* (2) **179**, 1121–1174. `http://dx.doi.org/10.4007/annals.2014.179.3.7`

21

Sieves of Fixed Dimension and Applications

21.1 Refresher on Sieves

In this chapter we return to the topic of Chapter 3, (small) sieves, which we now treat, at least initially, in some generality. However our objective is to give nothing more than an introduction and some applications to what has become a vast and complex subject. Readers who wish to see the many aspects of the subject in more detail are advised to consult the standard reference on the subject, Friedlander & Iwaniec (2010). Let $\mathcal{A} = \{a_n\}$ be a sequence of non-negative real numbers such that

$$A = \sum_{n \in \mathbb{Z}} a_n < \infty. \tag{21.1}$$

Usually this sequence has compact support and, most commonly, $a_n = 0$ or 1.

Let $\mathcal{P}$ be a set of primes, the *sifting range*, and define

$$P(z) = \prod_{\substack{p<z \\ p \in \mathcal{P}}} p. \tag{21.2}$$

Then we are concerned with estimates for the quantity

$$S(\mathcal{A}, \mathcal{P}, z) = \sum_{\substack{n \in \mathbb{Z} \\ (n, P(z))=1}} a(n).$$

Here z is often called the *sifting level* of the sifted set.

We find it useful to develop sieves with rather general weights. This facilitates applications. For example, suppose that $F(\boldsymbol{x})$ is a integral form of degree k in s variables and we are interested in the number of integer points $\boldsymbol{x}$ in a box such that $N - F(\boldsymbol{x})$ is prime, where N is a large positive integer.

As we saw in §3.1 in the special case of sifting an interval, it is natural to suppose that we have some information concerning

$$\mathcal{A}_m = \{a_m(n) : n \in \mathbb{Z}\},$$

where we define

$$a_m(n) = a(mn).$$

This is usually in the form of an approximation for

$$A(m) = \sum_{n \in \mathbb{Z}} a_m(n)$$

when m is square-free and has all its prime factors in $\mathscr{P}$, of the kind

$$A(m) = X\rho(m) + r(m), \tag{21.3}$$

where X is a large parameter and ρ is a non-negative multiplicative function. Hopefully $r(m)$ is relatively small compared with $X\rho(m)$, at least on average over some range of m. Often the $r(m)$ are not explicitly known, but we assume that there is a non-negative function $R(m)$ available such that $|r(m)| \le R(m)$.

Since

$$A = A(1) = X + r(1),$$

it is normal to expect that X is a good approximation to A. Hence if $a(n)$ is the characteristic function of the integers in an interval, then one would take X to be the length of the interval, and (21.3) holds with $\rho(m) = 1/m$ and $R(m) = 1$.

If we are interested in the twin prime conjecture, then we might take $a(n)$ to be the number of solutions of $r(r+2) = n$ in integers r with $1 \le r \le X$, and then (21.3) holds with $m\rho(m)$ the number of solutions of $x(x+2) \equiv 0 \pmod{m}$ and with $|R(m)| \le m\rho(m)$. Alternatively, we might take $a(n)$ to be the characteristic function of numbers of the form $p+2$ with $p \le x$. Then

$$A(m) = \pi(x; m, -2),$$

we take $\mathscr{P}$ to be the set of primes $p > 2$, and

$$X = \mathrm{li}(x), \quad \rho(m) = \frac{1}{\varphi(m)}.$$

A familiar way of writing the condition $(n, P(z)) = 1$ is to observe that

$$\sum_{m|q} \mu(m) = \begin{cases} 1 & (q = 1), \\ 0 & (q > 1). \end{cases}$$

However, as we saw in §3.1, the number of m with $m|P(z)$ grows too rapidly for us to make good use of this identity. Thus we seek functions $\lambda^\pm(m)$ that are one-sided approximations to $\mu(m)$ in the sense that

$$\sum_{m|q}\lambda^-(m) \le \sum_{m|q}\mu(m) \le \sum_{m|q}\lambda^+(m)$$

for all q, where the support of the $\lambda^\pm$ is controlled. Then

$$X\sum_{m|P(z)}\lambda^-(m)\rho(m) + \sum_{m|P(z)}\lambda^-(m)r(m) \le S(\mathcal{A},\mathcal{P},z)$$
$$\le X\sum_{m|P(z)}\lambda^+(m)\rho(m) + \sum_{m|P(z)}\lambda^+(m)r(m), \quad (21.4)$$

which gives

$$X\sum_{m|P(z)}\lambda^-(m)\rho(m) - \sum_{m|P(z)}|\lambda^-(m)|R(m) \le S(\mathcal{A},\mathcal{P},z)$$
$$\le X\sum_{m|P(z)}\lambda^+(m)\rho(m) + \sum_{m|P(z)}|\lambda^+(m)|R(m). \quad (21.5)$$

Suppose that, among all possible upper bound sifting functions λ^+, we take the one that minimizes the right-hand member above. By appealing to the fundamental duality theorem of linear programming, it can be shown that there exists a sequence of non-negative $a(n)$ satisfying $|A(m) - X\rho(m)| \le R(m)$ for all m, and which has the property that $S(\mathcal{A},\mathcal{P},z)$ is equal to the upper bound above. Similarly, if λ^- is chosen to maximize the lower bound on the left above, then there is a choice of the $a(n)$ for which $S(\mathcal{A},\mathcal{P},z)$ is equal to the lower bound above. Details of this will be discussed in §H.2. The beautiful thing about this is that when an optimal λ^+ can be found, and the worst case $a(n)$ is also constructed, then each one proves that the other is optimal. Unfortunately, we presently know of such optimal pairs in only a few isolated situations. See §21.4.

As described in §3.1, Brun's initial choice corresponds to taking, for a suitable positive integer r,

$$\mathcal{D}^- = \{n : \omega(n) \le 2r-1\}, \quad \mathcal{D}^+ = \{n : \omega(n) \le 2r\}; \quad (21.6)$$

thus $\mu(m)\lambda^\pm(m)$ is the characteristic function of $\mathcal{D}^\pm$.

We say that a set $\mathcal{D}$ of positive integers is *divisor closed* if for each $n \in \mathcal{D}$ all positive divisors of n are also members of $\mathcal{D}$. We now set $\mathcal{D} = \{d|P(z) : d \le z\}$. In §3.2 we saw that the Selberg sieve gives a superior choice of λ^+.

To construct the Selberg upper bound sifting function we take $\lambda(n)$ to be real-valued, supported on $\mathscr{D}$, with $\lambda(1) = 1$. Thus

$$\left(\sum_{l|q} \lambda(l)\right)^2 = \sum_{m|q} \sum_{\substack{l_1, l_2 \\ [l_1, l_2]=m}} \lambda(l_1)\lambda(l_2).$$

If $\lambda(l_1) \neq 0$ and $\lambda(l_2) \neq 0$, then $l_1 \leq z$, $l_2 \leq z$, and hence $m \leq l_1 l_2 \leq z^2$. This gives an upper bound sifting function

$$\lambda^+(m) = \sum_{\substack{l_1, l_2 \\ [l_1, l_2]=m}} \lambda(l_1)\lambda(l_2)$$

supported on the interval $\left[1, z^2\right]$.

Thus

$$\begin{aligned} S(\mathscr{A}, \mathscr{P}, z) &\leq \sum_{l} \sum_{m} \lambda(l)\lambda(m) \sum_{n} a(n[l, m]) \\ &= X \sum_{l} \sum_{m} \lambda(l)\lambda(m)\rho([l, m]) + r, \end{aligned}$$

where

$$r = \sum_{l} \sum_{m} \lambda(l)\lambda(m) r([l, m]).$$

The interesting part is the main term XF where

$$F = \sum_{d} \sum_{e} \lambda(d)\lambda(e)\rho([d, e]).$$

We want to minimize this subject to the condition $\lambda(1) = 1$, and in the special case $\rho(n) = 1/n$ we already did this in §3.2. The general case involves no new ideas.

It is helpful to view F as a quadratic form in the λ. Our first objective is to diagonalize F, and this can be done quite easily. It is natural to suppose that $\mathcal{D}$ is chosen so that

$$\rho(d) > 0 \quad (d \in \mathcal{D}). \tag{21.7}$$

Write $(d, e) = m$, $d = qm$, $e = rm$, so that $(q, r) = 1$. Since ρ is multiplicative and qrm is square-free we have $\rho([d, e]) = \rho(qrm) = \rho(qm)\rho(rm)/\rho(m)$ and

$$F = \sum_{m} \rho(m)^{-1} \sum_{q} \sum_{\substack{r \\ (q,r)=1}} \lambda(qm)\lambda(rm)\rho(qm)\rho(rm).$$

Now we use the Möbius function to remove the condition $(q, r) = 1$. Thus

$$F = \sum_{m} \rho(m)^{-1} \sum_{l} \mu(l) \left(\sum_{d} \lambda(dlm)\rho(dlm)\right)^2.$$

Next we collect the terms with $lm = n$ and observe that by multiplicativity that

$$\sum_{\substack{l,m\\ lm=n}} \rho(m)^{-1}\mu(l) = \prod_{p|n} \frac{1-\rho(p)}{\rho(p)}.$$

Denote this expression by $g(n)^{-1}$. Then we have

$$F = \sum_n g(n)^{-1} \left(\sum_d \lambda(dn)\rho(dn) \right)^2,$$

where

$$g(n) = \prod_{p|n} \frac{\rho(p)}{1-\rho(p)}. \tag{21.8}$$

Let

$$\nu(n) = \sum_d \lambda(dn)\rho(dn) \quad (n \in \mathscr{D}).$$

We have

$$F = \sum_n g(n)^{-1}\nu(n)^2,$$
$$\nu(n) = \sum_d \lambda(dn)\rho(dn) \quad (n \in \mathscr{D}).$$

There is a bijection between the λ and the ν. We can view the transformation from one to the other as being by an upper triangular matrix, which is obviously invertible. There is a standard number-theoretic way of expressing the inversion. Consider

$$\sum_n \mu(n)\nu(nm) = \sum_n \sum_d \lambda(dnm)\rho(dnm)\mu(n).$$

By collecting the terms with $nd = q$ this becomes, for $m \in \mathscr{D}$,

$$\sum_q \lambda(qm)\rho(qm) \sum_{n|q} \mu(n) = \lambda(m)\rho(m).$$

Hence

$$\lambda(m)\rho(m) = \sum_n \nu(nm)\mu(n) \quad (m \in \mathscr{D}).$$

Thus we are seeking to minimize

$$F = \sum_n g(n)^{-1}\nu(n)^2 \text{ under the condition } \sum_n \nu(n)\mu(n) = \lambda(1) = 1.$$

Let $\theta = 1/\sum_{n\in\mathscr{D}} g(n)$. Then

$$F = \sum_{n\in\mathscr{D}} \frac{(\nu(n) - \theta\mu(n)g(n))^2}{g(n)} + 2\theta\sum_n \nu(n)\mu(n) - \theta^2\sum_n g(n)$$
$$= \sum_{n\in\mathscr{D}} \frac{(\nu(n) - \theta\mu(n)g(n))^2}{g(n)} + \theta.$$

Obviously $F \geq \theta$, and the choice

$$\nu(n) = \theta\mu(n)g(n)$$

gives

$$\sum_n \nu(n)\mu(n) = 1 \text{ and } F = \theta.$$

We have just shown that the minimum of F is θ and the minimum is attained when

$$\nu(n) = \theta\mu(n)g(n).$$

We can now invert the transform to recover the minimizing $\lambda(m)$. Recall that

$$\lambda(m)\rho(m) = \sum_n \nu(nm)\mu(n) \quad (m \in \mathscr{D}).$$

Thus the minimizing $\lambda(m)$ are given by

$$\lambda(m) = \frac{\theta}{\rho(m)}\sum_n g(mn)\mu(mn)\mu(n) = \theta\mu(m)\frac{g(m)}{\rho(m)}\sum_{\substack{n\\ nm\in\mathscr{D}}} g(n).$$

We need to determine the $\lambda(m)$ because they occur in the remainder term. Write

$$\frac{g(m)}{\rho(m)} = \prod_{p|m}\frac{1}{1-\rho(p)} = \prod_{p|m}(1+g(p)) = \sum_{d|m} g(d).$$

Thus

$$|\lambda(m)| \leq \theta\sum_{d|m} g(d) \sum_{\substack{n\\ nd\in\mathscr{D}\\ (n,m/d)=1}} g(n) = \theta\sum_{d|m}\sum_{\substack{k\\ (k,m)=d}} g(k) = 1,$$

so

$$|\lambda(m)| \leq 1.$$

Theorem 21.1 (Selberg 1947) *Suppose that* (21.3), (21.2), *and* (21.8), *hold, and ρ is multiplicative and satisfies $0 \le \rho(p) < 1$. Let $\mathscr{D}$ be a divisor closed subset of the divisors of $P(z)$. Then*

$$S(\mathcal{A}, \mathcal{P}, z) \le \frac{X}{\sum_{n\in\mathscr{D}} g(n)} + \sum_{l\in\mathscr{D}} \sum_{m\in\mathscr{D}} \lambda(l)\lambda(m)r([l,m]).$$

Moreover

$$|\lambda| \le 1.$$

This bound is reminiscent of the arithmetical form of the large sieve, Theorem 19.13, but that, of course, is just an interval sieve.

Our main interest at this stage is to develop lower bound sieves, hopefully in tandem with upper bound sieves. For this purpose we introduce several new parameters. Let

$$V(z) = \prod_{\substack{p<z\\ p\in\mathcal{P}}} (1-\rho(p)).$$

Then it is natural to suppose that $XV(z)$ ought to give us the size of $S(\mathcal{A}, \mathcal{P}, z)$. It is normal at this point in the discussion of 'small' sieves to hypothesize that

$$\sum_{p\le z} \rho(p)\log p = \kappa \log z + O(1),$$

where $\kappa \ge 0$ is a constant, and this important number is usually referred to as the *sieve dimension*. By partial summation it follows that

$$\sum_{p\le z} \rho(p) = \kappa \log\log z + c + O\big((\log z)^{-1}\big),$$

where c is a constant. Then by Mertens' approximation, Theorem 2.7(e), we deduce that

$$V(z) = e^{-C_0\kappa}\mathfrak{S}(\log z)^{-\kappa}\big(1+O\big((\log z)^{-1}\big)\big),$$

where

$$\mathfrak{S} = \prod_p \frac{1-\rho(p)}{(1-1/p)^{\kappa}}.$$

For much of modern (small) sieve theory the weaker assumption that

$$\sum_{w\le p\le z} \rho(p)\log p \le \kappa\log(z/w) + \frac{C}{\log w}$$

suffices.

At this point it is convenient to introduce an identity that generalizes one used in the proof of Theorem 7.11.

Lemma 21.2 (Buchstab's Identity; Buchstab 1938) *Suppose that* $2 \le w \le z$. *Then*

$$S(\mathcal{A},\mathcal{P},w) = S(\mathcal{A},\mathcal{P},z) + \sum_{w\le p<z} S(\mathcal{A}_p,\mathcal{P},p).$$

Proof The identity is immediate on observing that the difference

$$S(\mathcal{A},\mathcal{P},w) - S(\mathcal{A},\mathcal{P},z)$$

is a sum over integers with at least one prime factor p with $w \le p < z$ and no prime factor $p' < w$. Hence the identity follows by sorting these terms according to their least prime factor. □

This identity has been very suggestive of a possible way to improve sieve estimates. Consider the special case $w = 1$, which asserts that

$$A - \sum_{p<z} S(\mathcal{A}_p,\mathcal{P},p) = S(\mathcal{A},\mathcal{P},z).$$

If we have an upper bound sieve estimate, we could insert it in the sum on the left and obtain a lower bound for $S(\mathcal{A},\mathcal{P},z)$. We could then use this lower bound in the sum on the left and obtain a new upper bound for $S(\mathcal{A},\mathcal{P},z)$. It was found that if one initially used a version of Brun's sieve then the new upper bound was stronger. This suggested an iterative procedure. Examination of the limit of the process suggested a more direct route, which was first discovered by Rosser in the 1950s (unpublished) and then rediscovered by Iwaniec (1971).

Suppose that we can control suitably the behaviour of

$$r(m)$$

when $m \le y$ (the *level of distribution* of $\mathcal{A}$). We might hope that in some generality there are smooth 'fudge factors' $f_\pm(s)$ with $s = \frac{\log y}{\log z}$ which satisfy

$$XV(z)f_-(s)(1+o(1)) \le S(\mathcal{A},\mathcal{P},z) \le XV(z)f_+(s)(1+o(1)). \tag{21.9}$$

We note that the Buchstab identity has enabled us to guess that

$$XV(z) \sim Xe^{-C_0\kappa}\mathfrak{S}(\log z)^{-\kappa} = \mathfrak{S}\frac{Xe^{-C_0\kappa}}{(\log X)^\kappa}\left(\frac{\log X}{\log z}\right)^\kappa$$

ought to be about the right size for the sifted set, at least when $y = X$. Thus we might imagine that, for suitable $f_\pm$,

$$\mathfrak{S}\frac{Xe^{-C_0\kappa}}{(\log X)^\kappa}s^\kappa f_-(s)(1+o(1)) \le S(\mathcal{A},\mathcal{P},z) \le \mathfrak{S}\frac{Xe^{-C_0\kappa}}{(\log X)^\kappa}s^\kappa f_+(s)(1+o(1))$$

are the limits of the Buchstab identity iteration. It is also reasonable to suppose that the sum over p can be replaced by an integral and retain asymptotic equality. Finally, put $s = \frac{\log X}{\log z}$ and $t = \frac{\log X}{\log w}$, divide by

$$\mathfrak{S}\frac{Xe^{-C_0\kappa}}{(\log X)^\kappa},$$

and let X, w, z go to infinity together so that the error terms tend to 0. We may need to suppose that $\beta < s < t$ where β is a positive constant. Then we find that the $f_\pm$ satisfy

$$t^\kappa f_\pm(t) - s^\kappa f_\pm(s) = \int_s^t \kappa u^{\kappa-1} f_\mp(u-1)du, \tag{21.10}$$

and hence that

$$\left(t^\kappa f_\pm(t)\right)' = \kappa t^{\kappa-1} f_\mp(t-1). \tag{21.11}$$

We also know from the Brun sieve that if s is large, then $f_\pm(s)$ should be asymptotically 1.

The analysis of the iterations can be quite complicated and instead we follow the Rosser–Iwaniec approach. To set this up, write

$$\lambda(m) = \mu(m)\sigma(m),$$

where we suppose that

$$\sigma(m) = 0 \text{ or } 1, \quad \sigma(1) = 1,$$

and define the least prime factor $l(m)$ of m, so that

$$l(1) = 1, \quad l(m) = \min\{p : p|m\} \quad (m > 1), \tag{21.12}$$

and then define

$$\tau(m) = \sigma(m/l(m)) - \sigma(m).$$

Theorem 21.3 *Suppose that $z > 1$. Then*

$$S(\mathcal{A}, \mathcal{P}, z) = \sum_{m|P(z)} \mu(m)\sigma(m)A(m) + \sum_{m|P(z)} \mu(m)\tau(m)S(\mathcal{A}_m, \mathcal{P}, l(m)).$$

Proof We substitute the definitions of $A(m)$ and $S(\mathcal{A}_m, \mathcal{P}, l(m))$. On interchanging the summation we find that

$$\sum_n a(n)\left(\sum_{m|(n,P(z))} \mu(m)\sigma(m) + \sum_{\substack{m|(n,P(z))\\(n,P(l(m)))=1}} \mu(m)\tau(m)\right).$$

When $(n, P(z)) = 1$ the sums reduce to $\sigma(1) = 1$. It remains to consider those n of the form

$$n = n' p_1^{k_1} \cdots p_r^{k_r}$$

with $(n', P(z)) = 1$ and $z > p_1 > \cdots > p_r$. Then the second inner sum is

$$\sum_{\substack{1<m|p_1\cdots p_r \\ (p_1\cdots p_r, P(l(m)))=1}} \mu(m)(\sigma(m/l(m)) - \sigma(m)).$$

The only m which satisfy these summation conditions and give a non-zero contribution have $l(m) = p_r$ and $m = jp_r$ with $j|p_1 \cdots p_{r-1}$. Thus the above sum is

$$\begin{aligned}\sum_{j|p_1\cdots p_{r-1}} \mu(jp_r)(\sigma(j) - \sigma(jp_r)) &= -\sum_{j|p_1\cdots p_{r-1}} (\mu(j)\sigma(j) + \mu(jp_r)\sigma(jp_r)) \\ &= -\sum_{m|p_1\cdots p_r} \mu(m)\sigma(m)\end{aligned}$$

and this cancels out the terms in the first sum. □

Suppose that $\sigma^{\pm}$ can be chosen so that

$$\mp\mu(m)\tau^{\pm}(m) \geq 0 \quad (m|P(z)). \tag{21.13}$$

Then

$$\sum_{m|P(z)} \mu(m)\sigma^{-}(m)A(m) \leq S(\mathcal{A}, \mathcal{P}, z) \leq \sum_{m|P(z)} \mu(m)\sigma^{+}(m)A(m)$$

and so

$$XS^{-} + r^{-} \leq S(\mathcal{A}, \mathcal{P}, z) \leq XS^{+} + r^{+}, \tag{21.14}$$

where

$$S^{\pm} = \sum_{m|P(z)} \mu(m)\sigma^{\pm}(m)\rho(m) \tag{21.15}$$

and

$$r^{\pm} = \sum_{m|P(z)} \mu(m)\sigma^{\pm}(m)r(m). \tag{21.16}$$

We can also use Theorem 21.3 to compute a suitable approximation to the main term. Suppose that

$$0 \leq \rho(p) < 1 \quad (p|P(z)).$$

Note that if $\rho(p) = 1$ for some p, then almost nothing will survive the sieving process and that would not be very interesting. Now define a to be the multiplicative function with

$$a(p^k) = \begin{cases} \rho(p)^k & (p \in \mathcal{P}, p < z, \text{ and } k \in \mathbb{N}), \\ 0 & (p \notin \mathcal{P} \text{ or } p \geq z, \text{ and } k \in \mathbb{N}). \end{cases}$$

Then

$$A = \prod_{p<z} (1 - \rho(p))^{-1} = V(z)^{-1},$$

and for $m|P(z)$

$$A(m) = a(m) \sum_n a(n) = \rho(m) A.$$

Moreover,

$$S(\mathcal{A}_m, \mathcal{P}, l(m)) = \sum_{\substack{n,\, m|n \\ (n, P(l(m))=1}} a(n) = \sum_{\substack{k \\ (k, P(l(m)))=1}} a(mk) = \rho(m) \frac{V(l(m))}{V(z)}.$$

Also $S(\mathcal{A}, \mathcal{P}, z)$. Thus by Theorem 21.3,

$$V(z) = \sum_{m|P(z)} \mu(m)\sigma^{\pm}(m)\rho(m) + \sum_{m|P(z)} \mu(m)\tau^{\pm}(m)\rho(m)V(l(m)).$$

By (21.14) and (21.15) this establishes Theorem 21.4.

Theorem 21.4 *Suppose that for every prime $p \in \mathcal{P}$ with $p < z$ we have $0 \leq \rho(p) < 1$ and for every $m|P(z)$ we have (21.13). Then*

$$XS^- + R^- \leq S(\mathcal{A}, \mathcal{P}, z) \leq XS^+ + R^+,$$

where

$$S^{\pm} = V(z) - \sum_{m|P(z)} \mu(m)\tau^{\pm}(m)\rho(m)V(l(m)),$$

$$R^{\pm} = \sum_{m|P(z)} \mu(m)\sigma^{\pm}(m)R(m),$$

$$\sigma^{\pm}(m) = 0 \text{ or } 1, \quad \sigma^{\pm}(1) = 1,$$

$$\mp\mu(m)\big(\sigma^{\pm}(m/l(m)) - \sigma^{\pm}(m)\big) \geq 0,$$

and (21.3) *holds.*

21.1.1 Exercises

1. Suppose that h is an even positive integer and

$$R(x;h) = \operatorname{card}\{p_1 \le x, p_2 \le x : p_1 - p_2 = h\}.$$

Let g be the multiplicative function with $g(2) = 0$, $g(p) = \frac{1}{p-2}$ when $p > 2$ and $g(p^k) = 0$ for all $k \ge 2$, and define

$$L = \sum_{\substack{n \le D \\ (n,2h)=1}} g(n).$$

Further, let $f(q)$ denote the number of pairs l, m of positive square-free integers $l \le D$, $m \le D$ such that $[l,m] = q$.

(a) Prove that

$$R(x;h) \le \operatorname{li}(x)L^{-1} + D + \sum_{\substack{q \le D \\ (q,h)=1}} f(q)\left|\pi(x;q,h) - \frac{\operatorname{li}(x)}{\varphi(q)}\right|.$$

(b) Prove that if n is square-free, then

$$g(n) = \frac{1}{\varphi(n)} \sum_{m|n} g(m)$$

and

$$L = \sum_{\substack{m \le D \\ (m,2h)=1}} \frac{g(m)}{\varphi(m)} \sum_{\substack{l \le D/m \\ (l,2hm)=1}} \frac{\mu(l)^2}{\varphi(l)}.$$

(c) Prove (cf. the argument after (3.18)) that if $Y \ge 1$, then

$$\frac{k}{\varphi(k)} \sum_{\substack{l \le Y \\ (l,k)=1}} \frac{\mu(l)^2}{\varphi(l)} \ge \sum_{m \le Y} \frac{\mu(m)^2}{\varphi(m)} \ge \log Y.$$

(d) Prove that

$$\begin{aligned} L &\ge \sum_{\substack{m \le D \\ (m,2h)=1}} \frac{g(m)\varphi(2hm)}{2hm\varphi(m)} \log \frac{D}{m} \\ &= (\log D)\frac{\varphi(2h)}{2h} \prod_{p \nmid 2h} \left(1 + \frac{1}{p(p-2)}\right) + O(1). \end{aligned}$$

2. Prove that, uniformly in x and h,

$$R(x;h) \le \frac{4\mathfrak{S}(h)x}{(\log x)^2}\left(1 + O\left(\frac{\log\log x}{\log x}\right)\right),$$

where

$$\mathfrak{S}(h) = c(h) = \frac{2h}{\varphi(2h)} \prod_{p \nmid 2h} \frac{p(p-2)}{(p-1)^2} = c \prod_{\substack{p \mid h \\ p>2}} \frac{p-1}{p-2},$$

$c(h)$ is the constant of Corollary 3.14, and c is the constant of Theorem 3.10 and (20.28) (cf. Exercise 19.2.1.10 with $k = 2$).

3. Suppose that $J \le H \le \log N$. By combining (20.36) and the previous question show that

$$\frac{1}{2}HN \log N + 2\sum_{h=1}^{J}(H-h)R(N,h) + 8\sum_{h=J+1}^{H}(H-h)\mathfrak{S}(h)N \geq H^2N + O\big(HN(\log\log N)^2\big).$$

   (a) Deduce that

$$\frac{1}{2}HN \log N + 2\sum_{h=1}^{J}(H-h)R(N,h) + 4(H-J)^2N \geq H^2N + O\big(HN(\log\log N)^2\big).$$

   (b) Let

$$J = \left(\frac{2+\sqrt{3}}{8} + \varepsilon\right)\log N, \qquad H = \frac{3+2\sqrt{3}}{12}\log N.$$

   Prove that if N is large, then

$$\sum_{h=1}^{J}(H-h)R(N,h) > 0.$$

   (c) Prove that

$$\liminf_{n\to\infty} \frac{p_{n+1}-p_n}{\log p_n} \le \frac{2+\sqrt{3}}{8} = 0.466506\ldots.$$

4. Let x be a large real number and define

$$R(n) = \sum_{3 \le p_1 \le x} \sum_{\substack{p_2 \ge 3 \\ p_1+p_2=n}} \log p_1,$$

and $f_n(q)$ to be the multiplicative function with $f_n(p^k) = 1/(p-2)$ when $k = 1$, $p \nmid n$, and p is odd, and $f_n(p) = 0$ otherwise. Let $y = x(\log x)^{-B}$ for a suitable constant B and write

$$L(n) = \sum_{q \le y^{1/2}} f_n(q).$$

(a) Prove that

$$\sum_{x<n\le 2x} R(n)L(n) \le \sum_{\substack{x<n\le 2x\\ R(n)>0}} \vartheta(x) + O\big(x^2(\log x)^{-A}\big).$$

(b) Prove that

$$L(n) \ge \Bigg(\prod_{\substack{p|n\\ p>2}} \frac{p-2}{p-1}\Bigg) \sum_{\substack{q\le y^{1/2}\\ 2\nmid q}} \frac{\mu(q)^2}{\varphi_1(q)},$$

where

$$\varphi_1(q) = \prod_{p|q}(p-2).$$

(c) Prove that

$$\begin{aligned}&\operatorname{card}\{n \in (x,2x] : R(n) > 0\}\\ &\qquad \ge \frac{\log y}{4\vartheta(x)} \int_3^{2x} \frac{\min(u, 2x-u)}{\log u}\,du + O\big(x(\log x)^{-1}\big).\end{aligned}$$

(d) Let $N(x)$ denote the number of even numbers $n \le x$ such that n is the sum of two odd primes. Deduce that

$$\liminf_{x\to\infty} \frac{N(x)}{x} \ge \frac{1}{4}.$$

5. Suppose that $s \ge 3$ and $k \ge 2$, and let $N(Y)$ denote the number of ordered s-tuples of integers $\boldsymbol{y} \in [1,Y]^s$ such that $y_1^k + \cdots + y_s^k$ is prime. Prove that

$$N(Y) \ll \frac{Y^s}{\log Y}.$$

6. Suppose that for a constant $C \ge 1$ we have $0 \le \rho(p) \le C/p$ and that for $m|P(z)$ we have $|R(m)| \le m\rho(m)$. Let $k \in \mathbb{N}$.

(a) Prove that

$$\Bigg|\sum_{\substack{m|P(z)\\ \omega(m)\le k}} \mu(m)R(m)\Bigg| \le C^k z^k,$$

and that

$$\begin{aligned}&\Bigg|\sum_{\substack{m|P(z)\\ \omega(m)=k+1}} \rho(m)\frac{V(l(m))}{V(z)}\Bigg|\\ &\qquad\le \frac{1}{(k+1)!}\Bigg(\sum_{p<z}\rho(p)\Bigg)^{k+1} \exp\Bigg(\sum_{p<z}\frac{1}{1-\rho(p)}\Bigg).\end{aligned}$$

(b) Suppose further that there are constants $\kappa \ge 0$ and $C_1 \ge 0$ such that for $z \ge 3$ we have

$$\sum_{p<z} \rho(p) < \kappa \log\log z + C_1.$$

Prove that

$$\left| \sum_{\substack{m|P(z)\\ \omega(m)=k+1}} \rho(m) \frac{V(l(m))}{V(z)} \right| \ll \frac{1}{(k+1)!} (\kappa \log\log z + C_1)^{k+1} (\log z)^{\kappa}.$$

(c) By taking $\sigma^{\pm}(m)$ to be the characteristic function of the sets $\mathscr{D}^{\pm}$ given by (21.6), or otherwise, prove that

$$XV(z) - E(2r-1) \le S(\mathcal{A}, \mathcal{P}, z) \le XV(z) + E(2r),$$

where

$$E(k) \ll \frac{XV(z)}{(k+1)!} (\kappa \log\log z + C_1)^{k+1} (\log z)^{\kappa} + (Cz)^k.$$

(d) Show that there is a positive constant C_2 such that if

$$3 \le z \le \exp\left(C_2 \frac{\log X}{\log\log X} \right),$$

then

$$S(\mathcal{A}, \mathcal{P}, z) = XV(z)\big(1 + O(\log^{-10} z)\big) + O(X^{1-C_2}).$$

21.2 The Rosser–Iwaniec Sieve

We are ultimately concerned with the one-dimensional sieve, but initially there is no need to distinguish any one particular value of κ. We will find that there is a point at which there is a tricky convergence problem. For simplicity we will give a treatment of this only when $\kappa = 1$. In principle, the method can be adapted for all κ, and gives the best results that are known when $0 < \kappa \le 1$. In particular, it can be shown to be optimal when $\kappa = \frac{1}{2}$ and 1. We would add that we are not aware of any interesting applications of dimension $\kappa > 1$ which cannot be treated more effectively by other methods.

In addition to the role played by z, we introduce two further parameters

$$\beta \ge 1 \tag{21.17}$$

and $y \ge 2$, which will give us some finer control of the lower bound and the error term. The quality of the final results will depend on β, and we will see that there is a choice for each κ which maximizes the range on which one can obtain a positive lower bound.

Let

$$\upsilon^{\pm}(m) = \begin{cases} 0 & \text{when } \mu(m) = \mp 1 \text{ and } l(m) \ge (y/m)^{1/\beta}, \\ 1 & \text{otherwise,} \end{cases}$$

where $l(m)$ is given by (21.12). We consider κ to be fixed. In the notation introduced in §21.1, let $\sigma^{\pm}(1) = 1$ and when $m = p_1 \cdots p_k$ with $p_1 > p_2 > \cdots > p_k$ let

$$\sigma^{\pm}(m) = \prod_{u=1}^{k} \upsilon^{\pm}(p_1 \cdots p_u),$$

$$\tau^{\pm}(m) = \sigma^{\pm}\big(m/l(m)\big) - \sigma^{\pm}(m) = \sigma^{\pm}(m/l(m))\big(1 - \upsilon^{\pm}(m)\big).$$

Clearly

$$\sigma^{\pm}(m) = 0 \text{ or } 1, \quad \tau^{\pm}(m) = 0 \text{ or } 1,$$

and it is readily checked that $\sigma^{\pm}(m) = 0$ when $m > y$. Moreover, if $\tau^{\pm}(m) = 1$, then $\upsilon^{\pm}(m) = 0$ and so $\mu(m) = \mp 1$. Hence

$$\mp\mu(m)\tau^{\pm}(m) \ge 0 \quad \text{for all } m|P(z). \tag{21.18}$$

Thus the hypothesis of Theorem 21.4 is satisfied.

There is an extremely useful way of describing the sets of m for which $\tau^{\pm}(m) = 1$. Let $\mathscr{D}_k(y, z)$ denote the set of divisors m of $P(z)$ of the form

$$m = p_1 \cdots p_k \quad \text{with}$$

$$p_1 > p_2 > \cdots > p_k, \quad p_1 \cdots p_{k-1} p_k^{\beta+1} \ge y, \text{ and } p_1 \cdots p_{k-2j-1} p_{k-2j}^{\beta+1} < y$$

whenever $1 \le j < k/2$. When $k \ge 3$ and k is odd the case $j = \lfloor \frac{k}{2} \rfloor$ is interpreted as $p_1^{\beta+1} < y$. Then define

$$S_k(y, z) = \sum_{m \in \mathscr{D}_k(y,z)} \rho(m) V(l(m)). \tag{21.19}$$

Note that $S_k(y, z) = 0$ when $k \ge 3$ and $2^{\beta+k-2} \ge y$, so the series given by (21.20) and (21.21) below are in fact finite.

Theorem 21.5 *Let $S_k(y, z)$ be as in* (21.19). *Then we have*

$$XS^{-}(y, z) + R^{-} \le S(\mathscr{A}, \mathscr{P}, z) \le XS^{+}(y, z) + R^{+},$$

where

$$S^+(y,z) = V(z) + \sum_{r=1}^{\infty} S_{2r-1}(y,z), \tag{21.20}$$

$$S^-(y,z) = V(z) - \sum_{r=1}^{\infty} S_{2r}(y,z), \tag{21.21}$$

and $R^\pm$ *satisfies* (21.16), *and so*

$$|R^\pm| \le R^* = \sum_{\substack{m<y \\ m|P(z)}} |R(m)|. \tag{21.22}$$

At this point we can see already that y can be used as a means of controlling the size of R^*. The initial problem is the convergence of the infinite series when we replace the terms by smooth approximations.

Proof The expressions for $S^\pm(y,z)$ follow from our discussion above concerning the m for which $\tau^\pm(m) = 1$.

To estimate $R^\pm$ we need only consider those $m = p_1 \cdots p_k$ with $p_1 > p_2 > \cdots > p_k$ for which $\sigma^\pm(m) \ne 0$. Then $\upsilon^\pm(q) = 1$ for $q|m$ and so for either $u = k$ or $u = k-1$ we have

$$p_u^{\beta+1} p_{u-1} \cdots p_1 < y.$$

By (21.17) we have $\beta \ge 1$. Hence $m < y$. □

We now have to investigate $S^\pm(y,z)$. It will surely be no great surprise to find that there is an iterative relationship between the S_k. To better understand it we introduce as an important parameter, namely the ratio

$$s = \frac{\log y}{\log z}.$$

For $s > 0$ and $k = 1, 2, \ldots$, we define

$$g_k(y,s) = V(y^{1/s})^{-1} S_k(y, y^{1/s}). \tag{21.23}$$

Suppose $k \ge 2$. Then for $m \in \mathscr{D}_k(y,z)$ we have $m = p_1 \cdots p_k$ with

$$p_1 > p_2 > \cdots > p_k, \quad p_1 \cdots p_{k-1} p_k^{\beta+1} \ge y, \text{ and } p_1 \cdots p_{k-1-2j} p_{k-2j}^{\beta+1} < y$$

whenever $j \ge 1$. These inequalities can be rewritten as

$$p_2 \cdots p_{k-1} p_k^{\beta+1} \ge y/p_1, \quad p_2 \cdots p_{k-1-2j} p_{k-2j}^{\beta+1} < y/p_1 \quad (1 \le j < k/2),$$

and also $p_1 < y^{1/(\beta+1)}$ when k is odd and $k \geq 3$. Thus m is of the form pm' with $m' \in \mathscr{D}_{k-1}(y/p, p)$ and, additionally, $p < y^{1/(\beta+1)}$ when k is odd, and every such m is in $\mathscr{D}_k(y, z)$. Thus, by (21.19),

$$S_{2r+1}(y,z) = \sum_{p<\min\{y^{1/(\beta+1)},z\}} \rho(p) S_{2r}(y/p, p),$$

$$S_{2r}(y,z) = \sum_{p<z} \rho(p) S_{2r-1}(y/p, p).$$

Thus, by (21.23), these relations can be rewritten as

$$g_{2r+1}(y,s) = \sum_{\substack{p \\ \max\{\beta+1,s\}<\frac{\log y}{\log p}}} \rho(p) \frac{V(p)}{V(y^{1/s})} g_{2r}\left(\frac{y}{p}, \frac{\log y}{\log p} - 1\right), \tag{21.24}$$

$$g_{2r}(y,s) = \sum_{\substack{p \\ s<\frac{\log y}{\log p}}} \rho(p) \frac{V(p)}{V(y^{1/s})} g_{2r-1}\left(\frac{y}{p}, \frac{\log y}{\log p} - 1\right). \tag{21.25}$$

Note that in (21.24), when $s \leq \beta + 1$,

$$g_{2r+1}(y,s) V(y^{1/s})$$

is independent of s, so

$$g_{2r+1}(y,s) = \frac{V(y^{1/(\beta+1)})}{V(y^{1/s})} g_{2r+1}(y, \beta+1). \tag{21.26}$$

Consider the case $k = 1$. Then

$$S_1(y,z) = \sum_{y^{1/(\beta+1)} \leq p < z} \rho(p) V(p),$$

and so

$$S_1(y,z) = 0 \quad \text{when } z \leq y^{1/(\beta+1)}.$$

Now suppose that $z > y^{1/(\beta+1)}$. The identity

$$\sum_{m=1}^{n} x_m \prod_{l=1}^{m-1} (1 - x_l) = 1 - (1 - x_1) \cdots (1 - x_n) \tag{21.27}$$

is easily proved by induction on n, and gives

$$S_1(y,z) = V(y^{1/(\beta+1)}) - V(z).$$

Thus

$$g_1(y,s) = \begin{cases} \dfrac{V(y^{1/(\beta+1)})}{V(y^{1/s})} - 1 & \text{when } 0 < s \leq \beta + 1, \\ 0 & \text{when } s > \beta + 1. \end{cases} \tag{21.28}$$

To make further progress we have to input some information which corresponds to the dimension of the sieve. Thus we assume that there is a positive constant C such that

$$\frac{V(w)}{V(z)} < \left(\frac{\log z}{\log w}\right)^{\kappa}\left(1+\frac{C}{\log w}\right) \quad (2 \le w < z). \tag{21.29}$$

Therefore

$$g_1(y,s) < G_1(s) + \frac{C(\beta+1)^{\kappa+1}}{s^{\kappa}\log y}, \tag{21.30}$$

where

$$G_1(s) = \begin{cases} (\beta+1)^{\kappa}s^{-\kappa}-1 & \text{when } 0 \le s < \beta+1, \\ 0 & \text{when } s \ge \beta+1. \end{cases} \tag{21.31}$$

Interestingly, G_1 is independent of y.

The form of (21.29) is not the most useful for all our purposes. Although we will only use it later, in the case $\kappa = 1$, it is convenient to establish here the following lemma.

Lemma 21.6 *Suppose that $0 \le \rho(p) < 1$ and that (21.29) holds.*

(a) *If $2 \le w < z$, then*

$$\sum_{w \le p < z} \rho(p) < \kappa \log\frac{\log z}{\log w} + \frac{C}{\log w}. \tag{21.32}$$

(b) *Suppose that $s < u$, $2 \le y^{1/u} < y^{1/s}$, that η is non-negative, continuous, and decreasing on $[s,u]$, and differentiable on (s,u) with a continuous and uniformly bounded derivative. Then*

$$\sum_{y^{1/u} \le p < y^{1/s}} \rho(p)\frac{V(p)}{V(y^{1/s})}\eta\left(\frac{\log y}{\log p}\right)$$
$$\le s^{-\kappa}\int_s^u \kappa t^{\kappa-1}\eta(t)dt + \frac{Cu^{\kappa+1}s^{-\kappa}\eta(s)}{\log y}. \tag{21.33}$$

Proof The bound (21.32) is immediate from (21.29) on observing that

$$\sum_{w \le p < z} \rho(p) \le \log\frac{V(w)}{V(z)}.$$

To prove (21.33), let

$$T(t) = \sum_{y^{1/t} \le p < y^{1/s}} \rho(p)\frac{V(p)}{V(y^{1/s})},$$

so that our sum is

$$T(u)\eta(u) - \int_s^u T(t)\eta'(t)\,dt. \tag{21.34}$$

Then, as in the proof of (21.28), combined with (21.29),

$$T(t) = \frac{V(y^{1/t})}{V(y^{1/s})} - 1 < \left(\frac{t}{s}\right)^\kappa - 1 + \frac{Ct^{\kappa+1}}{s^\kappa \log y}.$$

Since $\eta'(t) \le 0$, inserting this in (21.34) gives the upper bound

$$\left(\left(\frac{u}{s}\right)^\kappa \left(1 + \frac{Cu}{\log y}\right) - 1\right)\eta(u) - \int_s^u \left(\left(\frac{t}{s}\right)^\kappa \left(1 + \frac{Ct}{\log y}\right) - 1\right)\eta'(t)\,dt.$$

Then integration by parts gives

$$\frac{Cs\eta(s)}{\log y} + \int_s^u \left(\frac{\kappa t^{\kappa-1}}{s^\kappa} + \frac{C(\kappa+1)t^\kappa}{s^\kappa \log y}\right)\eta(t)\,dt.$$

Rearranging gives

$$s^{-\kappa}\int_s^u \kappa t^{\kappa-1}\eta(t)\,dt + \frac{Cs\eta(s)}{\log y} + s^{-\kappa}\int_s^u \frac{C(\kappa+1)t^\kappa}{\log y}\eta(t)\,dt.$$

In the second integral we replace $\eta(t)$ by its upper bound $\eta(s)$ and integrate. Part (b) follows. □

We need to consider what to do with g_k when $k > 1$. If for some k and suitably smooth $G_k(s)$ we have

$$g_k(y,s) \lesssim G_k(s),$$

then the relations (21.24) and (21.25) suggest that, at least for larger s,

$$\begin{aligned} g_{k+1}(y,s) &\lesssim \sum_{\substack{p \\ s<\frac{\log y}{\log p}}} \rho(p)\left(\frac{\log y}{s\log p}\right)^\kappa G_k\left(\frac{\log y}{\log p} - 1\right) \\ &\sim s^{-\kappa}\int_s^\infty \kappa t^{\kappa-1} G_k(t-1)\,dt. \end{aligned}$$

This in turn suggests that G_1 should be defined by (21.31) and that

$$G_{2r}(s) = s^{-\kappa}\int_s^\infty \kappa t^{\kappa-1} G_{2r-1}(t-1)\,dt \quad (s \ge \beta), \tag{21.35}$$

$$G_{2r+1}(s) = s^{-\kappa}\int_{\max(\beta+1,s)}^\infty \kappa t^{\kappa-1} G_{2r}(t-1)\,dt \quad (s > 0). \tag{21.36}$$

At this point we need to observe that, at least when $k = 1$, $G_1(t-1) \asymp (t-1)^{-\kappa}$ and so we need to suppose that

$$\beta \ge 1 \quad (\kappa < 1), \qquad \beta > 1 \quad (\kappa \ge 1).$$

By an easy induction on k we find that

$$G_k(s) = 0 \quad \text{when } s \geq \beta + k.$$

Let

$$h_{2q-1}(y, s) = \sum_{r=1}^{q} g_{2r-1}(y, s), \tag{21.37}$$

$$h_{2q}(y, s) = \sum_{r=1}^{q} g_{2r}(y, s), \tag{21.38}$$

$$H_{2q-1}(s) = \sum_{r=1}^{q} G_{2r-1}(s), \tag{21.39}$$

$$H_{2q}(s) = \sum_{r=1}^{q} G_{2r}(s). \tag{21.40}$$

The aim is to show that the first two sums can be approximated by the second two, with an error that can be controlled. With quite a lot of work it can be shown that there is a positive number δ such that

$$\begin{aligned} XV(z)\biggl(f_-(s) + O\biggl(\frac{s^{1-\kappa}e^{-s}}{(\log y)^{\delta}}\biggr)\biggr) - R^* \\ \leq S(\mathcal{A}, \mathcal{P}, z) \leq XV(z)\biggl(f_+(s) + O\biggl(\frac{s^{-\kappa}e^{-s}}{(\log y)^{\delta}}\biggr)\biggr) + R^*, \end{aligned} \tag{21.41}$$

where

$$f_+(s) = 1 + \sum_{r=1}^{\infty} G_{2r-1}(s) \quad (s > 0), \tag{21.42}$$

$$f_-(s) = 1 - \sum_{r=1}^{\infty} G_{2r}(s) \quad (s \geq \beta). \tag{21.43}$$

It then follows by (21.31), (21.35), and (21.36) that $f_+(s)$ is differentiable for $s > 0, \neq \beta + 1$ and continuous at $s = \beta + 1$, that f_- is differentiable for $s > \beta$, and continuous from the right at $s = \beta$, and that

$$f_+(s) = (\beta + 1)^{\kappa} f_+(\beta + 1)s^{-\kappa} \quad (0 < s < \beta + 1), \tag{21.44}$$

$$\bigl(s^{\kappa} f_+(s)\bigr)' = \kappa s^{\kappa-1} f_-(s-1) \quad (s > \beta + 1), \tag{21.45}$$

$$\bigl(s^{\kappa} f_-(s)\bigr)' = \kappa s^{\kappa-1} f_+(s-1) \quad (s > \beta). \tag{21.46}$$

It is perhaps not surprising that these are essentially the same relationships that we adduced from the Buchstab identity, *vide* (21.10) and (21.11).

21.2.1 Convergence

We assume hence forward that $\kappa = 1$ and that $1.75 \le \beta \le 3$.

Lemma 21.7 *When $s \ge \beta$ let*

$$\varpi(s,\beta) = e^s s^{-1} \int_s^\infty (t-1)^{-1} e^{-\max(\beta,t-2)}\, dt$$

and

$$\Upsilon(\beta) = \sup_{s \ge \beta} \varpi(s,\beta).$$

Then $0.27 < \Upsilon(\beta) < 0.98$.

Proof Suppose first that $s > \beta + 2$. Then

$$\varpi(s) = e^s s^{-1} \int_s^\infty (t-1)^{-1} e^{2-t}\, dt$$

and

$$\begin{aligned}\varpi'(s) &= e^s (s-1) s^{-2} \int_s^\infty (t-1)^{-1} e^{2-t}\, dt - e^2 s^{-1} (s-1)^{-1}\\ &< e^2 \left(\frac{1}{s^2} - \frac{1}{s(s-1)} \right)\\ &< 0.\end{aligned}$$

Hence

$$\begin{aligned}\varpi(s) \le \varpi(\beta+2) &= e^2 (\beta+2)^{-1} \int_0^\infty (\beta+1+u)^{-1} e^{-u} du\\ &< \frac{e^2}{(\beta+2)(\beta+1)} \le \frac{16e^2}{165}\\ &< 0.72.\end{aligned}$$

Also

$$\varpi(\beta+2) \ge \frac{e^2}{5} \int_0^\infty (u+4)^{-1} e^{-u} du > \frac{3e^2}{80} > 0.27.$$

Now suppose that $\beta \le s \le \beta + 2$. Then

$$\varpi(s) = \frac{e^{s-\beta}}{s} \left(\log \frac{\beta+1}{s-1} + I(\beta) \right),$$

where for brevity we have written

$$I(\beta) = \int_0^\infty \frac{e^{-u}}{\beta+1+u}\, du.$$

If

$$\sup_{\beta \le s \le \beta+2} \varpi(s) = \max\{\varpi(\beta), \varpi(\beta+2)\},$$

then in view of the bound above for $\varpi(\beta+2)$, it suffices to deal with $\varpi(\beta)$. It is readily checked that

$$\varpi(\beta) = \frac{1}{\beta}\log\left(1+\frac{2}{\beta-1}\right) + \frac{I(\beta)}{\beta}$$

is a decreasing function of β. Hence

$$\varpi(\beta) \le \frac{4}{7}\left(\log\frac{11}{3} + I(7/4)\right) < \frac{4}{7}\left(\log\frac{11}{3} + \frac{4}{11}\right) < 0.96.$$

It remains to deal with the possibility that

$$\sup_{\beta\le s\le\beta+2} \varpi(s) = \varpi(s_0) > \max\{\varpi(\beta), \varpi(\beta+2)\}$$

for some s_0 with $\beta < s_0 < \beta+2$. We have

$$\varpi'(s) = e^{s-\beta}\frac{s-1}{s^2}\left(\log\frac{\beta+1}{s-1} + I(\beta)\right) - \frac{e^{s-\beta}}{s(s-1)}$$

and

$$\varpi'(s_0) = 0.$$

Thus

$$\log\frac{\beta+1}{s_0-1} + I(\beta) = \frac{s_0}{(s_0-1)^2}. \tag{21.47}$$

We also have

$$\varpi''(s) = e^{s-\beta}\left(\frac{1}{s} - \frac{2}{s^2} + \frac{2}{s^3}\right)\left(\log\frac{\beta+1}{s-1} + I(\beta)\right) + e^{s-\beta}\left(\frac{2s-1}{s^2(s-1)^2} - \frac{1}{s^2} - \frac{1}{s(s-1)}\right).$$

Hence, substituting (21.47), when $s = s_0$ we have

$$\varpi''(s_0) = e^{s_0-\beta}\frac{s_0^2+1-(s_0-1)^2 - s_0(s_0-1)}{s_0^2(s_0-1)^2}$$

and this is

$$= e^{s_0-\beta}\frac{s_0(3-s_0)}{s_0(s_0-1)^2}.$$

Since $\varpi(s_0)$ is maximal we have $s_0 \ge 3$.

Now substituting (21.47) once more we obtain

$$\varpi(s_0) = \frac{e^{s_0-\beta}}{(s_0-1)^2} = e^{-\beta}\frac{e^{s_0}}{(s_0-1)^2}.$$

The function $e^y(y-1)^{-2}$ is an increasing function for $y \geq 3$. Also $s_0 \leq \beta + 2$. Hence

$$\varpi(s_0) \leq \frac{e^2}{(\beta+1)^2} \leq \frac{16e^2}{121} < 0.98,$$

and this completes the proof. □

Choose the positive constant δ so that

$$\left(\frac{7}{3}\right)^{\delta} = \Upsilon^{-1/6},$$

where for brevity we write Υ for $\Upsilon(\beta)$. For $s > 0$ define

$$E_+(s) = s^{-1}e^{-\max(s-1,\beta)}, \quad E_-(s) = \Upsilon^{1/2}e^{-s}. \tag{21.48}$$

Then by Lemma 21.7, $0.0039 < \delta < 0.29$,

$$s^{-1}\int_s^{\infty}\left(\frac{t}{t-1}\right)^{\delta} E_+(t-1)\,dt \leq \theta E_-(s) \quad (s \geq \beta), \tag{21.49}$$

$$s^{-1}\int_{\max(s,\beta+1)}^{\infty}\left(\frac{t}{t-1}\right)^{\delta} E_-(t-1)\,dt \leq \theta E_+(s) \quad (s > 0), \tag{21.50}$$

where

$$\theta = \Upsilon^{1/3}$$

satisfies $0 < \theta < 1$.

It is useful to define

$$E_{2r}(s) = E_-(s), \quad E_{2r-1}(s) = E_+(s). \tag{21.51}$$

By induction on k we have

$$G_k(s) \ll \theta^k E_k(s) \tag{21.52}$$

when k is odd and $s > 0$ and when k is even and $s \geq \beta$. Thus H_{2q-1} and H_{2q} converge locally uniformly for $s > 0$ and $s \geq \beta$, respectively. Therefore, if we can show that for some positive constant C_1

$$h_q(y,s) < H_q(s) + C_1E_q(s)(\log y)^{-\delta} \tag{21.53}$$

when q is odd and $s > 0$ and when q is even and $s \geq \beta$, then we have the following conclusion.

Lemma 21.8 *There is a positive constant δ such that if $\frac{7}{4} \le \beta \le 3$ and $s = (\log y)/\log z)$, then*

$$XV(z)\big(f_-(s) + O\big(e^{-s}(\log y)^{-\delta}\big)\big) - R^* \le S(\mathcal{A}, \mathcal{P}, z) \\ \le XV(z)\big(f_+(s) + O\big(e^{-s}(\log y)^{-\delta}\big)\big) + R^*,$$

where $f_\pm(s)$ satisfy (21.42), (21.43), *and R^* satisfies* (21.22). *Moreover, $f_+(s)$ is differentiable for $s > 0, \ne \beta + 1$, continuous at $\beta + 1$, f_- is differentiable for $s > \beta$, and continuous from the right at $s = \beta$, and*

$$f_+(s) = (\beta + 1) f_+(\beta + 1) s^{-1} \quad (0 < s \le \beta + 1), \tag{21.54}$$

$$f_-(s) = 0 \quad (0 < s < \beta), \tag{21.55}$$

$$\big(s f_+(s)\big)' = f_-(s - 1) \quad (s > \beta + 1), \tag{21.56}$$

$$\big(s f_-(s)\big)' = f_+(s - 1) \quad (s > \beta). \tag{21.57}$$

We also have

$$f_\pm(s) = 1 + O(e^{-s}) \quad \text{as } s \to \infty. \tag{21.58}$$

The utility of this conclusion depends on the finer details of the functions $f_\pm$, which we study in §21.2.2, and we give the ultimate conclusions in Theorem 21.9 of §21.3 below. The three equations (21.54), (21.56), and (21.57) are immediate from (21.44), (21.45), and (21.46), and (21.58) follows from (21.42), (21.43), (21.52), and (21.48). Note that we have extended the definition of f_- to the region $0 < s < \beta$ since the theorem remains true for trivial reasons with this extension. It is clear by (21.58) and continuity that $f_\pm(s) > 0$ for $s > s_0$ for some $s_0 \ge \beta$, and if $f_-(\beta) \ge 0$, then (21.54), (21.56), (21.57), and induction on k shows that $s f_\pm(s)$ is strictly increasing on $[\beta + k - 1, \beta + k]$, and hence is positive for $s > \beta$. We will eventually

$$\text{choose } \beta \text{ so } \inf\{t \ge 0 : f_-(s) > 0 \text{ for all } s > t\} \text{ is minimal.} \tag{21.59}$$

This optimal choice is known as the *sieving limit*.

Proof of (21.53) We now prove (21.53) by induction on q. The case $q = 1$ is immediate from (21.30), (21.37), and (21.39). Note that $\kappa = 1$, $1.75 \le \beta \le 3$, and $G_1(s) = 0$ when $s \ge \beta + 1$.

Suppose $q \ge 2$ and (21.53) holds with q replaced by $q - 1$. By (21.24) and (21.25) and induction, $g_k(y, s) = 0$ when $s \ge \beta + k$, and

$$g_k(y, s) \le V(y^{1/s})^{-1} \sum_{p_k < \cdots < p_1 < y^{1/s}} \rho(p_1 \cdots p_k).$$

Hence

$$g_k(y, s) = 0 \quad \text{for } s \ge \min\left(\beta + k, \frac{\log y}{\log 2}\right) \tag{21.60}$$

and

$$g_k(y, s) \le V(y^{1/s})^{-1} \frac{1}{k!}\left(\sum_{p<y^{1/s}} \rho(p)\right)^k. \tag{21.61}$$

Thus we may suppose that $s < \frac{\log y}{\log 2}$.

By (21.61), when $2 \le y \ll 1$ and $s \ge 1$ we have

$$g_k(y, s) \ll C_2^k/k!$$

for some positive constant C_2. Hence, by (21.48) and (21.51) we have

$$h_q(y, s) \ll \sum_{k>s-\beta} \frac{C_2^k}{k!}$$

whence

$$h_q(y, s) \le C_1 E_q(s)(\log y)^{-\delta}$$

provided that $2 \le y \le y_0$ and $C_1 = C_1(y_0)$. It follows in this case that we have (21.53). Thus we may now assume that

$$y > y_0, \quad s < \frac{\log y}{\log 2}. \tag{21.62}$$

For the time being we suppose that

$$s \ge \beta \text{ when } q \text{ is odd}, \quad s \ge \beta + 1 \text{ when } q \text{ is even}. \tag{21.63}$$

By (21.37), (21.38), (21.60), and (21.61),

$$h_{2r-1}(y, s) + h_{2r}(y, s) \le V(y^{1/s})^{-1} \sum_{s-\beta<k\le 2r} \frac{1}{k!}\left(\sum_{p<y^{1/s}} \rho(p)\right)^k.$$

Suppose that $Y \ge 1$. Then by (21.29) and Lemma 21.6,

$$\begin{aligned} h_q(y, s) &\le V(y^{1/s})^{-1} \sum_{s-\beta<k\le 2r} \frac{Y^{-k}}{k!}\left(\sum_{p<y^{1/s}} Y\rho(p)\right)^k \\ &\ll s^{-1}(\log y)Y^{\beta-s}\exp(Y(\log\log y - \log s + C)). \end{aligned}$$

Let $Y = s^{1/2}$. Then by (21.62),

$$h_q(y, s) < E_q(s)(\log y)^{-1} \quad \text{for } s \ge (\log\log y)^3,$$

which again gives (21.53). Thus it remains to consider s with

$$s < w, \quad w = (\log\log y)^3,$$

and we will deduce slightly more than (21.53), namely that there is a constant θ_0 with $0 < \theta_0 < 1$ such that

$$h_q(y, s) < H_q(s) + \theta_0 C_1 E_q(s)(\log y)^{-\delta}. \tag{21.64}$$

Suppose for now that $s \geq \beta$ when q is even and $s \geq \beta + 1$ when q is odd. By (21.24), (21.25), (21.37), and (21.38) with q replaced by $q - 1$,

$$\begin{aligned} h_q(y, s) &= \frac{V(y^{1/w})}{V(y^{1/s})} h_q(y, w) + \sum_{y^{1/w} \leq p < y^{1/s}} g(p) \frac{V(p)}{V(y^{1/s})} h_{q-1}\left(\frac{y}{p}, \frac{\log y}{\log p} - 1\right) \\ &\leq \left(1 + \frac{Cw}{\log y}\right) \frac{wE_q(w)}{s \log y} + \sum_{y^{1/w} \leq p < y^{1/s}} \frac{g(p)V(p)}{V(y^{1/s})} H_{q-1}\left(\frac{\log y}{\log p} - 1\right) \\ &\quad + \sum_{y^{1/w} \leq p < y^{1/s}} \frac{C_1 g(p) V(p)}{V(y^{1/s})\left(\log \frac{y}{p}\right)^{\delta}} E_{q-1}\left(\frac{\log y}{\log p} - 1\right). \end{aligned}$$

The functions $H_{q-1}(s)$, $E_{q-1}(s)$, and $sE_q(s)$ are decreasing functions of s for $s \geq \beta$. Hence by Lemma 21.6,

$$\begin{aligned} h_q(y, s) &\leq \frac{1}{s} \int_s^w \left(H_{q-1}(t-1) + \frac{C_1 E_{q-1}(t-1)}{(1 - 1/t)^{\delta} (\log y)^{\delta}}\right) dt \\ &+ \frac{Cw^2}{s \log y}\left(H_{q-1}(s-1) + \frac{C_1 E_{q-1}(s-1)}{(1-1/s)^{\delta}(\log y)^{\delta}}\right) + O\big(E_q(s)(\log y)^{-1}\big). \end{aligned}$$

By (21.35), (21.36), (21.39), and (21.40),

$$s^{-1} \int_s^w H_{q-1}(t-1)\, dt \leq H_q(s).$$

By (21.49), (21.50), and (21.51),

$$\frac{1}{s} \int_s^w C_1 E_{q-1}(t-1)(1 - 1/t)^{-\delta}\, dt \leq C_1 \theta E_q(s).$$

By (21.48), (21.51), and (21.52),

$$\frac{Cw^2}{s \log y}\left(H_{q-1}(s-1) + \frac{C_1 E_{q-1}(s-1)}{(1-1/s)^{\delta}(\log y)^{\delta}}\right) \ll E_q(s)(\log y)^{-1/2}.$$

This establishes (21.64) when (21.63) holds.

We now deduce (21.53) when $q = 2k - 1$ is odd and $0 < s \le \beta + 1$. We have established (21.64) when $s = \beta + 1$. By (21.26), (21.28), (21.25), and (21.37) we have

$$h_{2k-1}(y,s) = g_1(y,s) + \sum_{r=2}^{k} g_{2r-1}(y,s)$$
$$= \frac{V(y^{1/(\beta+1)})}{V(y^{1/s})}(1 + h_{2k-1}(y,\beta+1)) - 1.$$

Therefore, by (21.64),

$$h_{2k-1}(y,s)$$
$$< -1 + \frac{\beta+1}{s}\left(1 + \frac{C(\beta+1)}{\log y}\right)\left(1 + H_{2k-1}(\beta+1) + \frac{\theta_0 C_1 E_{2k-1}(\beta+1)}{(\log y)^{\delta}}\right)$$
$$< \frac{\beta+1}{s} + \frac{\beta+1}{s}H_{2k-1}(\beta+1) - 1 + \frac{C_1 E_+(s)}{(\log y)^{\delta}}.$$

By (21.31), (21.35), (21.36), and (21.39) we have

$$\frac{\beta+1}{s} + \frac{\beta+1}{s}H_{2k-1}(\beta+1) - 1 = H_{2k-1}(s),$$

which gives (21.53). □

21.2.2 The Differential Delay Equations

We now need to elicit the finer properties of the functions $f_{\pm}(s)$ when $\kappa = 1$. They satisfy (21.54), (21.56), (21.57), and (21.58). We can separate the functions by defining

$$S_+(s) = f_+(s) + f_-(s) - 2, \quad S_-(s) = f_+(s) - f_-(s)$$

so that

$$(sS_{\pm}(s))' = \pm S_{\pm}(s-1).$$

These functions $S_{\pm}(s)$ are differentiable for $s > \beta$, $s \ne \beta + 1$, and continuous at $\beta + 1$ and from the right at β. We have already encountered the equation for S_+. It is satisfied by Buchstab's function w, *vide* (7.38). The Dickman function (7.4) also has some similarities with S_-. Our initial concern is to optimize the choice of β. To that end we need to study the conjugate equations

$$s\phi_{\pm}'(s) = \mp\phi_{\pm}(s+1) \tag{21.65}$$

and the associated inner product

$$I_{\pm}(s) = sS_{\pm}(s)\phi_{\pm}(s) \pm \int_{s-1}^{s} S_{\pm}(t)\phi_{\pm}(t+1)\,dt.$$

It is clear that

$$\phi_-(s) = s - 1$$

is a solution of (21.65) in the − case.

In the contrary case we define

$$\phi_+(s) = \int_0^\infty \exp\left(-sx + \int_0^x \frac{e^{-y} - 1}{y}\, dy\right) dx.$$

This is differentiable for $s > 0$, and it is readily checked that then (21.65) holds in the + case and that

$$\frac{1}{s+1} < \phi_+(s) < \frac{1}{s}.$$

It follows from (21.58) that

$$S_\pm(s) \ll e^{-s} \quad \text{as } s \to \infty.$$

We also have

$$I'_\pm(s) = 0 \quad (s > \beta + 1)$$

so that $I_\pm(s)$ is constant for $s \geq \beta + 1$. Moreover, $I_\pm(s) \to 0$ as $s \to \infty$. Therefore

$$sS_\pm(s)\phi_\pm(s) \pm \int_{s-1}^{s} S_\pm(t)\phi_\pm(t+1)\, dt = I_\pm(s) = 0 \quad (s \geq \beta + 1). \quad (21.66)$$

Hence

$$\begin{aligned}(\beta+1)S_-(\beta+1)\beta &= \int_\beta^{\beta+1} tS_-(t)\, dt \\ &= \left[tS_-(t)\phi_-(t)\right|_\beta^{\beta+1} - \int_\beta^{\beta+1} (tS_-(t))'\phi_-(t)\, dt.\end{aligned}$$

When $\beta < t < \beta + 1$ we have

$$(tS_-(t))' = -f_+(t-1) = -(\beta+1)f_+(\beta+1)(t-1)^{-1}$$

and so

$$\beta S_-(\beta)(\beta - 1) = \int_\beta^{\beta+1} (\beta+1)f_+(\beta+1)dt = (\beta+1)f_+(\beta+1).$$

Moreover,

$$\beta S_-(\beta) = \beta(f_+(\beta) - f_-(\beta)) = (\beta+1)f_+(\beta+1) - \beta f_-(\beta).$$

Hence

$$(\beta - 1)\beta f_-(\beta) = (\beta - 2)(\beta+1)f_+(\beta+1).$$

By (21.66) and continuity we have $S_-(s) > 0$ for all $s \geq \beta + 1$, and hence for all $s > 0$. Thus $f_+(s) > f_-(s)$ for all $s > 0$. Also, by (21.54), $f_+'(s) < 0$ for $s \leq \beta + 1$. We show by contradiction that this holds for all $s > 0$. If this is false, then by continuity there will be an $s_0 > \beta + 1$ such that $f_+'(s_0) = 0$, and we can choose s_0 minimally. Hence, by (21.56), $0 = f_+(s_0) - f_-(s_0 - 1) > f_-(s_0) - f_-(s_0 - 1)$. Therefore, by the mean value theorem there is an s_1 with $\beta < s_0 - 1 < s_1 < s_0$ such that $f_-'(s_1) < 0$. Hence, by (21.57), $0 < f_(s_1) - f_+(s_1 - 1) < f_+(s_1) - f_+(s_1 - 1) = f_+'(s_2)$ where $0 < \beta - 1 < s_1 - 1 < s_2 < s_1 > s_0$, and so there is a change of sign before s_2 contradicting the minimality of s_0. Hence $f_+'(s) < 0$ for all $s > 0$. Therefore, by (21.57), when $s > \beta$ we have $sf_-'(s) = f_+(s-1) - f_-(s) > f_+(s-1) - f_+(s) > 0$. Thus

$$f_-(s) > f_-(\beta) \quad (s > \beta).$$

Hence, by (21.59) it is clear that the optimal choice of β is

$$\beta = 2,$$

which we assume hitherto.

We are finally concerned with evaluating $(\beta + 1)f_+(\beta + 1) = 3f_+(3)$. Let

$$\nu(s) = s\phi_+(s) + \int_s^{s+1} \phi_+(t)\, dt.$$

Then $\nu'(s) = 0$ $(s > 0)$ and $\nu(s) = \lim_{t\to\infty} \nu(t) = 1$, and so

$$2\phi_+(2) + \int_2^3 \phi(t)\, dt = 1.$$

By (21.66)

$$\begin{aligned} 3S_+(3)\phi_+(3) &= -\int_2^3 S_+(t)\phi_+(t+1)\, dt \\ &= \int_2^3 tS_+(t)\phi_+'(t)\, dt \\ &= \left[tS_+(t)\phi_+(t)\right|_2^3 - \int_2^3 (tS_+(t))'\phi_+(t)\, dt. \end{aligned}$$

Hence, as $(tf_+(t))' = 0$ when $t < 3$ we have

$$\begin{aligned} 2S_+(2)\phi_+(2) &= -\int_2^3 (tf_+(t) + tf_-(t))'\phi_+(t)\, dt + 2\int_2^3 \phi_+(t)\, dt \\ &= -\int_2^3 f_+(t-1)\phi_+(t)\, dt + 2(1 - 2\phi_+(2)). \end{aligned}$$

Thus

$$\begin{aligned}(2f_+(2)-4)\phi_+(2) &= -\int_2^3 \frac{3f_+(3)}{t-1}\phi_+(t) + 2 - 4\phi_+(2)\\ &= 3f_+(3)\int_2^3 \phi_+'(t-1)dt + 2 - 4\phi_+(2).\end{aligned}$$

Therefore

$$3f_+(3)\phi_+(2) = 3f_+(3)(\phi_+(2)-\phi_+(1)) + 2$$

and so

$$3f_+(3)\phi_+(1) = 2.$$

When $s > 0$, we have $(s\phi_+(s))' = \phi_+(s) - \phi_+(s+1)$ and so

$$\begin{aligned}\phi_+(1) &= \lim_{s\to 0+}\left(\phi_+(s) - (s\phi_+(s))'\right)\\ &= \lim_{s\to 0+}\int_0^\infty s\exp\left(\log x - sx - \int_0^x \frac{1-e^{-y}}{y}\,dy\right)dx.\end{aligned}$$

By (C.11) we have

$$C_0 = -\Gamma'(1) = -\int_0^\infty (\log y)e^{-y}\,dy.$$

Splitting the integral at x, writing it as

$$-\int_0^x (\log y)\,d(1-e^{-y}) + \int_x^\infty (\log y)\,d(e^{-y}),$$

and integrating each integral by parts, we obtain

$$C_0 = -\log x + \int_0^x \frac{1-e^{-y}}{y}\,dy - \int_x^\infty \frac{e^{-y}}{y}\,dy.$$

Hence

$$\begin{aligned}&\lim_{s\to 0+}\int_0^\infty s\exp\left(\log x - sx - \int_0^x \frac{1-e^{-y}}{y}\,dy\right)dx\\ &\qquad = \lim_{s\to 0+}\int_0^\infty s\exp\left(-sx - C_0 - \int_x^\infty \frac{e^{-y}}{y}\,dy\right)dx\\ &\qquad = \lim_{s\to 0+}\int_0^\infty \exp\left(-t - C_0 - \int_{t/s}^\infty \frac{e^{-y}}{y}\,dy\right)dt\\ &\qquad = e^{-C_0}.\end{aligned}$$

Therefore $3f_+(3) = 2e^{C_0}$.

21.2.3 Exercises

1. Suppose that $0 < \kappa \le 1$, that $1 \le \beta \le 1 + \kappa$ when $0 < \kappa \le \frac{1}{2}$ and $1 + (2\kappa - 1)^2 \le \beta \le 1 + \kappa$ when $\frac{1}{2} \le \kappa \le 1$, and let

$$\rho(s) = \kappa e^s s^{-\kappa}(s+1)^{\kappa-1} \int_s^\infty \kappa t^{\kappa-1}(t-1)^{-\kappa} e^{-\max(\beta, t-2)}\, dt.$$

Prove that

$$\sup_{s \ge \beta} \rho(s) < 1.$$

2. Suppose that $\kappa = \frac{1}{2}$, that (21.44), (21.45), (21.46), that

$$f_\pm(s) = 1 + O(e^{-s}) \qquad \text{as } s \to \infty,$$

and that $f_-(s) = 0$ when $s < \beta$ all hold. Let

$$S_+(s) = f_+(s) + f_-(s) - 2, \quad S_+(s) = f_+(s) - f_-(s).$$

(a) Prove that if $s > \beta$, then

$$\left(s^{1/2} S_\pm(s)\right)' = \pm \frac{1}{2} s^{-1/2} S_\pm(s-1)$$

and

$$s S_\pm'(s) = -\frac{1}{2} S_\pm(s) \pm \frac{1}{2} S_\pm(s-1).$$

(b) Prove that when $s > 0$ the equations

$$(s\phi_\pm(s))' = \frac{1}{2}\phi_\pm(s) \mp \frac{1}{2}\phi_\pm(s+1)$$

are satisfied by

$$\phi_+(s) = \int_0^\infty \exp\left(-sx - \frac{1}{2}\int_0^x \frac{1 - e^{-u}}{u}\, du\right) dx$$

and $\phi_-(s) = 1$, respectively, and that $\phi_+(s) \sim \frac{1}{s}$ as $s \to \infty$.

(c) Suppose that $s \ge \beta$. Let

$$I_\pm(s) = s S_\pm(s)\phi_\pm(s) \pm \frac{1}{2}\int_{s-1}^s S_\pm(t)\phi_\pm(t+1)\, dt.$$

Prove that $I_\pm(s) = 0$.

(d) Prove that

$$\beta f_-(\beta) = (\beta+1)^{1/2} f_+(\beta+1)(\beta-1)^{1/2}$$

and that the optimal choice of β is $\beta = 1$.

(e) Now assume that $\beta = 1$ and let

$$\nu(s) = s\phi_+(s) + \int_s^{s+1} \frac{1}{2}\phi_+(t)\, dt.$$

Show that $\nu(s) = 1$, that

$$s(f_+(s) + f_-(s))\phi_+(s) + \int_{s-1}^{s} \frac{1}{2}(f_+(t) + f_-(t))\varphi_+(t+1)\,dt = 2$$

and that

$$f_+(1) = \frac{2}{\phi_+(1)}.$$

(f) Prove that

$$\phi_+(1) = e^{-C_0/2}\Gamma(1/2)$$

and that

$$f_+(s) = s^{-1/2} 2\big(e^{C_0}/\pi\big)^{1/2} \quad (s \le 2).$$

21.3 The Linear Sieve

We can now state the linear sieve.

Theorem 21.9 *Suppose that* (21.3) *and* (21.29) *hold with* $\kappa = 1$. *Then there is a positive constant* δ *such that when* $s = (\log y)/\log z$ *we have*

$$\begin{aligned} XV(z)\big(f_-(s) + O\big(e^{-s}(\log y)^{-\delta}\big)\big) - R^* &\le S(\mathcal{A}, \mathcal{P}, z) \\ &\le XV(z)\big(f_+(s) + O\big(e^{-s}(\log y)^{-\delta}\big)\big) + R^*, \end{aligned}$$

where R^* *is given by* (21.22). *Moreover,* $f_+(s)$ *is differentiable for* $s > 0$ *and* f_- *is differentiable for* $s > 0, \neq 2$, *and*

$$f_+(s) = 2e^{C_0}s^{-1} \quad (0 < s \le 3), \tag{21.67}$$

$$f_-(s) = 0 \quad (0 < s \le 2), \tag{21.68}$$

$$f_-(s) = 2e^{C_0}s^{-1}\log(s-1) \quad (2 \le s \le 4), \tag{21.69}$$

$$(sf_+(s))' = f_-(s-1) \quad (s > 3), \tag{21.70}$$

$$(sf_-(s))' = f_+(s-1) \quad (s > 2). \tag{21.71}$$

We also have

$$f_\pm(s) = 1 + O(e^{-s}) \quad \textit{as } s \to \infty. \tag{21.72}$$

Equations (21.67), (21.68), and (21.70)–(21.72) are the analogues of (21.54)–(21.58) with $\beta = 2$. We remark that it is easily seen by integration that (21.69) holds, and then that f_+ is differentiable at $s = 3$. In addition,

$$f_+(s) = 2e^{C_0}s^{-1}\left(1 + \int_3^s \frac{\log(u-2)}{u-1}\,du\right) \quad (3 \le s \le 5) \tag{21.73}$$

and

$$f_-(s) = \frac{2e^{C_0}}{s}\left(\log(s-1) + \int_4^s \frac{\log(u-3)}{u-2}\log\frac{s-1}{u-1}\,du\right) \qquad (4 \le s \le 6). \quad (21.74)$$

There are some applications where we would like to have an asymptotic result rather than just upper and lower bounds. In §21.4 we will see that this is not possible when s is small, and that indeed Theorem 21.9 is best possible. However, when s is large (21.72) does permit an asymptotic conclusion.

Corollary 21.10 *Suppose that* (21.3) *and* (21.29) *hold with* $\kappa = 1$, *and that*

$$\sum_{m \le y} |R(m)| \ll XV(z)e^{-s}.$$

Let $s = \frac{\log y}{\log z}$ *and suppose that for some positive number* δ *we have* $s \ge 2 + \delta$. *Then*

$$S(\mathcal{A}, \mathcal{P}, z) = XV(z)(1 + O(e^{-s})).$$

21.3.1 Exercises

1. (Vaughan 1976, Iwaniec 1978b)
   Let $a \in \mathbb{Z}$ and $n \in \mathbb{N}$, $n > 1$, $X \ge 2$, $P = \prod_{p<X} p$,

$$T(h,n) = \sum_{\substack{m=a+1 \\ (m,n)=1}}^{a+h} 1, \quad S(h,X) = \sum_{\substack{m=a+1 \\ (m,P)=1}}^{a+h} 1.$$

   (a) Prove that

$$S(h,X) \le T(h,n) + \omega(n)(h/X + 1).$$

   (b) As in §7.3, let $g(n)$ denote the least integer so that amongst any $g(n)$ consecutive integers $a+1, \ldots, a+g(n)$ there is at least one coprime with n. Prove that $g(n) \ll \omega(n)^2(\log 2\omega(n))^4$.

2. Let p be an odd prime and let $G(p)$ denote the least positive primitive root modulo p. Prove that if the Generalized Riemann Hypothesis holds, then

$$G(p) \ll (\log p)^{6+\varepsilon}.$$

21.4 The Selberg Examples

Selberg showed that the inequalities in Theorem 21.9 are best possible by the presentation of a pair of extremal examples.

Theorem 21.11 *Let $X \geq 2$,*

$$a_{\pm}(n) = \begin{cases} 1 \mp \lambda(n) & n \leq X, \\ 0 & \textit{otherwise}, \end{cases}$$

$\mathcal{P}$ be the set of all primes, and

$$T_{\pm}(X, z) = S(\mathcal{A}_{\pm}, \mathcal{P}, z),$$

where λ is the Liouville function of §1.3. *Suppose that*

$$\exp\left((\log X)^{2/3}\right) \leq z \leq \frac{X}{\log X}.$$

Then

$$T_{\pm}(X, z) = XV(z) f_{\pm}\left(\frac{\log X}{\log z}\right) + O\left(X(\log z)^{-4/3}\right),$$

where

$$V(z) = \prod_{p<z} (1 - 1/p)$$

and the $f_{\pm}$ are as in Theorem 21.9.

We have

$$T_{-}(X, z) + T_{+}(X, z) = 2\Phi(X, z), \quad T_{-}(X, z) - T_{+}(X, z) = 2\Xi(X, z),$$

where

$$\Xi(X, z) = \sum_{\substack{n \leq X \\ (n, P(z))=1}} \lambda(n).$$

As discussed in §7.2, $\Phi(x, y)$ is the number of integers $\leq x$ composed entirely of primes $p \geq y$. Its asymptotics are described by Buchstab's function $w(u)$ where $u = (\log x)/\log y$. We could just appeal to Theorem 7.11 when $\frac{\log X}{\log z} \ll 1$, but here we push things a bit further. The function $\Xi(X, z)$ satisfies Buchstab's identity, Lemma 21.2, and consequently the Dickman function $\rho(u)$, which arose in §7.1 to determine the asymptotices of $\psi(x, y)$, is relevant. Both $w(u)$ and $\rho(u)$ are determined by differential–delay equations.

Note that by (21.58) we have

$$f_{\pm}\left(\frac{\log X}{\log z}\right) = 1 + O\left(\exp\left(-\frac{\log X}{\log z}\right)\right)$$

and so the upper and lower sieve bounds are anyway asymptotic when

$$z \leq \exp\left((\log X)^{2/3}\right).$$

When $s \geq 2 + \delta$, we have $f_{-}(s) \asymp 1$, and likewise for $f_{+}(s)$ when $s \geq 1$, so the above give asymptotic formulæ in those ranges.

This theorem illustrates one facet of the parity problem, namely that sieve methods generally cannot distinguish between numbers with an odd and an even number of prime factors.

Theorem 21.11 shows that essentially the linear sieve, as annunciated in Theorem 21.9, is best possible. To see this, let

$$y = X\exp(-(\log\log X)^3)$$

and

$$R(m) = \sum_{\substack{n\le X\\ m|n}} a_{\pm}(n) - X\rho(m)$$

with

$$\rho(m) = \frac{1}{m}.$$

Then for $m \le y$

$$\sum_{m\le y} |R(m)| \le y + \sum_{m\le y}\left|\sum_{l\le X/m} \lambda(l)\right|,$$

and by Exercise 6.2.1.1

$$\begin{aligned}\sum_{l\le X/m} \lambda(l) &\ll Xm^{-1}\exp\left(-c(\log(X/m))^{1/2}\right)\\ &\ll Xm^{-1}\exp\left(-c(\log\log X)^{3/2}\right).\end{aligned}$$

Thus

$$\sum_{m\le y} |R(m)| \ll X(\log X)^{-3}. \tag{21.75}$$

Moreover,

$$f_{\pm}(s) = f_{\pm}\left(\frac{\log X}{\log z}\right) + O\left((\log X)^{-1/2}\right).$$

Thus Theorem 21.9 would give

$$\begin{aligned}XV(z)f_-&\left(\frac{\log X}{\log z}\right)\left(1+O\left((\log X)^{-\delta}\right)\right) + O\left(X(\log X)^{-3/2}\right)\\ &\le T_{\pm}(X,z)\\ &\le XV(z)f_+\left(\frac{\log X}{\log z}\right)\left(1+O\left((\log X)^{-\delta}\right)\right) + O\left(X(\log X)^{-3/2}\right).\end{aligned}$$

Theorem 21.11 shows that the functions f_- and f_+ cannot be replaced by anything larger or smaller, respectively.

Proof To prove the theorem we use an inductive argument. When

$$X^{1/2} < z \le X/\log X \tag{21.76}$$

we have

$$T_\pm(X,z) = 1 \mp \lambda(1) + \sum_{z<p\le X} (1 \mp \lambda(p)). \tag{21.77}$$

Thus

$$\begin{aligned} T_\pm(X,z) &= (1 \pm 1)\frac{X}{\log X} + O\left(\frac{X}{(\log X)^2}\right) \\ &= (1 \pm 1)XV(z)e^{C_0}\frac{\log z}{\log X} + O\left(\frac{X}{(\log X)^2}\right), \end{aligned}$$

and so

$$T_\pm(X,z) = XV(z)f_\pm\left(\frac{\log X}{\log z}\right) + O\left(\frac{X}{(\log X)^2}\right), \tag{21.78}$$

where

$$V(z) = \prod_{p<z}(1 - 1/p).$$

Now the proof of Lemma 21.2 is readily adapted to show that if $2 \le z \le w$, then

$$T_\pm(X,z) = T_\pm(X,w) + \sum_{z\le p<w} T_\mp(X/p,p). \tag{21.79}$$

The plan now is to show that for a suitable positive constant C, when $k \in \mathbb{N}$, we have for every pair X, z satisfying $X > X_0$, $3 \le k \le (\log X)^{1/3} + 1$, and $X^{1/k} < z \le X^{1/(k-1)}$ the inequality

$$\left|T_\pm(X,z) - XV(z)f_\pm\left(\frac{\log X}{\log z}\right)\right| \le \frac{CXk^3 \log\log X}{(\log X)^2}. \tag{21.80}$$

The relationship in the inductive proof is that if $X^{1/(k+1)} < z \le p < X^{1/k}$, then $(X/p)^{1/k} \le p < (X/p)^{1/(k-1)}$ and the deduction will be routine when $k \ge 3$, but when $k = 2$ we have only established the necessary hypothesis when $X^{1/2} < z \le X/\log X$. That is, we have a problem when

$$\frac{X/p}{\log(X/p)} < p \le X/p.$$

We resolve this minor lacuna by a separate argument to establish the desired conclusion when

$$\left(\frac{X}{\log X}\right)^{1/2} \le z \le X^{1/2}.$$

Then $T_{\pm}(X,z)$ differs from the sum (21.77) considered previously in case (21.76) by just having an additional contribution

$$\sum_{\substack{p_1,p_2\\ z\le p_1\le p_2\le X/p_1}} (1\mp 1) \ll \sum_{z<p\le X^{1/2}} \frac{X}{p\log X} \ll \frac{X\log\log X}{(\log X)^2},$$

and we can argue much as before.

Consequentially we need only treat the case corresponding to $k=2$ above when $X^{1/3}\le z<(X/\log X)^{1/2}$.

In (21.79) let $w=(X/\log X)^{1/2}$ and $X^{1/3}\le z<w$. When $z\le p\le w$ we have

$$X^{1/3}\le p<\left(\frac{X}{\log(X/p)}\right)^{1/2}$$

and so

$$(X/p)^{1/2}\le p<\frac{X/p}{\log(X/p)}.$$

Thus we may appeal to the initial case (21.78) (with X replaced by X/p and z by p). Then we have to deal with

$$XV(w)f_{\pm}\left(\frac{\log X}{\log w}\right)+\sum_{z\le p<w} Xp^{-1}V(p)f_{\mp}\left(\frac{\log X}{\log p}-1\right) + O\left(\frac{X\log\log X}{(\log X)^2}\right). \quad (21.81)$$

We require an asymptotic version of Lemma 21.6.

Lemma 21.12 *Suppose that* $2\le z\le w\le X^{1/2}$ *and*

$$V(u)=\prod_{p<u}\left(1-\frac{1}{p}\right).$$

Then

$$V(w)f_{\pm}\left(\frac{\log X}{\log w}\right)+\sum_{z\le p<w} V(p)p^{-1}f_{\mp}\left(\frac{\log X}{\log p}-1\right) = V(z)f_{\pm}\left(\frac{\log X}{\log z}\right)\left(1+O\left(\frac{1}{\log z}\right)\right).$$

Proof The sum above is

$$\sum_{z\le p<w}\left(V(p)p^{-1}f_{\mp}\left(\frac{\log X}{\log w}-1\right)+\int_p^w f'_{\mp}\left(\frac{\log X}{\log u}-1\right)\frac{\log X}{u\log^2 u}\,du\right).$$

Then on interchanging the order of summation and integration and applying the identity (21.27), this is

$$(V(z) - V(w))f_{\mp}\left(\frac{\log X}{\log w} - 1\right)$$
$$+ \int_z^w (V(z) - V(u))f_{\mp}'\left(\frac{\log X}{\log u} - 1\right)\frac{\log X}{u \log u^2}\,du.$$

By Mertens' Theorem 2.7(e),

$$V(w) = V(z)\frac{\log z}{\log w}\left(1 + O\left(\frac{1}{\log z}\right)\right).$$

Hence the above is

$$V(z)\left(1 + O\left(\frac{1}{\log z}\right)\right)\left(\left(1 - \frac{\log w}{\log z}\right)f_{\mp}\left(\frac{\log X}{\log w} - 1\right)\right.$$
$$\left. + \int_z^w \left(1 - \frac{\log z}{\log u}\right)f_{\mp}'\left(\frac{\log X}{\log u} - 1\right)\frac{\log X}{u \log^2 u}\,du\right).$$

By the change of variables $t = \frac{\log X}{\log u}$ the integral above becomes

$$\int_{\frac{\log X}{\log w}}^{\frac{\log X}{\log z}} \left(1 - t\frac{\log z}{\log X}\right) f_{\mp}'(t-1)\,dt,$$

and then by integration by parts, (21.70), and (21.71) the above is

$$V(z)\left(1 + O\left(\frac{1}{\log z}\right)\right)\int_{\frac{\log X}{\log w}}^{\frac{\log X}{\log z}} \frac{\log z}{\log X} f_{\mp}(t-1)\,dt$$
$$= V(z)\left(1 + O\left(\frac{1}{\log z}\right)\right)\left(f_{\pm}\left(\frac{\log X}{\log z}\right) - \frac{\log z}{\log w}f_{\pm}\left(\frac{\log X}{\log w}\right)\right).$$

Hence

$$V(w)f_{\pm}\left(\frac{\log X}{\log w}\right) + \sum_{z \le p < w} V(p)p^{-1}f_{\mp}\left(\frac{\log X}{\log p} - 1\right)$$
$$= V(z)f_{\pm}\left(\frac{\log X}{\log z}\right)\left(1 + O\left(\frac{1}{\log z}\right)\right) + O\left(\frac{V(z)}{\log w}f_{\pm}\left(\frac{\log X}{\log w}\right)\right),$$

and the lemma follows from the monotonicity of $sf_{\pm}(s)$. □

The lemma applied to (21.81) completes the proof of (21.80) when $k = 3$. Now suppose that (21.80) holds for some k with $3 \le k \le (\log X)^{1/3}$ and suppose that $X^{1/(k+1)} < z \le X^{1/k}$. Let $w = X^{1/k}$ and consider (21.80). As we observed above, when $z \le p < w$ we have $(X/p)^{1/k} \le p < (X/p)^{1/(k-1)}$.

Hence we may insert (21.80) with x, z replaced by X, w or $X/p, p$ into (21.79). Thus

$$\left| T_\pm(X,z) - XV(w) f_\pm\left(\frac{\log X}{\log w}\right) - \sum_{z \le p < w} \frac{X}{p} V(p) f_\mp\left(\frac{\log X}{\log p} - 1\right) \right| \le \frac{CXk^3 \log\log X}{\log^2 X} + \sum_{z \le p < w} \frac{CXk^3 \log\log X}{p \log^2(X/p)}.$$

Now we apply the lemma and obtain

$$\left| T_\pm(X,z) - XV(z) f_\pm\left(\frac{\log X}{\log z}\right) \right| \le \frac{CXk^3 \log\log X}{\log^2 X} + \frac{C_1 X(k+1)^2}{\log^2 X} + \frac{C_1 k^5 \log\log X}{(k-1)^2 (\log X)^2}\left(\frac{1}{k} + \frac{k}{\log X}\right)$$

for an absolute constant C_1. Since $k \le (\log X)^{1/3}$, it follows that for a suitable positive constant C we have (21.80) with k replaced by $k+1$.

To complete the proof of the theorem, suppose

$$\exp\left((\log X)^{2/3}\right) \le z \le \frac{X}{\log X},$$

so that $(\log z)^{5/3} \ge (\log X)^{10/9}$ and choose k so that $k - 1 < \frac{\log X}{\log z} \le k$, whence $k < (\log X)^{1/3} + 1$. Hence, by (21.80), we obtain the theorem with an error term

$$\ll \frac{X(\log X)\log\log X}{(\log z)^3} \ll \frac{X}{(\log z)^{4/3}},$$

as required. □

21.4.1 Exercises

1. Let $a_\pm$, $S(\mathcal{A}_\pm, \mathcal{P}, z)$, $T_\pm(X,z)$, $V(z)$ be as in Theorem 21.11. Let $\kappa \in \mathbb{N}$ and define

$$a_\pm^{(\kappa)}(n) = \sum_{\substack{n_1,\cdots,n_\kappa \\ n_1\cdots n_\kappa = n}} a_\pm(n_1)\cdots a_\pm(n_\kappa),$$

$$\mathcal{A}_\pm^{(\kappa)} = \sum_{n\in\mathbb{Z}} a_\pm^{(\kappa)}(n),$$

$$X_\kappa = X^\kappa,$$

$$T_\pm^{(\kappa)}(X_\kappa, z) = S\left(\mathcal{A}_\pm^{(\kappa)}, \mathcal{P}, z\right).$$

(a) Let ρ_κ be the multiplicative function with

$$\rho_\kappa(p^r) = \begin{cases} 1-(1-1/p)^\kappa & (r=1), \\ 0 & (r>1). \end{cases}$$

Prove that

$$\sum_{p\le z} \rho_\kappa(p)\log p = \kappa \log z + O(1),$$

and that there is a positive constant C_κ such that whenever $2 \le w < z$ we have

$$\prod_{w\le p<z} (1-\rho_\kappa(p))^{-1} < \left(\frac{\log z}{\log w}\right)^\kappa \left(1+\frac{C_\kappa}{\log w}\right),$$

i.e., (21.29) holds.

(b) Prove that

$$T_\pm^{(\kappa)}(X_\kappa, z) = S(\mathcal{A}_\pm, \mathcal{P}, z)^\kappa = T_\pm(Z,z)^\kappa.$$

(c) Suppose further that

$$\exp\left((\log X)^{2/3}\right) \le z \le \frac{X}{\log X},$$

and the $f_\pm^{(1)}(s)$ are the functions $f_\pm(s)$ satisfying (21.67)–(21.71). Prove that

$$T_\pm^\kappa(X_\kappa, z) = X_\kappa V_\kappa(z) f_\pm^{(1)}\left(\frac{\log X_\kappa}{\kappa\log z}\right)^\kappa + O\big(X_\kappa(\log z)^{-\kappa-1/3}\big),$$

where

$$V_\kappa(z) = \prod_{p<z}(1-\rho_\kappa(p)).$$

(d) Let

$$A_\pm^{(\kappa)}(m) = \sum_{\substack{n\\ m|n}} a_\pm^{(\kappa)}(n).$$

Prove that if $\kappa > 1$ and m is square-free, then

$$A_\pm^{(\kappa)}(m) = \sum_{u|m}\sum_{v|m/u} \mu(v) A_\pm^{(1)}(uv) A_\pm^{(\kappa-1)}(m/u),$$

and

$$\rho_\kappa(m) = \sum_{u|m}\sum_{v|m/u} \mu(v)\rho_1(uv)\rho_{\kappa-1}(m/u).$$

(e) Let

$$R_\kappa(m) = \sum_{\substack{n\le X\\ m|n}} a_\pm^{(\kappa)}(n) - X_\kappa\rho_\kappa(m).$$

Prove that if $\kappa > 1$ and m is square-free, then

$$|R_\kappa(m)| \le \sum_{uv|m} \big(|R_1(uv)||R_{\kappa-1}(m/u)| + |R_1(uv)|X_{\kappa-1}\rho_{\kappa-1}(m/u) + X_1\rho_1(uv)|R_{\kappa-1}(m/u)|\big).$$

(f) Let

$$y = X\exp\big(-(\log\log X)^3\big).$$

Prove that there is a positive constant c_κ such that

$$\sum_{m\le y} \mu(m)^2|R_\kappa(m)| \ll_\kappa X_\kappa \exp\big(-c_\kappa(\log\log X)^{3/2}\big),$$

so that $X_\kappa\rho_\kappa(m)$ does indeed correspond to

$$\sum_{\substack{n\le X\\ m|n}} a_\pm^{(\kappa)}(n).$$

The bound (21.75) is useful here.

(g) Conclude that functions $f_\pm = f_\pm^{(\kappa)}$ that satisfy inequalities of the kind (21.9) when the dimension is $\kappa \in \mathbb{N}$ must of necessity satisfy

$$f_-^{(\kappa)}(s) \le f_-^{(1)}(s/\kappa)^\kappa, \quad f_+^{(1)}(s/\kappa)^\kappa \le f_+^{(\kappa)}(s)$$

when $s > \kappa$. In particular,

$$f_-^{(\kappa)}(s) = 0 \quad (s \le 2\kappa),$$
$$s^\kappa f_+^{(\kappa)}(s) \ge 2^\kappa e^{C_0\kappa}\kappa^\kappa \quad (\kappa < s \le 3\kappa),$$

where in the first inequality we have also used the fact that if $w \ge z$, then $S(\mathcal{A}_-^{(\kappa)}, \mathcal{P}, w) \le S(\mathcal{A}_-^{(\kappa)}, \mathcal{P}, z)$.

Thus the upper bound given by the Selberg sieve or by Theorem 19.13, as in Exercises 19.2.1.1 and 19.2.1.10, and likewise the range for which lower bound sieves are non-trivial, cannot be much improved in general, even for large dimension κ.

21.5 Some Applications of Sieve Theory

21.5.1 Almost Primes

Lower bound sieves are by themselves usually unable to establish primality, yet for a number of problems we have theorems that tell us that at any rate there are a plentiful supply of numbers of a particular kind which have a bounded number of prime factors. Such numbers are frequently called *almost primes.* More precisely, the notation P_k is sometimes used to denote a typical number having at most k prime factors.

Particular examples of this are the twin prime and Goldbach binary problems. Thus the lower bound sieve can be adapted readily without further ado to show that there are infinitely many primes p such that $p-2$ has at most four prime factors, and there is a very simple wrinkle using the Selberg sieve that shows that the four can be replaced by three. There are more sophisticated combinations of weights and upper and lower bounds that can give more substantial lower bounds for the number of primes $p \le x$ for which $p-2$ has at most three prime factors. This in combination with a clever idea of Chen can reduce the three to two. All of these results have analogues for the Goldbach binary problem. We start by establishing the following simple lower bound.

Theorem 21.13 *Suppose that ε is a small positive number, $x > x_0(\varepsilon)$, $a(n) = 1$ when $n+2$ is a prime $p \le x$ and 0 otherwise, $\mathcal{P}$ is the set of odd primes, and $A \ge 0$ is a constant. Let*

$$y = x^{1/2}(\log x)^{-A-4}$$

and suppose that

$$2 \le z \le y^{1/(2+\varepsilon)}.$$

Then there is a positive number δ such that

$$S(\mathcal{A}, \mathcal{P}, z) \ge X\frac{ce^{-C_0}f_-(s)}{\log z} + O\left(\frac{X}{(\log z)^{1+\delta}}\right),$$

where $X = \mathrm{li}(x)$, $s = \frac{\log y}{\log z}$, and c is the twin prime constant

$$c = 2\prod_{p>2}\frac{p(p-1)}{(p-1)^2}. \tag{21.82}$$

Proof For a given odd square-free number m we have

$$A(m) = \pi(x, m, 2).$$

Let

$$R(m) = A(m) - \frac{\mathrm{li}(x)}{\varphi(m)}.$$

Then, by the Bombieri–Vinogradov theorem in the form (20.24) we have

$$R^* = \sum_{\substack{m\le y\\ m|P(z)}} |R(m)| \ll x(\log x)^{-A-2}.$$

Also, by Mertens' theorem in the form Theorem 2.7(e) we have

$$\begin{aligned} V(z) &= 2\prod_{2<p\le z}\frac{p(p-2)}{(p-1)^2}\prod_{p<z}(1-1/p)\\ &= \frac{ce^{-C_0}}{\log z} + O\big((\log z)^{-2}\big). \end{aligned}$$

The theorem now follows from Theorem 21.9. □

When we take z to satisfy $x^{1/5} < z$ we see that the n remaining after sieving will have at most four prime factors. In particular, when we take $z = y^{1/(2+\varepsilon)}$ it follows from (21.69) that the number $N_4(x)$ of primes $p \le x$ such that $p-2$ has at most four prime factors all $\ge z$ satisfies

$$N_4(x) \ge \frac{2cx\log(1+\varepsilon)}{(\log x)^2} + O\big(x(\log x)^{-2-\delta}\big). \tag{21.83}$$

There are various sophisticated ways of reducing the four to three and even, as we shall see, to two. However there is a very simple way of deducing the following corollary.

Corollary 21.14 *The number $N_3(x)$ of primes $p \le x$ such that $p-2$ has at most three prime factors satisfies*

$$N_3(x) \gg \frac{x}{(\log x)^2}.$$

It suffices to bound the number $N_4^*(x)$ of primes p counted by $N_4(x)$ above such that

$$p-2 = p_1p_2p_3p_4 \quad\text{and}\quad z\le p_1\le p_2\le p_3\le p_4\le \frac{x}{p_1p_2p_3}.$$

We can do this quite easily via the Selberg sieve. Suppose we are given p_2, p_3, p_4 with $z \le p_2 \le p_3 \le p_4 \le \frac{x}{zp_2p_3}$. Then the number of choices of $p \le x$ with $p_2p_3p_4|p-2$, and $\frac{p-2}{p_2p_3p_4} \in [z, p_2]$ and prime, is equal to the number of primes l such that $z \le l \le p_2$ and $lp_2p_3p_4 + 2 \le x$ is a prime. This is bounded by

$$S(\mathcal{B}, \mathscr{P}, w),$$

where $\mathcal{B} = \{b(n)\}$ and $b(n)$ is the number of integers l with $z \le l \le \min\left(p_2, (x-2)/(p_2p_3p_4)\right)$, $l(lp_2p_3p_4+2) = n$, $w = x^{1/9}$, and $\mathcal{P}$ is the set of all primes.

Let $p\rho(p)$ denote the number of solutions of

$$l(lp_2p_3p_4+2) \equiv 0 \pmod{p}.$$

Since $w < p_2 \le p_3 \le p_4$, when $2 < p < w$ we have $\rho(2) = 1/2$ and $\rho(p) = 2/p$ for $p > 2$. Let

$$X = \sum_n b(n)$$

and

$$R(m) = B(m) - X\rho(m).$$

Then $|R(m)| \le d(m)$. Now let

$$g(n) = \prod_{p|n} \frac{\rho(p)}{1-\rho(p)}.$$

Then $X \le x/(p_2p_3p_4)$, and by Theorem 21.1 we have

$$S(\mathcal{B}, \mathcal{P}, w) \le \frac{x/(p_2p_3p_4)}{\sum_{\substack{n\le w\\ n|P(w)}} g(n)} + O\left(xp_2^{-1}p_3^{-1}.p_4^{-1}(\log x)^{-10}\right).$$

As in the proof of Theorem 3.10,

$$\sum_{\substack{n\le w\\ n|P(w)}} g(n) \gg (\log x)^2.$$

Hence

$$S(\mathcal{B}, \mathcal{P}, w) \ll \frac{x}{p_2p_3p_4(\log x)^2}.$$

We now obtain an upper bound for $N_4^*(x)$ by summing over the primes p_2, p_3, and p_4, which, crudely, satisfy $z \le p_j \le xz^{-3}$. Hence each sum over p_j is bounded by

$$\sum_{z\le p_j\le xz^{-3}} p^{-1} \le \log\frac{\log\left(xz^{-3}\right)}{\log z} + O\left((\log z)^{-1}\right) \ll \log(1+4\varepsilon).$$

Therefore

$$N_4^*(x) \ll \frac{\varepsilon^3 x}{(\log x)^2}.$$

Choosing ε small enough and comparing with (21.83) gives the desired conclusion.

21.5.2 Chen's Theorem

Let $N_2(x)$ denote the number of primes $p \le x$ such that $p - 2$ has at most two prime factors.

Theorem 21.15 (Chen 1973) *For every large x,*

$$N_2(x) > \frac{cx}{3\log^2 x},$$

where c is the twin prime constant (21.82).

Proof Let

$$z = x^{1/10}, \quad w = z^{1/3},$$

and consider

$$N^*(x) = \sum_{\substack{p \le x \\ (p-2, P(z))=1}} \left(1 - \sum_{\substack{z \le p_1 \le w \\ p_1 | p-2}} \frac{1}{2} - \sum_{\substack{p_1 p_2 p_3 = p-2 \\ z \le p_1 \le w \le p_2 \le p_3}} \frac{1}{2} \right).$$

The number $M_0(x)$ of primes $p \le x$ such that $p - 2$ is divisible by the square of a prime p_1 with $z \le p_1 \le w$ satisfies

$$M_0(x) \ll \sum_{z \le p_1 \le w} \frac{x}{p_1^2} \ll xz^{-1}.$$

If $p - 2$ has at least two distinct prime factors p_1 with $z \le p_1 \le w$, then the general term is non-positive. Also, if $p - 2$ has exactly one prime factor with $z \le p_1 \le w$, then it can have at most three prime factors in total, and if it has exactly three in total then again the general term is non-positive and at least one of those prime factors cannot exceed w. Hence

$$N_2(x) \ge N^*(x) = M_1(x) - M_2(x) - M_3(x) + O\left(xz^{-1}\right), \tag{21.84}$$

where

$$M_1(x) = \sum_{\substack{p \le x \\ (p-2, P(z))=1}} 1,$$

$$M_2(x) = \sum_{\substack{p \le x \\ (p-2, P(z))=1}} \sum_{\substack{z \le p_1 \le w \\ p_1 | p-2}} \frac{1}{2},$$

and

$$M_3(x) = \sum_{\substack{p \le x \\ (p-2, P(z))=1}} \sum_{\substack{p_1 p_2 p_3 = p-2 \\ z \le p_1 \le w \le p_2 \le p_3}} \frac{1}{2}.$$

We can read off a lower bound for $M_1(x)$ at once from Theorem 21.13. Thus

$$M_1(x) \geq 10ce^{-C_0} f_-(5)\frac{x}{(\log x)^2} + O\left(\frac{x}{(\log x)^{2+\delta}}\right),$$

and hence, by (21.74),

$$M_1(x) \geq \frac{cx}{(\log x)^2}\left(8\log 2 + \int_3^4 \frac{4}{u}\int_0^{u-3} \frac{\log(1+t)}{2+t}\,dt\,du\right) + O\left(\frac{x}{(\log x)^{2+\delta}}\right). \tag{21.85}$$

A similar argument can be applied to $M_2(x)$. We have

$$M_2(x) = \sum_{z\leq p_1\leq w} \sum_{\substack{p\leq x \\ p\equiv 2 \pmod{p_1} \\ (p-2,P(z))=1}} \frac{1}{2}.$$

Now we have

$$\sum_{\substack{p\leq x \\ p\equiv 2 \pmod{p_1 m}}} 1 = \pi(x; p_1 m, -2).$$

Let

$$R(l) = \pi(x; l, -2) - \frac{\mathrm{li}(x)}{\varphi(l)}.$$

Then we may apply the Bombieri–Vinogradov theorem as before to obtain

$$\sum_{z\leq p_1\leq w} \sum_{m\leq x^{1/2}p_1^{-1}(\log x)^{-A}} |R(p_1 m)| \ll \frac{x}{(\log x)^{2+\delta}}$$

for suitable positive A and δ. Thus

$$M_2(x) \leq \frac{5ce^{-C_0}x}{(\log x)^2} \sum_{z\leq p_1\leq w} \frac{1}{p_1-1} f_+\left(\frac{5\log x - 10\log p_1}{\log x}\right) + O\left(\frac{x}{(\log x)^{2+\delta}}\right).$$

Mertens' Theorem 2.7(e), partial summation, and a change of variables gives

$$\begin{aligned} M_2(x) &\leq \frac{5ce^{-C_0}x}{(\log x)^2}\int_z^w f_+\left(\frac{5\log x - 10\log u}{\log x}\right)\frac{du}{u\log u} + O\left(\frac{x}{(\log x)^{2+\delta}}\right) \\ &\leq \frac{5ce^{-C_0}x}{(\log x)^2}\int_3^{10} f_+\left(5-\frac{10}{t}\right)\frac{dt}{t} + O\left(\frac{x}{(\log x)^{2+\delta}}\right). \end{aligned}$$

Hence by (21.67) and (21.73),

$$M_2(x) \le \frac{cx}{(\log x)^2}\left(6\log 2 + \int_3^4 \frac{10}{u(5-u)}\int_0^{u-3}\frac{\log(1+t)}{2+t}\,dt\,du\right) + O\left(\frac{x}{(\log x)^{2+\delta}}\right).$$

Therefore, by (21.85),

$$M_1(x) - M_2(x) \ge \frac{2cx}{(\log x)^2}\left(\log 2 - \int_3^4 \frac{2u-5}{u(5-u)}\int_0^{u-3}\frac{\log(1+t)}{2+t}\,dt\,du\right) + O\left(\frac{x}{(\log x)^{2+\delta}}\right).$$

The double integral here is

$$\int_0^1 \frac{\log(1+t)}{2+t}\,dt\int_{t+3}^4\left(\frac{1}{5-u} - \frac{1}{u}\right)du = \int_0^1 \frac{\log(1+t)}{2+t}\log\frac{(2-t)(t+3)}{4}\,dt \le 0.029772694.$$

Hence

$$M_1(x) - M_2(x) \ge 1.3267489\frac{cx}{(\log x)^2} + O\left(\frac{x}{(\log x)^{2+\delta}}\right). \tag{21.86}$$

We now turn our attention to M_3. We have

$$M_3(x) \le \frac{1}{2}S(\mathcal{A},\mathcal{P},y) + O(x/z), \tag{21.87}$$

where

$$a(n) = \begin{cases} 1 & n = p_1p_2p_3 + 2 \le x + 2 \text{ with } z < p_1 \le w \le p_2 \le p_3, \\ 0 & \text{otherwise,} \end{cases}$$

$\mathcal{P}$ is the set of odd primes, and

$$y = x^{1/2}(\log x)^{-B}.$$

Here we have used the estimates

$$\sum_{w \le p_2 \le p_3 \le \frac{x}{zp_2}} 1 \ll xz^{-1}$$

(in case $p_1 = z$) and

$$\sum_{\substack{p_j \ge z \\ x-2 < p_1p_2p_3 \le x}} 1 \ll 1$$

(because x is tidier than $x - 2$).

Now we sieve the n for primeness. In order to apply the upper bound sieve effectively we need to deal with

$$\sum_n a(mn)$$

when $m|P(y)$, and so we need a variant of the Bombieri–Vinogradov theorem for triples of primes. In this situation the result is rather more straightforward, not requiring any identity similar to that used in the proof of the Bombieri–Vinogradov theorem since now the underlying bilinear form is already a good Type II form.

As is usual we need to input information about

$$\sum_n a(mn) = \sum_{\substack{z<p_1\le w\le p_2\le p_3\le x/(p_1p_2)\\ p_1p_2p_3\equiv -2 \pmod m}} 1$$

when $m|P(y)$ and $m \le y$. Since we automatically have $(m, p_1p_2p_3) = 1$ it follows that

$$\begin{aligned}\sum_n a(mn) &= \sum_{\chi \pmod m} \frac{\overline{\chi}(-2)}{\varphi(m)} \sum_{z<p_1\le w\le p_2\le p_3\le \frac{x}{p_1p_2}} \chi(p_1p_2p_3)\\ &= \sum_{q|m} \sum_{\chi \pmod q}^{\star} \frac{\overline{\chi}(-2)}{\varphi(m)} \sum_{\substack{z<p_1\le w\le p_2\le p_3\le \frac{x}{p_1p_2}\\ (p_1p_2p_3,m)=1}} \chi(p_1p_2p_3).\end{aligned}$$

Let

$$R(m) = \sum_n a(mn) - \frac{X}{\varphi(m)}, \tag{21.88}$$

where

$$X = \sum_{z<p_1\le w\le p_2\le p_3\le \frac{x}{p_1p_2}} 1 = \sum_n a(n). \tag{21.89}$$

Then

$$R(m) = R^*(m) - \frac{E(m)}{\varphi(m)}, \tag{21.90}$$

where

$$R^*(m) = \sum_{1<q|m} \sum_{\chi \bmod q}^{\star} \frac{\overline{\chi}(-2)}{\varphi(m)} \sum_{\substack{z<p_1\le w\le p_2\le p_3\le \frac{x}{p_1p_2}\\ (p_1p_2p_3,m)=1}} \chi(p_1p_2p_3)$$

and

$$E(m) = \sum_{\substack{z<p_1\le w\le p_2\le p_3\le \frac{x}{p_1p_2} \\ (p_1p_2p_3,m)>1}} 1.$$

When $m \le y$, the contribution from the p_1 dividing m is

$$\ll \sum_{w<p_2\le(x/z)^{1/2}} \frac{x}{p_2 z\log x} \ll \frac{x}{z\log x}$$

and from the p_2 or p_3 dividing m is

$$\ll \sum_{z<p_1\le w} \frac{x}{p_1 w\log x} \ll \frac{x}{w\log x}.$$

Thus

$$\sum_{m\le y} \frac{|E(m)|}{\varphi(m)} \ll xz^{-1}. \tag{21.91}$$

We also have

$$|R^*(m)| \le \sum_{1<q|m} \sum_{\chi \bmod q}^{\star} \frac{1}{\varphi(m)} \left| \sum_{\substack{z<p_1\le w\le p_2\le p_3\le \frac{x}{p_1p_2} \\ (p_1p_2p_3,m)=1}} \chi(p_1p_2p_3) \right|,$$

and so

$$\begin{aligned}\sum_{m\le y} |R^*(m)| &\le \sum_{1<q\le y} \sum_{l\le \frac{y}{q}} \frac{1}{\varphi(ql)} \sum_{\chi \bmod q}^{*} \left| \sum_{\substack{z<p_1\le w\le p_2\le p_3\le \frac{x}{p_1p_2} \\ (p_1p_2p_3,l)=1}} \chi(p_1p_2p_3) \right| \\ &\le \sum_{l\le y} \frac{1}{\varphi(l)} \sum_{1<q\le y/l} \frac{1}{\varphi(q)} \sum_{\chi \bmod q}^{*} \left| \sum_{\substack{z<p_1\le w\le p_2\le p_3\le \frac{x}{p_1p_2} \\ (p_1p_2p_3,l)=1}} \chi(p_1p_2p_3) \right|.\end{aligned}$$

Hence

$$\sum_{m\le y} |R^*(m)| \ll \max_{l\le y} \sum_{1<q\le y} \frac{\log y}{\varphi(q)} \sum_{\chi \bmod q}^{*} \left| \sum_{\substack{z<p_1\le w\le p_2\le p_3\le \frac{x}{p_1p_2} \\ (p_1p_2p_3,l)=1}} \chi(p_1p_2p_3) \right|.$$

The summation conditions imply that $p_2 \le (x/p_1)^{1/2}$ and so $p_1p_2 \le p_1^{1/2}x^{1/2} \le x^{2/3}$. Consequently, when $q \le (\log x)^A$ for a given constant A the Siegel–Walfisz theorem in the form of Corollary 11.18 applied to the sum over p_3 gives

$$\max_{l\le y} \sum_{1<q\le(\log x)^A} \frac{\log y}{\varphi(q)} \sideset{}{^*}\sum_{\chi \bmod q} \left| \sum_{\substack{z<p_1\le w\le p_2\le p_3\le \frac{x}{p_1p_2} \\ (p_1p_2p_3,l)=1}} \chi(p_1p_2p_3) \right| \ll x(\log x)^{-10}.$$

Let $a(u)$ be 1 when u is a prime p_1 with $z < p_1 \le w$ and $p_1 \nmid l$, and 0 otherwise, and let $b(v) = 1$ when $v = p_2p_3$ with $w \le p_2 \le p_3$ and $(p_2p_3, l) = 1$, and 0 otherwise. Then

$$\sum_{\substack{z<p_1\le w\le p_2\le p_3\le \frac{x}{p_1p_2} \\ (p_1p_2p_3,l)=1}} \chi(p_1p_2p_3) = \sum_{z<u\le w} \sum_{w<v\le x/u} a(u)b(v)\chi(uv).$$

By (19.35) and a division of the sum over u into dyadic intervals we have

$$\sum_{1<q\le Q} \frac{q}{\varphi(q)} \sideset{}{^*}\sum_{\chi \bmod q} \left| \sum_{z<u\le w} \sum_{w<v\le x/u} a(u)b(v)\chi(uv) \right| \ll (\log x)^2 \left(x + Qx^{19/20} + Q^2x^{1/2}\right)$$

and so, by partial summation, for a suitable choice of A we have

$$\max_{l\le y} \sum_{(\log x)^A<q\le y} \frac{\log y}{\varphi(q)} \sideset{}{^*}\sum_{\chi \bmod q} \left| \sum_{z<u\le w} \sum_{w<v\le x/u} a(u)b(v)\chi(uv) \right| \ll x(\log x)^{-10}.$$

Therefore

$$\sum_{m\le y} |R^*(m)| \ll x(\log x)^{-10}$$

and so, by (21.90) and (21.91),

$$\sum_{m\le y} |R(m)| \ll x(\log x)^{-10}.$$

Thus, by (21.87) and (21.88), and a by now familiar application of Theorem 21.9, we have

$$M_3(x) \le \frac{2Xc}{\log x} + O\big(X(\log x)^{-1-\delta} + x(\log x)^{-10}\big).$$

By (21.89) and multiple applications of Theorem 6.9 we have

$$\begin{aligned}2X &= \int_{x^{1/10}}^{x^{1/3}} \int_{x^{1/3}}^{(x/u)^{1/2}} \frac{2x\,dv}{(\log v)v\log(x/uv)} \frac{du}{u\log u} + O\big(x(\log x)^{-3/2}\big)\\ &= \frac{2x}{\log x}\int_{1/10}^{1/3} \frac{\log(2-3w)}{w(1-w)}dw + O\big(x(\log x)^{-3/2}\big)\\ &\le 0.98199041\frac{x}{\log x} + O\big(x(\log x)^{-3/2}\big).\end{aligned}$$

Therefore, by (21.84) and (21.86) for all large x,

$$N(x) \ge \frac{cx}{3(\log x)^2}.$$

□

21.5.3 Conjecture J of Hardy & Littlewood (1922)

Let

$$R(n) = \mathrm{card}\{p, x, y : p + x^2 + y^2 = n, x \in \mathbb{Z}, y \in \mathbb{Z}, p \text{ prime}\} \tag{21.92}$$

with p a prime and $x, y \in \mathbb{Z}$. Then Conjecture J states that

$$R(n) \sim \frac{\pi n}{\log n}\mathfrak{S}(n), \tag{21.93}$$

where, for $z \in \mathbb{Z} \setminus \{0\}$,

$$\mathfrak{S}(z) = \left(\prod_{p>2}\left(1 + \frac{(-1)^{(p-1)/2}}{(p-1)p}\right)\right)\prod_{\substack{p\mid z\\ p>2}}\left(1 - \frac{p(-1)^{(p-1)/2}}{p^2 - p + (-1)^{(p-1)/2}}\right). \tag{21.94}$$

As an aside, we observe that it is readily checked that

$$\mathfrak{S}(z) = \sum_{q=1}^{\infty} \sum_{\substack{a=1\\ (a,q)=1}}^{q} \frac{S(q,a)^2\mu(q)}{q^2\varphi(q)} e(-az/q),$$

where

$$S(q,a) = \sum_{x=1}^{q} e(ax^2/q).$$

Moreover, πn is the volume of the region $\{\alpha, \beta \in \mathbb{R} : \alpha^2 + \beta^2 \le n\}$. Thus one can see how Hardy and Littlewood read off the conjecture from their assumption that the major arcs for this question would dominate.

Following work of Hooley (1957), Linnik (1960, 1961, 1963) gave a proof of Conjecture J which was later simplified via the Bombieri–Vinogradov theorem (Elliott & Halberstam 1966). For $n \in \mathbb{N}$, let

$$r(n) = \mathrm{card}\{x, y : x^2 + y^2 = n, x \in \mathbb{Z}, y \in \mathbb{Z}\}, \tag{21.95}$$

and let

$$\chi(m) = \begin{cases} (-1)^{\frac{m-1}{2}} & 2 \nmid m, \\ 0 & 2|m. \end{cases} \tag{21.96}$$

Then, in view of the formula

$$r(n) = 4 \sum_{m|n} \chi(m) \quad (n \in \mathbb{N}) \tag{21.97}$$

(see, for example, (16.9.1) and Theorem 278 of Hardy & Wright 2008), one might hope to imitate the method used to prove Theorem 20.7. However, now the main term is a factor of $\log n$ smaller. Whilst this is not a problem for the divisors m not near $\sqrt{n}$, those near $\sqrt{n}$ require a more delicate treatment than a crude application of the Brun–Titchmarsh theorem. In particular, some use needs to be made of possible cancellation arising from changes of sign of $\chi(m)$. Hooley's idea to overcome this is, when this range occurs, to replace the primes by a larger set which, whilst including the primes, now satisfies the asymptotics of (21.10) and so the resulting main terms are cancelling. The delicacy of the situation is such that this idea also needs to be combined with a careful accounting of the primes p for which $n - p$ has slightly more than its normally expected number of prime factors.

Theorem 21.16 (Hooley–Linnik) *Let $R(n)$ be as in* (21.92) *and $\mathfrak{S}(n)$ be as in* (21.94). *Then for any number ν with $1 < \nu < \frac{3}{2} - \frac{e \log 2}{4} = 1.028957\ldots$ we have*

$$R(n) = \frac{\pi n}{\log n}\mathfrak{S}(n) + O\left(\frac{n}{(\log n)^{\nu}}\right)$$

as $n \to \infty$.

We initiate the proof by eliminating the easy parts. Let B be a constant at our disposal to be fixed later and define

$$M = n^{1/2}(\log n)^{-B}, \quad N = n^{1/2}(\log n)^{B}. \tag{21.98}$$

By (21.97),

$$R(n) = 4X(0, M) + 4X(M, N) + 4X(N, n) + O(1), \tag{21.99}$$

where

$$X(U, V) = \sum_{p<n} \sum_{\substack{U<m\le V \\ m|(n-p)}} \chi(m). \tag{21.100}$$

Note that if n should be prime then there is a contribution to the error term of $r(0) = 1$.

If $p < n$, $m|(n-p)$ and $(m, n) > 1$, then $p|n$ and there can be at most $\ll \log n$ such primes p. Moreover, then m/p divides $n/p - 1$. Thus

$$X(0, M) = \sum_{\substack{m \le M \\ (m,n)=1}} \chi(m) \sum_{\substack{p<n \\ p \equiv n \pmod m}} 1 + O(n^{\varepsilon}).$$

Hence, by (21.98) and the form of the Bombieri–Vinogradov theorem stated in Corollary 20.5 we have

$$X(0, M) = \mathrm{li}(n) \sum_{\substack{m \le M \\ (m,n)=1}} \frac{\chi(m)}{\varphi(m)} + O\big(n(\log n)^{3-B}\big).$$

By a simple elementary argument we show that the main term here gives the main term of our theorem.

Lemma 21.17 *Let* χ *be the quadratic character modulo* 4, *as given in* (21.96), *and let* (n) *be the singular series given in* (21.94). *Then*

$$\sum_{\substack{m \le M \\ (m,n)=1}} \frac{\chi(m)}{\varphi(m)} = \frac{\pi}{4}\mathfrak{S}(n) + O\big(d(n)M^{-1}\log M\big).$$

Proof We note that

$$\frac{1}{\varphi(m)} = \frac{1}{m}\sum_{l|m} \frac{\mu(l)^2}{\varphi(l)}.$$

Thus the subject of interest is

$$\sum_{\substack{l \le M \\ (l,n)=1}} \frac{\chi(l)\mu(l)^2}{l\varphi(l)} \sum_{k \le M/l} \frac{\chi(k)}{k} \sum_{j|(k,n)} \mu(j)$$
$$= \sum_{\substack{l \le M \\ (l,n)=1}} \frac{\chi(l)\mu(l)^2}{l\varphi(l)} \sum_{\substack{j|n \\ j \le M/l}} \frac{\mu(j)\chi(j)}{j} \sum_{m \le M/lj} \frac{\chi(m)}{m}.$$

By an explicit version of the alternating series test we have

$$\sum_{m \le M/lj} \frac{\chi(m)}{m} = L(1,\chi) + O(lj/M) = \frac{\pi}{4} + O(lj/M).$$

Thus the multiple sum becomes

$$\frac{\pi}{4} \sum_{\substack{j|n \\ j \le M}} \frac{\mu(j)\chi(j)}{j} \sum_{\substack{l \le M/j \\ (l,n)=1}} \frac{\chi(l)\mu(l)^2}{l\varphi(l)} + O\big(d(n)M^{-1}\log M\big).$$

We then complete in turn each sum, which gives

$$\frac{\pi}{4}\sum_{j|n}\frac{\mu(j)\chi(j)}{j}\sum_{\substack{l=1\\(l,n)=1}}^{\infty}\frac{\chi(l)\mu(l)^2}{l\varphi(l)}+O\big(d(n)M^{-1}\log M\big).$$

The main term here is

$$\frac{\pi}{4}\left(\prod_{p|n}\left(1-\frac{\chi(p)}{p}\right)\right)\prod_{p\nmid n}\left(1+\frac{\chi(p)}{p(p-1)}\right),$$

and the Euler products here match (21.94). □

We now turn to $X(N,n)$. In the inner sum in (21.100) we replace m by $\frac{n-p}{l}$ so that

$$X(N,n)=X(N,n,1)-X(N,n,-1),$$

where

$$X(N,n,\pm 1)=\sum_{p<n}\sum_{\substack{l|(n-p)\\ l<(n-p)/N\\ (n-p)/l\equiv\pm 1\ (\mathrm{mod}\ 4)}}1=\sum_{l<n/N}\sum_{\substack{p<n-lN\\ p\equiv n\mp l\ (\mathrm{mod}\ 4l)}}1.$$

Should $d=(n\mp l,4l)$ be >1, then $p=2$ or $p|l|n$ and so

$$X(N,n,\pm 1)=\sum_{\substack{l<n/N\\ (n\mp l,4l)=1}}\sum_{\substack{p<n-lN\\ p\equiv n\mp l\ (\mathrm{mod}\ 4l)}}1+O(n^{\varepsilon}).$$

Thus on applying Corollary 20.5 we find that

$$X(N,n,\pm 1)=\sum_{\substack{l<n/N\\ (n\mp l,4l)=1}}\frac{\mathrm{li}(n-lN)}{\varphi(4l)}+O\big(n(\log n)^{3-B}\big).$$

Now $(n\mp l,4l)=1$ if and only if $(n\mp l,2l)=1$, and this holds in turn if and only if $(n\pm l,2l)=1$. Thus the main terms are independent of the sign of $\pm l$, and so cancel, giving the bound

$$X(N,n)\ll n(\log n)^{3-B}.$$

We now come to the more delicate part of the argument: the treatment of $X(M,N)$. As with $X(N,n)$ we expect this to contribute something which is smaller than the main term, but unfortunately our knowledge of the distribution of primes in this case lacks the desired precision. There are two features of the situation which come to our aid. The divisors m of $n-p$ are now in a very restricted range, $M<m\le N$, and their occurrence is likely to be relatively infrequent. The heuristic here is that the normal number of divisors of $n-p$

is about $(\log n)^{\log 2}$. Thus of the approximately $\log n$ intervals $[e^{k-1}, e^k)$ with $1 \le k \le \log n$ only a proportion $(\log n)^{\log 2-1}$ can be expected to contain a divisor of $n-p$, and so the probability that $(M, N]$ contains such a divisor is about $(\log n)^{\log 2-1} \log\log n$. Thus the sum over the primes p can be expected to be bounded by something like $n(\log n)^{\log 2-2} \log\log n$. The second feature is that we can embed the primes in a somewhat larger set for which we can establish a suitable distribution into residue classes for the appropriate modulus m. Moreover, we can separate and enable these two features by the usual process in analytic number theory when we cannot think of anything better to do, namely apply the Cauchy–Schwarz inequality. Let

$$\Delta(k) = \sum_{\substack{m|k \\ M<m\le N}} 1$$

and

$$\Xi(k) = \sum_{\substack{m|k \\ M<m\le N}} \chi(m).$$

Then

$$X(M,N) = \sum_{\substack{p<n \\ \Delta(p-n)>0}} \Xi(n-p) \le Y(M,N)^{1/2} Z(M,N)^{1/2}, \tag{21.101}$$

where

$$Y(M,N) = \sum_{\substack{p<n \\ \Delta(n-p)>0}} 1, \qquad Z(M,N) = \sum_{p<n} \Xi(n-p)^2. \tag{21.102}$$

To manage $Y(M,N)$ we divide the sum into two parts dependent on the number of prime factors of $n-p$. To mark the division we introduce a real number α which satisfies

$$1 < \alpha \le \frac{3}{2} \tag{21.103}$$

and is otherwise at our disposal. Thus

$$Y(M,N) = Y_1(M,N) + Y_2(M,N),$$

where

$$Y_1(M,N) = \sum_{\substack{p<n \\ \Omega(p-n)\le\alpha\log\log n}} \Delta(n-p), \quad Y_2(M,N) = \sum_{\substack{p<n \\ \Omega(n-p)>\alpha\log\log n}} 1.$$

The sum $Y_1(M,N)$ is bounded by the number of pairs p and m with $p < n$, $M < m \le N$, $m|(p-n)$, and $\Omega(n-p) \le \alpha \log\log n$, or equivalently triples p, l, m with

$$p < n,\ lm = n-p,\ M < m \le N,\ \Omega(n-p) \le \alpha \log\log n.$$

We have $l < n/M = N$ and the number of such triples with $m \le N$ and $l \le M(\log n)^{-2}$ is $\ll n(\log n)^{-2}$. Thus, when $M(\log n)^{-2} < l$ we certainly have

$$M(\log n)^{-2} < l \le N \quad \text{and} \quad M(\log n)^{-2} < m \le N.$$

We also have

$$\min\big(\Omega(m), \Omega(l)\big) \le \frac{1}{2}\alpha \log\log n.$$

Hence

$$Y_1(M,N) \ll Y_3(M,N) + \frac{n}{(\log n)^2}, \tag{21.104}$$

where

$$Y_3(M,N) = \\ \operatorname{card}\left\{m, p : \frac{M}{\log^2 n} < m \le N, p < n, m|(n-p), \Omega(m) \le \frac{\alpha}{2}\log\log n\right\}.$$

By the Brun–Titchmarsh theorem, Theorem 3.9, we have

$$Y_3(M,N) \ll \sum_{\substack{M(\log n)^{-2} < m \le N \\ \Omega(m) \le \frac{\alpha}{2}\log\log n}} \frac{n}{\varphi(m)\log n}.$$

Here and later we need some estimates concerning Ω.

Lemma 21.18 *For a real number $\theta > 0$ define $\gamma(\theta) = \theta - \theta\log\theta$.*
(i) *Suppose that X is a real number with $X \ge 1$ and ϖ is a real number with $1/2 \le \varpi \le 7/4$. Then*

$$\sum_{m \le X} \varpi^{\Omega(m)} \ll X(\log 2X)^{\varpi - 1}.$$

(ii) *Suppose that $1/2 \le \beta < 1$ and $n \ge e^e$. Then*

$$\sum_{\substack{n^{1/2}(\log n)^{-B-2} < m \le n^{1/2}(\log n)^B \\ \Omega(m) \le \beta \log\log n}} \frac{1}{m} \ll (\log n)^{\gamma(\beta)-1}\log\log n.$$

(iii) *Suppose that* $1 < \beta \le 3/2$ *and* $n \ge e^e$. *Then*

$$\sum_{\substack{m\le n\\ \Omega(m)\ge\beta\log\log n-1}} \frac{1}{m} \ll (\log n)^{\gamma(\beta)}.$$

Proof (i) This can be proved in the same way as Theorem 7.17. Here the Dirichlet series generating function is $\zeta(s)^{\varpi}\eta(s)$ where $\eta(s)$ is a series that converges absolutely in a halfplane $\operatorname{Re} s > 1-\delta$ for some $\delta > 0$.

(ii) By (i) and partial summation when $1/2 \le \varpi \le 1$,

$$\sum_{n^{1/2}(\log n)^{-B-2}<m\le n^{1/2}(\log n)^B} \frac{\varpi^{\Omega(m)}}{m} \ll (\log n)^{\varpi-1}\log\log n.$$

For those terms with $\Omega(m) \le \beta\log\log n$, since $0<\beta<1$, we have

$$1 \le (\log n)^{-\beta\log\beta}\beta^{\Omega(m)}$$

and the result follows on taking $\varpi = \beta$.

(iii) This follows from the bound

$$\sum_{m\le n}\frac{\varpi^{\Omega(m)}}{m} \ll (\log n)^{\varpi}$$

by a concomitant argument. □

By (21.103) we have $1<\alpha<2$. Hence, by Lemma 21.18 (ii) with $\beta = \alpha/2$ we have

$$Y_3(M,N) \ll \frac{n(\log\log n)^2}{(\log n)^{2-\gamma(\alpha/2)}},$$

whence, as $\gamma(\alpha/2) > 0$, by (21.104),

$$Y_1(M,N) \ll \frac{n(\log\log n)^2}{(\log n)^{2-\gamma(\alpha/2)}}. \tag{21.105}$$

We now turn to $Y_2(M,N)$. We divide this sum into two further parts so that

$$Y_2(M,N) \le Y_4(M,N) + Y_5(M,N),$$

where

$$Y_4(M,N) = \operatorname{card}\{m<n : \Omega(m) > 10\log\log n\},$$

$$Y_5(M,N) = \operatorname{card}\{p<n : \alpha\log\log n < \Omega(n-p) \le 10\log\log n\}.$$

By Lemma 21.18(i),

$$Y_4(M,N) < \sum_{m<n} e^{\Omega(m)/2}(\log n)^{-5} \ll n(\log n)^{\sqrt{e}-6} \ll \frac{n}{(\log n)^2}.$$

When $p < n$, $\Omega(n-p) \le 10 \log\log n$ and $n-p$ has no prime factors exceeding $\exp\big((\log n)/(20\log\log n)\big)$ we have $n - p \le \sqrt{n}$. Otherwise, for p counted by Y_5 there exist numbers r and p' so that $n - p = rp'$, $p' > \exp\big((\log n)/(20\log\log n)\big)$ and $\Omega(s) > \alpha\log\log n - 1$. Moreover, $r < n \exp\big(-(\log n)/(20\log\log n)\big)$. Thus

$$Y_5(M,N) \le \sum_{\substack{r<n\exp((\log n)/20\log\log n))\\ \Omega(r)>\alpha\log\log n-1}} \sum_{\substack{p,p'\\ p+rp'=n}} 1 + n^{1/2}.$$

Let $N(n;r)$ denote the inner sum here. Then the bound

$$N(n;r) \ll \frac{n^2}{\varphi(nr)\big(\log(n/r)\big)^2} \ll \frac{n(\log\log n)^3}{r(\log n)^2}$$

is trivial when $(n,r) > 1$ and when $(n,r) = 1$ follows by the methods of §3.4, or from Theorem 19.13 or Corollary 21.10 by sifting out the $y < n/r$ with $p|y(n-yy)$ for $p \le (n/r)^\theta$ for some $\theta \le 1/2$. Thus

$$Y_5(M,N) \ll n^{1/2} + \frac{n(\log\log n)^3}{(\log n)^2} \sum_{\substack{r\le n\\ \Omega(r)>\alpha\log\log n-1}} \frac{1}{s}.$$

By Lemma 21.18(iii) this is

$$\ll \frac{n(\log\log n)^3}{(\log n)^{2-\gamma(\alpha)}}.$$

Since $\gamma(\alpha) > 0$ this gives

$$Y_2(M,N) \ll \frac{n(\log\log n)^3}{(\log n)^{2-\gamma(\alpha)}}.$$

Comparing this with (21.105) we see that the optimal choice of α occurs when $\gamma(\alpha/2) = \gamma(\alpha)$, i.e., $\alpha = e/2$. Thus

$$Y(M,N) \ll \frac{n(\log\log n)^3}{(\log n)^{2-e(\log 2)/2}} \tag{21.106}$$

and

$$2 - e(\log 2)/2 = 1.057915\ldots.$$

The final stage of the proof is the examination of $Z(M,N)$, given by (21.102). Let

$$z = \exp\big((\log n)/(\log\log n)^2\big). \tag{21.107}$$

Multiplying out $Z(M,N)$ would yield sums over p, m_1, m_2 with $m_j|(n-p)$.

We want to replace the primes by an essentially larger set of $r < n$ such that $(r, P) = 1$ for a suitable choice of P. Let

$$\mathscr{P} = \{p < z\}, \quad P(z) = \prod_{p \in \mathscr{P}} p,$$

and

$$\mathscr{N} = \{r < n : (r, P(z)) = 1\}.$$

Then the primes $p \le z$ contribute $\ll n^{\varepsilon/2}$ to $Z(M, N)$ and $\Xi(n-p) \ll n^{\varepsilon/4}$. Thus

$$Z(M, N) \le Z_1(M, N) + O(n^{\varepsilon}), \tag{21.108}$$

where

$$Z_1(M, N) = \sum_{r \in \mathscr{N}} \Xi(n-r)^2 = \sum_{\substack{m_1, m_2 \\ M < m_j \le N}} \chi(m_1 m_2) \sum_{\substack{r \in \mathscr{N} \\ m_1 | (n-r),\, m_2 | (n-r)}} 1.$$

If $m_1 m_2$ and

$$n_1 = \prod_{p | n, p < z} p \tag{21.109}$$

were to have a prime factor in common, then it would divide r which we have excluded, so such an inner sum would be empty. Thus

$$Z_1(M, N) = \sum_{\substack{(m_1 m_2, n_1) = 1 \\ M < m_j \le N}} \chi(m_1 m_2) \sum_{\substack{r \in \mathscr{N} \\ m_1 | (n-r),\, m_2 | (n-r)}} 1.$$

We have $m_1 | (n-r)$ and $m_2 | (n-r)$ if and only if $[m_1, m_2] | (n-r)$. Put $d = (m_1, m_2)$ and $k_j = m_j / d$. Then

$$(k_1, k_2) = 1, M < dk_j \le N, dk_1 k_2 | (n-r).$$

Now we split the sum according as to whether $d \le D$ or $d > D$ where

$$D = n^{1/8}.$$

Thus

$$Z_1(M, N) = Z_2(M, N) + Z_3(M, N), \tag{21.110}$$

where

$$Z_2(M, N) = \sum_{\substack{(k_1, k_2) = 1, (dk_1 k_2, n_1) = 1 \\ M < dk_j \le N,\, d > D}} \chi(d^2 k_1 k_2) \sum_{\substack{r \in \mathscr{N} \\ dk_1 k_2 | (n-r)}} 1$$

and

$$Z_3(M,N) = \sum_{\substack{(k_1,k_2)=1,(dk_1k_2,n_1)=1\\ M<dk_j\le N,\, d\le D}} \chi(d^2k_1k_2) \sum_{\substack{r\in\mathcal{N}\\ dk_1k_2|(n-r)}} 1. \tag{21.111}$$

In the sum $Z_2(M,N)$ we have

$$dk_1k_2 \le N^2d^{-1} \le n^{7/8}(\log n)^{2B}.$$

Given such d, k_1, k_2, the r in the inner sum arise by removing the a with $a \equiv n \pmod{dk_1k_2}$ with $a < n$ which have a prime factor that divides $P(z)$. Thus we can apply Corollary 21.10. For $q = dk_1k_2$ with $(q,n_1) = 1$ let

$$\mathscr{A} = \{a < n : a \equiv n \pmod q\}.$$

Then

$$\sum_{\substack{r\in\mathcal{N}\\ q|(n-r)}} 1 = S(\mathscr{A}, \mathscr{P}, z).$$

When $m|P(z)$, in the notation of (21.3) we have

$$A(m) = \sum_{\substack{a<n\\ a\equiv n \pmod q\\ a\equiv 0 \pmod m}} 1 = \sum_{\substack{x=1\\ x\equiv n \pmod q\\ x\equiv 0 \pmod m}}^{[q,m]} \left(\frac{n}{[q,m]} + O(1)\right).$$

Moreover, since $m|P(z)$ and $(q,n_1) = 1$ it follows that

$$\sum_{\substack{y=1\\ ym\equiv n \pmod q}}^{q/(q,m)} 1 = \begin{cases} 1 & (m,q) = 1,\\ 0 & (m,q) > 1.\end{cases}$$

Thus

$$A(m) = X\rho(m) + O(1),$$

where

$$X = \frac{n}{q}, \quad \rho(m) = \begin{cases} \frac{1}{m} & (m,q) = 1,\\ 0 & (m,q) > 1.\end{cases}$$

Hence with z as in (21.107), $y = n^{1/16}$, and $s = \frac{\log y}{\log z}$ we have

$$V(z) = \prod_{\substack{p<z\\ p\nmid q}} (1 - 1/p)$$

and

$$\sum_{m\le y}|R(m)| \ll XV(z)e^{-s}.$$

Moreover,

$$\prod_{\substack{p\ge z\\ p|q}}(1-1/p)^{-1} \le \exp\left((\log n)z^{-1}\right) = 1 + O\left((\log n)^{-10}\right)$$

and

$$e^{-s} \ll (\log n)^{-10}.$$

Hence

$$Z_2(M,N) = n\left(\prod_{p<z}\left(1-\frac{1}{p}\right)\right)\sum_{\substack{(k_1,k_2)=1\\ (dk_1k_2,n_1)=1\\ M<dk_j\le N,\, d>D}}\frac{\chi(d^2k_1k_2)}{\varphi(dk_1k_2)} + O\left(\frac{n}{(\log n)^5}\right).$$

We first consider the contribution from the d with $d > M$. In that case we have $k_j \le N/M$. We also know that

$$\varphi(dk_1k_2) \ll (dk_1k_2)^{-1}\log\log n$$

and

$$\prod_{p<z}\left(1-\frac{1}{p}\right) \ll (\log n)^{-1}(\log\log n)^2.$$

Thus the total contribution from the terms with $d > M$ is

$$\ll n\frac{(\log\log n)^3}{\log n}\sum_{M<d\le N}\frac{1}{d}\left(\sum_{k\le N/M}\frac{1}{k}\right)^2 \ll n\frac{(\log\log n)^6}{\log n}.$$

Now consider the terms with $d \le M$. In the multiple sum we replace $\varphi(dk_1k_2)^{-1}$ by

$$\frac{1}{\varphi(dk_1)k_2}\sum_{\substack{m|k_2\\ (m,dk_1)=1}}\frac{\mu(m)^2}{\varphi(m)}.$$

Then we replace k_2 by lm so that the contribution from k_2 and m becomes

$$\sum_{\substack{(k_1,lm)=1,(dk_1lm,n_1)=1\\ M<dlm\le N,D<d\le M\\ M<dk_1\le N}}\frac{\chi(d^2k_1lm)\mu(m)^2}{\varphi(dk_1)lm\varphi(m)}.$$

Next we replace the condition $(l, k_1 n_1) =$ by

$$\sum_{u|(k_1 n_1, l)} \mu(u)$$

and then write $l = uv$. Thus the multiple sum becomes

$$\sum_{\substack{m\le N\\(m,n_1)=1}} \frac{\mu(m)^2\chi(m)}{m\varphi(m)} \sum_{\substack{D<d\le M\\(d,n_1)=1}} \sum_{\substack{M/d<k_1\le N/d\\(k_1,n_1)=1}} \frac{\chi(d^2k_1)}{\varphi(dk_1)} \sum_{u|k_1n_1} \frac{\mu(u)\chi(u)}{u} \sum_{M/(dmu)<v\le N/(dmu)} \frac{\chi(v)}{v}.$$

Since $(k_1, n_1) = 1$, the sum over u can be rewritten as a sum over $u = hj$ with $h|k_1$ and $j|n_1$ and then k_1 can be replaced by hr. Thus we obtain

$$\sum_{\substack{m\le N\\(m,n_1)=1}} \frac{\mu(m)^2\chi(m)}{m\varphi(m)} \sum_{j|n_1} \frac{\mu(j)\chi(j)}{j} \sum_{\substack{D<d\le M\\(d,n_1)=1}} \sum_{(h,n_1)=1} \frac{\mu(h)\chi(h)}{h} \sum_{\substack{M/(dh)<r\le N/(dh)\\(r,n_1)=1}} \frac{\chi(d^2hr)}{\varphi(dhr)} \sum_{M/(dmhj)<v\le N/(dmhj)} \frac{\chi(v)}{v}.$$

The contribution from the terms with $dmhj > M$ is

$$\ll \sum_{m\le N} \frac{1}{m\varphi(m)} \sum_{j|n_1} \frac{1}{j} \sum_{h} \frac{1}{h\varphi(h)} \sum_{M/(mhj)<d\le N/(mhj)} \frac{1}{d} \sum_{M/(dh)<r\le N/(dh)} \frac{\log\log n}{r} \sum_{M/(dmhj)<v\le N/(dmhj)} \frac{1}{v} \ll (\log\log n)^5$$

and the contribution from those with $dmhj \le M$ is

$$\ll \sum_{m\le N} \frac{1}{m\varphi(m)} \sum_{j|n_1} \frac{1}{j} \sum_{h} \frac{1}{h\varphi(h)} \sum_{d\le M/(mhj)} \frac{1}{d} \sum_{M/(dh)<r\le N/(dh)} \frac{\log\log n}{r} \left| \sum_{M/(dmhj)<v\le N/(dmhj)} \frac{\chi(v)}{v} \right|.$$

Here the innermost sum is $\ll dmhj/M$. Thus the above is

$$\ll \sum_{m\le N} \frac{1}{m\varphi(m)} \sum_{j|n_1} \frac{1}{j} \sum_{h} \frac{1}{h\varphi(h)} \sum_{d\le M/(mhj)} \frac{mhj}{M} (\log\log n)^2 \ll (\log\log n)^3.$$

Thus it follows that

$$Z_2(M,N) \ll n\frac{(\log\log n)^6}{\log n}. \tag{21.112}$$

It remains to consider $Z_3(M,N)$ given by (21.111). Here we replace the condition $(k_1,k_2)=1$ by the sum

$$\sum_{l|(k_1,k_2)} \mu(l),$$

interchange the order of summation, and replace k_j by lh_j. The dummy variable l plays a similar role to that of d in $Z_2(M,N)$. Thus

$$Z_3(M,N) = \sum_{\substack{(dlh_1h_2,n_1)=1\\ M<dlh_j\le N,d\le D}} \chi(d^2l^2h_1h_2)\mu(l) \sum_{\substack{r\in\mathcal{N}\\ dl^2h_1h_2|n-r}} 1.$$

We now again divide the summation according as to whether $l\le D$ or not. The contrary case can be readily dismissed since the total contribution from such terms is

$$\ll \sum_{l>D}\sum_{M/l<dh_j\le N/l}\sum_{\substack{r\in\mathcal{N}\\ dl^2h_1h_2|n-r}} 1 \ll \sum_{l>D}\frac{n}{l^2}(\log n)^3 \ll \frac{n}{(\log n)^2}.$$

We are left with

$$\sum_{d\le D}\sum_{l\le D}\chi(d^2l^2)\mu(l) \sum_{\substack{M/dl<h_j\le N/dl\\ (h_1h_2dl^2,n_1)=1}} \chi(h_1h_2) \sum_{\substack{r\in\mathcal{N},(r,P(z))=1\\ dl^2h_1h_2|(n-r)}} 1.$$

The condition $(h_2,n_1)=1$ is a nuisance and it is convenient to remove it by the usual resort to the formula

$$\sum_{v|(h_2,n_1)} \mu(v)$$

so that, on writing $h_2=vw$, our sum is

$$\sum_{d\le D}\sum_{l\le D}\chi(d^2l^2)\mu(l) \sum_{\substack{M/dl<h_1\le N/dl\\ (h_1dl^2,n_1)=1}} \sum_{v|n_1}\mu(v)\chi(h_1v)$$
$$\sum_{M/dlv<w\le N/dlv} \chi(w) \sum_{\substack{r\in\mathcal{N},\ (r,P(z))=1\\ dl^2h_1vw|n-r}} 1.$$

Given d, l, h_1, v, w, r let

$$u = \frac{n-r}{dl^2h_1vw}.$$

Then the sum becomes

$$\sum_{d\le D}\sum_{l\le D}\chi(d^2l^2)\mu(l)\sum_{\substack{M/dl<h_1\le N/dl\\(h_1dl^2,n_1)=1}}\sum_{v|n_1}\mu(v)\chi(h_1v)$$
$$\sum_{u}\sum_{M/dlv<w\le N/dlv}\chi(w)\sum_{\substack{r\in\mathcal{N},\ (r,P(z))=1\\ dl^2h_1vwu=n-r}}1.$$

Then for a given q we collect the terms with $dl^2h_1vu=q$. Thus $w=\frac{n-r}{q}$ and the multiple sum becomes

$$\sum_{q\le\frac{nD^2n_1}{M}}\sum_{\substack{d,l,h_1,v\\ dl^2h_1v|q}}\chi(d^2l^2)\mu(l)\mu(v)\chi(h_1v)\sum_{\substack{r\in\mathcal{N},(r,P(z))=1\\ n-\frac{Nq}{dlv}\le r<n-\frac{Mq}{dlv}\\ 1\le r<n\\ q|n-r}}\chi\left(\frac{n-r}{q}\right),$$

where d, l, h_1, v satisfy

$$d\le D, l\le D, \frac{M}{dl}<h_1<\frac{N}{dl}, v|n_1, Ndlv>q.$$

Let

$$X=\max\{0,\lceil n-Nq/dlv\rceil-1\},\quad Y=\max\{0,\lceil n-Mq/dlv\rceil-1\},$$

so that $Y<r\le X$. We also have

$$q\le n^{3/4}n_1(\log n)^B.$$

Thus if

$$Y-X\ll\frac{n}{(\log n)^7},$$

then the innermost sum is

$$\sum_{\substack{r\in\mathcal{N},\ (r,P(z))=1\\ X<r\le Y\\ 1\le r<n\\ q|n-r}}\chi\left(\frac{n-r}{q}\right)\ll\frac{n}{q(\log n)^7}\tag{21.113}$$

and the total contribution from such terms is

$$\ll\frac{n}{(\log n)^2}.$$

Thus we may suppose that

$$Y-X\gg\frac{n}{(\log n)^7}.$$

The general term in the sum over r is ± 1 according as

$$\frac{n-r}{q} \equiv \pm 1 \pmod 4,$$

that is

$$r \equiv n \mp q \pmod{4q}$$

and so is

$$\sum_{\substack{X<r\le Y\\ (r,P(z))=1\\ r\equiv n-q \pmod{4q}}} 1 - \sum_{\substack{X<r\le Y\\ (r,P(z))=1\\ r\equiv n+q \pmod{4q}}} 1.$$

Given $m|P(z)$ and $m \le y = n^{1/8}$, so that $mq \le n^{7/8}n_1(\log n)^B$, we have

$$\sum_{\substack{X<a\le Y\\ m|a\\ a\equiv n\pm q \pmod{4q}}} 1 = \sum_{\substack{x=1\\ x\equiv n\pm q \pmod{4q}\\ x\equiv 0 \pmod m}}^{[4q,m]} \left(\frac{Y-X}{[4q,m]} + O(1)\right).$$

Moreover,

$$\sum_{\substack{x=1\\ x\equiv n\pm q \pmod{4q}\\ x\equiv 0 \pmod m}}^{[4q,m]} 1 = \begin{cases} 1 & (4q,m)|n\pm q,\\ 0 & (4q,m)\nmid n\pm q.\end{cases}$$

Since m is square-free we have $(4q,m) = (2q,m)$, and so $(4q,m)|n \pm q$ if and only if $(2q,m)|n+q$ if and only if $(2q,m)|n-q$. Hence we may apply Corollary 21.10 and the main term will be independent of the sign $\pm$. Thus the main terms will cancel and we obtain (21.113) once more, and so we can conclude that

$$Z_3(M,N) \ll \frac{n}{(\log n)^2}.$$

Hence, by (21.108), (21.110), and (21.112)

$$Z(M,N) \ll n\frac{(\log\log n)^6}{\log n}.$$

Hence, by (21.101) and (21.106) we have

$$Y(M,N) \ll n\frac{(\log\log n)^5}{(\log n)^\lambda},$$

where

$$\lambda = \frac{3}{2} - \frac{e\log 2}{4} = 1.028957\ldots,$$

and this completes the proof of the Hooley–Linnik theorem.

21.5.4 Exercises

1. Let $R(N)$ denote the number of solutions of $p + P_2 = N$ with p prime and P_2 a number having at most two prime factors. Prove that if N is even and large, then

$$R(N) > \frac{N\mathfrak{S}(N)}{3(\log N)^2},$$

where

$$\mathfrak{S}(N) = c \prod_{\substack{p|N \\ p>2}} \frac{p-1}{p-2}$$

and c is the twin prime constant.

2. (Vaughan 1973) Prove that at least one of the following two statements is valid.

   (a) For infinitely many primes p, $3p + 2$ is prime.
   (b) For infinitely many n, $d(n) = d(n+2)$.

3. (a) Let $S(X)$ denote the number of $n \le X$ such that $n^2 + 1$ is prime. Prove that there is a positive constant δ such that

$$S(X) \le \frac{2X\mathfrak{S}}{\log X} + O\big(X(\log X)^{-1-\delta}\big),$$

where

$$\mathfrak{S} = \frac{4}{\pi} \prod_{p>2} \left(1 - \frac{(-1)^{\frac{p-1}{2}}}{\big(p - (-1)^{\frac{p-1}{2}}\big)(p-1)}\right).$$

   (b) Prove that the number $S_4(X)$ of $n \le X$ such that $n^2 + 1$ has at most four prime factors satisfies

$$S_4(n) \gg \frac{X}{\log X}.$$

4. (Halberstam & Richert 1974, §9.5) Let $a(n)$ be the number of solutions of $n = l^2 + 1$ with $1 \le l \le X$, and define for $z = X^{1/4}$, $w = X^{7/10}$,

$$W(\mathcal{A}, \mathcal{P}, z, w) = \sum_{\substack{n \\ (n, P(z))=1}} a(n) \left(1 - \sum_{\substack{z \le p < w \\ p|n}} \frac{1}{2}\right).$$

   (a) Show that

$$W(\mathcal{A}, \mathcal{P}, z, w) \ge$$
$$XV(z)\left(f_-(4) - \int_z^w f_+\left(\frac{\log(X/u)}{\log z}\right) \frac{du}{2u \log u}\right) + O\big(X(\log X)^{-1-\delta}\big),$$

where $V(z) = \prod_{p<z}\left(1-\rho(p)\right)$, $\rho(2) = \frac{1}{2}$ and

$$p\rho(p) = 1 + (-1)^{\frac{p-1}{2}} \quad (p > 2).$$

(b) Show that

$$V(z) = \frac{\mathfrak{S}e^{-C_0}}{\log z} + O\left((\log z)^{-2}\right),$$

where $\mathfrak{S}$ is as in Exercise 21.5.4.3.

(c) Let $S_3(X)$ denote the number of $n \le X$ such that n^2+1 has at most three prime factors. Prove that

$$\begin{aligned} S_3(X) &\ge \frac{2\mathfrak{S}X}{\log X}\left(\frac{\log 3}{4} - \int_{1/4}^{7/10} \frac{dv}{8v(1-v)}\right) + O\left(X(\log X)^{-1-\delta}\right) \\ &= \frac{\mathfrak{S}X}{8\log X}\log\frac{9}{7} + O\left(X(\log X)^{-1-\delta}\right). \end{aligned}$$

5. Let a be a given non-zero integer and

$$N(x) = \sum_{0<p+a\le x} r(p+a),$$

where r is as in (21.95). Prove that

$$N(x) \sim \frac{\pi x}{\log x}\mathfrak{S}(a),$$

where $\mathfrak{S}(a)$ is as in (21.94).

21.6 Almost Primes in Polynomial Sequences

In the exercises of §25.5.4 we sieved a thin set, namely the sequence n^2+1, to limit the number of prime factors. Thus there are n for which n^2+1 is an almost prime; in this case a P_4 or a P_3. Suppose we have a sequence $c(n)$ of positive integers, increasing for large n, roughly of size n^d where $d \in \mathbb{N}$, $d > 1$, and we want to consider those $c(n)$ with $n \le X$ which remain when those terms with a prime factor $p < w$ are removed. Except possibly in very special cases one cannot expect sieving techniques to deal with prime factors p significantly larger than X and the expectation is that one will need to restrict to the situation when $w = X^\theta$ with $\theta < 1$. We can hope to cope with a θ satisfying

$$\frac{d}{d+1} < \theta < 1, \tag{21.114}$$

but not with a larger θ. Thus the best we could conclude is that there are n so that $c(n)$ has at most d prime factors p with $p \ge w$. It remains to see what can be said for smaller prime factors. In Exercise 21.5.4.4, with a suitable set of weights and $\theta = \frac{7}{10}$, it can be shown that some of these n have at most one prime factor $p < w$ and so there are infinitely many n such that $n^2 + 1$ has at most three prime factors. This argument is readily modified to deal with general irreducible quadratic polynomials. For polynomials of higher degree, $d > 2$, the requirement that (21.114) holds becomes too demanding for the system of weights to give a concomitant conclusion. We require more sophisticated weights and to this end we use those introduced by Richert (1969). These lead to an elegant conclusion.

Theorem 21.19 *Suppose that $d \in \mathbb{N}$, $d \geq 2$, and that $g \in \mathbb{Z}[x]$ is irreducible over $\mathbb{Q}$ and has positive leading coefficient. Let*

$$\mathcal{N}_g(X) = \{m \leq X : \Omega(g(m)) \leq d+1 \quad \text{and} \quad (g(m), P(x^{1/4})) = 1\}.$$

If for every prime p there exist integers m such that $(g(m), p) = 1$, then

$$\mathcal{N}_g(X) \gg_g \frac{X}{\log X}.$$

Let $a(n)$ denote the number of $m \leq X$ such that $g(m) = n$, let $r(m)$ denote the number of solutions of $g(x) \equiv 0 \pmod{m}$, and let $\rho(m) = r(m)/m$. Then in the notation of §21.1 we have (21.3) and

$$|R(m)| \leq r(m).$$

Thus we need to understand the behaviour of $r(m)$, certainly when m is square-free. It is also useful to understand $r(p^2)$. A substantial part of the input here is from algebraic number theory.

Theorem 21.20 *Suppose that $g \in \mathbb{Z}[x]$ is irreducible over $\mathbb{Q}$ and r is as above.*

(i) *There is a positive constant c_g such that whenever $Y \geq 2$ we have*

$$\sum_{p \leq Y} r(p) = \operatorname{li}(Y) + O_g\big(Y \exp(-c_g\sqrt{\log Y})\big). \tag{21.115}$$

(ii) *Suppose that p does not divide the discriminant D_g of g. Then for every k we have*

$$r(p^k) \leq d. \tag{21.116}$$

(iii) *Suppose that m is square-free. Then*

$$r(m) \leq d^{\omega(m)}. \tag{21.117}$$

Proof We begin by dismissing the second and third statements. The function r is multiplicative and since the polynomial g is irreducible it does not have a fixed prime divisor. Hence by Lagrange's theorem $r(p) \le d$, and (iii) follows. Moreover, when $p \nmid D_g$ and there is an x such that $g(x) \equiv 0 \pmod{p}$ we have $g'(x) \not\equiv 0 \pmod{p}$, so by Hensel's lemma x lifts to a unique solution modulo p^2, and likewise modulo p^3, and so on. Thus (ii) holds.

To prove the first part we require a classical result on prime ideals.

Lemma 21.21 (Dedekind–Kummer) *Let K be a number field of the form $K = \mathbb{Q}(\theta)$ with $\theta \in \mathcal{O}_K$ and suppose that f is the minimal polynomial of θ over $\mathbb{Z}[x]$. For any prime p not dividing $[\mathcal{O}_K : \mathbb{Z}[\theta]]$ consider the factorization in $\mathbb{F}_p[x]$*

$$h(x) = h_1(x)^{d_1} \cdots h_k(x)^{d_k},$$

where the $h_j(x)$ are monic irreducible polynomials and each $d_j \in \mathbb{N}$. Then the ideal (p) factors into prime ideals $\mathfrak{p}_j$

$$(p) = \mathfrak{p}_1^{d_1} \cdots \mathfrak{p}_k^{d_k}$$

and $N(\mathfrak{p}_j) = p^{\deg h_j}$, where N is the norm of K.

This can be hard to pin down in standard expositions of algebraic number theory, but see Neukirch (1999, Chapter 1, Proposition 8.3), or Lang (1970, Chapter 1, Proposition 25).

We note also that

$$d_1 \deg h_1 + \cdots + d_k \deg h_k = d$$

and so $k \le d$.

Let a_k be the leading coefficient of g and let $h(x) = a_k^{k-1} g(x a_k^{-1})$. Then $h \in \mathbb{Z}[x]$ is monic and irreducible over $\mathbb{Q}$. Moreover, when $p \nmid a_k$ we have

$$r(p; g) = r(p; h).$$

Let θ be a root of h and let $K = \mathbb{Q}(\theta)$. Since f is monic we have $\theta \in \mathcal{O}_K$. Hence by the lemma we have

$$N(\mathfrak{p}_j) = p^{\deg h_j}$$

for each prime ideal $\mathfrak{p}_j$ factor of (p). Moreover $r(p; h)$ is the number of h_j in the factorization of h over $\mathbb{F}_p$ which are linear, i.e., $\deg h_j = 1$. Hence

$$r(p; h) = \sum_{\substack{\mathfrak{p} \mid (p) \\ N(\mathfrak{p}) = p}} 1.$$

Therefore if we choose Y_0 so that when $p > Y_0$ we have $p \nmid a_k[\mathcal{O}_K : \mathbb{Z}[\theta]]$, then

$$\sum_{Y_0 < p \le Y} r(p; g) = \sum_{Y_0 < N(\mathfrak{p}) \le Y} 1 - \sum_{\substack{\mathfrak{p} | (p) \\ Y_0 < N(\mathfrak{p}) = p^r \le Y \\ 2 \le r \le d}} 1.$$

By our observations immediately after the statement of Lemma 21.21, the second sum on the right is

$$\le \sum_{\substack{p^r \le Y \\ r \ge 2}} d \ll Y^{1/2}.$$

The theorem then follows from the prime ideal theorem, Theorem 8.9. □

Proof of Theorem 21.19 Let X be large and

$$z = X^{1/4},\ w = X^{1/v},\ v \in (1, 4),\ \lambda \in \left(\frac{1}{2}, \frac{\log 3}{\log 4} \right). \tag{21.118}$$

There is some flexibility in the choice of v and λ, and determining what values are possible is quite instructive. Whilst we leave the exact values of v and λ open at the moment, we will ultimately choose v close to 1 and then any λ satisfying the above will work. If we wish to maximize the lower bound in our theorem then we should take λ close to $1/2$.

With $a(n)$ as defined immediately after Theorem 21.19 we let

$$W(X; g) = \sum_{(n, P(z))=1}^{*} a(n) \left(1 - \sum_{\substack{z \le p < w \\ p | n}} \frac{\lambda \log(w/p)}{\log w} \right), \tag{21.119}$$

where Σ^* indicates that we exclude n with a repeated prime factor p in the range $z \le p < w$. Note that no n counted by W can have more than $4d$ prime factors.

First we observe that the sum over n with the condition Σ^* can be replaced by the sum over n without this condition with an error

$$\ll \sum_n \sum_{\substack{z \le p < w \\ p^2 | n}} a(n) d = \sum_{z \le p < w} \sum_{\substack{m \le X \\ g(m) \equiv 0 \ (\mathrm{mod}\ p^2)}} d$$

$$\ll \sum_{z \le p \le \sqrt{X}} \frac{X r(p^2)}{p^2} d + \sum_{\sqrt{X} < p < w} r(p^2) d.$$

Since $p \ge z > D_g$, by Theorem 21.20(ii) the above is

$$\ll d^2 X z^{-1} + d^2 w.$$

Hence

$$W(X;g) = S(\mathcal{A},\mathcal{P},z) - \sum_{z\le p<w} \frac{\lambda\log(w/p)}{\log w} S(\mathcal{A}_p,\mathcal{P},z) + O\big(X(\log X)^{-2}\big), \tag{21.120}$$

where $\mathcal{P}$ is the set of all primes. Now we have

$$A(m) = X\rho(m) + R(m),$$

where

$$\rho(m) = \frac{r(m)}{m}$$

and

$$|R(m)| \le r(m).$$

Let

$$y = \frac{X}{(\log X)^{d+3}} = X^{\frac{1}{1+\eta}},$$

say, so that

$$0 < \eta \ll \frac{\log\log X}{\log X}, \tag{21.121}$$

and assume that

$$v > 1 + \eta$$

so that $w < y$.

By Theorem 21.20(iii),

$$\sum_{m\le y} \mu(m)^2|R(m)| \ll y\prod_{p\le y}\left(1 + \frac{r(p)}{p}\right) \ll y(\log y)^d,$$

and by parts (ii) and (iii),

$$\sum_{z\le p<w}\sum_{pm\le y} \mu(m)^2|R(pm)| \ll y(\log y)^{d+1}.$$

Thus, by Theorem 21.9,

$$S(\mathcal{A},\mathcal{P},z) \ge XV(z)f_-\left(\frac{\log y}{\log z}\right) + O\big(X(\log X)^{-1-\delta}\big). \tag{21.122}$$

Here

$$V(z) = \prod_{p<z}(1-\rho(p)), \tag{21.123}$$

and by (21.115) and partial summation this satisfies

$$V(z) \gg_g \frac{1}{\log X}. \tag{21.124}$$

Note that as g has no fixed prime divisor we have $r(p) < p$. Similarly, by Theorem 21.9,

$$\sum_{z \le p < w} \frac{\lambda \log(w/p)}{\log w} S(\mathcal{A}_p, \mathcal{P}, z)$$
$$\le XV(z) \sum_{z \le p < w} \frac{\rho(p)\lambda \log(w/p)}{\log w} f_+\left(\frac{\log(y/p)}{\log z}\right) + O\big(X(\log X)^{-1-\delta}\big).$$

By (21.115), the properties of f_+ given in §21.3, and by partial summation we have

$$\sum_{z \le p < w} \frac{r(p)\lambda \log(w/p)}{p \log w} f_+\left(\frac{\log(y/p)}{\log z}\right)$$
$$= \int_z^w \frac{\lambda \log(w/t)}{t(\log t)\log w} f_+\left(\frac{\log(y/t)}{\log z}\right) dt + O\big((\log X)^{-2}\big).$$

By the change of variable $t = X^{1/\alpha}$ we have

$$\int_z^w \frac{\lambda \log(w/t)}{t(\log t)\log w} f_+\left(\frac{\log(y/t)}{\log z}\right) dt = \int_v^4 \frac{\lambda(\alpha - v)}{\alpha^2} f_+\left(\frac{4}{1+\eta} - \frac{4}{\alpha}\right) d\alpha.$$

For $\alpha \in [v, 4]$ we have

$$0 < \frac{4}{1+\eta} - \frac{4}{\alpha} \le \frac{4}{1+\eta} - 1 < 3.$$

Hence, by (21.67),

$$\int_v^4 \frac{\lambda(\alpha - v)}{\alpha^2} f_+\left(\frac{4}{1+\eta} - \frac{4}{\alpha}\right) d\alpha = \frac{2e^{C_0}\lambda}{4} \int_v^4 \frac{(\alpha - v)(1+\eta)}{\alpha(\alpha - 1 - \eta)} d\alpha.$$

The integrand here is

$$\frac{v}{\alpha} - \frac{v - 1 - \eta}{\alpha - 1 - \eta},$$

so the integral is

$$v \log \frac{4}{v} - (v - 1 - \eta) \log \frac{3 - \eta}{v - 1 - \eta}.$$

Now we advert to (21.122). We have

$$\frac{\log y}{\log z} = \frac{4}{1+\eta}.$$

Hence, by (21.69),

$$f_-\left(\frac{\log y}{\log z}\right) = 2e^{C_0}(1+\eta)\frac{\log \frac{3-\eta}{1+\eta}}{4}.$$

Thus, by (21.120), (21.122), and (21.121) we have established that

$$W(X;g) \geq \frac{e^{C_0}}{2} XV(z)\Xi + O\big(X(\log X)^{-1-\delta}\big),$$

where

$$\Xi = \log 3 - \lambda\left(v\log\frac{4}{v} - (v-1-\eta)\log\frac{3-\eta}{v-1-\eta}\right).$$

We can choose v so that $v-1-\eta$ is as small as we please, so the term

$$(v-1-\eta)\log\frac{3-\eta}{v-1-\eta}$$

can be made as small as we please, and then, by (21.121), v will also be close to 1. Thus we can make Ξ close to

$$\log 3 - \lambda \log 4,$$

which will be positive as long as we choose the constant λ to satisfy

$$\lambda < \frac{\log 3}{\log 4} = 0.7924\ldots.$$

We now examine W from a different direction. Consider an n counted by (21.119) for which the term

$$b(n) = 1 - \sum_{\substack{z\leq p<w\\ p|n}} \frac{\lambda\log(w/p)}{\log w}$$

is non-negative. Let $p_1,\ldots,p_j$ be the primes counted in this sum and let k be the number of prime factors q of n with $q \geq w$. Then the expression above is

$$1-\lambda j + \lambda\frac{\log(p_1\cdots p_j)}{\log w} \geq 0.$$

We also have

$$p_1\cdots p_j \leq nw^{-k} \leq CX^d w^{-k}$$

for some constant $C \geq 2$. Hence

$$1-\lambda j + \lambda\frac{\log(CX^d w^{-k})}{\log w},$$

so that

$$\Omega(n) = j + k \le \frac{1}{\lambda} + \nu d + \frac{\nu \log C}{\log X}.$$

By (21.118) we have

$$\frac{\log 4}{\log 3} < \lambda^{-1} < 2.$$

Since X is large the last term is negligible, and as ν can be taken arbitrarily close to 1 it follows that given any $\varepsilon > 0$ we have

$$\Omega(n) < \frac{1}{\lambda} + d + \varepsilon < d + 2.$$

Thus

$$\Omega(n) \le d + 1,$$

and so, as $b(n) \le 1$,

$$\sum_{\substack{(n,P(z))=1 \\ \Omega(n)\le d+1}}^{*} a(n) \ge \sum_{\substack{(n,P(z))=1 \\ b(n)\ge 0}}^{*} a(n)b(n) = W(X;g) \gg XV(z) \gg \frac{X}{\log X},$$

as required. □

21.7 Notes

Section 21.1. For an overview of Selberg's seminal contributions to sieve theory see Selberg (1952a,b, 1991). For background on the Exercises 21.1.1.1–3 see the notes to Chapter 20, and for later work see Chapter 22.

Section 21.2. The argument of §21.2.1 is readily extended to all dimensions κ. There are considerable complications of detail, although no new ideas of principle are required. In some cases the analogues of $\phi_\pm$ are most conveniently represented by a contour integral. For the full details see the standard work on the subject, Friedlander & Iwaniec (2010). It is not clear that questions requiring dimension $\kappa > 1$ are not better served by other methods. For example, the core method applied in Chapter 22 uses a form of the Selberg idea.

Section 21.3. For the best bound for $G(p)$ in Exercise 21.3.1.2, and some history, see Shoup (1992).

Section 21.4. Exercise 21.4.1.1. It seems quite possible that this is the extremal example for sieves of arbitrary dimension $\kappa \in \mathbb{N}$. This observation does not seem to be in the extant literature. However, it may not be extremal for sieves which are *not* products of sieves of lower dimension.

Section 21.5. The account of Chen's theorem, Theorem 21.15, is based on Ross (1974). The weights are essentially those of Kuhn (1941, 1954). Hooley (1957) established Theorem 21.16 by assuming the Riemann Hypothesis for Dirichlet L-functions formed from Dirichlet characters. This requirement was removed by Linnik (1963) using his dispersion method. Then it was observed by Elliott & Halberstam (1966) that this can be simplified by using the Bombieri–Vinogradov theorem, Theorem 20.2. See also Theorem 5 of Hooley (1976). Hooley's use of the asymptotic sieve has never been superseded and is combinatorially quite intense, which in the interests of digestibility we have expanded somewhat. Hooley's paper was also the first appearance of a Δ function, which is now usually written in the form

$$\Delta(n) = \max_u \operatorname{card}\{m|n : u < m \le eu\}. \tag{21.125}$$

It appears in this modern form in Erdős (1974), and then in Hooley (1979), about 20 years after Hooley's work described here, and was developed then either for its own interest or by Hooley for applications in additive number theory. See also Vaughan (1986a,b). This led to a substantial body of work, for which see Hall & Tenenbaum (2008), and is still of ongoing interest. See the Wikipedia article on the Hooley Delta function (`https://en.wikipedia.org/wiki/Hooley%27s_delta_function`) or the section of Hooley's obituary on the Δ-function at `https://royalsocietypublishing.org/doi/10.1098/rsbm.2020.0027`.

Concerning Exercise 21.5.4.2, for some history of questions related to the equation $d(n) = d(n+1)$, see Erdős, Pomerance, & Sárközy (1987).

Section 21.6. The first part of Theorem 21.20 in many expositions of sieves is rather airily said to follow from the prime ideal theorem, Theorem 8.9. However, as we show here there is more to it than that. Some details were given by Erdős (1952). For other sources and background to Lemma 21.21 see

`https://encyclopediaofmath.org/wiki/Kummer_theorem` or
`https://en.wikipedia.org/wiki/Dedekind-Kummer_theorem` or
`https://kconrad.math.uconn.edu/blurbs/gradnumthy/dedekindf.pdf`.

Weights improving on Kuhn's were first introduced by Ankeny & Onishi (1964) and developed further by Richert (1969) (see also Buchstab (1967), Halberstam & Richert (1974), and Greaves (2001)).

In the special case $n^2 + 1$, Iwaniec (1978a) has pushed things further and shown that there are infinitely many n such that this polynomial has at most two prime factors. This depends significantly on a deeper analysis of the error term (21.16), leading to an expression as a bilinear form where ideas can be employed similar to those in Chapter 17.

We have only touched the surface of possible applications of sieves, even the linear sieve and applications to almost primes. There is a long and complicated history of such results going back to the 1920s. Let P_r denote a number having at most r prime factors. Then it can be shown that for each large x there are $y_r(x)$ so that the interval $(x - y_r(x), x]$ contains a P_r. In the case $r = 2$ the current best result is due to Wu (2010), where it is shown that

$$y_2(x) = \frac{101}{232}$$

is possible. A variant of this is to establish such a result just for almost all x. Thus in Matomäki (2022), it is shown that there exists a constant $c > 0$ such that the following holds. Suppose that $x \geq 2$ and $2 \leq h \leq X^{1/100}$. Then

$$\sum_{\substack{x-h\log x<n\leq x\\ (n,P(X^{1/8}))=1\\ \Omega(n)\leq 2}} 1 \geq ch$$

for all $x \in (X/2, X]$ apart from an exceptional set of s of measure $\ll X/h$.

Another class of questions which has been studied is, given $k \in \mathbb{N}$ and $l \in \mathbb{Z}$ with $(l, k) = 1$, to find exponents e_r such that there are $P_r \equiv l \pmod k$ with $P_r \leq k^{e_r}$. Thus in Cai, Li, & Zhang (2023) it is shown that $e_2 = 1.8345$ is possible, improving on Iwaniec (1982) who had $e_2 = 1.845$. For a somewhat different view of sieves and their applications, see Harman (2007).

21.8 References

Ankeny, N. & Onishi, H. (1964). The general sieve, *Acta Arith.* **10**, 31–62.

Buchstab, A. A. (1938). New improvements in the method of the sieve of Eratosthenes, *Mat. Sb.* (N. S.) **4** (46), 1239–1246.

Buchstab, A. A. (1967). A combinatorial strengthening of the Eratosthenian sieve method, *Usp. Mat. Nauk* **22**, 199–226.

Cai, Y., Li, J. & Zhang, M. (2023). On the least almost-prime in arithmetic progression, *Czech. Math. J.* **73**, 177–188. `https://doi.org/10.21136/CMJ.2022.0478-21`.

Chen, J.-R. (1973). On the representation of a large even integer as the sum of a prime and the product of at most two primes, *Scientia Sinica* **16**, 157–176.

Elliott, P. D. T. A. & Halberstam, H. (1966). Some applications of Bombieri's theorem, *Mathematika* **13**, 196–203.

Erdős, P. (1952). On the sum $\sum_{k=1}^{x} d(f(x))$, *J. London Math. Soc.* **27**, 7–15.

(1974). On abundant-like numbers, *Canadian Math. Bull.* **17**, 599–602. doi:10.4153/CMB-1974-108-5.

Erdős, P., Pomerance, C. & Sárkőzy, A. (1987). On locally repeated values of certain arithmetic functions, III, *Proc. Amer. Math. Soc.* **101**, 1–7.

Friedlander, J. B. & Iwaniec, H. (2010). *Opera de Cribro*, AMS Colloquium Publications 57, Providence: Amer. Math. Soc., xx+527 pp.

Greaves, G. (2001). *Sieves in Number Theory*, Ergeb. Math. Grenzgeb (3) **43**, Berlin Heidelberg: Springer, xii+304 pp.

Halberstam, H. & Richert, H.-E. (1974). *Sieve Methods*, London Math. Soc. Monograph No. 4, London: Academic Press, xiv+364 pp.

Hall, R. R. & Tenenbaum, G. (2008). *Divisors*, Cambridge Tracts in Math. 90, Reissue edition, Cambridge: Cambridge University Press, 184 pp.

Hardy, G. H. & Littlewood, J. E. (1922). Some problems of 'partito numerorum'; III: On the expression of a number as a sum of primes, *Acta Math.* **44**, 1–70.

Hardy, G. H. & Wright, E. M. (2008). *An Introduction to the Theory of Numbers*, Sixth edition, Oxford: Oxford University Press.

Harman, G. (2007). *Prime-Detecting Sieves*, London Math. Soc. Monograph 33, Princeton: Princeton University Press, xiv+362 pp.

Hooley, C. (1957). On the representation of a number as the sum of two squares and a prime, *Acta Math.* **97**, 189–210.

(1976). *Applications of Sieve Methods to the Theory of Numbers*, Cambridge Tracts in Math. 70. Cambridge: Cambridge University Press.

(1979). On a new technique and its applications to the theory of numbers, *Proc. London Math. Soc.* (3) **38**, 115–151.

Iwaniec, H. (1971). On the error term in the linear sieve, *Acta Arith.* **s19**, 1–30.

(1978a). Almost-primes represented by quadratic polynomials, *Invent. Math.* **47**, 171–188.

(1978b). On the problem of Jacobsthal, *Demonstratio Math.* **11**, 225–231.

(1982). On the Brun–Titchmarsh theorem, *J. Math. Soc. Japan* **34**, 95–123.

Kuhn, P. (1941). Zur Viggo Brun'schen Siebmethode I, *Norske Vid. Selsk. Forh. Trondhjem* **14**, 145–148.

(1954). Neue abschätzungen auf grund der Viggo Brunschen siebmethode, *Tolfte Skandinaviska Matematikerkongressen, Lund*, 160–168.

Lang, S. (1970). *Algebraic Number Theoery*, Massachusetts: Addison-Wesley, xi+354 pp.

Linnik, Y. V. (1960). All large numbers are sums of a prime and two squares (A problem of Hardy and Littlewood) I, *Mat. Sb. (N.S.)* **52**(94), 661–700.

(1961), All large numbers are sums of a prime and two squares (A problem of Hardy and Littlewood) II, *Mat. Sb. (N.S.)* **53**(95), 3–38.

(1963). *The Dispersion Method for Binary Additive Problems*, Amer. Math. Soc., Trans. of Math. Monographs, Vol. 4, 186 pp.

Matomäki, K. (2022). Almost primes in almost all very short intervals, *J. London Math. Soc.* (2) **106**, 1061–1097.

Neukirch, J. (1999). *Algebraic Number Theory*, Grundlehren der mathematischen Wissenschaften, 322, Springer, xvii+571 pp.

Richert, H.-E. (1969). Selberg's sieve with weights, *Mathematika* **16**, 1–22.

Ross, P. M. (1974). On Chen's theorem that each large even integer has the form $p_1 + p_2$ or $p_1 + p_2p_3$, *J. London Math. Soc.* (2) **10**, 500–506.

Selberg, A. (1947). On an elementary method in the theory of primes, *Norske Vid. Selsk. Forh., Trondhjem* **19**(18), 64–67; *Collected Papers*, Vol. I, Berlin: Springer-Verlag, pp. 363–368.

(1952a). On elementary methods in prime-number theory and their limitations. In *C. R. Onzième Contrè Math. Scandiaves* (Trondheim 1949), Oslo: Johan Grundt Tanums Forlag, pp. 13–22; *Collected Papers*, Vol. I, Berlin: Springer-Verlag, pp. 388–397.

(1952b). The general sieve-method and its place in prime-number theory. In *Proc. Int. Cong. Math.* (Cambridge, Mass. 1950), Vol. I, Providence: Amer. Math. Soc. pp. 411–417.

(1991). *Collected Papers*, Vol. II, Berlin: Springer, viii+253 pp.

Shoup, V. (1992). Searching for primitive roots in finite fields, *Math. Comp.* **58**, 369–380. `www.shoup.net/papers/primroots.pdf`.

Vaughan, R. C. (1973). A remark on the divisor function, *Glasgow Math. J.* **14**, 54–55.

(1976). On the order of magnitude of Jacobsthal's function, *Proc. Edinburgh Math. Soc.* **29**, 329–331.

(1986a). On Waring's problem for cubes, *J. Reine Angew. Math.* **365**, 122–170.

(1986b). On Waring's problem for smaller exponents, *Mathematika* **33**, 6–22.

Wu, J. (2010). Almost primes in short intervals, *Sci. China Math.* **53**, 2511–2524. `https://doi.org/10.1007/s11425-010-4039-y`.

22

Bounded Gaps between Primes

22.1 The GPY Sieve

An important role is played in the most recent developments on gaps between primes by suitable sets of prime k-tuples. Thus before proceeding with this chapter the reader would be well advised to review the contents of §18.5. The principal idea is to use artefacts from sieve theory, especially the Selberg sieve, not directly in the form of a sieve but as a means to increase the likelihood that certain constellations of k-tuples have relatively few prime factors.

As a preliminary observation consider the starting point for the Selberg upper bound sieve (see §3.2 or the discussion on page 228),

$$\sum_{\substack{n\in\mathbb{Z}\\(n,P(z))=1}} a(n) \le \sum_{n\in\mathbb{Z}} a(n)\left(\sum_{l|n}\lambda(l)\right)^2 .$$

On following the argument above Theorem 21.1, one sees that the plan is to minimize this under the assumptions that $\lambda(1) = 1$ and that

$$A(d) = \sum_{\substack{n\in\mathbb{Z}\\d|n}} 1$$

can be approximated by an expression of the form

$$CX\rho(d).$$

Here X is a good approximation to $A(1)$, and ρ is multiplicative. The minimizing choice of $\lambda(q)$ is given by

$$\lambda(q) = \mu(q)\frac{S(R,q)}{S(R,1)}\prod_{p|q}\left(\frac{1}{1-\rho(p)}\right),$$

where

$$S(R,q) = \sum_{\substack{r\le R/q\\(r,q)=1}} \mu(q)^2 \prod_{p|q} \frac{\rho(p)}{1-\rho(p)}.$$

Typically, we apply this when the sieve is of dimension κ, i.e., when

$$\sum_{p\le y} \rho(p)\frac{\log p}{p} = \kappa \log y + O(1).$$

Under this kind of condition one might expect that

$$S(R,q) \sim C(\log R/q)^\kappa \prod_{p|q}(1-\rho(p)),$$

and so $\lambda(q)$ could be replaced by

$$\lambda(q) = \mu(q)\frac{\log^\kappa(R/q)}{\log^\kappa R} = \mu(q)\Big(1-\frac{\log q}{\log R}\Big)^\kappa.$$

Indeed, this is correct, and whilst we encounter some loss in precision in the final conclusion, there is one significant advantage, namely that this choice of $\lambda(q)$ can be applied quite effectively to any sieving question where the dimension is κ.

Let $\mathbb{1}_{\mathbb{P}}$ denote the characteristic function of the set of primes $\mathscr{P}$. Then our basic idea is to construct an expression of the form

$$\sum_{N\le n\le 2N} \left(\sum_{j=1}^{k} \mathbb{1}_{\mathbb{P}}(n+h_j) - \varrho\right)\left(\sum_{\substack{q\le R\\ q|Z(n;\boldsymbol{h})}} \lambda(q)\right)^2, \tag{22.1}$$

where $Z = \prod_{i=1}^k (n+h_i)$. Since our objective is to construct a large number of primes in a short interval, the k-tuples $\boldsymbol{h}$ that we consider will always be admissible in the sense defined in §18.5. A wrinkle introduced by Goldston, Pintz, & Yıldırım (2006) is to use a more general $\lambda(q)$ of the form

$$\lambda(q) = \mu(q) f\left(\frac{\log q}{\log R}\right),$$

where f is at our disposal. If one can show that the expression in (22.1) is positive, then it follows that there are n such that the number of primes amongst the $n+h_j$ is at least $\lfloor\varrho\rfloor+1$. Such a sieve application is now known as the GPY sieve.

Following Maynard (2015) we will use a more sophisticated version of this. Let $n+\boldsymbol{h}$ denote the k-tuple $(n+h_1,\ldots,n+h_k)$ and let $\boldsymbol{d}$ denote the k-tuple $(d_1, \ldots, d_k)$. We generally use the notation $\boldsymbol{z}$ to denote the k-tuple $(z_1, \ldots, z_k)$. Moreover, given two k-tuples $\boldsymbol{d}$ and $\boldsymbol{r}$ of integers, we define $\boldsymbol{d}|\boldsymbol{r}$ to mean that $d_j|r_j$ for all j. We also let $[\boldsymbol{d}, \boldsymbol{e}]$ denote the k-tuple $(\operatorname{lcm}[d_1, e_1], \ldots, \operatorname{lcm}[d_k, e_k])$. Finally, whenever we name a k-tuple, say $\boldsymbol{a} = (a_1, a_2, \ldots, a_k)$, we are implicitly setting $a = a_1 a_2 \cdots a_k$.

First, we perform some initial sieving for small primes so as to simplify some later expressions. A simple way to do this is to restrict our attention to a given residue class a modulo q where

$$q = \prod_{p \le Q} p, \qquad Q = \log\log\log N, \tag{22.2}$$

and N is a large integer parameter. Since the k-tuple $\boldsymbol{h}$ is admissible, there exist residue classes $a \pmod q$ for which $(a + h_j, q) = 1$ for $1 \le j \le k$. If we restrict n to the arithmetic progression $n \equiv a \pmod q$, and look for primes among the k-tuples $n + \boldsymbol{h}$, then the heuristic approach we used in §18.5 would predict that the frequency of k-tuples of primes encountered would be governed by the singular series

$$\mathfrak{S}(\boldsymbol{h}) = \prod_{p > Q} \left(1 - \frac{k}{p}\right)\left(1 - \frac{1}{p}\right)^{k} \sim 1$$

for large N.

Now we consider

$$\sum_{\substack{N < n \le 2N \\ n \equiv a \ (\mathrm{mod}\ q)}} \left(\sum_{j=1}^{k} \mathbb{1}_{\mathbb{P}}(n + h_j) - \varrho\right)\left(\sum_{\substack{d \le R \\ \boldsymbol{d}|n+\boldsymbol{h} \\ (d,q)=1}} \lambda(\boldsymbol{d})\right)^{2}. \tag{22.3}$$

In the first instance we ought to consider

$$\lambda(\boldsymbol{d}) = \mu(d) g(\boldsymbol{d})$$

for some suitable g. However, we shall be carrying out diagonalization of quadratic forms in the λ and this leads to a natural representation of the $\lambda(\boldsymbol{d})$, when d is square-free with $(d, q) = 1$, in the form

$$\lambda(\boldsymbol{d}) = \mu(d) d \sum_{\substack{\boldsymbol{r} \\ \boldsymbol{d}|\boldsymbol{r} \\ (r,q)=1}} \frac{\mu(r)^2}{\varphi(r)} f\left(\frac{\log r_1}{\log R}, \ldots, \frac{\log r_k}{\log R}\right). \tag{22.4}$$

We further suppose that

$$\operatorname{supp} f = \mathscr{R} = \{\boldsymbol{x} \in [0,1]^k : x_1 + \cdots + x_k \le 1\}. \tag{22.5}$$

There are two major tasks to be undertaken. The first is to obtain a good approximation to (22.3) with (22.4) for a wide class $\mathscr{F}$ of f. In practice, this means good approximations $S^*(f)$ and $T^*(f)$ to $S(f)$ and $T(f)$ where

$$S(f) = \sum_{j=1}^{k} S_j(f)$$

with

$$S_j(f) = S_j = \sum_{\substack{N<n\leq 2N \\ n\equiv a \pmod q}} \mathbb{1}_{\mathbb{P}}(n+h_j) \left(\sum_{\substack{d\leq R \\ \boldsymbol{d}|n+\boldsymbol{h} \\ (d,q)=1}} \lambda(\boldsymbol{d}) \right)^2, \tag{22.6}$$

$$T(f) = T = \sum_{\substack{N\leq n\leq 2N \\ n\equiv a \pmod q}} \left(\sum_{\substack{d\leq R \\ \boldsymbol{d}|n+\boldsymbol{h} \\ (d,q)=1}} \lambda(\boldsymbol{d}) \right)^2. \tag{22.7}$$

The second is then to maximize the ratio

$$\frac{S^*(f)}{T^*(f)}$$

over the class f. The optimal solution to this latter task is not known, although the former can be carried out for a very wide class, for example for f for which the partial derivatives are continuous on $\mathscr{R}$, and even this requirement can be relaxed somewhat.

Since we have to deal with $T(f)$ as well as the $S_j(f)$, we are pretty much forced to choose $\lambda(\boldsymbol{d})$ corresponding to a k-dimensional sieve, although in $S_j(f)$ since one of the variables is prescribed to be prime we would only need a $(k-1)$-dimensional sieve. On the other hand, the normalization we choose means that the logarithmic powers are essentially the same, and since the prime factors p of the d satisfy $p > Q = \log\log\log N$, any factors such as

$$\prod_{p|d} \frac{p^k - kp^{k-1}}{(p-1)^k}$$

will be close to 1, at least on average, and so will not differ in any important way from the $k-1$ version.

A major input into the approximation for $S_j(f)$ will be the Bombieri–Vinogradov theorem (Corollary 20.5) or a variant thereof. We define the *level* θ of distribution for the prime numbers to be the assumption that for every sufficiently small positive δ and every $A > 0$ we have

$$\sum_{r\leq x^{\theta-\delta}} \max_{(a,r)=1} \sup_{y\leq x} \left| \pi(y;r,a) - \frac{\mathrm{li}(y)}{\varphi(r)} \right| \ll_{\delta,A} x(\log x)^{-A}.$$

The Bombieri–Vinogradov theorem asserts that $\theta = \frac{1}{2}$ is permissible. However, it is useful to be able to see at once the consequence of the Elliott–Halberstam conjecture ($\theta = 1$) (Conjecture 20.4) or some intermediate improvement in the Bombieri–Vinogradov theorem.

Let $\mathscr{R}_j$ denote the set of $(k-1)$-tuples $(t_1, \ldots, t_{j-1}, t_{j+1}, \ldots, t_k)$ with $\boldsymbol{t} \in \mathscr{R}$ for some t_j. We define $\mathscr{F}$ to be the class of functions f, not identically 0, defined on $\mathscr{R}$ such that for each j, if $\boldsymbol{t}^* = (t_1, \ldots, t_{j-1}, t_{j+1}, \ldots, t_k)$ with $t_i \geq 0$ and $t_1 + \cdots + t_{j-1} + t_{j+1} + \cdots + t_k \leq 1$, then the function $f_j^*(t_j) = f(\boldsymbol{t})$ is absolutely continuous on $[0, 1 - t_1 - \cdots - t_{j-1} - t_{j+1} - \cdots - t_k]$. Given an $f \in \mathscr{F}$ it is useful first to extend its definition to $[0,1]^k$ by taking it to be 0 outside $\mathscr{R}$ and then to presume that

$$F(f) = \sup_{\boldsymbol{t} \in \mathscr{R}} |f(\boldsymbol{t})| + \sum_{j=1}^{k} \sup_{\boldsymbol{t}^* \in \mathscr{R}_j} \int_0^1 \left| \frac{\partial f}{\partial t_j}(\boldsymbol{t}) \right| dt_j \tag{22.8}$$

is bounded.

Theorem 22.1 (Maynard 2015) *Let $k \geq 2$. Suppose that the primes have level of distribution θ where $0 < \theta \leq 1$, let δ be a sufficiently small positive number, and let N be a large positive integer. Put $R = N^{\frac{\theta}{2} - \delta}$, define Q and q as in (22.2) and f and $\mathscr{R}$ as in (22.5), and assume $f \in \mathscr{F}$. Let $\boldsymbol{h}$ be an admissible set and choose a modulo q so that $(a + h_j, q) = 1$ for each j. Let*

$$I_j = \int_{[0,1]^{k-1}} \left(\int_0^1 f(\boldsymbol{t}) dt_j \right)^2 dt_1 \cdots dt_{j-1} dt_{j+1} \cdots dt_k,$$

$$J = \int_{[0,1]^k} f(\boldsymbol{t})^2 d\boldsymbol{t},$$

and let $S(f)$ and $T(f)$ be as in (22.6) and (22.7). Then as $N \to \infty$,

$$S(f) = \frac{(1 + o(1))\varphi(q)^k N (\log R)^{k+1}}{q^{k+1} \log N} \sum_{j=1}^{k} I_j$$

and

$$T(f) = \frac{(1 + o(1))\varphi(q)^k N (\log R)^k}{q^{k+1}} J.$$

In particular,

$$\frac{S(f)}{T(f)} = (1 + o(1)) \left(\frac{\theta}{2} - \delta \right) \frac{\sum_{j=1}^{k} I_j}{J}.$$

22.1.1 Exercises

1. (Graham 1978; see also Barban & Vehov 1968 and Motohashi 1974) Let $\mathscr{P}$ be the set of all primes, let $X \ge 2$, and let $\mathscr{A} = \{n : Y < n \le Y + X\}$. Further suppose that $z \ge 2$,

$$P(z) = \prod_{\substack{p\in\mathscr{P}\\ p\le z}} p,$$

and

$$\lambda(k) = \begin{cases} \mu(k)\frac{\log(z/k)}{\log z} & (k \le z),\\ 0 & (k > z).\end{cases}$$

(a) Prove that $|\lambda(k)| \le 1$ for every $k \ge 1$, and that

$$\sum_{k\le z} |\lambda(k)| \ll \frac{z}{\log z}.$$

(b) Prove that, in the notation of Chapter 21,

$$S(\mathcal{A}, \mathscr{P}, z) \le X \sum_{k,l} \frac{\lambda(k)\lambda(l)}{[k,l]} + O\left(\frac{z^2}{(\log z)^2}\right).$$

(c) Prove that if $Q \ge 1$ and $A > 0$, then

$$\sum_{\substack{n\le Q\\ (n,r)=1}} \frac{\mu(n)}{n} \log\frac{Q}{n} = \frac{r}{\varphi(r)} + O\big(\sigma_{-1/2}(r)(\log 2Q)^{-A}\big).$$

(d) Prove that

$$(\log z)^2 \sum_{k,l} \frac{\lambda(k)\lambda(l)}{[k,l]} = \sum_{r\le z} \frac{\mu(r)^2}{\varphi(r)} + O(1).$$

(e) Prove that

$$S(\mathcal{A}, \mathscr{P}, x) \le \frac{X}{\log z} + O\left(\frac{X + z^2}{(\log z)^2}\right).$$

(f) Let $P(z) = \prod_{p\le z} p$. Conclude, in the notation of §3.1, that

$$S\big(X, Y, P(z)\big) \le \frac{2X}{\log X}\left(1 + O\left(\frac{1}{\log X}\right)\right),$$

and that

$$\pi(X+Y) - \pi(Y) \le \frac{2X}{\log X} + O\left(\frac{X}{(\log X)^2}\right).$$

Compare with this with Theorem 3.3 and Corollary 3.4. Thus, although the λ used here are not exactly optimal for the method, they yield the same estimates.

2. Let $\mathscr{P}$ be the set of all odd primes, let $x \ge 2$, $\mathcal{A} = \{p - 2 : p \le x\}$, and $X = \mathrm{li}(x)$. Further suppose that

$$3 \le z \le x^{1/4}(\log x)^{-B}$$

for a suitable positive constant B, and where $P(z)$ and λ are as in the preceding exercise. (Note that 2 is now omitted from $\mathscr{P}$.)

(a) Prove that, in the notation of Chapter 21,

$$S(\mathcal{A}, \mathscr{P}, z) \le X \sum_{k,l} \frac{\lambda(k)\lambda(l)}{\varphi([k,l])} + O\big(X(\log X)^{-2}\big).$$

(b) Prove that if $Q \ge 1$, $r|P(z)$, and $A > 0$, then

$$\sum_{\substack{n \le Q \\ (n,2r)=1}} \frac{\mu(n)}{\varphi(n)} \log \frac{Q}{n} = c \prod_{\substack{p|r \\ p>2}} \frac{p-1}{p-2} + O\big(\sigma_{-1/2}(r)(\log 2Q)^{-A}\big),$$

where

$$c = 2 \prod_{p>2} \left(\frac{p(p-2)}{(p-1)^2} \right)$$

is the twin prime constant.

(c) Prove that

$$(\log z)^2 \sum_{k,l} \frac{\lambda(k)\lambda(l)}{\varphi([k,l])} = c^2 \sum_{\substack{r \le z \\ 2 \nmid r}} \frac{\mu(r)^2}{\varphi_2(r)} + O(1),$$

where

$$\varphi_2(n) = \sum_{m|n} \mu(m)\varphi(n/m),$$

and so, for square-free r,

$$\varphi_2(r) = r \prod_{p|r} (1 - 2/p).$$

(d) Prove that

$$\sum_{\substack{r \le z \\ 2 \nmid r}} \frac{\mu(r)^2}{\varphi_2(r)} = \frac{\log z}{c} + O(1).$$

(e) Prove that

$$S(\mathcal{A}, \mathcal{P}, x) \leq \frac{4cx}{(\log x)^2} + O\left(\frac{x(\log\log x)}{(\log x)^2}\right).$$

(f) Conclude that the number $N_2(x)$ of primes $p \leq x$ such that $p - 2$ is prime satisfies

$$N_2(x) \leq \frac{4cx}{(\log x)^2} + O\left(\frac{x(\log\log x)}{(\log x)^3}\right).$$

Compare with Exercise 21.1.1.2. Again, these λ work quite nicely for another sieving problem of dimension 1.

22.2 The Proof of Maynard's Theorem

The proof of Theorem 22.1 is divided into several stages. Fortunately, the treatments of $S(f)$ and $T(f)$ are similar. Initially we do not assume (22.4) but suppose only that the $\lambda(\boldsymbol{d})$ are general real-valued functions with support satisfying $d_1 \cdots d_k = d \leq R$, $(d, q) = 1$ where q satisfies (22.2), and d square-free. Thus it can be supposed that $(d_i, d_j) = 1$ when $i \neq j$. We begin with the normal diagonalization process. To this end it is useful to define the multiplicative function $\varphi_2(n)$ by

$$\varphi_2(n) = \sum_{m|n} \mu(m)\varphi(n/m),$$

so that

$$\varphi(n) = \sum_{m|n} \varphi_2(m)$$

and, in particular, $\varphi_2(p) = p - 2$ and $\varphi_2(p^t) = (p-1)^2 p^{t-2}$ when $t \geq 2$. When $\varphi_2(n)$ appears below, n will be odd and square-free.

Lemma 22.2 *For $j = 1, \ldots, k$ let*

$$\kappa_j(\boldsymbol{r}) = \mu(r)\varphi_2(r) \sideset{}{^j}\sum_{\substack{\boldsymbol{d} \\ \boldsymbol{r}|\boldsymbol{d}}} \frac{\lambda(\boldsymbol{d})}{\varphi(d)},$$

where $\sum^j$ indicates that the summation variable is a k-tuple, say $\boldsymbol{d}$, which is restricted by $d_j = 1$, and let

$$\kappa(\boldsymbol{r}) = \mu(r)\varphi(r) \sum_{\substack{\boldsymbol{d} \\ \boldsymbol{r}|\boldsymbol{d}}} \frac{\lambda(\boldsymbol{d})}{d}.$$

Then

$$\lambda(\boldsymbol{d}) = \mu(d)\varphi(d) \sideset{}{^j}\sum_{\substack{\boldsymbol{r} \\ \boldsymbol{d}|\boldsymbol{r}}} \frac{\kappa_j(\boldsymbol{r})}{\varphi_2(r)} \tag{22.9}$$

and

$$\lambda(\boldsymbol{d}) = \mu(d)d \sum_{\substack{\boldsymbol{r} \\ \boldsymbol{d}|\boldsymbol{r}}} \frac{\kappa(\boldsymbol{r})}{\varphi(r)}. \tag{22.10}$$

Proof This is Möbius inversion. Consider

$$\sideset{}{^j}\sum_{\substack{\boldsymbol{r} \\ \boldsymbol{d}|\boldsymbol{r}}} \frac{\kappa_j(\boldsymbol{r})}{\varphi_2(r)}.$$

On substituting the definition of κ_j this becomes

$$\sideset{}{^j}\sum_{\substack{\boldsymbol{r} \\ \boldsymbol{d}|\boldsymbol{r}}} \mu(r) \sideset{}{^j}\sum_{\substack{\boldsymbol{s} \\ \boldsymbol{r}|\boldsymbol{s}}} \frac{\lambda(\boldsymbol{s})}{\varphi(s)} = \sideset{}{^j}\sum_{\boldsymbol{s}} \frac{\lambda(\boldsymbol{s})}{\varphi(s)} \sum_{\substack{\boldsymbol{r} \\ \boldsymbol{d}|\boldsymbol{r}|\boldsymbol{s}}} \mu(r).$$

The innermost sum is a sum over $r_1, \ldots, r_{j-1}, r_{j+1}, \ldots, r_k$ with $d_i|r_i|s_i$, and the general term is $\mu(r) = \mu(r_1)\cdots\mu(r_{j-1})\mu(r_{j+1})\cdots\mu(r_k)$. Thus the sum over r_i is $\mu(d_i)\sum_{t_i|s_i/d_i}\mu(t_i) = 0$ unless $s_i = d_i$ in which case it is $\mu(d_i)$. Thus $d_j = 1$ and

$$\sideset{}{^j}\sum_{\substack{\boldsymbol{r} \\ \boldsymbol{d}|\boldsymbol{r}}} \frac{\kappa_j(\boldsymbol{r})}{\varphi_2(r)} = \mu(d)\frac{\lambda(\boldsymbol{d})}{\varphi(d)},$$

which is equivalent to (22.9).

The inversion formula (22.10) follows in the same way. □

At this point we observe that if (22.4) were to hold, then it follows in the same way that

$$\kappa(\boldsymbol{r}) = f\left(\frac{\log r_1}{\log R}, \ldots, \frac{\log r_k}{\log R}\right), \tag{22.11}$$

and then any bound for f predicated on (22.8) with $f \in \mathcal{F}$ will hold for κ also.

Lemma 22.3 *Let*

$$K_j = \max_{\boldsymbol{r}} |\kappa_j(\boldsymbol{r})|, \quad K = \max_{\boldsymbol{r}} |\kappa(\boldsymbol{r})|.$$

Then for any fixed $A > 0$,

$$S_j = \frac{N}{\varphi(q)\log N} \sideset{}{^j}\sum_{\boldsymbol{r}} \frac{\kappa_j(\boldsymbol{r})^2}{\varphi_2(r)} + O\left(\frac{K_j^2\varphi(q)^{k-2}N(\log R)^{k-2}}{q^{k-1}Q}\right)$$

and

$$T = \frac{N}{q}\sum_{\boldsymbol{r}} \frac{\kappa(\boldsymbol{r})^2}{\varphi(\boldsymbol{r})} + O\left(\frac{K^2 N(\log R)^k}{qQ}\right).$$

Proof We set the pattern with S_j. Not only do we need to substitute κ_j for λ in the main term, but we need suitable bounds for the $\lambda(\boldsymbol{d})$ in any error terms which arise. Moreover, we need to do so in terms of κ and κ_j rather than λ.

We square out and invert the order of summation. Thus

$$S_j = \sum_{\substack{\boldsymbol{d},\boldsymbol{e}\\ d_j=e_j=1}} \lambda(\boldsymbol{d})\lambda(\boldsymbol{e}) \sum_{\substack{N<n\le 2N\\ [\boldsymbol{d},\boldsymbol{e}]|n+\boldsymbol{h}\\ n\equiv a \bmod q}} \mathbb{1}_{\mathbb{P}}(n+h_j).$$

We recall that for $\lambda(\boldsymbol{d}) \neq 0$ we have d square-free and $(d,q) = 1$. Therefore $(d_u, d_v) = 1$ when $u \neq v$. Likewise for $\boldsymbol{e}$. Also, if $p|n + h_u$ and $p|n + h_v$, then $p|h_v - h_u$ and this is impossible since $p > \log\log\log N > \max|h_v - h_u|$. Thus, when $u \neq v$, $([d_u, e_u], [d_v, e_v]) = 1$, whence $(d_u, e_v) = 1$. Since $d_j = e_j = 1$ we have $[d_j, e_j] = 1$. Hence in the inner sum we are left with the system of congruences $n \equiv -h_i \pmod{[d_i, e_i]}$ $i \neq j$ and $n \equiv a \pmod q$. Then the innermost sum can be rewritten as

$$\sum_{\substack{N+h_j<p\le 2N+h_j\\ p\equiv h_j-h_i \bmod [d_i,e_i]\ (i\neq j)\\ p\equiv a+h_j \bmod q}} 1\,.$$

By construction, $(a + h_j, q) = 1$ and $(h_j - h_i, de) = 1$ when $i \neq j$. Let

$$m = q\prod_{i=1}^{k}[d_i, e_i], \tag{22.12}$$

$$X = \int_{N+h_j}^{2N+h_j} \frac{dt}{\log t},$$

and

$$E = \sum_{\boldsymbol{d},\boldsymbol{e}}{}^{\star} |\lambda(\boldsymbol{d})\lambda(\boldsymbol{e})| \max_{(b,m)=1} \sup_{x\le 2N+H} \left|\pi(x; m, b) - \frac{\mathrm{li}(x)}{\varphi(m)}\right|,$$

where $\sum^*$ indicates the restrictions $d_j = e_j = 1$ and $(d_u, e_v) = 1$ when $u \neq v$, and $H = \max_j h_j$. Then

$$S_j = X\sum_{\boldsymbol{d},\boldsymbol{e}}{}^{*} \frac{\lambda(\boldsymbol{d})\lambda(\boldsymbol{e})}{\varphi(m)} + O(E).$$

By (22.9), on taking the maximum over $\boldsymbol{d}$ with $d_j = 1$ we have

$$\max_{\boldsymbol{d}, d_j=1} |\lambda(\boldsymbol{d})| \le \max_{\boldsymbol{d}, d_j=1} \varphi(d) \sum_{\substack{\boldsymbol{r} \\ \boldsymbol{d}|\boldsymbol{r} \\ (d,q)=1}}^{\star} \frac{K_j \mu(r)^2}{\varphi_2(r)} = K_j \max_{\boldsymbol{d}} \frac{\varphi(d)}{\varphi_2(d)} \sum_{\substack{\boldsymbol{s} \\ (s,dq)=1}}^{\star} \frac{\mu(s)^2}{\varphi_2(s)}.$$

Thus

$$\max_{\boldsymbol{d}, d_j=1} |\lambda(\boldsymbol{d})| \le K_j \max_{\boldsymbol{d}} \frac{\varphi(d)}{\varphi_2(d)} \prod_{\substack{Q<p\le R \\ p\nmid d}} \left(1 + \frac{1}{p-2}\right)^{k-1} \ll K_j (\log R)^{k-1}.$$

A concomitant argument shows that

$$\max_{\boldsymbol{d}} |\lambda(\boldsymbol{d})| \ll K(\log R)^k. \tag{22.13}$$

Now consider the number of ways that the modulus m/q can arise in E. By (22.12) this is square-free, and so a prime p dividing m/q can divide exactly one of the $[d_i, e_i]$. Since $i \ne j$, there are $k-1$ choices of i, and for any one choice there are three possibilities. These are $p|(d_i, e_i)$, $p|d_i$ and $p \nmid e_i$, and $p \nmid e_i$ and $p|e_i$. Thus there are at most $(3(k-1))^{\omega(m/q)} \le (3k)^{\omega(m)}$ possible $\boldsymbol{d}$, $\boldsymbol{e}$ which give rise to m. Therefore

$$E \ll K_j^2 (\log R)^{2k} \sum_{m \le qR^2} \mu(m)^2 (3k)^{\omega(m)} \max_{(b,m)=1} \sup_{x \le 2N} \left| \pi(x; m, b) - \frac{\mathrm{li}(x)}{\varphi(m)} \right|.$$

Crudely, we have

$$\sum_{m \le qR^2} \mu(m)^2 (3k)^{2\omega(m)} \max_{(b,m)=1} \sup_{x \le 2N} \left| \pi(x; m, b) - \frac{\mathrm{li}(x)}{\varphi(m)} \right|$$
$$\ll \sum_{m \le qR^2} \mu(m)^2 (3k)^{2\omega(m)} N m^{-1} \ll N (\log N)^{(3k)^2}.$$

Thus, by Cauchy's inequality and the form of the Bombieri–Vinogradov theorem for $\pi(x; m, b)$ given by Corollary 20.5, we have

$$E \ll K_j^2 N (\log N)^{-A}.$$

It remains to deal with the main term for S_j, and it is desirable to rid ourselves of the condition that $(d_u, e_v) = 1$ when $u \ne v$. That this is possible without undue effect on the main term is due to the prior sieving resulting from the choice of the residue class a modulo q. Thus any primes p which can potentially divide (d_u, e_v) satisfy $p > Q$.

We have

$$\frac{1}{\varphi([d_i,e_i])} = \frac{\varphi((d_i,e_i))}{\varphi(d_i)\varphi(e_i)}, \quad \varphi((d_i,e_i)) = \sum_{n_i|d_i, n_i|e_i} \varphi_2(n_i),$$

and $\varphi(m) = \varphi(q)\prod_{i\neq j}\varphi([d_i,e_i])$. Hence

$$\frac{1}{\varphi(m)} = \frac{1}{\varphi(q)\varphi(d)\varphi(e)} \sum_{\boldsymbol{n}|\boldsymbol{d},\boldsymbol{n}|\boldsymbol{e}} \varphi_2(n).$$

We substitute this in the main term and invert the order of summation to obtain

$$\frac{X}{\varphi(q)} \sum_{\mathbf{n}}{}^{j} \varphi_2(n) \sum_{\substack{\boldsymbol{d},\boldsymbol{e} \\ \boldsymbol{n}|\boldsymbol{d},\boldsymbol{n}|\boldsymbol{e}}}^{\star} \frac{\lambda(\boldsymbol{d})\lambda(\boldsymbol{e})}{\varphi(d)\varphi(e)}.$$

We now take the first step in dealing with the condition $(d_u, e_v) = 1$ for $u \neq v$. We replace it by

$$\sum_{s_{uv}|d_u, s_{uv}|e_v} \mu(s_{uv}).$$

There are various observations we can make with regard to the s_{uv}. We have $n_u|d_u$. Thus $(d_v, n_u) = 1$. Hence $(s_{uv}, n_u) = 1$. Likewise, $(s_{uv}, n_v) = 1$. Also, when $w \neq v$, $s_{uw}|e_w$ and $(e_v, e_w) = 1$. Hence $(s_{uv}, s_{uw}) = 1$. Likewise, $(s_{uv}, s_{wv}) = 1$ when $w \neq u$, and so in summary

$$(s_{uv}, n_u) = 1, \ (s_{uv}, n_v) = 1, \ (s_{uv}, s_{uw}) = 1, \ (s_{uv}, s_{wv}) = 1. \tag{22.14}$$

Thus

$$\sum_{\mathbf{n}}{}^{j} \varphi_2(n) \sum_{\substack{\boldsymbol{d},\boldsymbol{e} \\ \boldsymbol{n}|\boldsymbol{d},\boldsymbol{n}|\boldsymbol{e}}}^{\star} \frac{\lambda(\boldsymbol{d})\lambda(\boldsymbol{e})}{\varphi(d)\varphi(e)}$$

$$= \sum_{\mathbf{n}}{}^{j} \varphi_2(n) \sum_{\substack{s_{uv} \\ u\neq v}}^{\dagger} \prod_{u\neq v} \mu(s_{uv}) \left(\sum_{\substack{\boldsymbol{d} \\ \boldsymbol{n}|\boldsymbol{d} \\ s_{uv}|d_u}}^{j} \frac{\lambda(\boldsymbol{d})}{\varphi(d)} \right) \left(\sum_{\substack{\boldsymbol{e} \\ \boldsymbol{n}|\boldsymbol{e} \\ s_{uv}|e_v}}^{j} \frac{\lambda(\boldsymbol{e})}{\varphi(e)} \right),$$

where $\sum^{\dagger}$ indicates that (22.14) holds. We now substitute the κ_j, defined in Lemma 22.2, for the λ. Thus the above becomes

$$\sum_{\mathbf{n}}{}^{j} \frac{1}{\varphi_2(n)} \sum_{\substack{s_{uv} \\ u\neq v}}^{\dagger} \left(\prod_{u\neq v} \frac{\mu(s_{uv})}{\varphi_2(s_{uv})^2} \right) \kappa_j(\boldsymbol{a})\kappa_j(\boldsymbol{b}),$$

where $\boldsymbol{a} = (a_1, \ldots, a_k)$, $\boldsymbol{b} = (b_1, \ldots, b_k)$, and

$$a_u = n_u \prod_{\substack{v \\ v\neq u}} s_{uv}, \quad b_v = n_v \prod_{\substack{u \\ u\neq v}} s_{uv}.$$

In particular, $a = b = ns$ where $s = \prod_{u \neq v} s_{uv}$. Thus the main term is

$$\frac{X}{\varphi(q)} \sum_{\mathbf{n}}{}^{j} \frac{1}{\varphi_2(n)} \sum_{\substack{s_{uv} \\ u \neq v}}{}^{\dagger} \frac{\mu(s)}{\varphi_2(s)^2} \kappa_j(\boldsymbol{a}) \kappa_j(\boldsymbol{b}).$$

Since $n_j = 1$, the terms with $s > 1$ contribute

$$\ll \frac{K_j^2 N}{\varphi(q) \log N} \sum_{\substack{n \leq R \\ (n,q)=1}} \frac{d_{k-1}(n) \mu(n)^2}{\varphi_2(n)} \sum_{\substack{s>1 \\ (s,q)=1}} \frac{d_{k(k-1)}(s) \mu(s)^2}{\varphi_2(s)^2}.$$

The inner sum is

$$\ll -1 + \prod_{p>Q} \left(1 + \frac{k(k-1)}{(p-2)^2}\right) \ll \frac{1}{Q \log Q},$$

and the sum over n is

$$\ll \prod_{Q<p\leq R} (1 + (k-1)/(p-2)) \ll (\varphi(q)(\log R)/q)^{k-1}.$$

Thus the total contribution from the terms with $s > 1$ is

$$\frac{K_j^2 \varphi(q)^{k-2} N (\log R)^{k-2}}{q^{k-1} Q}.$$

For the remaining terms we have $a = b = n$. Thus they give

$$\frac{X}{\varphi(q)} \sum_{\mathbf{n}}{}^{j} \frac{\kappa_j(\boldsymbol{n})^2}{\varphi_2(n)}.$$

We recall that

$$X = \int_{N+h_j}^{2N+h_j} \frac{dt}{\log t} = \frac{N}{\log N} + O\left(\frac{N}{(\log N)^2}\right).$$

Moreover,

$$\begin{aligned} \frac{1}{\varphi(q)} \sum_{\mathbf{n}}{}^{j} \frac{\kappa_j(\boldsymbol{n})^2}{\varphi_2(n)} &\ll \frac{K_j^2}{\varphi(q)} \prod_{Q<p\leq R} \left(1 + \frac{1}{p-2}\right)^{k-1} \\ &\ll \frac{K_j^2 (\varphi(q))^{k-2} (\log R)^{k-1}}{q^{k-1}}. \end{aligned}$$

This completes the proof of the approximation for S_j.

The proof of the approximation for T is essentially the same, except that we do not use Bombieri's theorem and we do not have the restriction that $d_j = 1$ to contend with. Thus on the initial application of the Chinese remainder theorem, the main term is $\frac{N}{m}$, and the error term is $O(1)$. By (22.13) we see that the total contribution arising from this error is

$$\ll K^2R^2(\log R)^{4k-2},$$

which is acceptable. Then, just as the function φ now plays the role that φ_2 played earlier, so the κ_j is replaced by its understudy κ. Then the process of replacing λ by κ is identical, as is the elimination of the restriction $(d_u, e_v) = 1$.

□

The functions κ_j and κ introduced in Lemma 22.2 are clearly related to each other, as can be seen explicitly by (22.9) and (22.10). Thus when we insert (22.9) into the definition of κ_j and invert the order of summation, we obtain (when $r_j = 1$)

$$\kappa_j(\boldsymbol{r}) = \mu(r)\varphi_2(r)\sum_{\substack{\boldsymbol{s}\\ \boldsymbol{r}|\boldsymbol{s}}}\frac{\kappa(\boldsymbol{s})}{\varphi(s)}\sum_{\substack{\boldsymbol{d}\\ \boldsymbol{r}|\boldsymbol{d}|\boldsymbol{s}}}^{j}\frac{\mu(d)d}{\varphi(d)}.$$

Write $e_i = d_i/r_i$ and $t_i = s_i/r_i$. Then the inner sum is

$$\frac{\mu(r)r}{\varphi(r)}\sum_{\substack{\boldsymbol{e}\\ \boldsymbol{e}|\boldsymbol{t}\\ e_j=1}}\frac{\mu(e)e}{\varphi(e)} = \frac{\mu(r)r\mu(t/t_j)}{\varphi(r)\varphi(t/t_j)} = \frac{r\mu(s/s_j)}{\varphi(s/s_j)}.$$

On using the notation $\boldsymbol{rt}$ for $(r_1t_1, \ldots, r_kt_k)$ we find that

$$\kappa_j(\boldsymbol{r}) = \frac{r\varphi_2(r)}{\varphi(r)^2}\sum_{\boldsymbol{t}}\kappa(\boldsymbol{rt})\frac{\mu(t)\varphi(t_j)\mu(t_j)}{\varphi(t)^2}.$$

The terms with $t > t_j$ contribute

$$\ll K\sum_{\substack{t_j\le R\\ (t_j,q)=1}}\frac{\mu(t_j)^2}{\varphi(t_j)}\sum_{\substack{n>1\\ (n,q)=1}}\frac{(k-1)^{\omega(n)}\mu(n)^2}{\varphi(n)^2}.$$

We have

$$\sum_{\substack{t_j\le R\\ (t_j,q)=1}}\frac{\mu(t_j)^2}{\varphi(t_j)} \ll \left(\prod_{Q<p\le R}\frac{p}{p-1}\right) \ll \frac{\varphi(q)}{q}\log R$$

and

$$\left(-1+\prod_{Q>p}\left(1+\frac{k-1}{(p-1)^2}\right)\right) \ll Q^{-1}.$$

Since also

$$\frac{r\varphi_2(r)^2}{\varphi(r)} = 1 + O(1/Q),$$

it follows that, when $r_j = 1$,

$$\kappa_j(\boldsymbol{r}) = \sum_{t_j} \frac{\kappa(\boldsymbol{r}')}{\varphi(t_j)} + O\left(\frac{K\varphi(q)\log R}{qQ}\right), \tag{22.15}$$

where $\boldsymbol{r}' = (r_1, \ldots, r_{j-1}, t_j, r_{j+1}, \ldots, r_k)$.

Having come this far, we should take stock. The ultimate aim is to maximize the ratio

$$\frac{\sum_{j=1}^{k} \sum_{\boldsymbol{r}}^{j} \frac{\kappa_j(\boldsymbol{r})^2}{\varphi_2(r)}}{\sum_{\boldsymbol{r}} \frac{\kappa(\boldsymbol{r})^2}{\varphi(r)}}.$$

We henceforward make the assumption that (22.11) holds with $f \in \mathscr{F}$ which, by (22.10), gives (22.4).

The final step of the proof of Theorem 22.1 is to obtain smooth approximations to the main terms in Lemma 22.2. We have standard methods of carrying this out when $k = 1$, i.e., $\boldsymbol{r} = r_1$. We adopt the simple expedient of establishing a suitable one-dimensional approximation and then applying it k times.

Suppose that $g\colon [0,1] \to \mathbb{R}$. Then we say that g is *l-piecewise absolutely continuous on* $[0,1]$ when, associated with g, there is a partition $a_0 = 0 < a_1 < \cdots < a_l = 1$ of $[0,1]$ such that for $1 \le j \le l$:

1. $g_+(a_{j-1}) = \lim_{x\to a_{j-1}+} g(x)$ and $g_-(a_j) = \lim_{x\to a_j-} g(x)$ both exist, and
2. g is absolutely continuous on $[a_{j-1}, a_j]$ when we replace $g(a_{j-1})$ and $g(a_j)$ by $g_+(a_{j-1})$ and $g_-(a_j)$, respectively.

We define $\mathscr{G}(l, G)$ to be the class of l-piecewise absolutely continuous functions g on $[0,1]$ such that

$$\sup_{v\in[0,1]} |g(v)| + \int_0^1 |g'(v)|dv \le G.$$

We observe in passing that, in practice, it suffices for our application that g' is continuous except for at most one x in $[0,1]$ where g and g' have jump discontinuities.

Lemma 22.4 *Suppose that* $\eta\colon \mathbb{N} \to \mathbb{R}$ *is multiplicative with its support on the square-free numbers, that* $0 \le \eta(p) \le 2$*, and that there is a constant* C *such that whenever* $p > C$ *we have*

$$\left|\eta(p) - \frac{1}{p}\right| \le \frac{C}{p^2}.$$

Suppose also that $g \in \mathscr{G}(l, G)$ *and* $m \in \mathbb{N}$*. Then*

$$\sum_{\substack{n \le x\\ (n,m)=1}} \eta(n) g\left(\frac{\log n}{\log x}\right)$$

$$= \mathfrak{S}(m) \int_0^1 g(v)\,dv \log x + O\left(lG\left(1 + \sum_{p|m} \frac{\log p}{p}\right)\prod_{p|m}\left(1 + \frac{1}{p}\right)\right),$$

where

$$\mathfrak{S}(m) = \frac{\varphi(m)}{m} \prod_{p\nmid m} (1 + \eta(p))\left(1 - \frac{1}{p}\right).$$

We also have

$$\mathfrak{S}(m) \ll \frac{\varphi(m)}{m}.$$

In order to make a comparison with the main term, which is of size

$$\asymp \frac{\varphi(m)}{m} \log x \int_0^1 g(v)dv,$$

it is useful to observe that the error term is

$$\ll G\frac{\varphi(m)}{m}(\log\log 3m)^3.$$

Proof We begin with the case when g is identically 1. Also, we may suppose that $\eta(p) = 0$ when $p|m$. Let ρ be the multiplicative function with $\rho(p) = \eta(p) - 1/p$, $\rho(p^2) = -\eta(p)/p$, $\rho(p^t) = 0$ $(t \ge 3)$, and let $\nu = 0$ or 1. Then

$$\sum_{u|n} \frac{\rho(n/u)}{u} = \eta(n)$$

and

$$\sum_{l \le y} (\log 2l)^\nu |\rho(l)| \ll \sum_{\substack{rst^2 \le y\\ r|m,(st,m)=1}} (\log 2rst)^\nu \frac{\mu(rst)^2}{rs^2t^2} C^{\omega(s)} \sum_{u|t} \frac{C^{\omega(u)}}{u}$$

$$\ll \left(1 + \sum_{p|m} \frac{\log p}{p}\right)\prod_{p|m}\left(1 + \frac{1}{p}\right).$$

Also,

$$\sum_{x<l\le y} |\rho(l)| \ll \frac{1}{\log x} \sum_l (\log l)|\rho(l)|.$$

Therefore

$$\sum_{n\le x} \eta(n) = \sum_{\substack{u,v\\ uv\le x}} \frac{\rho(v)}{u} = \sum_{v\le x} \rho(v)\Big(\log\frac{x}{v} + O(1)\Big)$$

$$= \mathfrak{S}(m)\log x + O\left(\left(1 + \sum_{p|m} \frac{\log p}{p}\right)\prod_{p|m}\left(1+\frac{1}{p}\right)\right).$$

Now we apply this to general $g \in \mathscr{G}(l, G)$. Let

$$E(x) = \sum_{n\le x} \eta(n) - \mathfrak{S}(m)\log x,$$

and choose a_j as provided by the definition of $\mathscr{G}(l, G)$. When $x^{a_{j-1}} < n \le x^{a_j}$ we have

$$g\left(\frac{\log n}{\log x}\right) = g_-(a_j) - \int_{\frac{\log n}{\log x}}^{a_j} g'(v)\,dv$$

except possibly when $n = x^{a_j}$, in which case the two sides differ by $\ll G$. We multiply by $\eta(n)$, sum over the $n \in (x^{a_{j-1}}, x^{a_j}]$, interchange the order of summation and integration, and apply the formula for E to obtain

$$\big(\mathfrak{S}(m)(\log x)(a_j - a_{j-1}) + E(x^{a_j}) - E(x^{a_{j-1}})\big)g_-(a_j)$$
$$- \int_{a_{j-1}}^{a_j} \big(\mathfrak{S}(m)(\log x)(v - a_{j-1}) + E(x^v) - E(x^{a_{j-1}})\big)g'(v)\,dv + O(G).$$

We integrate the main term, i.e., the parts not containing E, by parts to obtain

$$\int_{a_{j-1}}^{a_j} \mathfrak{S}(m)(\log x)g(v)\,dv$$

which on summing over j gives the desired main term. We insert the bound for E given by the first part of the proof and sum over j. This completes the proof of the lemma. □

We are now in a position to complete the proof of Theorem 22.1. We make the choice (22.11) for some f in $\mathscr{F}$. To simplify some of the formulæ we then extend the definition of f to $[0, 1]^k$ by taking f to be 0 outside $\mathscr{R}$. Again we concentrate on S_j rather than T. We recall that $\kappa_j(\boldsymbol{r}) = 0$ unless $r_j = 1$, $(r, q) = 1$ and r is square-free, in which case, by (22.15) and (22.11), we have

$$\kappa_j(\boldsymbol{r}) = \sum_{t_j} \frac{\mu(t_j)^2}{\varphi(t_j)} f\left(\frac{\log r_1}{\log R}, \ldots, \frac{\log r_{j-1}}{\log R}, \frac{\log t_j}{\log R}, \frac{\log r_{j+1}}{\log R}, \ldots, \frac{\log r_k}{\log R}\right) + O\left(\frac{F\varphi(q)\log R}{qQ}\right),$$

where $\boldsymbol{r}' = (r_1, \ldots, r_{j-1}t_j, r_{j+1}, \ldots, r_k)$. Thus

$$K_j \ll F\frac{\varphi(q)}{q}\log R\,.$$

Moreover, by Lemma 22.4, with $\eta(p) = 1/p$ and $m = qr$ we have

$$\kappa_j(\boldsymbol{r}) = (\log R)\frac{\varphi(qr)}{qr} f_j(\boldsymbol{r}) + O\left(\frac{F\varphi(q)\log R}{qQ}\right),$$

where

$$f_j(\boldsymbol{r}) = \int_0^1 f\left(\frac{\log r_1}{\log R}, \ldots, \frac{\log r_{j-1}}{\log R}, u_j, \frac{\log r_{j+1}}{\log R}, \ldots, \frac{\log r_k}{\log R}\right) du_j\,.$$

This holds when $r_j = 1$, $(r, q) = 1$ and r is square-free, and otherwise $\kappa_j(\boldsymbol{r}) = 0$. Thus, by Lemma 22.3,

$$S_j = \frac{\varphi(q)N(\log R)^2}{q^2\log N} \sideset{}{^j}\sum_{\substack{\boldsymbol{r}\\(r,q)=1}} \frac{\mu(r)^2\varphi(r)^2}{\varphi_2(r)r^2} f_j(\boldsymbol{r})^2 + O\left(\frac{F^2\varphi(q)^k N(\log R)^k}{q^{k+1}Q}\right).$$

The general arithmetical factor in the main term in the sum can be rewritten as

$$\prod_{i=1}^{k} \frac{\mu(r_i)^2\varphi(r_i)^2}{\varphi_2(r_i)r_i^2}$$

provided that the sum over $\boldsymbol{r}$ is restricted to $\boldsymbol{r}$ with $(r_u, r_v) = 1$ when $u \neq v$. However, if $(r_u, r_v) > 1$, then there is a prime $p > Q$ such that $p|r_u$ and $p|r_v$. Therefore when we remove the condition $(r_u, r_v) = 1$, the total error in so doing is

$$\ll \frac{F^2\varphi(q)N(\log R)^2}{q^2\log N} \sum_{p>Q} \frac{\varphi(p)^4}{\varphi_2(p)^2p^4} \left(\sum_{\substack{n<R\\(n,q)=1}} \frac{\mu(n)^2\varphi(n)^2}{\varphi_2(n)n^2}\right)^{k-1}$$
$$\ll \frac{F^2\varphi(q)^k N(\log R)^k}{q^{k+1}Q}\,.$$

Thus the sum in the main term can be replaced by

$$\sum_{\substack{\boldsymbol{r}\\(r,q)=1}}^{j} f_j(\boldsymbol{r})^2 \prod_{i=1}^{k} \frac{\mu(r_i)^2 \varphi(r_i)^2}{\varphi_2(r_i) r_i^2}.$$

Here we apply Lemma 22.4 to each variable r_i in turn, i.e., $k-1$ times, with

$$\eta(p) = \frac{(p-1)^2}{(p-2)p^2} = \frac{1}{p} + \frac{1}{p^2(p-2)}$$

and $m = q$. In each case we have

$$\mathfrak{S}(q) = 1 + O(1/Q).$$

Thus

$$S_j = \frac{\varphi(q)^k N (\log R)^{k+1}}{q^{k+1} \log N} I_j + O\left(\frac{F^2 \varphi(q)^k N (\log R)^k}{q^{k+1} Q} \right),$$

where I_j is as in Theorem 22.1. This gives the first part of that theorem. The second part follows in the same way.

22.2.1 Exercises

1. Prove (22.10).
2. Prove (22.13).
3. Prove the last part of Lemma 22.3.

22.3 Consequences of Maynard's Theorem

Theorem 22.5 (Maynard 2015) *Suppose that when $k \geq 2$, we take $f \in \mathcal{F}$ and then $I_j = I_j(f)$ and $J = J(f)$ are as in Theorem 22.1. Let*

$$\varsigma = \sup_{f \in \mathcal{F}} \frac{\sum_{j=1}^{k} I_j(f)}{J(f)}.$$

Then for k sufficiently large,

$$\varsigma > \log k - \log\log k - 1.$$

Corollary 22.6 (Zhang 2014) *There are bounded gaps in the sequence of primes.*

Corollary 22.7 (Maynard 2015, Tao (unpublished)) *For each $m \in \mathbb{N}$ we have*

$$\liminf_{n\to\infty} (p_{n+m} - p_n) \ll m^2 e^{4m}.$$

Corollary 22.8 (Maynard 2015) *Let $m \in \mathbb{N}$ and let $\mathscr{G} = \{g_1, \ldots, g_l\}$ be a set of l distinct non-negative integers. Let $M(m, l, \mathscr{G})$ be the number of admissible m-tuples contained in $\mathscr{G}$ and let $N(m, l, \mathscr{G})$ be the number of admissible m-tuples $\boldsymbol{h}$ contained in $\mathscr{G}$ such that there are infinitely many n for which each member of the m-tuple $n + \boldsymbol{h}$ is prime. Then for $l > l_0(m)$,*

$$l^m \geq M(m, l, \mathscr{G}) \gg_m l^m$$

and

$$\frac{N(m, l, \mathscr{G})}{M(m, l, \mathscr{G})} \gg_m 1.$$

The conjecture of de Polignac's (1849) asserts that every even integer is the difference of infinitely many pairs of primes. That the conjecture holds for a positive proportion of all even integers follows on taking $m = 2$ and $g_j = 2j - 2$ in Corollary 22.7, for then the number of solutions of $g_{j_2} - g_{j_1} = 2d$ is at most l and so there must be $\gg l^2/l = 1$ different differences $g_{j_2} - g_{j_1}$ arising from the admissible pairs counted by $N(2, l, \mathscr{G})$.

Corollary 22.9 *There is an infinite subset $\mathbb{D}$ of $\mathbb{N}$ with positive lower asymptotic density such that for each $d \in \mathbb{D}$ there are infinitely many pairs of primes p_1, p_2 such that $p_2 - p_1 = d$.*

Proof of Theorem 22.5 We have to construct a suitable f. For simplicity of construction we will take f to be essentially a product of single variable functions. That is, we separate the variables. In part, this is motivated by putting most of the mass of f near the axes. This has the effect of minimizing the importance of the boundary condition $t_1 + \cdots + t_k \leq 1$. It also means that f is symmetric, which one might suspect would be true for an extremal f.

The function

$$\nu\colon (1, \infty] \to \mathbb{R} : \nu(\alpha) = \alpha/\log\alpha$$

has its minimum at $\alpha = e$ and is increasing for $\alpha > e$. Thus for $k \geq 2$ we have

$$\frac{k}{\log k} \geq e,$$

and

$$x = \frac{k/\log k}{\log(k/\log k)} \tag{22.16}$$

satisfies $x \geq e > 1$. Hence we can define ξ to be the positive solution to

$$1 + \xi x = e^{\xi}. \tag{22.17}$$

Then

$$\nu(e^{\xi}) > x = \nu(k/\log k),$$

and so by monotonicity,

$$\log k - \log\log k < \xi.$$

Also, for large k,

$$\xi = \log \xi + \log x + \log(1 + 1/(x\xi)) \sim \log \xi + \log k - 2\log\log k$$

and so

$$\xi < \log k.$$

Let $g\colon [0, \infty) \to \mathbb{R}$ be defined by

$$g(y) = \begin{cases} \frac{1}{1+\xi y} & 0 \le y \le x, \\ 0 & x < y. \end{cases}$$

We need to compute various integrals, which we denote by $\alpha, \beta, \gamma, \tau$, as follows:

$$\alpha = \int_0^\infty g(y)\, dy = 1, \tag{22.18}$$

$$\beta = \int_0^\infty g(y)^2\, dy = \frac{1}{\xi} - \frac{1}{\xi e^{\xi}}, \tag{22.19}$$

$$\gamma = \int_0^\infty y g(y)^2\, dy = \frac{1}{\xi} - \frac{1}{\xi^2} + \frac{1}{\xi^2 e^{\xi}}, \tag{22.20}$$

$$\tau = \int_0^\infty y^2 g(y)^2\, dy = \frac{x}{\xi^2} - \frac{2}{\xi^2} + \frac{1}{\xi^3} - \frac{1}{\xi^3 e^{\xi}}. \tag{22.21}$$

We now take

$$f(\boldsymbol{t}) = \begin{cases} \prod_{i=1}^k g(kt_i) & \boldsymbol{t} \in \mathscr{R}, \\ 0 & \boldsymbol{t} \notin \mathscr{R}. \end{cases}$$

Since f is symmetric we have $I_j(f) = I_k(f)$ for every $j \le k$. Thus

$$\varsigma \ge \frac{kI_k(f)}{J(f)} \tag{22.22}$$

and we now proceed to estimate $I_k(f)$ and $J(f)$. Since we are concerned with only a lower bound for ς, lower and upper bounds for $I_k(f)$ and J, respectively, will suffice. An upper bound for $J(f)$ is easy. We have

$$J(f) \le \int_{[0,\infty)^k} \prod_{i=1}^k g(kt_i)^2\, d\boldsymbol{t} = k^{-k}\beta^k. \tag{22.23}$$

Thus we can concentrate on $I_k(f)$. Let $\mathcal{S}$ denote the set of $(k-1)$-tuples $(y_1, \ldots, y_{k-1})$ with $y_i \geq 0$ and $y_1 + \cdots + y_{k-1} \leq k - x$. Then we have

$$kI_k(f) = k \int_{\mathcal{R}_{k-1}} \left(\prod_{i=1}^{k-1} g(kt_i)^2 \right) \left(\int_0^{1-t_1-\cdots-t_{k-1}} g(kt_k)\, dt_k \right)^2 dt_1 \cdots dt_{k-1}$$
$$\geq k^{-k} \alpha^2 \int_{\mathcal{S}} \prod_{i=1}^{k-1} g(y_i)^2\, d\boldsymbol{y},$$

and so

$$kI_k(f) \geq k^{-k} \alpha^2 \beta^{k-1} - E, \tag{22.24}$$

where

$$E = k^{-k} \alpha^2 \int_{\mathcal{S}^*} \prod_{i=1}^{k-1} g(y_i)^2\, d\boldsymbol{y}$$

and

$$\mathcal{S}^* = [0, \infty)^{k-1} \setminus \mathcal{S}.$$

Let

$$\sigma = \gamma/\beta = \frac{1 - \xi^{-1} + \xi^{-1} e^{-\xi}}{1 - e^{-\xi}} = 1 - \frac{1}{\xi} + \frac{1}{e^{\xi} - 1}. \tag{22.25}$$

The condition $\boldsymbol{y} \in \mathcal{S}^*$ is equivalent to $y_1 + \cdots + y_{k-1} > k - x$ and this in turn is equivalent to

$$\frac{y_1 + \cdots + y_{k-1}}{k-1} - \sigma > \frac{k - x - \sigma(k-1)}{k-1} = \frac{(1-\sigma)(k-1) - x + 1}{k-1}.$$

For k sufficiently large we have

$$(1-\sigma)(k-1) - x + 1 > 0$$

and

$$1 - \sigma - \frac{x-1}{k-1} = \xi^{-1} + O(\xi^{-2}),$$

so that

$$\zeta = \left(1 - \sigma - \frac{x-1}{k-1}\right)^{-1} = \xi + O(1). \tag{22.26}$$

In particular, if $\boldsymbol{y} \in \mathcal{S}^*$, then

$$\left(\frac{y_1 + \cdots + y_{k-1}}{k-1} - \sigma\right)^2 \zeta^2 \geq 1.$$

Hence

$$E \le k^{-k}\alpha^2\zeta^2 \int_{[0,\infty)^{k-1}} \left(\frac{y_1 + \cdots + y_{k-1}}{k-1} - \sigma\right)^2 \prod_{i=1}^{k-1} g(y_i)^2\, d\mathbf{y}.$$

We now square out the expression

$$\left(\frac{y_1 + \cdots + y_{k-1}}{k-1} - \sigma\right)^2 = \frac{1}{(k-1)^2}\sum_{i=1}^{k-1} y_i^2 + \frac{2}{(k-1)^2}\sum_{1\le i<j\le k-1} y_i y_j - \frac{2\sigma}{k-1}\sum_{i=1}^{k-1} y_i + \sigma^2,$$

and evaluate the various integrals with reference to the integrals evaluated above. Thus

$$E \le k^{-k}\alpha^2\zeta^2 \left(\frac{1}{k-1}\tau\beta^{k-2} + \frac{k-2}{k-1}\gamma^2\beta^{k-3} - 2\sigma\gamma\beta^{k-2} + \sigma^2\beta^{k-1}\right).$$

By the definition of σ, (22.25),

$$E \le k^{-k}\alpha^2\zeta^2\beta^{k-3}\frac{\tau\beta - \gamma^2}{k-1} < k^{-k}\alpha^2\zeta^2\beta^{k-2}\frac{\tau}{k-1}.$$

Thus by (22.22), (22.23), and (22.24),

$$\varsigma > \beta^{-1}\left(1 - \frac{\zeta^2\tau}{\beta(k-1)}\right).$$

By (22.16) and (22.17),

$$\log k - \log\log k < \xi = \log k - \log\log k + O(1),$$

by (22.19), (22.26), and (22.21),

$$\begin{aligned} \beta^{-1} &= \xi + O(\xi k^{-1}\log k), \\ \zeta^2 &= \xi^2 + O(\xi), \\ \tau &= x\xi^{-2} + O(\xi^{-2}), \end{aligned}$$

respectively, and we have

$$\frac{1}{k-1} = \frac{1}{k} + O(k^{-2}).$$

Thus

$$\frac{\zeta^2\tau}{\beta(k-1)} = (\xi + O(1))xk^{-1} = \frac{\xi + O(1)}{(\log k)\log(k/\log k)}$$
$$= \frac{1}{\log k} + O\left(\frac{1}{(\log k)^2}\right).$$

Hence

$$\varsigma > \xi\left(1 + O\left(\frac{\log k}{k}\right)\right)\left(1 - \frac{1}{\log k} + O\left((\log k)^{-2}\right)\right) > \log k - \log\log k - 1 \tag{22.27}$$

if k is sufficiently large. □

Proof of Corollary 22.6 By Theorem 18.18, for every large k there exist admissible k-tuples. Then by (22.6), (22.7), and Theorems 22.1 and 22.5 we have

$$\sum_{N\le n\le 2N}\left(\sum_{j=1}^{k} \mathbb{1}_{\mathbb{P}}(n+h_j) - \varrho\right)\left(\sum_{\substack{q\le R\\ q\mid Z(n;\boldsymbol{h})}} \lambda(q)\right)^2 > 0,$$

where

$$\varrho = \left(\frac{\theta}{2} - \delta\right)\varsigma \tag{22.28}$$

and δ is arbitrarily small, θ is the level of distribution of the primes in arithmetic progressions, and ς is as in Theorem 22.5. Since we know that $\theta \ge \frac{1}{2}$, and ς is large for large k, it follows that for sufficiently large k there are admissible k-tuples $\boldsymbol{h}$ for which there are arbitrarily large N such that for some n with $N \le n \le 2N$ the k-tuple $n + \boldsymbol{h}$ contains at least two primes. This establishes the first corollary. □

Proof of Corollary 22.7 Let C be a constant chosen so that for every $m \in \mathbb{N}$ we have

$$\frac{Cme^{4m}}{4m + \log m + \log C} > e^{2+4m}.$$

Hence for $k \ge \max(3, Cme^{4m})$ we have

$$\frac{k}{\log k} \ge e^{2+4m}$$

and so

$$\log k - \log\log k - 1 > 4m + 1.$$

Thus if k is large enough, then

$$\left(\frac{1}{4} - \frac{1}{k}\right)(\log k - \log\log k - 1) > m.$$

Taking the level of distribution θ to be $\frac{1}{2}$ and choosing $\delta = \frac{1}{k}$, we see by (22.27) and (22.28) that $\varsigma > m$, and so every admissible k-tuple $\boldsymbol{h}$ has the property that there are infinitely many n such that the k-tuple $n + \boldsymbol{h}$ contains at least m primes. By Theorem 18.18 there is an admissible k-tuple of diameter $\ll k \log k \ll m^2 e^{4m}$. □

Proof of Corollary 22.8 Let $k = \lceil \max(3, Cme^{4m}) \rceil$ with C suitably large as in the proof of Corollary 22.7 and let $\boldsymbol{h}$ be an admissible k-tuple. By considering all possible m-tuples $\boldsymbol{h}' = (h_1', \dots, h_m')$ that are subsets of $\boldsymbol{h}$ we see that at least one has the property that there are infinitely many n such that $n + h_1', \dots, n + h_m'$ are simultaneously prime, i.e., the prime m-tuple conjecture holds for this m-tuple.

Starting from $\mathscr{G}$, we construct a subset $\mathscr{G}'$ by successively removing elements from $\mathscr{G}$. Given a prime p and a finite set $\mathscr{L}$ of integers we can construct a subset as follows. Let $\mathscr{L}(p; h) = \{n \in \mathscr{L} : n \equiv h \pmod{p}\}$ and $L(p; h) = \operatorname{card} \mathscr{L}(p; h)$. Choose an h for which $L(p; h)$ is minimal and take $\mathscr{L}' = \mathscr{L} \setminus \mathscr{L}(p; h)$. Then $\operatorname{card} \mathscr{L}' \ge (1 - 1/p) \operatorname{card} \mathscr{L}$. We apply this operation successively to $\mathscr{G}$ for $p \le k$, giving a subset $\mathscr{G}'$ that satisfies

$$\operatorname{card} \mathscr{G}' \ge \operatorname{card} \mathscr{G} \prod_{p \le k} \left(1 - \frac{1}{p}\right) \gg_m l.$$

Thus on taking l to be sufficiently large we have $s = \operatorname{card} \mathscr{G}' > k$. Every subset $\boldsymbol{h}$ of $\mathscr{G}'$ of cardinality k is an admissible set since it omits a residue class modulo p for every $p \le k$. There are $\binom{s}{k}$ such $\boldsymbol{h}$ and, from above, each one contains at least one m-tuple $\boldsymbol{h}'$ for which the prime m-tuples conjecture holds. Subsets $\boldsymbol{b}$ of $\mathscr{G}'$ of cardinality k that contain $\boldsymbol{h}'$ are exactly those in which the $k - m$ remaining elements of $\boldsymbol{b}$ are chosen at random from the $s - m$ remaining elements of $\mathscr{G}'$. Thus there are precisely $\binom{s-m}{k-m}$ such $\boldsymbol{b}$. Hence there are at least

$$\frac{\binom{s}{k}}{\binom{s-m}{k-m}} = \frac{(s-m+1)\cdots(s-1)s}{(k-m+1)\cdots(k-1)k} \gg_m s^m \gg_m l^m$$

admissible subsets of $\mathscr{G}$ of cardinality m that satisfy the prime m-tuple conjecture. On the other hand, there are $\binom{l}{m} \le l^m$ subsets $\boldsymbol{h}$ of $\mathscr{G}$ of cardinality m, and this completes the proof of Corollary 22.8. □

22.3.1 Exercises

1. Suppose that $k \ge 2$. Let $\mathscr{R}_k \subset [0,1]^k$ be defined by $\mathscr{R}_k = \{\boldsymbol{t}\colon t_i \ge 0, t_1 + \cdots + t_k \le 1\}$, and let $m \in \mathbb{N}$ and $f(\boldsymbol{t}) = (1 - t_1 - \cdots - t_k)^m$. Given $(t_1, \ldots, t_{j-1}, t_{j+1}, \ldots, t_k) \in [0,1]^{k-1}$ with $t_1 + \cdots + t_{j-1} + t_{j+1} + \cdots + t_k \le 1$, let A_j denote the interval $[0, 1 - t_1 - \cdots - t_{j-1} - t_{j+1} - \cdots - t_k]$ (and take it to be the empty set otherwise), and define

$$I_j(f) = \int_0^1 \cdots \int_0^1 \left(\int_{A_j} f(\boldsymbol{t})\, dt_j \right)^2 dt_1 \cdots dt_{j-1} dt_{j+1} \cdots dt_k$$

and

$$J(f) = \int_{\mathscr{R}_k} f(\boldsymbol{t})^2 \, d\boldsymbol{t} \, .$$

(a) Prove that $\displaystyle\sum_{j=1}^{k} I_j(f) = \frac{k(2m+2)!}{(2m+1+k)!(m+1)^2}$.

(b) Prove that $J(f) = \dfrac{(2m)!}{(2m+k)!}$.

(c) Prove that

$$\frac{\sum_{j=1}^{k} I_j(f)}{J(f)} = 4\left(1 - \frac{1}{2m+2}\right)\left(1 - \frac{2m+1}{2m+1+k}\right).$$

(d) (Goldston, Pintz & Yıldırım 2009) Prove that if the level θ of distribution of the primes satisfies $\theta > \frac{1}{2}$, then there are infinitely many bounded gaps in the sequence of primes.

2. Let $\mathscr{R}_k$ be as in the previous exercise. For $\boldsymbol{t} \in \mathscr{R}_k$ let $\alpha_k(\boldsymbol{t}) = t_1 + \cdots + t_k$ and $\beta_k(\boldsymbol{t}) = t_1^2 + \cdots + t_k^2$.

(a) Suppose that a and a_j are non-negative integers. Prove, by induction on k or otherwise, that

$$\int_{\mathscr{R}_k} (1 - \alpha_k(\boldsymbol{t}))^a \prod_{j=1}^{k} t_j^{a_j} \, d\boldsymbol{t} = \frac{a! \prod_{j=1}^{k} a_j!}{(k + a + \sum_{j=1}^{k} a_j)!}.$$

(b) Suppose that a and b are non-negative integers. Prove that

$$\int_{\mathscr{R}_k} (1 - \alpha_k(\boldsymbol{t}))^a \beta_k(\boldsymbol{t})^b \, d\boldsymbol{t} = \frac{a!b!}{(k+a+2b)!} \sum_{\substack{\boldsymbol{b} \\ b_1 + \cdots + b_k = b}} \prod_{j=1}^{k} \frac{(2b_j)!}{b_j!}.$$

(The multinomial theorem applied to β_k^b is useful here.)

3. (Maynard 2015)

(a) Let $k = 5$. In the notation of the preceding exercise, when $\boldsymbol{t} \in \mathscr{R}_5$, let

$$f(\boldsymbol{t}) = (1-\alpha_5(\boldsymbol{t}))\beta_5(\boldsymbol{t}) + \frac{7}{10}(1-\alpha_5(\boldsymbol{t}))^2 + \frac{1}{14}\beta_5(\boldsymbol{t})^2 - \frac{3}{14}(1-\alpha_5(\boldsymbol{t})).$$

Prove that

$$\frac{\sum_{j=1}^{5} I_j(f)}{J(f)} = \frac{1,417,255}{708,216}.$$

(b) Prove that if the Elliott–Halberstam conjecture (Conjecture 20.4) holds, then

$$\liminf_{n\to\infty} p_{n+1} - p_n \le 12.$$

22.4 Notes

Section 22.1. In the first couple of decades of the twenty-first century there were a series of major advances. In a seminal paper, Goldston, Pintz & Yıldırım (2009) proved that

$$\liminf_{n\to\infty} \frac{p_{n+1} - p_n}{\log p_n} = 0,$$

and Goldston, Pintz, & Yıldırım (2010) showed that

$$p_{n+} - p_n \ll (\log p_n)^{1/2}(\log\log p_n)^2 \tag{22.29}$$

for infinitely many n. They also showed that if the level of distribution exceeds $\frac{1}{2}$, then there are infinitely many bounded gaps between primes. Indeed, if the level of distribution can be taken to be 1 (as in Conjecture 20.4), they were able to show that infinitely often $p_{n+1} - p_n \le 16$. All subsequent work is based on their method. There have been two sensational developments. Zhang (2014) proved a version of the Bombieri–Vinogradov theorem in which the moduli of the arithmetic progressions are restricted to being numbers with only relatively small prime factors but, crucially, the level of distribution exceeds $\frac{1}{2}$ by a small amount. Then, although the moduli are restricted, nevertheless the modified Bombieri–Vinogradov theorem contains enough information to enable an adaptation of the Goldston, Pintz, Yıldırım machinery to work. Thus Zhang showed that

$$\liminf_{n\to\infty} p_{n+1} - p_n \le 70{,}000{,}000. \tag{22.30}$$

Then Maynard (2015), by returning to an earlier version of the GPY method that predates their 2009 paper and which had been abandoned due to being as unsuccessful, was able to adapt their method to establish that infinitely many bounded gaps between the primes exist even if one only assumes a positive level of distribution for the primes. In particular, by using the Bombieri–Vinogradov theorem Maynard showed that

$$\liminf_{n\to\infty} p_{n+1} - p_n \le 600\,. \tag{22.31}$$

Apparently Tao independently also discovered similar methods to Maynard but as far as we have been able to ascertain they have never appeared in a refereed journal. These most recent methods involve quite heavy computations to obtain the sharpest bounds. For example, in the notation of Exercise 22.3.1.2, Maynard considers Theorem 22.1 with

$$f(\boldsymbol{t}) = \sum_{i=1}^{d} a_i(1-\alpha_k(\boldsymbol{t}))^{b_i}\beta_k(\boldsymbol{t})^{c_i}$$

and finds that (cf. Exercise 22.3.1.2)

$$\frac{\sum_{j=1}^{k} I_j(f)}{J(f)} = \frac{\boldsymbol{a}^T \mathcal{M}\boldsymbol{a}}{\boldsymbol{a}^T \mathcal{N}\boldsymbol{a}},$$

where the $d \times d$ positive definite matrices $\mathcal{M}$, $\mathcal{N}$ depend on the exponents b_i, c_i. He shows that this ratio is maximized when $\boldsymbol{a}$ is an eigenvector of $\mathcal{M}\mathcal{N}^{-1}$ corresponding to the largest eigenvalue. He then takes $k = 105$ and considers all choices of b_i, c_i with $b_i + 2c_i \le 11$, so that $d = 42$. It transpires that the largest eigenvalue is

$$4.0020697\ldots,$$

and so an appeal to Theorem 22.1 establishes that for any admissible 105-tuple $\boldsymbol{h}$ there are infinitely many n such that $n + \boldsymbol{h}$ contains at least two primes. He then displays a known admissible 105-tuple of diameter 600 discovered by T. Engelsma to establish (22.31). Maynard also showed on the assumption of the Elliott–Halberstam conjecture (Conjecture 20.4) that

$$\liminf_{n\to\infty} p_{n+1} - p_n \le 12, \tag{22.32}$$

for which see Exercise 22.3.1.3.

Following Zhang's work, Tao instituted a polymath project (Polymath 2014) to improve the bound. This led to the upper bound 4680. Then using Maynard's new ideas, the Polymath project improved this futher to

$$\liminf_{n\to\infty} p_{n+1} - p_n \le 246. \tag{22.33}$$

Assuming a more general form of the Elliott–Halberstam conjecture (see Theorem 20 of Polymath 2014) the bound (22.33) has been reduced further to 6.

The methods described here are very flexible and offer many potential applications. One is to a conjecture made by Dickson (1904), which states that if the g_i, h_i are integers and $\prod_{i=1}^{k}(g_i n + h_i)$ has no fixed prime divisor, then there are infinitely many n such that the $g_i n + h_i$ are simultaneously prime. Pintz (2016) has investigated questions involving consecutive primes in arithmetic progressions. In yet another application, Goldston, Graham, Pintz, & Yıldırım (2011) have considered n for which $d(n) = d(n+1)$, $\omega(n) = \omega(n+1)$, and $\Omega(n) = \Omega(n+1)$ simultaneously. There are also applications to cognate problems in algebraic number fields.

In the opposite direction, Maynard (2016) has developed the GPY sieve so as to show that there are exceptionally large gaps in the primes. In Theorem 7.15 we established Rankin's estimate

$$\limsup_{n\to\infty} \frac{p_{n+1} - p_n}{\left(\frac{(\log p_n)(\log\log p_n)(\log\log\log\log p_n)}{(\log\log\log p_n)^2}\right)} \ge c$$

for a suitable positive constant c, and in the Notes to §7.3 described the state of play as of 2007. Maynard (2016) showed that c can be made arbitrarily large. This was also established independently by a different method by Ford, Green, Konyagin, & Tao (2016). Although Erdős died in 1996, Ron Graham, who managed Erdős' finances honored the \$10,000 prize mentioned in the Notes to §7.3 . See Hartnett (2017). Ford, Green, Konyagin, Maynard, & Tao (2018) then showed that

$$\limsup_{n\to\infty} \frac{p_{n+1} - p_n}{\left(\frac{(\log p_n)(\log\log p_n)(\log\log\log\log p_n)}{\log\log\log p_n}\right)} \ge c$$

for some positive constant c. In the spirit of Erdős, Tao has offered \$10,000 for a proof that this c may be taken arbitrarily large.

22.5 References

Barban, M. B. & Vehov, P. P. (1968). On an extremal problem, *Trudy Moscov. Obšč* **18**, 83–90. See also: *Trans. Moscow Math. Soc.* **18**, 91–99.

Dickson, L. E. (1904). A new extension of Dirichlet's theorem on prime numbers, *Messenger of Math.* **33**, 155–161.

Ford, K., Green, B., Konyagin, S., & Tao, T. (2016). Large gaps between consecutive prime numbers, *Ann. Math.* **183**, 935–974. arXiv:1408.4505. doi:10.4007/annals.2016.183.3.4.

Ford, K., Green, B., Konyagin, S., Maynard, J. A., & Tao, T. (2018). Long gaps between primes, *J. Amer. Math. Soc.* **31**, 65–105. doi:10.1090/jams/876.

Goldston, D. A., Pintz, J., & Yıldırım, C. Y. (2006). Primes in tuples. III. On the difference $p_{n+\nu} - p_n$, *Funct. Approx. Comment. Math.* **35**, 79–89.

(2009). Primes in tuples. I, *Ann. Math.* **170**, 819–862.

(2010). Primes in tuples. II, *Acta Math.* **204**, 1–47.

(2011). Positive proportion of small gaps between consecutive primes, *Publ. Math. Debrecen* **79**, 433–444.

(2011). Small gaps between almost primes, the parity problem and some conjectures of Erdős on consecutive integers, *Int. Math. Res. Not.* **2011**, no. 7, 1439–1450.

Graham, S. W. (1978). An asymptotic estimate related to Selberg's sieve, *J. Number Theory* **10**, 83–94.

Hartnett, K. (2017), Cash for Math: The Erdős Prizes Live On, *Quanta Magazine*, `www.quantamagazine.org/cash-for-math-the-erdos-prizes-live-on-20170605/`

Maynard, J. A. (2015). Small gaps between primes, *Ann. Math.* (2) **181**, 383–413.

(2016). Large gaps between primes, *Ann. Math.* (3) **183**, 915–933. arXiv:1408.5110. `doi:10.4007/annals.2016.183.3.3`.

Motohashi, Y. (1974). On a problem in the theory of sieve methods (in Japanese), *Res. Inst. Math. Sci. Kyoto Univ. Kōkyūroko* **222**, 9–50.

Pintz, J. (2016). Polignac numbers, conjectures of Erdős on gaps between primes, arithmetic progressions in primes, and the bounded gap conjecture. In *From Arithmetic to Zeta-Functions*, Cham: Springer, pp. 367–384.

de Polignac, A. (1849). Recherches nouvelles sur les nombres premiers, *Comptes Rendus Paris* **29**, 397–401 & 738–739.

Polymath, D. H. J. (2014). Variants of the Selberg sieve, and bounded intervals containing many primes, *Res. Math. Sci.* **1**, 12. arXiv:1407.4897. doi:10.1186/s40687-014-0012-7.

Zhang, Y. (2014). Bounded gaps between primes, *Ann. Math.* **179**, 1121–1174.

Appendix E
Topics in Harmonic Analysis II

E.1 Uniform Approximation of Continuous Functions

Let $C(\mathbb{T})$ denote the set of continuous functions with period 1. Our objective in this section is to show that if $f \in C(\mathbb{T})$ and $\varepsilon > 0$, then there is a trigonometric polynomial $T(x)$ such that $|f(x) - T(x)| < \varepsilon$ for all x. This is elegantly achieved by using the Cesàro partial sums of the Fourier series of f, namely

$$\sigma_N(x) = \sigma_N(f, x) = \sum_{n=-N}^{N} (1 - |n|/N)\widehat{f}(n)e(nx). \tag{E.1}$$

Here $e(x) = e^{2\pi i x} = \cos 2\pi x + i \sin 2\pi x$ is the complex exponential with period 1, and the numbers $\widehat{f}(n)$ are the Fourier coefficents of f, which are defined to be

$$\widehat{f}(n) = \int_0^1 f(x)e(-nx)\,dx$$

for integers n. The functions $e(nx)$ form an orthonormal system, and the integral above is an inner product where $\langle f, g\rangle = \int_0^1 f(x)\overline{g(x)}\,dx$.

From the formula for the sum of a geometric progression we see that

$$\begin{aligned}\sum_{n=0}^{N-1} e(nx) &= \frac{1 - e(Nx)}{1 - e(x)} = e((N-1)x/2)\frac{e(Nx/2) - e(-Nx/2)}{e(x/2) - e(-x/2)}\\ &= e((N-1)x/2)\frac{\sin \pi N x}{\sin \pi x}.\end{aligned}$$

Hence

$$\left|\sum_{n=0}^{N-1} e(nx)\right|^2 = \left(\frac{\sin \pi N x}{\sin \pi x}\right)^2.$$

On the other hand, the left-hand side above is

$$= \sum_{m=0}^{N-1}\sum_{n=0}^{N-1} e((m-n)x) = \sum_{n=-N}^{N} (N-|n|)e(nx).$$

We divide through by N and set

$$\Delta_N(x) = \sum_{n=-N}^{N} (1-|n|/N)e(nx) = \frac{1}{N}\left(\frac{\sin \pi N x}{\sin \pi x}\right)^2. \tag{E.2}$$

This is the *Fejér kernel.* ('Fejér' is pronounced fay-air, because he was Hungarian, not French.) We note that if $f \in L^1(\mathbb{T})$, then

$$\begin{aligned}\int_0^1 f(u)\Delta_N(x-u)\,du &= \sum_{n=-N}^{N} (1-|n|/N)\int_0^1 f(u)e(n(x-u))\,du\\ &= \sum_{n=-N}^{N} (1-|n|/N)\widehat{f}(n)e(nx) = \sigma_N(x).\end{aligned}$$

Since $\int_0^1 \Delta_N(x)\,dx = 1$ and $\Delta_N(x) \ge 0$ for all x, it follows that $\sigma_N(x)$ is a weighted average of the values of f. Also, $\max \Delta_N(x) = \Delta_N(0) = N$. Let $\|x\| = \min_{n\in\mathbb{Z}} |x-n|$ be the distance from x to the nearest integer. (This is the natural distance function, when working modulo 1.) As $|\sin \pi x| \ge 2\|x\|$, it follows that

$$0 \le \Delta_N(x) \le \min\left(N, \frac{1}{4N\|x\|^2}\right). \tag{E.3}$$

It is useful to note that the pointwise estimate above implies that if $0 < \delta \le 1/2$, then

$$\begin{aligned}\int_\delta^{1-\delta} \Delta_N(u)\,du = 2\int_\delta^{1/2} \Delta_N(u)\,du &< \frac{1}{2N}\int_\delta^{1/2} \frac{1}{u^2}\,du\\ &< \frac{1}{2N}\int_\delta^{\infty} \frac{1}{u^2}\,du = \frac{1}{2N\delta}.\end{aligned} \tag{E.4}$$

Theorem E.1 *If f is a continuous function with period 1 and $\sigma_N(f,x)$ is defined as above, then $\sigma_N(f,x) \to f(x)$ uniformly in x, as $N \to \infty$.*

Proof We note that

$$\begin{aligned} f(x) - \sigma_N(x) &= \int_0^1 \Delta_N(x-u)(f(x) - f(u))\,du \\ &= \int_0^1 \Delta_N(u)(f(x) - f(x-u))\,du \\ &= \int_{-\delta}^{\delta} + \int_{\delta}^{1-\delta} = I_1 + I_2, \end{aligned}$$

say. Hence by the triangle inequality, $|f(x) - \sigma_N(x)| \le |I_1| + |I_2|$. Since f is continuous, it follows by compactness that f is uniformly continuous, which is to say that for any $\varepsilon > 0$ there is a $\delta > 0$ such that $|f(x) - f(y)| < \varepsilon$ whenever $\|x - y\| < \delta$. By the triangle inequality it follows that

$$|I_1| \le \int_{-\delta}^{\delta} \Delta_N(u)|f(x) - f(x-u)|\,du < \varepsilon \int_{-\delta}^{\delta} \Delta_N(u)\,du < \varepsilon.$$

Since f is continuous, it also follows by compactness that f is bounded, say $|f(x)| \le M$ for all x. Hence $|f(x) - f(x-u)| \le |f(x)| + |f(x-u)| \le 2M$. Thus from (E.4) we deduce that

$$|I_2| \le 2M \int_{\delta}^{1-\delta} \Delta_N(u)\,du < \frac{M}{N\delta}.$$

This quantity is $< \varepsilon$ if $N > M/(\delta\varepsilon)$. Then $|f(x) - \sigma_N(x)| < 2\varepsilon$ for all x, as desired. □

E.2 Quantitative Trigonometric Approximation

For $f \in L^1(\mathbb{R})$, we let $\widehat{f}(t)$ denote its Fourier transform,

$$\widehat{f}(t) = \int_{\mathbb{R}} f(x)e(-tx)\,dx.$$

Let $I = [\alpha, \beta]$ be an interval of $\mathbb{R}$ with χ_I its characteristic function, and suppose that $\delta > 0$ is given. Our objective is to construct functions $S_+(x)$ and $S_-(x)$ such that

$$\begin{aligned} \widehat{S_\pm}(t) &= 0 \quad \text{when } |t| \ge \delta, \\ S_-(x) \le \chi_I(x) &\le S_+(x) \quad \text{for all } x, \end{aligned}$$

and such that the integrals

$$\int_{\mathbb{R}} S_+(x) - \chi_I(x)\,dx, \qquad \int_{\mathbb{R}} \chi_I(x) - S_-(x)\,dx$$

are small. We do not attempt to determine exactly the extreme values of these integrals, but the functions we construct are elegant and close to optimal. With S_+ and S_- in hand, we use the Poisson summation formula to derive corresponding trigonometric polynomials $T_\pm$ that approximate closely the characteristic function of an arc of $\mathbb{T} = \mathbb{R}/\mathbb{Z}$. These $T_\pm$ are useful in a number of connections. We employ them in discussing the large sieve (in §19.1), in discussing quantitative measures of uniform distribution (in §F.2), and in proving Kronecker's theorem (in §F.3).

We begin by defining *Beurling's function*,

$$B(z) = \left(\frac{\sin \pi z}{\pi}\right)^2 \left(\frac{2}{z} + \sum_{n=0}^{\infty} \frac{1}{(z-n)^2} - \sum_{n=1}^{\infty} \frac{1}{(z+n)^2}\right), \tag{E.5}$$

whose basic properties are as follows.

Theorem E.2 *The function $B(z)$ given by* (E.5) *is an entire function such that*

(i) $B(n) = 1$ *for all integers* $n \ge 0$, $B(n) = -1$ *for all integers* $n < 0$;
(ii) $B'(n) = 0$ *for all integers* $n \ne 0$, $B'(0) = 2$;
(iii) $B(x) \ge \operatorname{sgn}(x)$ *for all real* x;
(iv) $B(x) - \operatorname{sgn}(x) \ll \min(1, x^{-2})$ *for all real* x;
(v) $B'(x) \ll \min(1, x^{-2})$ *for all real* x;
(vi) $B(z) - \operatorname{sgn}(x) \ll |z|^{-2} e^{2\pi|y|}$ *where* $z = x + iy$;
(vii) $\int_{-\infty}^{\infty} B(x) - \operatorname{sgn}(x)\, dx = 1$.

An entire function $f(z)$ belongs to the class E^σ of *functions of exponential type* σ if for every constant $\varepsilon > 0$ the inequality $|f(z)| < \exp((\sigma + \varepsilon)|z|)$ holds for all z with $|z|$ large. Thus we see that $B(x) \in E^{2\pi}$. Other examples of functions of exponential type are provided by observing that if $f \in L^1([-c, c])$, then its Fourier transform

$$\widehat{f}(z) = \int_{-c}^{c} f(u) e^{-2\pi i z u}\, du$$

is an entire function of the class $E^{2\pi c}$. In the case of $B(z)$, we note that $B \notin L^1(\mathbb{R})$, and also that there is no $f \in L^1(\mathbb{R})$ of which $B(z)$ is the Fourier transform (since $B(x) \not\to 0$ as $x \to \infty$). Nevertheless, the estimate (vi) above may be thought of as asserting that $\operatorname{supp} \widehat{B} \subseteq [-1, 1]$.

Proof We first establish further formulæ for $B(z)$. We recall the partial fraction formula

$$\left(\frac{\pi}{\sin \pi z}\right)^2 = \sum_{n=-\infty}^{\infty} \frac{1}{(z-n)^2}. \tag{E.6}$$

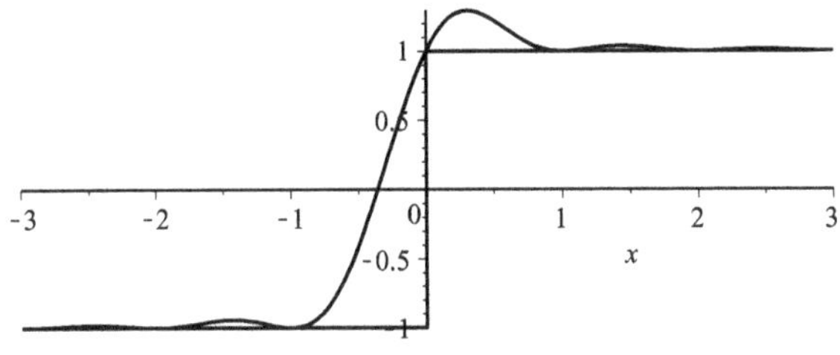

Figure E.1 Graph of Beurling's function $B(x)$ for $-3 \le x \le 3$.

(This may be proved by noting that the difference between the two sides is a bounded entire function that tends to 0 as $z \to i\infty$.) On combining this with (E.5) we find that

$$B(z) = 1 + 2\Big(\frac{\sin \pi z}{\pi}\Big)^2 \left(\frac{1}{z} - \sum_{n=1}^{\infty} \frac{1}{(z+n)^2}\right). \tag{E.7}$$

Suppose that $z \notin (-\infty, 0]$. The integral test suggests that the sum above is approximately

$$\int_0^\infty (u+z)^{-2}\, du = \frac{1}{z}.$$

Hence the second factor on the right-hand side is the difference between this approximation and the sum. To express this quantity more explicitly, we observe that if f has continuous first derivative on an interval $[\alpha, \beta]$, then

$$\int_\alpha^\beta f(u)\, du = f(\beta)(\beta - \alpha) - \int_\alpha^\beta f'(u)(u-\alpha)\, du$$

by integration by parts. By taking $\alpha = n-1$, $\beta = n$, $f(u) = (u+z)^{-2}$, it follows that

$$\int_{n-1}^n (u+z)^{-2}\, du = (z+n)^{-2} + 2\int_{n-1}^n (z+u)^{-3}\{u\}\, du$$

provided that $z \notin [-n, -n+1]$. If $z \notin (-\infty, 0]$, then we may sum over $n = 1, 2, \ldots$, and thus we deduce from (E.6) that

$$B(z) = 1 + 4\Big(\frac{\sin \pi z}{\pi}\Big)^2 \int_0^\infty \frac{\{u\}}{(u+z)^3}\, du. \tag{E.8}$$

Similarly, from (E.5) and (E.6) we also have

$$B(z) = -1 + 2\Big(\frac{\sin \pi z}{\pi}\Big)^2 \left(\frac{1}{z} + \sum_{n=0}^{\infty} \frac{1}{(z-n)^2}\right), \tag{E.9}$$

and that if $z \notin [0, \infty)$, then

$$B(z) = -1 + 4\Big(\frac{\sin \pi z}{\pi}\Big)^2 \int_0^\infty \frac{1 - \{u\}}{(u - z)^3}\, du. \tag{E.10}$$

The assertions (i) and (ii) are immediate from the definition (E.5) of $B(z)$. For $x > 0$, the inequality (iii) and the estimate (iv) follow from (E.8) since the value of the integral lies between 0 and $\frac{1}{2}x^{-2}$. For $x < 0$, these assertions follow similarly from (E.10). Since $B(x)$ is continuous, these relations therefore also hold when $x = 0$. To obtain the estimate (v) it suffices to differentiate the formulæ (E.8), (E.10), and then estimate the quantities that arise. As for (vi), we note that $(\sin \pi z)^2 \ll e^{2\pi|y|}$, and that if $\operatorname{Re} z \ge 0$, then $|u + z| \ge \max(u, |z|) \ge (u + |z|)/2$, so

$$\int_0^\infty \frac{\{u\}}{(u + z)^3}\, du \ll \int_0^\infty \frac{du}{(u + |z|)^3} \ll |z|^{-2}.$$

Thus we obtain (vi) from (E.8) when $\operatorname{Re} z \ge 0$, and similarly from (E.10) when $\operatorname{Re} z < 0$. As for (vii), let

$$V(z) = \Big(\frac{\sin \pi z}{\pi}\Big)^2 \left(\frac{2}{z} + \sum_{n=-\infty}^{\infty} \frac{\operatorname{sgn}(n)}{(z - n)^2}\right), \tag{E.11}$$

so that $B(z) = V(z) + (\sin \pi z)^2/(\pi z)^2$. Since $V(x)$ and $\operatorname{sgn}(x)$ are odd functions, we know that

$$\int_{-X}^{X} V(x) - \operatorname{sgn}(x)\, dx = 0$$

for any X. Hence

$$\begin{aligned}
\int_{-\infty}^{\infty} B(x) - \operatorname{sgn}(x)\, dx &= \lim_{X\to\infty} \int_{-X}^{X} B(x) - \operatorname{sgn}(x)\, dx \\
&= \lim_{X\to\infty} \int_{-X}^{X} V(x) - \operatorname{sgn}(x) + (\sin \pi x)^2/(\pi x)^2\, dx \\
&= \lim_{X\to\infty} \int_{-X}^{X} \Big(\frac{\sin \pi x}{\pi x}\Big)^2 dx \\
&= \int_{-\infty}^{\infty} \Big(\frac{\sin \pi x}{\pi x}\Big)^2 dx = 1.
\end{aligned}$$

The final definite integral can be evaluated by means of the calculus of residues.

□

Although the proof is now complete, it is instructive to note that (iii) can be derived from (E.7) and (E.9) by appealing to the integral test. For example, if $x > 0$, then

$$\sum_{n=1}^{\infty} \frac{1}{(x+n)^2} < \int_0^{\infty} \frac{du}{(x+u)^2} = \frac{1}{x}.$$

We now use the function $B(z)$ to construct approximations to the characteristic function χ_I of an interval $[\alpha, \beta]$.

Theorem E.3 *Let $I = [\alpha, \beta]$ be a finite interval, and suppose that $\delta > 0$ is given. Then there exist entire functions $S_+(z)$ and $S_-(z)$ such that*

(i) $S_\pm(x) \ll_{\alpha,\beta,\delta} \min(1, x^{-2})$ *for real* x;
(ii) $S_-(x) \le \chi_I(x) \le S_+(x)$ *for real* x;
(iii) $\int_{-\infty}^{\infty} S_\pm(x)\, dx = \beta - \alpha \pm 1/\delta$;
(iv) $\widehat{S_\pm}(t) = 0$ *when* $|t| \ge \delta$;
(v) $S_\pm(x)$ *is of bounded variation on* $\mathbb{R}$;
(vi) $\left|\widehat{S_\pm}(t)\right| \le \beta - \alpha + 1/\delta$ *for all real* t.

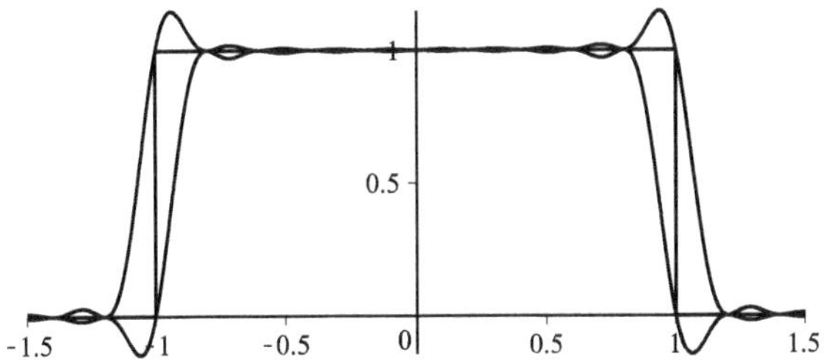

Figure E.2 Selberg's functions $S_\pm(x)$ and $\chi_I(x)$ for $I = [-1, 1]$ and $\delta = 5$.

Proof We take

$$S_+(z) = \frac{1}{2}B(\delta(z-\alpha)) + \frac{1}{2}B(\delta(\beta - z)),$$
$$S_-(z) = -\frac{1}{2}B(\delta(\alpha - z)) - \frac{1}{2}B(\delta(z-\beta));$$

these are the *Selberg functions*. Then the assertion (i) follows immediately from Theorem E.2(iv). To obtain the inequalities (ii) we note that

$$S_+(x) \ge \frac{1}{2}\operatorname{sgn}(\delta(x-\alpha)) + \frac{1}{2}\operatorname{sgn}(\delta(\beta - x))$$

by Theorem E.2(iii). Here the right-hand side is $\chi_I(x)$ unless $x = \alpha$ or $x = \beta$.

If $\alpha < \beta$, then we may conclude that $S_+(\alpha) \ge 1$, $S_+(\beta) \ge 1$, because S_+ is continuous. If $\alpha = \beta$, then $S_+(\alpha) = 1$ because $B(0) = 1$. Similarly, we see that $S_-(x) \le \chi_I(x)$ for all x. As for (iii), we note that

$$\begin{aligned}\int_{-\infty}^{\infty} S_+(x)\,dx &= \int_{-\infty}^{\infty} \chi_I(x)\,dx + \int_{-\infty}^{\infty} S_+(x) - \chi_I(x)\,dx \\ &= \beta - \alpha + \frac{1}{2}\int_{-\infty}^{\infty} B(\delta(x-\alpha)) - \operatorname{sgn}(\delta(x-\alpha))\,dx \\ &\quad + \frac{1}{2}\int_{-\infty}^{\infty} B(\delta(\beta - x)) - \operatorname{sgn}(\delta(\beta - x))\,dx \\ &= \beta - \alpha + 1/\delta,\end{aligned}$$

by Theorem E.2(iv),(vii), and similarly for S_-. Since the functions $S_\pm$ are in $L^1(\mathbb{R})$, we can define their Fourier transforms,

$$\widehat{S_\pm}(t) = \int_{-\infty}^{\infty} S_\pm(x)e(-tx)\,dx.$$

Here $S_\pm(z)e^{-2\pi i t z}$ is an entire function, and if $t \ge \delta$, then by Theorem E.2(vi) we see that this function is $\ll_{\alpha,\beta,\delta} |z|^{-2}$ in the lower halfplane $\operatorname{Im} z \le 0$. We consider the integral above to be a contour integral in the complex plane, and on replacing this path by a semicircle in the lower halfplane we conclude that $\widehat{S_\pm}(t) = 0$ if $t \ge \delta$. Similarly, $S_\pm(t) = 0$ if $t \le -\delta$, so we have (iv). Also, from Theorem E.2(v) we see that $B(x)$ is of bounded variation on $\mathbb{R}$, and hence the same is true of $S_\pm$. Finally, $\widehat{S_\pm}(t) = \widehat{\chi_I}(t) + \big(\widehat{S_\pm}(t) - \widehat{\chi_I}(t)\big)$, so by the triangle inequality

$$\begin{aligned}\big|\widehat{S_\pm}(t)\big| &\le \big|\widehat{\chi_I}(t)\big| + \big|\widehat{S_\pm}(t) - \widehat{\chi_I}(t)\big| \le \|\chi_I\|_{L^1(\mathbb{R})} + \|S_\pm - \chi_I\|_{L^1(\mathbb{R})} \\ &= \beta - \alpha + \delta^{-1}.\end{aligned}$$

□

We now derive analogous results for approximations in $\mathbb{T} = \mathbb{R}/\mathbb{Z}$ by trigonometric polynomials.

Theorem E.4 *For any arc $I = [\alpha, \beta]$ in $\mathbb{T}$ with length $\beta - \alpha < 1$, and for any positive integer N, there are trigonometric polynomials*

$$T_\pm(x) = \sum_{k=-N}^{N} \widehat{T_\pm}(k)e(kx) \tag{E.12}$$

of degree at most N such that:

(i) *$T_-(x) \le \chi_I(x) \le T_+(x)$ for all real x;*
(ii) *$\int_0^1 T_\pm(x)\,dx = \beta - \alpha \pm 1/(N+1)$;*
(iii) *$\big|\widehat{T_\pm}(k)\big| \le \beta - \alpha + \frac{1}{N+1}$ for all integers k.*

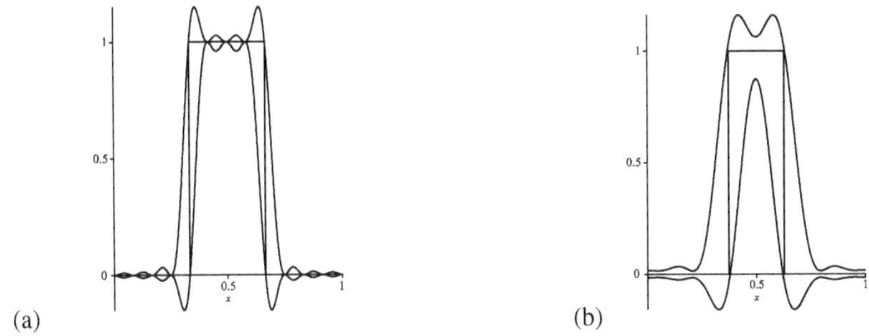

Figure E.3 (a) Graph of $\chi_I(x)$ and $T_\pm(x)$ for $I = [1/3, 2/3]$ with $K = 11$; (b) $I = [3/8, 5/8]$, $K = 5$.

Proof Take $\delta = N + 1$, and let $S_\pm$ be the functions described in Theorem E.3. Put

$$T_\pm(x) = \sum_n S_\pm(x + n).$$

From Theorem E.3(i) we see that this series is uniformly convergent for x in a compact set, so that $T_\pm(x)$ is continuous. The inequalities (i) follow from Theorem E.3(ii). From Theorem E.3(i),(v) we see that the Poisson summation formula, in the form given in Theorem D.3, applies to $S_\pm$. Thus

$$T_\pm(x) = \lim_{K\to\infty} \sum_{k=-K}^{K} \widehat{S}_\pm(k)e(kx).$$

But $\widehat{S}_\pm(k) = 0$ for $|k| \geq \delta = N + 1$, and $\widehat{T}_\pm(k) = \widehat{S}_\pm(k)$ for all k, so we find that $T_\pm$ is a trigonometric polynomial, as in (E.12). Also, the integral in (ii) is

$$\widehat{T}_\pm(0) = \widehat{S}_\pm(0) = \int_{-\infty}^{\infty} S_\pm(x)\,dx,$$

and the stated result follows from Theorem E.3(iii). The final assertion follows from Theorem E.3(vi). □

Majorants and minorants constructed as above are optimal if and only if $(K + 1)|I|$ is an integer. Hence the estimates in (i) are optimal, while those in (ii) are not.

In the above situation, the interval I is short, δ is large, $T_\pm(x)$ has period 1, and the $\widehat{S}_\pm(k)$ become Fourier coefficients. With an alternative application of the Poisson summation formula we reverse this, so that I is long, δ is small, $\widehat{S}_\pm$ has period 1, and the $S_\pm$ are Fourier coefficients.

Theorem E.5 *Let M and N be integers, $N \ge 1$. Suppose that $0 < \delta \le 1/2$. There exist functions $W_\pm(x)$ with period 1 and absolutely convergent Fourier expansions $W_\pm(x) = \sum_n w_\pm(n)e(nx)$ such that*

(i) $w_-(n) \le \chi_{[M+1,M+N]}(n) \le w_+(n)$ *for all integers n;*

(ii) $W_\pm(x) = 0$ *if* $\|x\| \ge \delta$;

(iii) $\sum_n w_\pm(n) = W_\pm(0) = N - 1 \pm 1/\delta$.

Proof Let $S_\pm(u)$ be the Selberg functions for the interval $I = [M+1, M+N]$, and set $w_\pm(u) = S_\pm(u)$. Thus we have (i). We apply the Poisson summation formula to $f(u) = S_\pm(u)e(ux)$. Hence by Theorem D.3 we see that

$$\sum_{n=-\infty}^{\infty} w_\pm(n)e(nx) = \sum_{k=-\infty}^{\infty} \widehat{S}_\pm(k-x),$$

and then properties (ii) and (iii) are immediate. □

E.2.1 Exercises

1. Suppose that $I = [\alpha, \beta]$ is an interval on the real line, put $K = (\beta - \alpha)\delta$, and suppose that K is a positive integer. Suppose that $f \in L^1(\mathbb{R})$, that f is continuous, that $f(x) \ge \chi_I(x)$ for all x, that f has bounded variation on $\mathbb{R}$, and that $\widehat{f}(t) = 0$ when $|t| \ge \delta$.

   (a) Show that

   $$\sum_{n=-\infty}^{\infty} f(n/\delta + x) = \delta \widehat{f}(0)$$

   for all x.

   (b) Show that x can be chosen so that $n/\delta + x \in I$ for $K + 1$ values of n.

   (c) Deduce that

   $$\int_{-\infty}^{\infty} f(u)\,du \ge \beta - \alpha + 1/\delta.$$

   That is, the function S_+ described in Theorem E.3 is optimal when $(\beta - \alpha)\delta$ is an integer.

2. Prove the following identities:

(a) $\left(\frac{\sin \pi x}{\pi x}\right)^2 = \int_{-1}^{1} (1 - |t|)e(tx)\,dt;$

(b) $\frac{(\sin \pi x)^2}{x} = \pi \int_0^1 \sin 2\pi t x\,dt;$

(c) $\sum_{n=-N}^{N} \operatorname{sgn}(n)e(-nt) = -i\cot \pi t + i\frac{\cos \pi(2N+1)t}{\sin \pi t};$

(d) $\operatorname{sgn}(x) = \frac{2}{\pi}\int_0^\infty \frac{1}{t}\sin 2\pi t x\,dt.$

3. Let $V(z)$ be *Vaaler's function* as defined in (E.11), and put

$$V_N(z) = \left(\frac{\sin \pi z}{\pi}\right)^2 \left(\frac{2}{z} + \sum_{-N}^{N} \frac{\operatorname{sgn}(n)}{(z-n)^2}\right).$$

(a) Using the identities in the previous exercise, or otherwise, show that

$$\begin{aligned} V_N(x) = 2&\int_0^1 ((1-t)\cot \pi t + 1/\pi)\sin 2\pi t x\,dt \\ &- 2\int_0^1 \frac{\cos \pi(2N+1)t}{\sin \pi t}(1-t)\sin 2\pi t x\,dt. \end{aligned}$$

(b) By using the Riemann–Lebesgue lemma, show that

$$V(x) = 2\int_0^1 \left((1-t)\cot \pi t + \frac{1}{\pi}\right)\sin 2\pi t x\,dt.$$

(c) Let

$$\phi(t) = \begin{cases} 1 & \text{if } t = 0, \\ \pi(1-|t|)t\cot \pi t + |t| & \text{if } 0 < |t| \le 1, \\ 0 & \text{if } |t| > 1. \end{cases} \tag{E.13}$$

Show that

$$V'(x) = 2\int_{-1}^{1} \phi(t)e(xt)\,dt.$$

(d) Show that $\phi(t)$ is non-negative, continuously differentiable on $\mathbb{R}$, and that is is strictly decreasing on $[0, 1]$.

(e) Show that $V(z)$ is an odd entire function, and that

$$V(z) = 1 - 6\left(\frac{\sin \pi z}{\pi}\right)^2 \int_0^\infty \frac{\{u\}(1-\{u\})}{(z+u)^4}\,du$$

provided that $z \notin (-\infty, 0]$.

(f) Show that $V(n) = \operatorname{sgn}(n)$ for all integers n, that $V'(n) = 0$ for all integers $n \neq 0$, that $V'(0) = 2$, and that $0 \leq V(x) \leq 1$ for $x > 0$.

(g) Show that if $x > 0$, then

$$V(x) - 1 \ll \min(1, x^{-3}),$$
$$V'(x) \ll \min(1, x^{-3}).$$

(h) Show that all zeros of $V'(z)$ lie on the real axis.

(i) Show that

$$V(x) - \operatorname{sgn}(x) = \int_{-\infty}^{\infty} \frac{\phi(t) - 1}{\pi i t} e(tx)\, dt.$$

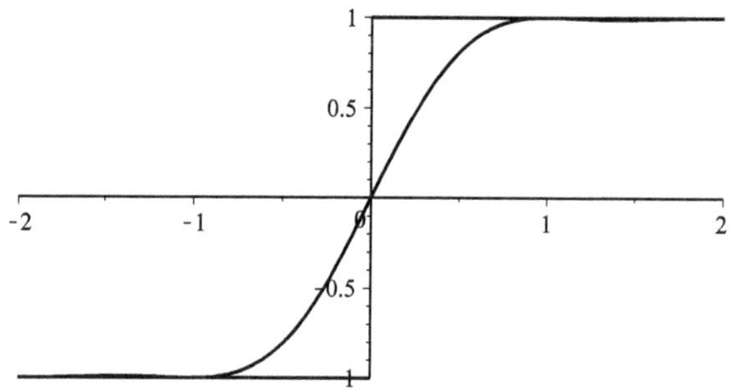

Figure E.4 Graph of Vaaler's function $V(x)$ for $-2 \leq x \leq 2$.

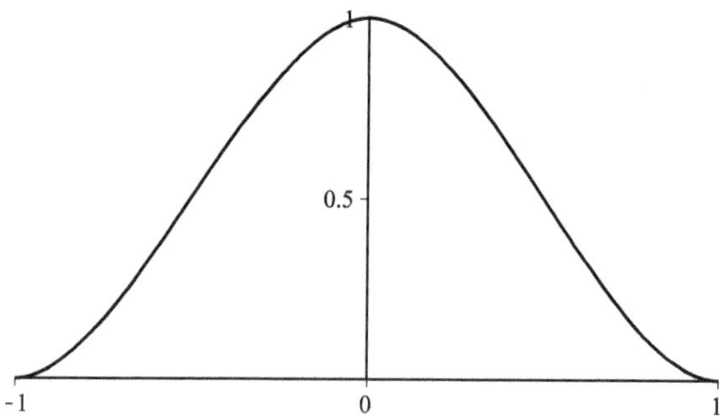

Figure E.5 Graph of $\phi(t)$ for $-1 \leq t \leq 1$.

4. Let

$$P(x) = \frac{K+1}{2} \sum_{n=-\infty}^{\infty} V'((K+1)(n+x)),$$

$$Q(x) = \frac{1}{2} \sum_{n=-\infty}^{\infty} V((K+1)(n+x)) - \operatorname{sgn}(n+x),$$

$$R(x) = Q(x) - \{x\} + 1/2.$$

(a) Show that $P(x)$ is a trigonometric polynomial of degree K, with coefficients $\widehat{P}(k) = \phi(k/(K+1))$ where $\phi(t)$ is defined as in (E.13).

(b) Show that $Q(x)$ has Fourier coefficients

$$\widehat{Q}(k) = \frac{\phi\left(\frac{k}{K+1}\right) - 1}{2\pi i k}$$

for $k \neq 0$, and that $\widehat{Q}(0) = 0$.

(c) Show that $R(x)$ is a trigonometric polynomial of degree K with coefficients

$$\widehat{R}(k) = \frac{\phi\left(\frac{k}{K+1}\right)}{2\pi i k}$$

for $k \neq 0$, $\widehat{R}(0) = 0$, and that $R'(x) = P(x) - 1$.

(d) Show that for all x,

$$R(x) - \frac{\Delta_{K+1}(x)}{2(K+1)} \le 1/2 - \{x\} \le R(x) + \frac{\Delta_{K+1}(x)}{2(K+1)}.$$

5. Let $P(x)$ and $Q(x)$ be as in the previous exercise. Suppose that f is of bounded variation on $\mathbb{T}$.

(a) Show that if f is continuous at x, then

$$f(x) = \int_0^1 f(x+u)P(u)\, du + \int_0^1 Q(u)\, df(x+u).$$

(b) Suppose that f is a real-valued function of bounded variation on $\mathbb{T}$. Show that

$$-\int_0^1 \frac{\Delta_{K+1}(x-u)}{2(K+1)} |df(u)| \le f(x) - \int_0^1 f(x+u)P(u)\, du$$
$$\le \int_0^1 \frac{\Delta_{K+1}(x-u)}{2(K+1)} |df(u)|$$

for all x.

(c) Show that $\int_0^1 f(x+u)P(u)\, du$ is a trigonometric polynomial of degree at most K with coefficients $\phi(k/(K+1))\widehat{f}(k)$.

(d) Show that $\int_0^1 \Delta_{K+1}(x-u)\,|df(u)|$ is a trigonometric polynomial of degree at most K with coefficients

$$\frac{1-\frac{|k|}{K+1}}{2(K+1)}\int_0^1 e(-ku)\,|df(u)|.$$

(e) Let

$$T_{\pm}(x) = \int_0^1 f(x+u)P(u)\,du \pm \int_0^1 \frac{\Delta_{K+1}(x-u)}{2(K+1)}\,|df(u)|.$$

Show that $T_{\pm}$ is a trigonometric polynomial of degree at most K such that $T_-(x) \le f(x) \le T_+(x)$ for all x, and that

$$\int_0^1 T_{\pm}(u)\,du = \int_0^1 f(u)\,du \pm \frac{\mathrm{Var}_{\mathbb{T}}(f)}{2(K+1)}.$$

(f) Show that if $f = \chi_{[\alpha,\beta]}$, then the $T_{\pm}$ in part (e) are the same as in Theorem E.4, and hence that the trigonometric polynomials in that theorem have coefficients

$$\widehat{T}_{\pm}(k) = \left(\phi\left(\frac{k}{K+1}\right)\frac{\sin\pi k(\beta-\alpha)}{\pi k} \pm \frac{1-\frac{|k|}{K+1}}{K+1}\cos\pi k(\beta-\alpha)\right) \times e\left(-k(\beta+\alpha)/2\right)$$

for $0 < |k| \le K$, $\widehat{T}_{\pm}(0) = \beta - \alpha \pm 1/(K+1)$.

6. (a) Suppose that $T(x)$ is a trigonometric polynomial of degree at most K, and that $N > K$. Show that for any real α,

$$\frac{1}{N}\sum_{n=1}^{N} T(\alpha + n/N) = \int_0^1 T(x)\,dx.$$

(b) Suppose that $I = [\alpha,\beta]$ is an arc of $\mathbb{T}$, and that $(\beta-\alpha)N$ is an integer $< N$. Show that if a function $T \in L^1(\mathbb{T})$ has the property that $T(x) \ge \chi_I(x)$ for all $x \in \mathbb{T}$, then

$$\sum_{n=1}^{N} T(\alpha + n/N) \ge (\beta-\alpha)N + 1.$$

(c) Suppose that $T(x)$ is a trigonometric polynomial of degree at most K, that $(\beta-\alpha)(K+1)$ is an integer $< K+1$, and that $T(x) \ge \chi_{[\alpha,\beta]}(x)$ for all $x \in \mathbb{T}$. Show that $\int_0^1 T(x)\,dx \ge \beta - \alpha + 1/(K+1)$.

(d) Suppose that $T(x)$ is a trigonometric polynomial of degree at most K, that $(\beta-\alpha)(K+1)$ is an integer $< K+1$, and that $T(x) \le \chi_{[\alpha,\beta]}(x)$ for all $x \in \mathbb{T}$. Show that $\int_0^1 T(x)\,dx \le \beta - \alpha - 1/(K+1)$.

7. (Barton, Montgomery, & Vaaler 2001, Lemma 4) Let $B(x)$ be the Beurling function, as defined in (E.5). Suppose that M is a positive integer and that α and β are real numbers such that $\beta-\alpha = M$. Show that $B(x-\alpha)+B(\beta-x) \ge 0$ for all real x.

E.3 An Additional Trigonometric Majorant

Let $s(x)$ denote the sawtooth function

$$s(x) = \begin{cases} \{x\} - 1/2 & (x \notin \mathbb{Z}), \\ 0 & (x \in \mathbb{Z}). \end{cases} \tag{E.14}$$

In Lemma D.1 we showed that

$$s(x) = -\sum_{0<|k|\le K} \frac{e(kx)}{2\pi ik} + O\left(\min\left(1, \frac{1}{K\|x\|}\right)\right). \tag{E.15}$$

In Exercise E.2.1.4 we find sharp trigonometric majorants and minorants for $s(x)$. These, as well as the estimate (E.15), apply equally to $s(x)$, to $\{x\} - 1/2$, and to $-\{-x\} + 1/2$, since these functions differ only in the value taken at 0, which is either 0, $-1/2$, or $1/2$, while our approximants are continuous. To estimate expressions of the sort

$$\sum_k a_k s(x_k),$$

the majorants and minorants are applicable if the a_k are real and of one sign, but are useless if the a_k are complex or of indeterminate sign. From Lemma 16.4 we see that

$$\sum_{k=1}^{K} e(kx) \ll \min\left(K, \frac{1}{\|x\|}\right). \tag{E.16}$$

Thus we encounter the same expression, but now divided by K, in the current context. Let

$$f_K(x) = \min\left(1, \frac{1}{K\|x\|}\right). \tag{E.17}$$

When $f_K(x)$ occurs in an expression (perhaps repeatedly with various values of x), one may derive an estimate by expanding f in its Fourier series, and then estimating the contribution of each Fourier coefficient. We now show that it suffices to consider the contribution of the Fourier coefficients $\widehat{f_K}(k)$ for $-K \le k \le K$.

Theorem E.6 *Let K be a given integer, $K \ge 2$, let $f_K(x)$ be defined as in* (E.17), *and put*

$$g_K(x) = \sum_{k=-K}^{K} \widehat{f_K}(k)(1 - |k|/K)e(kx).$$

Then

$$\widehat{f_K}(k) \ll \frac{1}{K} \log \frac{3K}{|k| + 1}$$

uniformly for $|k| \le K$, and $f_K(x) \ll g_K(x)$ uniformly in x and K.

From this we see that the error term in (E.15) can be replaced by $g_K(x)$, and that the right-hand side of (E.16) can be replaced by $Kg_K(x)$. The advantage here is not so much that we expect to obtain stronger results, but rather that we need not consider the contribution of $\widehat{f}(k)$ for larger k.

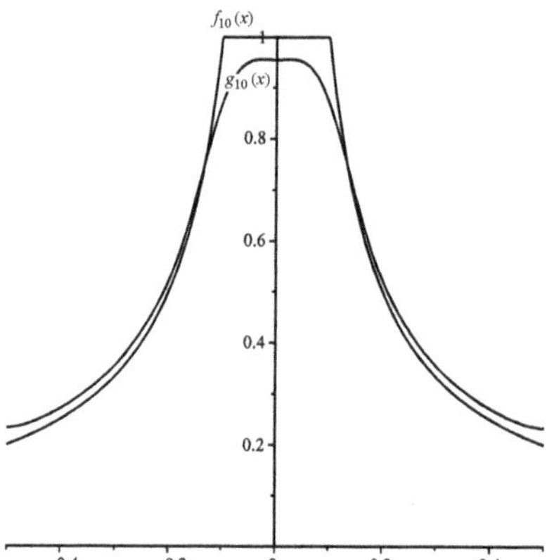

Figure E.6 Graphs of $f_{10}(x)$ and $g_{10}(x)$ for $-1/2 \le x \le 1/2$.

Proof Clearly $\widehat{f_K}(0) \asymp (\log K)/K$. Since f_K is real-valued and even, we know that $\widehat{f_K}(-k) = \widehat{f_K}(k)$, so it suffices to estimate $|\widehat{f_K}(k)|$ for $k > 0$. If $0 < k \le K$, then

$$\widehat{f_K}(k) \ll \frac{1}{K}\left(1 + \int_{1/K}^{1/k} \frac{1}{x}\,dx + \left|\int_{1/k}^{1/2} \frac{e(kx)}{x}\,dx\right|\right).$$

By integration by parts we see that the second integral above is

$$= \left[\frac{e(kx)}{2\pi ikx}\right|_{1/k}^{1/2} + \frac{1}{2\pi ik}\int_{1/k}^{1/2} \frac{e(kx)}{x^2}\,dx \ll 1 + \frac{1}{k}\int_{1/k}^{1/2} \frac{1}{x^2}\,dx \ll 1.$$

Thus we have the stated bound for $|\widehat{f_K}(k)|$. In establishing the second assertion, we may suppose that $0 \le x \le 1/2$ since f_K and g_K are even functions with period 1. Let

$$\Delta_K(x) = \sum_{k=-K}^{K} \left(1 - \frac{|k|}{K}\right) e(kx) = \frac{1}{K}\left(\frac{\sin \pi K x}{\sin \pi x}\right)^2$$

be the Fejér kernel. Then

$$g_K(x) = (f * \Delta_K)(x) = \int_0^1 \Delta_K(u) f_K(x-u)\, du. \tag{E.18}$$

Since $\Delta_K(x)$ is decreasing for $0 \le x \le 1/K$, it follows that

$$\int_0^{1/(2K)} \Delta_K(u)\, du \ge \frac{\Delta_K(1/(2K))}{2K} = \frac{1}{2K^2 \sin^2 \pi/(2K)} \ge \frac{2}{\pi^2}$$

because $\sin \delta \le \delta$ for $\delta \ge 0$. Since f_K and Δ_K are non-negative and f_K is (weakly) decreasing in $[-1/K, 1/2]$, if follows from (E.18) that

$$\begin{aligned} g_K(x) = \int_0^1 \Delta_K(u) f_K(x-u)\, du &\ge \int_0^{1/(2K)} \Delta_K(u) f_K(x-u)\, du \\ &\ge f_K(x) \int_0^{1/(2K)} \Delta_K(u)\, du \gg f_K(x). \end{aligned}$$

□

E.3.1 Exercises

1. Suppose that $K \ge 2$, and that $f_K(x)$ is defined as in (E.17).

   (a) Show that

   $$\widehat{f_K}(0) = \frac{2}{K}\left(1 + \log \frac{K}{2}\right).$$

   (b) Write

   $$\widehat{f_K}(k) = 2\int_0^{1/K} \cos 2\pi k x\, dx + \frac{2}{K}\int_{1/K}^{1/2} \frac{\cos 2\pi k x}{x}\, dx = T_1 + T_2,$$

   say. Suppose that $k \ne 0$. Show that

   $$\begin{aligned} T_1 &= \frac{\sin 2\pi k/K}{\pi k}, \\ T_2 &= -\frac{\sin 2\pi k/K}{\pi k} + \frac{1}{\pi k K}\int_{1/K}^{1/2} \frac{\sin 2\pi k x}{x^2}\, dx. \end{aligned}$$

   (c) Deduce that if $k \ne 0$, then

   $$\widehat{f_K}(k) = \frac{1}{\pi k K}\int_{1/K}^{1/2} \frac{\sin 2\pi k x}{x^2}\, dx.$$

(d) Conclude that $\widehat{f_K}(k) \ll K/k^2$ for $k \ge K$.

(e) Show that if $(\beta - \alpha)\delta$ is not an integer, then $S_+(x) > \chi_I(x)$, and hence that S_+ is not optimal, because there is a $c < 1$ such that $cS_+(x) \ge \chi_I(x)$ for all x.

E.4 Maximal Inequalities

Sometimes we may have an estimate for the size of a sum, say $\left|\sum_{n=1}^N c_n\right| \le M_N$, but it would be convenient to have a similar upper bound for the maximum size of its subsums, $\max_{\nu \le N} \left|\sum_{n=1}^{\nu} c_n\right| \le M_N^\star$, hopefully with $M_N^\star$ not much larger than M_N. Such an upper bound $M_N^\star$ is known as a *maximal inequality.*

E.4.1 Elementary Estimates

As in Appendix D, if $f \in L^1(\mathbb{T})$, then its Fourier coefficients are $\widehat{f}(n) = \int_{\mathbb{T}} f(x)e(-nx)\,dx$, the partial sums of its Fourier series are

$$s_N(x) = \sum_{n=-N}^{N} \widehat{f}(n)e(nx),$$

and the Dirichlet kernel is

$$D_K(x) = \sum_{n=-K}^{K} e(kx) = \frac{\sin(2K+1)\pi x}{\sin \pi x}.$$

Thus $s_K(x) = (f * D_K)(x) = \int_{\mathbb{T}} f(u)D_K(x-u)\,du$. Unfortunately, $|D_K(x)|$ decays only like an inverse first power, with the result that $\int_{\mathbb{T}} |D_K(x)|\,dx \asymp \log 2N$. Let

$$E_K(x) = \min(2K+1, 1/\|x\|). \tag{E.19}$$

(Note that this is a totally different function than the one with the same name discussed in Appendix D.) The letter 'E' is suggested here because $E_K(x)$ provides an *envelope* of $D_K(x)$: $|D_K(x)| \le E_K(x)$ and E_K is monotonically decreasing for $0 \le x \le 1/2$. Thus

$$|s_K(x)| \le \int_{\mathbb{T}} |f(u)|E_K(x-u)\,du.$$

Put

$$s_K^\star(x) = \max_{1 \le k \le K} |s_k(x)|. \tag{E.20}$$

Since $E_k(x) \le E_K(x)$ if $1 \le k \le K$, it follows that

$$s_K^\star(x) \le \int_{\mathbb{T}} |f(u)| E_K(x-u)\, du. \tag{E.21}$$

Hence

$$\max_x s_K^\star(x) \ll \|f\|_\infty \log 2K, \tag{E.22}$$

which is best possible since it might happen that $f(x) = \operatorname{sgn} D_K(x)$, in which case $s_K(0) = \int_{\mathbb{T}} |D_K(x)|\, dx \asymp \log 2K$. By Cauchy's inequality we see that

$$\begin{aligned}\left(\int_{\mathbb{T}} |f(u)| E_K(x-u)\, du\right)^2 &\le \int_{\mathbb{T}} |f(u)|^2 E_K(x-u)\, du \int_{\mathbb{T}} E_K(x-u)\, du \\ &\ll \int_{\mathbb{T}} |f(u)|^2 E_K(x-u)\, du \log 2K.\end{aligned}$$

By integrating this with respect to x we find that

$$\int_{\mathbb{T}} s_K^\star(x)^2\, dx \ll (\log 2K)^2 \int_{\mathbb{T}} |f(u)|^2\, du. \tag{E.23}$$

Finally, it is also evident from (E.21) that

$$\int_{\mathbb{T}} s_K^\star(x)\, dx \ll (\log 2K) \int_{\mathbb{T}} |f(u)|\, du. \tag{E.24}$$

We turn now to additive characters. Let f be an arithmetic function with period q. Our convention is to define the discrete Fourier transform by setting

$$\widehat{f}(k) = \frac{1}{q} \sum_{n=1}^{q} f(n) e(-nk/q).$$

This yields the discrete Fourier expansion

$$f(n) = \sum_{k=1}^{q} \widehat{f}(k) e(kn/q),$$

as in (4.3). Hence if $0 < N \le q$, then

$$\sum_{0<n\le N} f(n) = \sum_{k=1}^{q} \widehat{f}(k) \sum_{0<n\le N} e(kn/q).$$

Here $\widehat{f}(0)$ is the mean value of f, so

$$\sum_{0<n\le N} f(n) - N\widehat{f}(0) = \sum_{0<k<q} \widehat{f}(k) \sum_{0<n\le N} e(kn/q).$$

It is easy to write the sum on the right over n in closed form, but it suffices to observe that it is $\ll \min\left(N, \|k/q\|^{-1}\right)$, by (16.4). Thus the above is

$$\ll \sum_{0<k<q} \left|\widehat{f}(k)\right| \min\left(N, \|k/q\|^{-1}\right).$$

We note that this estimate is much more sensitive to the size of $\widehat{f}(k)$ when k is near a multiple of q (i. e., 0 or q) than otherwise. In any case,

$$\max_{0<N\le q} \left| \sum_{0<n\le N} f(n) - N\widehat{f}(0) \right| \ll (q \log 2q) \max_{0<k<q} \left|\widehat{f}(k)\right|.$$

Suppose that $f(n) = \chi(n)$ where χ is a non-principal character modulo q. Then $\widehat{f}(k)$ can be expressed in terms of Gauss sums, and from Theorems 9.7 and 9.10 we see that $\widehat{f}(k) \ll q^{-1/2}$, and then the above is the Pólya–Vinogradov inequality, as found in Theorem 9.18. The reasoning above is just a generalization of the proof we gave of that theorem. In Exercise E.4.5.2 it is shown that if χ is a primitive character modulo q, then

$$\sum_{N=1}^{q} \left| \sum_{0<n\le N} \chi(n) \right| \gg q^{3/2}.$$

Thus the bound provided by the Pólya–Vinogradov inequality is never more that a factor $\log q$ larger than the truth.

Let

$$D(s) = \sum_{n=1}^{N} a_n n^{-s}. \tag{E.25}$$

In a manner analogous to the above arguments, we now bound the maximal partial sum of $D(0)$ by an integral involving $|D(iu)|$. We begin by noting that

$$\begin{aligned} \int_{-U}^{U} e^{i\beta u} \frac{\sin \alpha u}{u}\, du &= \int_{-U}^{U} \frac{\cos \beta u \, \sin \alpha u}{u}\, du \\ &= \frac{1}{2} \int_{-U}^{U} \frac{\sin(\alpha+\beta)u + \sin(\alpha-\beta)u}{u}\, du \\ &= \operatorname{sgn}(\alpha+\beta) \int_0^{|\alpha+\beta|U} \frac{\sin u}{u}\, du \\ &\quad + \operatorname{sgn}(\alpha-\beta) \int_0^{|\alpha-\beta|U} \frac{\sin u}{u}\, du. \end{aligned} \tag{E.26}$$

We recall that $\int_0^\infty \frac{\sin u}{u}\, du = \pi/2$, and that the *sine integral* $\operatorname{si}(x)$ is defined to be

$$\operatorname{si}(x) = -\int_x^\infty \frac{\sin u}{u}\, du.$$

Thus the expression (E.26) is

$$\operatorname{sgn}(\alpha+\beta)\Big(\frac{\pi}{2}+\operatorname{si}(|\alpha+\beta|U)\Big)+\operatorname{sgn}(\alpha-\beta)\Big(\frac{\pi}{2}+\operatorname{si}(|\alpha-\beta|U)\Big).$$

Let χ_I denote the characteristic function of the interval $I=[-\alpha,\alpha]$, and note that $\operatorname{si}(x)\ll\min(1,1/x)$ for $x\ge 0$, as was recorded in (5.6). Thus the above is

$$=\pi\chi_I(\beta)+O\left(\min\left(1,\frac{1}{U|\alpha-\beta|}\right)\right)+\left(\min\left(1,\frac{1}{U|\alpha+\beta|}\right)\right).$$

For integers K, $0\le K<N$, we take $\alpha=\log(K+1/2)$, $\beta=-\log n$, multiply by a_n, and sum over n. Thus we find that

$$\sum_{n=1}^{K}a_n=\int_{-U}^{U}D(iu)\frac{\sin\alpha u}{u}\,du \\ +O\left(\sum_{n=1}^{N}|a_n|\min\left(1,\frac{1}{U|\log n/(K+1/2)|}\right)\right).$$

Now $(\sin\alpha u)/u\ll\min(|\alpha|,1/|u|)$, and $|\log n/(K+1/2)|\gg 1/N$. Hence

$$\max_{y\le N}\left|\sum_{n\le y}a_n\right|\ll\int_{-U}^{U}|D(iu)|\min(\log N,1/|u|)\,du \\ +\frac{N}{U}\sum_{n=1}^{N}|a_n|. \tag{E.27}$$

Here we can replace a_n by $a_n n^{-it}$ and integrate with respect to t, with or without squaring, depending on the objective. The above is used in §19.4.

E.4.2 The Hardy–Littlewood Maximal Inequality

Suppose that $f\in L^1(\mathbb{T})$. The *Hardy–Littlewood maximal function* of f is

$$M_f(x)=\sup_{0<|y|\le 1/2}\frac{1}{y}\int_x^{x+y}|f(u)|\,du. \tag{E.28}$$

Thus $M_f(x)$ is the maximum of two suprema, namely

$$\sup_{0<y\le 1/2}\frac{1}{y}\int_{x-y}^{x}|f(u)|\,du,\qquad \sup_{0<y\le 1/2}\frac{1}{y}\int_{x}^{x+y}|f(u)|\,du.$$

At first sight, it would seem remarkable that we consider such a non-linear operator, but its value is immediately apparent when we consider the following theorem.

Theorem E.7 (The Hardy–Littlewood maximal theorem) *Suppose that $f \in L^1(\mathbb{T})$ and that M_f is defined as in* (E.27). *If $r > 1$ and $\int_{\mathbb{T}} |f(x)|^r \, dx < \infty$, then*

$$\int_{\mathbb{T}} M_f(x)^r \, dx \le r\Big(\frac{r}{r-1}\Big)^r \int_{\mathbb{T}} |f(x)|^r \, dx.$$

To exhibit how this theorem is useful, recall from §E.1 that the Cesàro partial sums $\sigma_N(f, x)$ of a Fourier series, given in (E.1), are obtained by convolving f with the Fejér kernel (E.2). Let

$$F_N(x) = \min\left(1, \frac{1}{4N\|x\|^2}\right).$$

Then (E.3) asserts that

$$0 \le \Delta_N(x) \le F_N(x).$$

Here $F_N(x)$ is even and monotonically decreasing for $0 \le x < 1/2$, so

$$\sigma_N(x) \ll \int_{-1/2}^{1/2} |f(x-u)| F_N(u) \le M_f(x) \int_{-1/2}^{1/2} F_N(u) \, du \ll M_f(x).$$

Hence $\sup_N |\sigma_N(x)| \ll M_f(x)$. Thus if $r > 1$ and $f \in L^r(\mathbb{T})$, then

$$\| \sup_N |\sigma_N| \|_{L^r(\mathbb{T})} \ll_r \|f\|_{L^r(\mathbb{T})}.$$

This line of reasoning succeeds when we have an even envelope that decreases on $[0, 1/2]$, and has a finite integral. Abelian weights give rise to the Poisson kernel, which is monotonic, so there is no need to construct an envelope. See Exercise E.4.5.5.

E.4.3 The Rademacher–Menchov Device

We now seek to bound the quantity

$$\max_{1 \le \nu \le N} \left| \sum_{n=1}^{\nu} c_n \right|$$

by breaking the sum into short subsums. Let $R = \lceil (\log N)/(\log 2) \rceil$. Numbers of the form $\frac{d}{2^R} N$ form an arithmetic progression with common difference $N/2^R \le 1$, so each interval of the form $[n, n+1)$ contains at least one number of this form. Let $\mathscr{X}$ denote the set of all dyadic rationals of the form $x = \sum_{r=1}^{R} \varepsilon_r(x) 2^{-r}$ where $\varepsilon_r(x) = 0$ or 1. Hence

$$\max_{1 \le \nu \le N} \left| \sum_{n=1}^{\nu} c_n \right| = \max_{x \in \mathscr{X}} \left| \sum_{1 \le n \le xN} c_n \right|.$$

For $x \in \mathscr{X}$ and $1 \le r \le R+1$ we set $d_r = d_r(x) = \sum_{s<r} \varepsilon_s(x) 2^{-s}$. Then

$$\sum_{1 \le xN} c_n = \sum_{r=1}^{R} \sum_{N d_r < n \le N d_{r+1}} c_n.$$

By Cauchy's inequality,

$$\Big|\sum_{1\le xN} c_n\Big|^2 \le R\sum_{r=1}^{R}\Big|\sum_{Nd_r<n\le Nd_{r+1}} c_n\Big|^2 .$$

Here d_r is of the form $s/2^{r-1}$, and either $d_{r+1} = d_r$ or $d_{r+1} = d_r + 1/2^r$. Here s depends on x, but since we do not know its value, we sum over all 2^{r-1} possible values of s. It is somewhat astounding that this can lead to anything useful. In any case,

$$\max_{1\le\nu\le N}\Big|\sum_{n=1}^{\nu} c_n\Big|^2 \le R\sum_{r=1}^{R}\sum_{s=0}^{2^{r-1}-1}\Big|\sum_{\frac{Ns}{2^{r-1}}<n\le\frac{Ns}{2^{r-1}}+\frac{N}{2^r}} c_n\Big|^2 . \qquad \text{(E.29)}$$

For fixed r, n runs through intervals $I_s = (N2^{1-r}s, N2^{1-r}(s+1/2)]$. These intervals I_s are disjoint, their union is a subset of $(0, N]$, and the sum of their lengths is $N/2$.

To see how the above might be applied, replace c_n by $c_n e(nx)$, and integrate. It is immediate that

$$\int_0^1 \max_{1\le\nu\le N}\Big|\sum_{n=1}^{\nu} c_n e(nx)\Big|^2 \ll (\log N)^2\sum_{n=1}^{N}|c_n|^2.$$

Here the $e(nx)$ are orthonormal, but other families of functions for which we have a Bessel-like or bilinear form inequality can be introduced. The bounds obtained in this way are typically weaker than optimal by a factor of R^2. See (E.38).

While (E.29) is interesting and useful, it does not reveal the potential of the Rademacher–Menchov device. We now consider an application in which the power of the approach is fully realized.

Theorem E.8 (Montgomery & Vaughan 1979) *For Dirichlet characters χ modulo q, let $M(\chi) = \max_{1\le N\le q}\big|\sum_{n=1}^{N}\chi(n)\big|$. Then*

$$\sum_{\chi\ne\chi_0} M(\chi)^{2k} \ll_k \varphi(q)q^k$$

for any positive real k.

Thus $M(\chi) \ll q^{1/2}$ for most χ mod q, in the sense that if C is large, then $M(\chi) \le Vq^{1/2}$ with the exception of $\ll_k \varphi(q)/V^{2k}$ characters χ.

Proof By Hölder's inequality we see that the assertion becomes stronger as k increases through real values. Hence it suffices to prove the assertion for a sequence of k tending to infinity. We consider integral $k \ge 2$. In the proof we allow implicit constants to depend on k. We shall show that for $q > 1$ we have

$$\sum_{\chi}{}^{\star} M(\chi)^{2k} \ll \varphi(q)q^k. \tag{E.30}$$

To deduce the theorem from this, let χ be a character modulo q, let $\chi^\star$, modulo r, be the primitive character that induces χ, and let $s = q/r$. Then

$$\begin{aligned}\sum_{\chi \ne \chi_0} M(\chi)^{2k} &\ll \sum_{\substack{r\mid d\\ r>1}} d(q/r)^{2k} \sum_{\chi \bmod r}{}^{\star} M(\chi)^{2k} \\ &\ll \sum_{r\mid q} d(q/r)^{2k} r^k \varphi(r) \ll q^k\varphi(q) \sum_{s\mid q} d(s)^{2k}/s^k \\ &\ll q^k\varphi(q).\end{aligned}$$

Let

$$\mathscr{A} = \{a2^{-R} \, : \, a \in \mathbb{Z}, 0 \le a < 2^R\},$$

where R is an integer to be chosen later. For $\alpha \in \mathscr{A}$ we write $\alpha = \sum_{r=1}^{R} \varepsilon_r 2^{-r}$ with $\varepsilon_r = \varepsilon_r(\alpha) = 0$ or 1. Let $\nu_1 = 0$, and for $r > 1$ let

$$\nu_r = \nu_r(\alpha) = 2^r \sum_{m=1}^{r-1} \varepsilon_m 2^{-m}.$$

Then $\nu_r < 2^r$ and the interval $(0, \alpha]$ is a disjoint union of intervals $(\nu_r 2^{-r}, (\nu_r + \varepsilon_r)2^{-r}]$ for $1 \le r \le R$. Choose $N = N(\chi)$ so that $N < q$ and $\left|\sum_{n=1}^{N} \chi(n)\right| = M(\chi)$. Then there is an $\alpha = \alpha(\chi) \in \mathscr{A}$ such that $N \le \alpha q < N + q2^{-R}$. Hence

$$M(\chi) \le \left| \sum_{1\le n\le \alpha q} \chi(n) \right| + q2^{-R}. \tag{E.31}$$

We take $R = \lfloor (\log q)/(2\log 2) \rfloor$. Thus to prove (E.30) it suffices to show that

$$\sum_{\chi}{}^{\star} \left| \sum_{1\le n\le \alpha q} \chi(n) \right|^{2k} \ll \varphi(q)q^k \tag{E.32}$$

(where of course, $\alpha = \alpha(\chi)$, as above). By Hölder's inequality

$$\begin{aligned}\left| \sum_{1\le n\le \alpha q} \chi(n) \right|^{2k} &= \left| \sum_{r=1}^{R} \sum_{\nu_r 2^{-r} q < n \le (\nu_r+\varepsilon_r)2^{-r} q} \chi(n) \right|^{2k} \\ &\le \left(\sum_{r=1}^{R} r^{-2k/(2k-1)} \right) \left(\sum_{r=1}^{R} r^{2k} \left| \sum_{\nu_r 2^{-r} q < n \le (\nu_r+\varepsilon_r)2^{-r} q} \chi(n) \right|^{2k} \right).\end{aligned} \tag{E.33}$$

In our discussion of the Pólya–Vinogradov inequality in §9.4 (note, especially, pages 309–311), we showed that if χ is a primitive character modulo q, $q > 1$, then for real u and v with $u < v$ we have

$$\sum_{uq<n\le vq} \chi(n) = \tau(\chi) \sum_{0<|h|\le H} \overline{\chi}(h)\frac{e(-hu)-e(-hv)}{2\pi i h} + O\big(1 + qH^{-1}\log q\big).$$

Thus

$$\begin{aligned}\sum_{v_r 2^{-r} q<n\le (v_r+\varepsilon_r)2^{-r}q} \chi(n) \ll{}& q^{1/2}\Big|\sum_{0<h\le H}\chi(h)e(hv_r/2^r)a(h)\Big| \\ &+ q^{1/2}\Big|\sum_{0<h\le H}\overline{\chi}(h)e(hv_r/2^r)a(h)\Big| \\ &+ 1 + qH^{-1}\log q,\end{aligned}$$

where

$$a(h) = a(h,r) = \frac{e(h/2^r)-1}{h} \ll \min\big(2^{-r}, h^{-1}\big). \tag{E.34}$$

Thus by (E.33),

$$\begin{aligned}\sum_{\chi}{}^{\star}\Big|\sum_{n\le\alpha q}\chi(n)\Big|^{2k} \ll{}& \sum_{\chi}{}^{\star}\sum_{r=1}^{R} r^{2k}q^k\Big|\sum_{0<h\le H}\chi(h)e(hv_r/2^r)a(h)\Big|^{2k} \\ &+ \sum_{\chi}{}^{\star}\sum_{r=1}^{R} r^{2k}\big(1+\big(qH^{-1}\log q\big)^{2k}\big).\end{aligned}$$

Here the second sum over χ is

$$\ll \varphi(q)R^{2k+1}\big(1+\big(qH^{-1}\log q\big)^{2k}\big).$$

This is acceptable provided that $H \asymp q^{1/2}(\log q)^3$.

In order to obviate the dependence of v_r on χ, we sum over all possible v. We make no further use of χ being primitive, so we also permit χ to run over all characters modulo q. Therefore, to establish (E.32) it suffices to show that

$$\sum_{\chi}\sum_{r=1}^{R}\sum_{v=0}^{2^r-1} r^{2k}\Big|\sum_{0<h\le H}\chi(h)e(hv2^{-r})a(h)\Big|^{2k} \ll \varphi(q). \tag{E.35}$$

We now write

$$\Big(\sum_{0<h\le H}\chi(h)e(hv2^{-r})a(h)\Big)^k = \sum_{0<h\le H^k}\chi(h)b(h), \tag{E.36}$$

where by (E.34),

$$b(h) = b_k(h;r,v) \ll d_k(h)\min\big(2^{-kr}, h^{-1}\big). \tag{E.37}$$

In Exercise 4.2.1.2 we used the orthogonality property (4.15) to show that

$$\sum_{\chi}\left|\sum_{n=1}^{q} c_n\chi(n)\right|^2 = \varphi(q)\sum_{\substack{n=1\\(n,q)=1}}^{q}|c_n|^2$$

for arbitrary complex numbers c_n. Hence

$$\sum_{\chi}\left|\sum_{n=M+1}^{M+N} c_n\chi(n)\right|^2 = \varphi(q)\sum_{\substack{h=1\\(h,q)=1}}^{q}\left|\sum_{n\equiv h\,(\mathrm{mod}\,q)} c_n\right|^2,$$

so

$$\begin{aligned}\sum_{\chi}&\left|\sum_{0<h\le H^k}\chi(b)b(h)\right|^2\\ &\ll \varphi(q)\sum_{h=1}^{q}\left(\sum_{m=0}^{q^k} d_k(h+mq)\min\left(2^{-kr},(h+mq)^{-1}\right)\right)^2.\end{aligned}$$

For $m\le q^k$ we have $d_k(h+mq)\ll q^{\varepsilon}$. On considering separately the cases $m=0$ and $m>0$, we obtain

$$\begin{aligned}\sum_{\chi}\left|\sum_{0<h\le H^k}\chi(h)b(h)\right|^2 &\ll \varphi(q)\sum_{h=1}^{q} d_k(h)^2\min\left(2^{-2kr},h^{-2}\right)\\ &\quad+\varphi(q)\sum_{h=1}^{q}\left(q^{-1+\varepsilon}\sum_{m=1}^{q^k}1/m\right)^2\\ &\ll \varphi(q)2^{-kr}r^{k^2-1}+q^{3\varepsilon}\end{aligned}$$

since

$$\sum_{s\le x} d_k(s)^2 \ll_k x(\log 2x)^{k^2-1}.$$

We have assumed that $k\ge 2$ and we have chosen R so that $2^R\le q^{1/2}$. Thus the left-hand side of (E.35) is

$$\ll \sum_{r=1}^{R} r^{2k}2^r\left(\varphi(q)2^{-kr}r^{k^2-1}+q^{3\varepsilon}\right) \ll \varphi(q)+q^{4\varepsilon}2^R \ll \varphi(q)$$

as required. □

E.4.4 The Carleson–Hunt Theorem

The most memorable form of the Carleson–Hunt theorem states that if $p>1$ and $f\in L^p(\mathbb{T})$, then the Fourier series of f converges to f almost everywhere.

However, this is in fact a corollary of a much more fundamental result, namely that if $p > 1$, $f \in L^p(\mathbb{T})$ and

$$s^\star(x) = \sup_{K\ge 1}\left|\sum_{k=-K}^{K} \widehat{f}(k)e(kx)\right|,$$

then

$$\int_{\mathbb{T}} |s^\star(x)|^p \, dx \ll_p \int_{\mathbb{T}} |f(x)|^p \, dx.$$

We note in particular that the case $p = 2$ implies that there is an absolute constant C_H such that

$$\int_{\mathbb{T}} \max_{1\le\nu\le N}\left|\sum_{n=1}^{\nu} a_n e(nx)\right|^2 dx \le C_H \sum_{n=1}^{N} |a_n|^2 \tag{E.38}$$

for any choice of the complex numbers a_n. In Chapter 19 this is used to derive maximal versions of the large sieve.

E.4.5 Exercises

1. Let $E_K(x)$ and $s_K(x)$ be defined as in (E.19) and (E.20). Suppose that p and q are real numbers with $1 < p < \infty$ and $1/p + 1/q = 1$. Use Hölder's inequality to show that

$$\|s_K^\star\|_{L^p(\mathbb{T})} \ll (\log 2K)\|f\|_{L^p(\mathbb{T})}.$$

2. Suppose that χ is a primitive character modulo $q > 1$. Then $\widehat{\chi}(-1) = \tau(\chi)$, so $|\widehat{\chi}(-1)| = q^{1/2}$.

   (a) Let $s(u) = \sum_{0<n\le u} \chi(n)$. By Riemann–Stieltjes integration by parts, or otherwise, show that

$$\widehat{\chi}(-1) = \frac{2\pi i}{q^2}\int_1^q s(u)e(u/q)\,du.$$

   (b) Deduce that

$$\frac{1}{q}\int_0^q |s(u)|\,du \ge \frac{q^{1/2}}{2\pi}.$$

   (c) Let $M(\chi)$ be defined as in Theorem E.8. Conclude that $M(\chi) \ge q^{1/2}/(2\pi)$ for all primitive characters modulo q.

3. Let f be an arithmetic function with period q.
   (a) Suppose that M and N are integers, with $0 < N \le q$. Explain why
$$\sum_{n=M+1}^{M+N} f(n)e(an/q) = \sum_{k=1}^{q} \widehat{f}(k) \sum_{n=M+1}^{M+N} e(n(a+k)/q).$$
   (b) Show that the above is
$$\ll \sum_{k=1}^{q} |\widehat{f}(k)| \min\left(N, 1/\|(a+k)/q\|\right).$$
   (c) Deduce that
$$\max_{\substack{1\le M\le q\\1\le N\le q}} \left|\sum_{n=M+1}^{M+N} f(n)e(an/q)\right| \ll \sum_{k=1}^{q} |\widehat{f}(k)| \min\left(q, 1/\|(a+k)/q\|\right). \tag{E.39}$$
   (d) Show that
$$\max_{\substack{1\le M\le q\\1\le N\le q\\1\le a\le q}} \left|\sum_{n=M+1}^{M+N} f(n)e(an/q)\right| \ll q(\log 2q) \max_k |\widehat{f}(k)|. \tag{E.40}$$
   Note that by taking $M = 0$, $N = q$, and a suitably, the left-hand side can be made as large as $q \max |\widehat{f}(k)|$, so the above is within a factor $\log 2q$ of being best possible.
   (e) Show that
$$\sum_{a=1}^{q} \max_{\substack{1\le M\le q\\1\le N\le q}} \left|\sum_{n=M+1}^{M+N} f(n)e(an/q)\right| \ll q(\log 2q) \sum_{k=1}^{q} |\widehat{f}(k)|. \tag{E.41}$$
   (f) Show that
$$\sum_{k=1}^{q} |\widehat{f}(k)| \min\left(q, 1/\|(a+k)/q\|\right)$$
$$\ll (q\log 2q)^{1/2} \left(\sum_{k=1}^{q} |\widehat{f}(k)|^2 \min\left(q, 1/\|(a+k)/q\|\right)\right)^{1/2}.$$
   (g) Deduce that
$$\sum_{a=1}^{q} \max_{\substack{1\le M\le q\\1\le N\le q}} \left|\sum_{n=M+1}^{M+N} f(n)e(an/q)\right|^2 \ll (q\log 2q)^2 \sum_{k=1}^{q} |\widehat{f}(k)|^2. \tag{E.42}$$
   Note that if $M = 0$ and $N = q$, then the left-hand side is $q^2 \sum |\widehat{f}(k)|^2$, so the upper bound is never larger than the truth by more than a factor of $(\log 2q)^2$.

4. Let β be a real number, and set $f(x) = \|x\|^{-1}(-\log \|x\|)^{\beta}$.
   (a) Show that $f \in L^1(\mathbb{T})$ if $\beta < -1$.
   (b) Define $M_f(x)$ as in (E.28). Show that if $\beta \neq -1$, then
   $$M_f(x) \asymp_\beta \|x\|^{-1}(-\log \|x\|)^{1+\beta}.$$
   (c) Conclude that if $-2 < \beta < -1$, then $f \in L^1(\mathbb{T})$, but that $M_f \notin L^1(\mathbb{T})$.
5. For $0 \le r < 1$, the *Poisson kernel* is
   $$P_r(x) = \sum_{k=-\infty}^{\infty} r^{|k|} e(kx) = 1 + 2\sum_{k=1}^{\infty} r^k \cos 2\pi kx. \tag{E.43}$$
   In this context, $r \to 1^1$ corresponds to $K \to \infty$ for a discretely indexed kernel.
   (a) Let r be fixed, $0 \le r < 1$. Show that the series defining P_r is absolutely and uniformly convergent, that $P_r(x)$ is a continuous function of x, and that $\widehat{P_r}(k) = r^{|k|}$ for all integers k.
   (b) Show that
   $$P_r(x) = \frac{1 - r^2}{1 - 2r\cos 2\pi x + r^2}.$$
   (c) Show that
   $$P_r(x) = \frac{1 - r^2}{(1 - r)^2 + 4r\sin^2 \pi x}.$$
   (d) Show that $\int_0^1 P_r(x)\,dx = 1$.
   (e) Show that $P_r(x) \ge 0$ for all x.
   (f) Show that if $1/2 \le r < 1$, then
   $$P_r(x) \le \min\left(\frac{1+r}{1-r}, \frac{1-r}{\sin^2 \pi x}\right).$$
   (g) Show that if $f \in L^1(\mathbb{T})$, then
   $$(f * P_r)(x) = \sum_{k=-\infty}^{\infty} r^{|k|}\widehat{f}(k)e(kx).$$
   (h) Show that if f is continuous and has period 1, then $(f * P_r)(x) \to f(x)$ uniformly as $r \to 1^-$.
   (i) Show that for fixed r, the function $P_r(x)$ is decreasing for $0 \le x \le 1/2$.
   (j) Suppose that $f \in L^1(\mathbb{T})$, and let M_f be defined as in (E.28). Show that $|(P_r * f)(x)| \le M_f(x)$ for all x.
   (k) Show that if $f \in L^r(\mathbb{T})$ with $r > 1$, then $\|\sup_{r<1}|(P_r * f)|\|_{L^r(\mathbb{T})} \ll \|f\|_{L^r(\mathbb{T})}$.

6. (a) Show that

$$\sum_{\chi}\left|\sum_{n=M+1}^{M+N} c_n\chi(n)\right|^2 \le \varphi(q)\left(1+\left\lfloor\frac{N-1}{q}\right\rfloor\right)\sum_{\substack{n=M+1\\(n,q)=1}}^{M+N}|c_n|^2,$$

where the χ run over characters modulo q and the c_n are arbitrary complex numbers.

(b) Show that for any integers $M, N > 1$, $q > 1$, and complex numbers c_n,

$$\sum_{\chi}\max_{1\le\nu\le N}\left|\sum_{M+1}^{M+\nu} c_n\chi(n)\right|^2 \ll (\varphi(q)(\log N)^2 + N\log N)\sum_{\substack{n=M+1\\(n,q)=1}}^{M+N}|c_n|^2.$$

E.5 Notes

Section E.1. The notation $e(x)$ was introduced by the Russian number theorist I. M. Vinogradov. It is particularly useful in analytic number theory where x is often a complicated expression with superscripts and subscripts, which become scriptsize in $e^{2\pi i x}$ but are larger in $e(x)$.

As the nineteenth century drew to a close, it was already clear that many functions in $L^1(\mathbb{T})$ have Fourier series that fail to converge, and the prospects for the future of Fourier analysis looked bleak. But there was a Hungarian teenager, Lipót Fejér, studying in Berlin, who submitted a manuscript in December 1899. The Cesàro partial sums $\sigma_N(x)$ have all the lovely properties that one wishes the unweighted partial sums $s_N(x)$ would have (but generally do not). For example:

1. If $f \in L^1(\mathbb{T})$, then $\|f(x) - \sigma_N(f,x)\|_1 \to 0$ as $N \to \infty$.
2. If $f \in L^1(\mathbb{T})$, then $\sigma_N(f,x) \to f(x)$ a. e. (a theorem of Lebesgue 1905).

Additional useful kernels were invented, and the entire subject was reborn. See Kahane (1981).

Section E.2. In the late 1930s, Arne Beurling showed that if $F \in E^{2\pi}$, $F(x) \ge \operatorname{sgn}(x)$ for all real x, then

$$\int_{\mathbb{R}} F(x) - \operatorname{sgn}(x)\,dx \ge 1,$$

and that equality is attained only when $F(z) = B(z)$ as defined in (E.5). He also showed that if $G \in E_{2\pi}$, then

$$\int_{\mathbb{R}} |G(x) - \operatorname{sgn}(x)|\,dx \ge 1/2$$

with equality if and only if

$$G(z) = \frac{\sin 2\pi z}{\pi}\left(\log 4 + \sum_{n=-\infty}^{\infty} (-1)^n \operatorname{sgn}(n) \left(\frac{1}{n} + \frac{1}{2z - n} \right) \right).$$

This function gives a better approximation in the L^1 norm, but does not lend itself to one-sided approximations. Beurling never published his work on this subject, and Selberg rediscovered Beurling's function $B(z)$ in the early 1970s; see Selberg (1991, p. 226). For full proofs of Beurling's theorems see Vaaler (1985).

Beurling's work has since been extended to find optimal L^1 majorants and minorants for various weights, often with applications to inequalities occurring in analytic number theory. Let λ be a positive real number, and put $E(\lambda, x) = e^{-\lambda x}$ for $x \ge 0$, and $E(\lambda, x) = 0$ for $x < 0$. Graham & Vaaler (1981) found the unique L^1 majorants and minorants (whose Fourier transforms are supported on $[-1, 1]$) for $E(\lambda, x)$, $\operatorname{sgn}(x)e^{-\lambda|x|}$, and $e^{-\lambda|x|}$, and derived a precise form of the Wiener–Ikehara Tauberian theorem, which improves on a less precise version of Heilbronn & Landau (1933a,b). Holt & Vaaler (1996) generalized Beurling's analysis by finding bandlimited functions $S^{\pm}$ such that $S^-(x) \le \operatorname{sgn}(x) \le S^+(x)$ and that $\int_{-\infty}^{\infty}(S^+(x) - S^-(x))|x|^{2\nu+1}\,dx$ is minimized. Here ν is a real parameter, $\nu > -1$. They also used de Brange's theory of Hilbert spaces of entire functions to construct approximations to the characteristic function of a ball in Euclidean space. Carneiro & Vaaler (2010a,b) give best possible bounds for some Hermitian forms, and they determine the unique trigonometric polynomial $u_N(x)$ of degree N and period 1 such that $\log|e(x) - 1| \le u_N(x)$ for all x with $\int_{\mathbb{T}} u_N(x)\,dx$ as small as possible. The least such value is $(\log 2)/(N + 1)$. Suppose that $F_N(z)$ is a monic polynomial of degree N whose roots lie on the unit circle. Then $\max_{|z|=1} \log|F_N(z)|$ is small if the roots of F_N are approximately equally spaced. An upper bound for this maximum is given with sharp constants, in terms of the power sums of the zeros. This situation is the harmonic conjugate of discrepancy as discussed in §F.2. Carneiro & Vaaler (2010b) determine best possible $L^1(\mathbb{R})$ approximations to a wide class of even functions by entire functions of exponential type. Corresponding results are then derived for functions with period 1; in particular, the best approximation in $L^1(\mathbb{T})$ by a trigonometric polynomial of degree at most N to the function $\log|1 - e(x)|$. Carneiro & Chandee (2011) used extremal approximations to refine work of Littlewood concerning the size of the zeta function, assuming RH. Chandee & Soundararajan (2011) give an improved estimate for $|\zeta(1/2 + it)|$ assuming RH. Carneiro, Littmann, & Vaaler (2013) find extremal functions for majorizing, minorizing, and approximating the function $e^{-\pi\lambda x^2}$ by entire functions of exponential type, and provide numerous applications. Carneiro, Chandee, & Milinovich (2013) give two proofs that the estimate $|S(t)| \le (1/4 + o(1))(\log t)/(\log\log t)$ follows from RH. Carneiro & Finder (2015) extend bounds for the zeta function to a wide class of L-functions, assuming the relevant Riemann Hypothesis. Carneiro, Chandee, & Milinovich (2013) give a new and simple proof of the best known bound for $|S(t)|$ assuming RH, and give generalizations to L-functions. Carneiro & Chirre (2018) give sharp bounds for $S_n(t)$ assuming RH.

Suppose that $F_\pm \in L^1(\mathbb{R})$ are functions such that $F_-(x) \le \chi_{[-L/2,L/2]}(x) \le F_+(x)$ for all real x, and supp $\widehat{F}_\pm \subseteq [-\delta, \delta]$. Then

$$\max \int_{\mathbb{R}} F_-(x) - F_-(x)\,dx = L - \delta^{-1} f_-(L\delta),$$

$$\min \int_{\mathbb{R}} F_+(x) - \chi_{[-L/2,L/2]}(x)\,dx = L + \delta^{-1} f_+(L\delta)$$

for some functions $f_\pm$ whose values we would like to know. Selberg's construction using Beurling's function demonstrates that $0 \le f_-(x) \le 1$ and $0 < f_+(x) \le 1$ for all x, that they are both equal to 1 when x is a positive integer, and that they are less than 1 when x is not an integer. Logan (1977) announced that he had identified the function f_+, but he never published his proof. Donoho & Logan (1992) settled the issue when $0 < L\delta < 1$; they showed that

$$\max \int_{\mathbb{R}} F_-(x)\,dx = 0, \quad \min \int_{\mathbb{R}} F_+(x)\,dx = \frac{2}{\delta}\left(1 + \frac{\sin \pi L\delta}{\pi L\delta}\right)^{-1}.$$

Littmann (2013) has identified the extremal $F_\pm$, and has shown that

$$\int_{\mathbb{R}} F_+(x) - F_-(x)\,dx = \frac{2}{\delta}\left(1 + \left|\frac{\sin \pi L\delta}{\sin \pi L\delta}\right|\right)^{-1}$$

when $L\delta \ge 1$, but it seems to be difficult to derive useful formulæ for the $f_\pm$ from his analysis.

Section E.3. As far as we are aware the material in this section is novel.

Section E.4. The proof of the Hardy–Littlewood maximal inequality involves considering the equidistributed rearrangement of a given function. While we speak of *the* Hardy–Littlewood maximal inequality, in fact it is a family of seven theorems, three for an interval $[a, b]$, three for $\mathbb{T}$, and one for the real line. Zygmund (1968, pp. 29–33) gives detailed proofs of all of them.

The Rademacher–Menchov device has its origins in Rademacher (1922) and Menchov (1923). Theorem E.8 originates in Montgomery & Vaughan (1979) where it is also shown that

$$\sum_{2<p\leq P} \max_N \left|\sum_{n=1}^{N} \left(\frac{n}{p}\right)\right|^{2k} \ll_k \pi(P)P^k$$

for all real numbers $k > 0$.

The papers of Carleson (1966) and Hunt (1068) are quite difficult to read. Lacey (2004) has given a more accessible account of the L^2 case.

E.6 References

Barton, J. T., Montgomery, H. L., & Vaaler, J. D. (2001). Note on a Diophantine inequality in several variables, *Proc. Amer. Math. Soc.* **129**(2) 337–345.

Carleson, L. (1966). On convergence and growth of partial sums of Fourier series, *Acta Math.* **116**, 135–157.

Carneiro, E. & Chandee, V. (2011). Bounding $\zeta(s)$ in the critical strip, *J. Number Theory* **131**, 363–384.

Carneiro, E., Chandee, V. & Milinovich, M. B. (2013). Bounding $S(t)$ and $S_1(t)$ on the Riemann hypothesis, *Math. Ann.* **356**, 939–968.

(2015). A note on the zeros of zeta and L-functions, *Math. Z.* **281**, 315–332.

Carneiro, E. & Chirre, A. (2018). Bounding $S_n(t)$ on the Riemann hypothesis, *Math. Proc. Camb. Phil. Soc.* **164**, 259–283.

Carneiro, E. & Finder, R. (2015). On the argument of L-functions, *Bull. Braz. Math. Soc. N. S.* **46**, 601–620.

Carneiro, E., Littmann, F. & Vaaler, J. D. (2013). Gaussian subordination for the Beurling–Selberg extremal problem, *Trans. Amer. Math. Soc.* **365**, 3493–3534.

Carneiro, E. & Vaaler, J. D. (2010a). Some extremal functions in Fourier analysis, II, *Trans. Amer. Math. Soc.* **362**, 5803–5843.

(2010b). Some extremal functions in Fourier analysis, III, *Constr. Appro.* **31**, 259–288.

Chandee, V. & Soundararajan, K. (2011). Bounding $|\zeta(\frac{1}{2}+it)|$ on the Riemann hypothesis, *Bull. London Math. Soc.* **43**, 243–250.

Donoho, D. L. & Logan, B. F. (1992). Signal recovery and the large sieve, *SIAM J. App. Math.* **52**, 577–591.

Graham, S. D. & Vaaler, J. D. (1981). A class of extremal functions for the Fourier transform, *Trans. Amer. Math. Soc.* **265**, 283–302.

Heilbronn, H. & Landau, E. (1933a). Bermerkungen zur vorstehenden Arbeit von Herrn Bochner, *Math. Z.* **37**, 10–16; *Collected Works of Edmund Landau* Vol. 9, Essen: Thales Verlag, 215–221.

(1933b). Anwendungen der N. Wienerschen Methode, *Math. Z.* **37**, 18–21; *Collected Works of Edmund Landau* Vol. 9, Essen: Thales Verlag, 223–226.

Holt, J. J. & Vaaler, J. D. (1996). The Beurling–Selberg extremal functions for a ball in Euclidean space, *Duke Math. J.* **83**, 203–248.

Hunt, R. A. (1968). On the convergence of Fourier series, *Orthogonal Expansions and Their Continuous Analogues (Proc. Conf. Edwardsville, IL 1967)*, Carbondale: Southern Illinois University Press, pp. 235–255.

Kahane, J.-P. (1981). Leopold Fejér et l'analyse mathématique au début du XXe siècle. In *Cahiers du Séminaire d'Histoire des Mathématiques, 2*, pp. 67–84, Paris: Institut Henri Poincaré.

Lacey, M. T. (2004). Carleson's theorem: Proof, complements, variations, *Publ. Mat.* **48** (2), 251–307.

Lebesgue, H. (1905). Recherches sur la convergence des séries de Fourier, *Math. Ann.* **61**, 184–210.

Littmann, F. (2013). Quadrature and extremal bandlimited functions, *SIAM J. Math. Anal.* **45**, 732–747.

Logan, B. F. (1977). Bandlimited functions bounded below over an interval, *Notices Amer. Math. Soc.* **24**, A-331.

Menchov, D. (1923). Sur les séries de fonctions orthogonales (Première Partie. La convergence), *Fund. Math.* **4**, 82–105.

Montgomery, H. L. & Vaughan, R. C. (1979). Mean values of character sums, *Can. J. Math.* **31**, 476–487.

Rademacher, H. (1922). Einige Sätze über Reihen von allgemeinen Orthogonalfunktionen, *Math. Ann.* **87**, 112–138.

Selberg, A. (1991). *Collected Papers*, Vol. II, Berlin: Springer.

Vaaler, J. D. (1985). Some extremal functions in Fourier analysis, *Bull. Amer. Math. Soc. N. S.* **12**, 183–216.

Zygmund, A. (1968). *Trigonometric Series, Volumes I and II*, Cambridge: Cambridge University Press.

Appendix F
Uniform Distribution

F.1 Uniform Distribution (mod 1)

In this appendix we consider the uniform distribution of various quantities, the simplest being that of a sequence of real numbers considered modulo 1. We find that the distribution modulo 1 of a sequence $\{u_n\}$ can be described in terms of the asymptotic size of the associated exponential sums $\sum_{n=1}^{N} e(ku_n)$. Here k runs over integral values, and $e(\theta) = e^{2\pi i\theta}$ is the complex exponential with period 1. This motivates us to develop (in Chapter 16) methods for estimating exponential sums.

Let $u_1, u_2, \ldots$ be a sequence of real numbers, and for $0 \le \alpha \le 1$ let $Z(N, \alpha)$ denote the number of n, $1 \le n \le N$, such that $0 \le u_n \le \alpha \pmod 1$. We say that the sequence $\{u_n\}$ is *uniformly distributed* (mod 1) if

$$\lim_{N\to\infty} \frac{1}{N} Z(N, \alpha) = \alpha \tag{F.1}$$

for all $\alpha \in [0, 1]$. To characterize uniformly distributed sequences we have Weyl's criterion.

Theorem F.1 (Weyl's criterion) *The following are equivalent*:

(i) *The sequence* $\{u_n\}$ *is uniformly distributed*;
(ii) *For every integer* $k \neq 0$,

$$\lim_{N\to\infty} \frac{1}{N} \sum_{n=1}^{N} e(ku_n) = 0\,;$$

(iii) *For each function* f *with period* 1 *that is properly Riemann-integrable on* $[0, 1]$,

$$\lim_{N\to\infty} \frac{1}{N} \sum_{n=1}^{N} f(u_n) = \int_0^1 f(x)\,dx. \tag{F.2}$$

Proof We note that (i) is equivalent to the assertion that (F.2) holds whenever f is the characteristic function χ_I of the interval $I = [0, \alpha] \pmod 1$ where $0 \le \alpha \le 1$. Similarly, (ii) is equivalent to the assertion that (F.2) holds when $f(x) = e(kx)$ for all integers k (including $k = 0$ since (F.2) holds trivially when $f \equiv 1$). Moreover, the relation (F.2) is linear, so that if it holds for several functions, then it holds for any linear combination of (finitely many of) them. Hence (i) asserts that (F.2) holds for step functions with period 1, and (ii) asserts that (F.2) holds for trigonometric polynomials with period 1. We complete the proof by showing that (iii) $\Rightarrow$ (ii) $\Rightarrow$ (i) $\Rightarrow$ (iii).

The implication (iii) $\Rightarrow$ (ii) is trivial since trigonometric polynomials are Riemann-integrable.

To show that (ii) implies (i), we appeal to Theorem E.1. Of course the characteristic function of the arc $[\alpha, \beta]$ is not continuous, so we first construct a continuous one-sided approximation to the characteristic function, whose integrals are close to $\beta - \alpha$. Specifically, let $L_+(x)$ be the piecewise linear function with period 1 whose graph has the vertices $(0, 1+\varepsilon)$, $(\alpha, 1+\varepsilon)$, $(\alpha+\varepsilon, \varepsilon)$, $(1-\varepsilon, \varepsilon)$, $(1, 1+\varepsilon)$, and similarly let $L_-(x)$ be the piecewise linear function with period 1 whose graph has the vertices $(0, -\varepsilon)$, $(\varepsilon, 1-\varepsilon)$, $(\alpha-\varepsilon, 1-\varepsilon)$, $(\alpha, -\varepsilon)$, $(1, -\varepsilon)$. (We may suppose that $0 < \alpha < 1$ and that ε is so small that $2\varepsilon \le \alpha \le 1-2\varepsilon$.) Then the $L_\pm$ are continuous, $L_-(x)+\varepsilon \le \chi_I(x) \le L_+(x)-\varepsilon$ for all x, and the $L_\pm$ are good approximations to χ_I in the L^1-norm since $\int_0^1 L_\pm(x)\,dx = \alpha \pm 2\varepsilon$. By Theorem E.1 there exist trigonometric polynomials $T_\pm(x)$ such that $|L_\pm(x) - T_\pm(x)| < \varepsilon$ for all x. Hence $T_-(x) \le \chi_I(x) \le T_+(x)$ for all x, $\int_0^1 T_-(x)\,dx \ge \alpha - 3\varepsilon$, and $\int_0^1 T_+(x)\,dx \le \alpha + 3\varepsilon$. But then

$$Z(N, \alpha) = \sum_{n=1}^{N} \chi_I(u_n) \le \sum_{n=1}^{N} T_+(u_n),$$

and by the hypothesis (ii) we know that

$$\lim_{N\to\infty} \frac{1}{N} \sum_{n=1}^{N} T_+(u_n) = \int_0^1 T_+(x)\,dx \le \alpha + 3\varepsilon,$$

so it follows that $\limsup_{N\to\infty} Z(N, \alpha)/N \le \alpha + 3\varepsilon$. By arguing similarly with $T_-(x)$, we see that $\liminf_{N\to\infty} Z(N, \alpha)/N \ge \alpha - 3\varepsilon$. Since ε may be taken arbitrarily small, we have (i).

Finally, we show that (i) implies (iii); our method is the same as the one just completed. If $f(x)$ is properly Riemann-integrable on $[0, 1]$, then for any $\varepsilon > 0$ there exist step functions $S_\pm(x)$ such that $S_-(x) \le f(x) \le S_+(x)$, $\int_0^1 f(x) - S_-(x)\,dx < \varepsilon$, and $\int_0^1 S_+(x) - f(x)\,dx < \varepsilon$. By proceeding as above, but with χ_I replaced by f and $T_\pm$ replaced by $S_\pm$, we see that

$$\limsup_{N\to\infty} \frac{1}{N}\sum_{n=1}^{N} f(u_n) \le \lim_{N\to\infty} \frac{1}{N}\sum_{n=1}^{N} S_+(u_n)$$
$$= \int_0^1 S_+(x)\,dx < \int_0^1 f(x)\,dx + \varepsilon,$$

and similarly for the lim inf. Hence we see that (i) implies (iii), and the proof is complete. □

When we consider a real number x (mod 1), or equivalently the fractional part $\{x\}$ of x, we are treating x as a representative of a member of the circle group $\mathbb{T} = \mathbb{R}/\mathbb{Z}$. Similarly, a function with period 1 may be thought of as having domain $\mathbb{T}$. Thus Weyl's criterion can be considered to be a statement concerning the distribution of points $u_1, u_2, \ldots$ in $\mathbb{T}$.

We find it fruitful to cast Weyl's criterion in the language of measure theory. We call a measure μ on $\mathbb{T}$ a *probability measure* if both $\mu(\mathcal{S}) \ge 0$ for all measurable sets $\mathcal{S} \subseteq \mathbb{T}$ and also $\mu(\mathbb{T}) = 1$. Let δ, the *Dirac delta*, denote the probability measure that assigns unit mass to the point 0. Thus $\delta(\mathcal{S}) = 1$ or 0 according to whether $0 \in \mathcal{S}$ or not. The measure δ and the Lebesgue measure λ are examples of probability measures on $\mathbb{T}$. If $u_1, u_2, \ldots$ is a sequence of points in $\mathbb{T}$, then for each N put

$$\mu_N(x) = \frac{1}{N}\sum_{n=1}^{N} \delta(x - u_n). \tag{F.3}$$

Thus μ_N is a probability measure that places mass $1/N$ at each of the points $u_1, u_2, \ldots, u_N$, and hence

$$\int_{\mathbb{T}} f(x)\,d\mu_N = \frac{1}{N}\sum_{n=1}^{N} f(u_n).$$

In general, if μ is a measure on $\mathbb{T}$, then for integers k we define its *Fourier coefficient* $\widehat{\mu}(k)$ to be

$$\widehat{\mu}(k) = \int_{\mathbb{T}} e(-kx)\,d\mu. \tag{F.4}$$

Thus for the special measures μ_N we see that $\widehat{\mu}_N(k) = \frac{1}{N}\sum_{n=1}^{N} e(-ku_n)$. Hence Weyl's criterion asserts that the following three statements are equivalent:

1. $\mu_N([0,\alpha]) \to \alpha$ as $N \to \infty$ for all $\alpha \in [0,1]$ (i.e., $\mu_N \to \lambda$ weakly);
2. For each integer $k \neq 0$, $\widehat{\mu}_N(k) \to 0$ as $N \to \infty$;
3. If f is properly Riemann-integrable on $\mathbb{T}$, then $\int_{\mathbb{T}} f\, d\mu_N \to \int_{\mathbb{T}} f\, d\lambda$ as $N \to \infty$.

Here the restriction to measures of the special shape (F.3) may be dropped since it is easy to see that the proof of Theorem F.1 applies equally to any sequence of probability measures $\mu_1, \mu_2, \ldots$.

As a first application of Weyl's criterion we have the following theorem.

Theorem F.2 *If θ is irrational, then the numbers $n\theta$ are uniformly distributed* (mod 1).

We note that the converse of this theorem is obvious.

Proof From Lemma 16.4 we know that

$$\left|\sum_{n=1}^{N} e(n\alpha)\right| \le \min\left(N, \frac{1}{2\|\alpha\|}\right).$$

On taking $\alpha = k\theta$ in the above, we see that

$$\frac{1}{N}\sum_{n=1}^{N} e(kn\theta) \ll \frac{1}{\|k\theta\| N}.$$

Since θ is irrational, it follows that $k\theta$ is not an integer, so that the above is $\ll 1/N$ with an implicit constant that depends on k and on θ. Thus we have Theorem F.1(ii), and hence the sequence $n\theta$ is uniformly distributed (mod 1). □

Suppose that θ is irrational. Since the numbers $n\theta$ are uniformly distributed modulo 1, it follows that they are dense. That is, for any real β, and any $\varepsilon > 0$, there exist n (even infinitely many n) such that $\|n\theta + \beta\| < \varepsilon$.

F.1.1 Exercises

1. Suppose that $\{u_n\}$ is uniformly distributed (mod 1), and let c be a real number. Put $v_n = u_n + c$. Show that $\{v_n\}$ is uniformly distributed.
2. (a) Suppose that $f \in L^1(\mathbb{T})$. Show that for every $\varepsilon > 0$ there is a trigonometric polynomial $T(x)$ such that $\int_{\mathbb{T}} |f(x) - T(x)|\, dx < \varepsilon$.

   (b) Suppose that f is real-valued, has period 1, and that for every $\varepsilon > 0$ there exist trigonometric polynomials T_+ and T_- such that $T_-(x) \le f(x) \le T_+(x)$ for all x and $\int_{\mathbb{T}} T_+(x) - T_-(x)\, dx < \varepsilon$. Show that f is Riemann-integrable on $[0,1]$.

3. Suppose that f has period 1 and that $\lim_{N\to\infty} \frac{1}{N}\sum_{n=1}^{N} f(u_n)$ exists whenever $\{u_n\}$ is uniformly distributed. Show that f is properly Riemann-integrable.
4. (a) Show that

$$\limsup_{N\to\infty} \frac{1}{N} \operatorname{card}\{n : 1 \le n \le N, \{\log n\} \in [0, 1/2]\} = \frac{e - e^{1/2}}{e - 1}.$$

(b) Show that

$$\liminf_{N\to\infty} \frac{1}{N} \operatorname{card}\{n : 1 \le n \le N, \{\log n\} \in [0, 1/2]\} = \frac{e^{1/2} - 1}{e - 1}.$$

(c) Show that

$$\frac{1}{N}\sum_{n=1}^{N} e(k \log n) = \frac{N^{2\pi i k}}{2\pi i k + 1} + O\left((|k| + 1)\frac{\log N}{N}\right).$$

(d) Show that the sequence $\{\log n\}$ is not uniformly distributed (mod 1).

5. Suppose that $\{u_n\}$ is a sequence such that $\lim_{n\to\infty} u_{n+1} - u_n = \alpha$. Show that if α is irrational, then $\{u_n\}$ is uniformly distributed (mod 1).
6. Let I and $\mathcal{J}$ be arcs of $\mathbb{T}$, and suppose that α is an irrational number. Put $u_n = n\alpha$. Show that for each non-negative integer m the limit

$$d_m = \lim_{N\to\infty} \frac{1}{N} \operatorname{card}\{n \in [1, N] : u_n \in I,\ u_{n-m} \in \mathcal{J}\}$$

exists. Prove that

$$\lim_{M\to\infty} \frac{1}{M}\sum_{m=1}^{M} d_m = |I||\mathcal{J}|.$$

F.2 Quantitative Estimates

Suppose that a sequence $\{u_n\}$ is given, and let $Z(N, \alpha)$ be defined as in §F.1. For $0 \le \alpha \le 1$ put

$$D(N, \alpha) = Z(N, \alpha) - N\alpha.$$

The *discrepancy* of the sequence $\{u_n\}$ is the quantity

$$D^\star(N) = \sup_{0\le\alpha\le 1} |D(N, \alpha)|.$$

As a companion to (i)–(iii) of Theorem F.1, consider the assertion

(iv) $D^\star(N) = o(N)$ as $N \to \infty$.

On the one hand, this is equivalent to asserting that (F.1) holds uniformly in α. Hence (iv) implies part (i) of Theorem F.1. To establish the converse, we first observe that $D(N,\alpha)$ has a sort of one-sided Lipschitz property: If $0 \le \alpha \le \beta \le 1$, then

$$D(N,\beta) - D(N;\alpha) \ge -N(\beta - \alpha)$$

because $Z(N,\alpha)$ is an increasing function of α. Hence if $|D^\star(N, m/M)| < \varepsilon N$ for $m = 0, 1, 2, \dots, M$, then $|D(N,\alpha)| < (\varepsilon + 1/M)N$ for all $\alpha \in [0, 1]$. Thus we see that if (F.1) holds everywhere pointwise, then it holds uniformly, and hence the assertion (iv) above is equivalent to the assertions of Theorem F.1.

Since the discrepancy of a sequence provides a measure of the rate at which the limit (F.1) is attained, it is reasonable to ask for quantitative connections between the size of the discrepancy and of the exponential sums considered in part (ii) of Theorem F.1. Such links can be established in both directions, but since we shall shortly be developing methods for estimating exponential sums, the most useful tool is a bound for the discrepancy in terms of exponential sums.

Theorem F.3 (The Erdős–Turán inequality) *Let $u_1, u_2, \dots, u_N$ be N numbers in $\mathbb{T}$, let $I = [\alpha,\beta]$ be an arbitrary arc of $\mathbb{T}$ of length $\beta - \alpha \le 1$, and let K be an arbitrary positive integer. Then*

$$|\,\mathrm{card}\{n \in [1,N] : u_n \in I\} - (\beta - \alpha)N| \le \frac{N}{K+1} + 3\sum_{k=1}^{K}\frac{1}{k}\left|\sum_{n=1}^{N} e(ku_n)\right|.$$

By taking $\alpha = 0$ and allowing β to vary, we may make the left-hand side as close as we like to $D^\star(N)$, and hence this inequality provides the desired bound for the discrepancy.

Proof We proceed in the same manner as in the proof of Theorem F.1, but now we employ quantitative one-sided trigonometric approximations to χ_I that are sharp in the L^1-norm. Specifically, suppose that $T_-(x)$ is chosen as in Theorem E.4. Then

$$\begin{aligned}\mathrm{card}\{n \in [1,N] : u_n \in I\} &= \sum_{n=1}^{N}\chi_I(u_n) \ge \sum_{n=1}^{N} T_-(u_n)\\ &= \sum_{k=-K}^{K}\widehat{T}_-(k)\sum_{n=1}^{N} e(ku_n).\end{aligned}$$

By Theorem E.4(ii) we know that $\widehat{T}_-(0) = \beta - \alpha - 1/(K+1)$. To estimate the Fourier coefficients $\widehat{T}_-(k)$ for $k \neq 0$, we recall that if f is in $L^1(\mathbb{T})$, then

$$|\widehat{f}(k)| = \left|\int_{\mathbb{T}} f(x)e(-kx)\,dx\right| \leq \int_{\mathbb{T}} |f(x)|\,dx = \|f\|_{L^1}.$$

Since

$$\widehat{\chi}_I(k) = e(-k(\alpha+\beta)/2)\frac{\sin \pi k(\beta-\alpha)}{\pi k} \qquad (k \neq 0),$$

by taking $f = \chi_I - T_-$ we see by Theorem E.4(ii) that

$$\left|e(-k(\alpha+\beta)/2)\frac{\sin \pi k(\beta-\alpha)}{\pi k} - \widehat{T}_-(k)\right| \leq \frac{1}{K+1} \qquad (k \neq 0),$$

and hence that

$$|\widehat{T}_-(k)| \leq \frac{1}{K+1} + \left|\frac{\sin \pi k(\beta-\alpha)}{\pi k}\right| \qquad (k \neq 0).$$

Thus

$$\begin{aligned} \operatorname{card}\{n \in [1,N] : u_n \in I\} &\geq (\beta-\alpha)N - \frac{N}{K+1} \\ &- 2\sum_{k=1}^{K}\left(\frac{1}{K+1} + \left|\frac{\sin \pi k(\beta-\alpha)}{\pi k}\right|\right)\left|\sum_{n=1}^{N} e(ku_n)\right|. \end{aligned} \tag{F.5}$$

But $|\sin u| \leq 1$ and $2\left(\frac{1}{K+1} + \frac{1}{\pi k}\right) \leq 2\left(\frac{1}{k} + \frac{1}{3k}\right) < 3/k$, so this gives the desired lower bound. The corresponding upper bound is proved similarly, using the approximation $T_+(x)$ from Theorem E.4. □

The Erdős–Turán inequality provides a good estimate for the discrepancy in terms of exponential sums, but for short intervals we can do better.

Theorem F.4 *Let $u_1, u_2, \ldots, u_N$ be given, and suppose that K is an integer such that*

$$\sum_{k=1}^{K}\left|\sum_{n=1}^{N} e(ku_n)\right| < N/8.$$

Then any arc $I = [\alpha, \beta]$ of $\mathbb{T}$ of length $\beta - \alpha \geq 3/(K+1)$ contains at least $\frac{1}{3}(\beta-\alpha)N$ of the points u_n.

Proof Since $|\sin u| \leq |u|$, the lower bound in (F.5) is

$$\geq (\beta-\alpha)N - \frac{1}{3}(\beta-\alpha)N - \frac{8}{3}(\beta-\alpha)\sum_{k=1}^{K}\left|\sum_{n=1}^{N} e(ku_n)\right|,$$

which gives the result. □

F.2.1 Exercises

1. Let $\mathscr{A}$ be a dense subset of $[0, 1]$, and suppose that a sequence $\{u_n\}$ is given. Show that if the relation (F.1) holds for all $\alpha \in \mathscr{A}$, then (F.1) holds for all α, and hence $\{u_n\}$ is uniformly distributed (mod 1).
2. Let p be an odd prime, and put $u_n = n^2/p$ for $1 \le n \le p$.

   (a) Show that
   $$\left|\sum_{n=1}^{p} e(ku_n)\right| = \begin{cases} p & \text{if } k \equiv 0 \pmod p, \\ \sqrt{p} & \text{if } k \not\equiv 0 \pmod p. \end{cases}$$
   (b) Show that $D^\star(p) \ll p^{1/2} \log p$.

   (c) This is a special case of what familiar inequality?
3. (a) Show that $D^\star(N) \ge 1/2$ for any sequence $\{u_n\}$ and any $N \ge 1$.

   (b) Show that if N points $u_1, u_2, \ldots, u_N$ are equally spaced (mod 1), then $D^\star(N) \le 1$.
4. (a) Let $u_1, u_2, \ldots, u_M$ and $v_1, v_2, \ldots, v_N$ be two sequences with discrepancies $D^\star(M; \boldsymbol{u})$ and $D^\star(N; \boldsymbol{v})$, respectively. Let $w_1, w_2, \ldots, w_{M+N}$ be the concatenation of these two sequences (i.e., $w_m = u_m$ for $1 \le m \le M$, $w_{M+n} = v_n$ for $1 \le n \le N$), and let $D^\star(M+N; \boldsymbol{w})$ be its discrepancy. Show that $D^\star(M+N; \boldsymbol{w}) \le D^\star(M; \boldsymbol{u}) + D^\star(N; \boldsymbol{v})$.

   (b) Show that if $\|u_n - v_n\| \le \delta$ for all n, then $|D^\star(N; \boldsymbol{u}) - D^\star(N; \boldsymbol{v})| \le \delta N$ for all $N \ge 1$.

   (c) Suppose that $u_n = n\theta + \beta$. Show that if $|\theta - a/q| \le A/q^2$ and $(a, q) = 1$, then $D^\star(q) \le A + 1$.

   (d) In the remaining parts of this exercise, let θ denote the 'golden ratio', $\theta = (1 + \sqrt{5})/2$, and let F_h denote the hth Fibonacci number. Show that $F_h\theta = F_{h+1} + (-1)^{h+1}\theta^{-h}$. Deduce that $|\theta - F_{h+1}/F_h| \le F_h^{-2}$.

   (e) Show also that $(F_h, F_{h+1}) = 1$.

   (f) Deduce that if $N = F_h$ for some h and if $u_1, u_2, \ldots, u_N$ are any N consecutive members of the sequence $\{n\theta\}$, then $D^\star(N) \le 2$.

   (g) Show that any positive integer N may be written in the form $N = F_{h_1} + F_{h_2} + \cdots + F_{h_R}$ where $h_i \ge h_{i+1} + 2 (1 \le i \le R - 1)$.

   (h) Show that if $u_n = n\theta$ where θ is the golden ratio, then $D^\star(N) \ll \log N$.

   (i) Show that
   $$\sum_{k=1}^{F_h} \frac{1}{\|k\theta\|} \asymp F_h \log F_h.$$

(j) Deduce that

$$\sum_{k=1}^{K} \frac{1}{k\|k\theta\|} \asymp (\log K)^2.$$

(k) Conclude that the Erdős–Turán inequality gives a bound weaker than in part (h), namely $D^\star(N) \ll (\log N)^2$.

5. Suppose that a sequence $\{u_n\}$ is given, and let

$$D(N) = \sup_{0\le\alpha\le\beta\le 1} |\operatorname{card}\{n \in [1,N] : \alpha \le u_n \le \beta \pmod 1\} - (\beta-\alpha)N|.$$

(a) Show that $D(N) = \sup_{0\le\alpha\le 1} D(N,\alpha) - \inf_{0\le\alpha\le 1} D(N,\alpha)$.
(b) Show that $D^\star(N) \le D(N) \le 2D^\star(N)$.
(c) Suppose that $v_n = u_n + c$ for all n. Show that $D(N;\boldsymbol{u}) = D(N;\boldsymbol{v})$.

6. Let μ denote a probability measure on $\mathbb{T}$. Show that if $I = [\alpha,\beta]$ is an arc of $\mathbb{T}$, $0 \le \beta-\alpha \le 1$, then for any positive integer K

$$|\mu([\alpha,\beta]) - (\beta-\alpha)| \le \frac{1}{K} + 3\sum_{k=1}^{K}\frac{1}{k}|\widehat{\mu}(k)|,$$

where $\widehat{\mu}(k)$ is defined as in (F.4).

In the next exercise we establish a quantitative version of the implication (iv) $\Rightarrow$ (iii) for a restricted – but important – class of functions.

7. Suppose that $\{u_n\}$ is a given sequence, and that f has bounded variation on $\mathbb{T}$. Show that

$$\left|\sum_{n=1}^{N} f(u_n) - \int_0^1 f(x)\,dx\right| \le D^\star(N)\operatorname{Var}_{\mathbb{T}}(f).$$

(Careful! The Riemann–Stieltjes integral $\int f(\alpha)\,dD(N,\alpha)$ does not exist if f is has a jump discontinuity at any of the points u_n.)

Next we establish a quantitative version of the implication (iv) $\Rightarrow$ (ii).

8. Let $\{u_n\}$ be a given sequence.

(a) Show that if $k \ne 0$, then

$$\sum_{n=1}^{N} e(ku_n) = -2\pi i k\int_0^1 D(N,\alpha)e(k\alpha)\,d\alpha.$$

(b) Show that if $k \ne 0$, then

$$\left|\sum_{n=1}^{N} e(ku_n)\right| \le 2\pi|k|D^\star(N).$$

(c) Now show that the constant 2π in part (b) can be improved. Write $\sum_{n=1}^N e(ku_n) = \rho e(\theta)$ in polar coordinates, so that $\rho = |\sum_{n=1}^N e(ku_n)| = \sum_{n=1}^N \cos 2\pi(ku_n - \theta)$. Show that

$$\left|\sum_{n=1}^N e(ku_n)\right| = 2\pi k \int_0^1 D(N,\alpha)\sin 2\pi(k\alpha-\theta)\,d\alpha \le 4|k|D^\star(N).$$

(d) Construct examples to show that the conclusion of part (c) would be false if the constant 4 were replaced by a number < 4. Suggestion: Consider sequences of N terms where N is even, $N \ge 4$, $1 \le k < N/2$, ε is sufficiently small, and the sequence starts with k repetitions of ε, followed by $\frac{1}{N}, \frac{2}{N}, \dots, \frac{1}{2}-\frac{k}{N}, \frac{1}{2}+\frac{k}{N}, \frac{1}{2}+\frac{k+1}{N}, \dots, \frac{N-2}{N}, \frac{N-1}{N}$, and then k entries of $1-\varepsilon$.

9. (a) Suppose that the points u_n are distinct from 0 (mod 1) and from α (mod 1). Show that

$$D(N,\alpha) = \sum_{n=1}^N s(u_n-\alpha) - s(u_n),$$

where $s(x)$ is the 'sawtooth function' as in Lemma D.1, namely

$$s(x) = \begin{cases} \{x\} - 1/2 & (x \notin \mathbb{Z}), \\ 0 & (x \in \mathbb{Z}). \end{cases} \tag{F.6}$$

(b) Show that

$$\int_0^1 D(N,\alpha)\,d\alpha = \sum_{n=1}^N \frac{1}{2} - \{u_n\}.$$

(c) By using Lemma D.1, or otherwise, show that if α is distinct (mod 1) from the points u_n, then

$$D(N,\alpha) = \left(\sum_{n=1}^N \frac{1}{2} - \{u_n\}\right) + \frac{1}{2\pi i}\lim_{K\to\infty}\sum_{0<|k|\le K}\frac{1}{k}\left(\sum_{n=1}^N e(-ku_n)\right)e(k\alpha).$$

(d) Deduce that

$$\int_0^1 D(N,\alpha)^2\,d\alpha = \left(\sum_{n=1}^N \frac{1}{2} - \{u_n\}\right)^2 + \frac{1}{2\pi^2}\sum_{k=1}^\infty \frac{1}{k^2}\left|\sum_{n=1}^N e(ku_n)\right|^2.$$

10. (a) Suppose that δ is given. Show that if α and $\alpha+\delta$ are both distinct from the u_n (mod 1), then

$$D(N,\alpha+\delta)-D(N,\alpha) = \frac{1}{2\pi i}\lim_{K\to\infty}\sum_{0<|k|\le K}\frac{1}{k}\left(\sum_{n=1}^{N} e(-ku_n)\right)(e(k\delta)-1)e(k\alpha).$$

(b) Deduce that

$$\int_0^1 (D(N,\alpha+\delta)-D(N,\alpha))^2\,d\alpha = \sum_{k\ne 0}\left(\frac{\sin\pi\delta k}{\pi k}\right)^2\left|\sum_{n=1}^{N} e(ku_n)\right|^2.$$

(c) Show that

$$\sum_{k=1}^{K}\left|\sum_{n=1}^{N} e(ku_n)\right|^2 \le 2\pi^2K^2D^\star(N)^2.$$

11. (a) Show that

$$\sum_{|k|<K}\left(1-\frac{|k|}{K}\right)\left|\sum_{n=1}^{N} e(ku_n)\right|^2 = \sum_{m=1}^{N}\sum_{n=1}^{N}\Delta_K(u_m-u_n)\ge NK,$$

where $\Delta_K(\alpha)$ is Fejér's kernel,

$$\Delta_K(\alpha)=\sum_{|k|<K}\left(1-\frac{|k|}{K}\right)e(k\alpha)=\frac{1}{K}\left(\frac{\sin\pi K\alpha}{\sin\pi\alpha}\right)^2.$$

(b) Show that

$$\max_{1\le k\le 2N}\left|\sum_{n=1}^{N} e(ku_n)\right|\ge (N/2)^{1/2}.$$

12. Suppose that $0<u_1\le u_2\le\cdots\le u_N=1$, and put $\delta_n=u_n-n/N$.

(a) Show that

$$\max_{0\le\alpha\le 1} D(N,\alpha) = -N\min_{1\le n\le N}\delta_n.$$

(b) Show that

$$\inf_{0\le\alpha\le 1} D(N,\alpha) = -1-\max_{1\le n\le N}\delta_n.$$

(c) Show that

$$\int_0^1 D(N,\alpha)^2\,d\alpha=\sum_{n=1}^{N}\int_{u_{n-1}}^{u_n}(n-1-N\alpha)^2\,d\alpha,$$

where $u_0=0$.

(d) Deduce that

$$\sum_{n=1}^{N} \delta_n^2 = \frac{1}{N}\int_0^1 D(N,\alpha)^2\,d\alpha + \frac{1}{N}\int_0^1 D(N,\alpha)\,d\alpha + \frac{1}{6N}.$$

(e) Show that if $N > 1$, then

$$\sum_{n=1}^{N} e(u_n) = \sum_{n=1}^{N} (e(\delta_n) - 1)e(n/N).$$

(f) Deduce that

$$\left|\sum_{n=1}^{N} e(u_n)\right| \le 2\pi \sum_{n=1}^{N} |\delta_n|.$$

13. Take the u_n to be the Farey fractions $a/q \in [0,1)$ of order Q. Thus $(a,q) = 1$, $q \le Q$, and

$$N = N(Q) = \sum_{q=1}^{Q} \varphi(q) \sim \frac{3}{\pi^2}Q^2.$$

(a) By considering the contribution of the interval $[1 - 1/Q, 1)$, show that

$$\int_0^1 D(N,\alpha)^2\,d\alpha \ge \frac{N^2}{3Q^3} \asymp Q.$$

(b) Use properties of Ramanujan's sum $c_q(k)$ (as defined in §4.1) to show that

$$\sum_{q=1}^{Q} \sum_{\substack{a=1\\(a,q)=1}}^{q} e(ak/q) = \sum_{d|k} dM(Q/d),$$

where $M(x) = \sum_{n \le x} \mu(n)$ is the summatory function of the Möbius function.

(c) Show that $D^\star(N) = o(N)$ as $N \to \infty$.

(d) Show that for every $Q \ge 1$,

$$\int_0^1 D(N,\alpha)^2\,d\alpha \ge \frac{M(Q)^2}{2\pi^2}.$$

(e) Show that if $Q \ge 1$, then

$$\int_0^1 D(N,\alpha)^2\,d\alpha$$

$$= \frac{1}{4} + \frac{1}{12}\sum_{r \le Q}\left(\prod_{p|r}(1 - p^{-2})\right)\left(\sum_{s \le Q/r} \frac{1}{s}M(Q/(rs))\right)^2.$$

(f) (Franel 1924) Show that the Riemann Hypothesis is equivalent to the assertion that

$$\int_0^1 D(N,\alpha)^2\,d\alpha \ll_\varepsilon Q^{1+\varepsilon}$$

for every $\varepsilon > 0$.

(g) (Franel 1924) Let the numbers δ_n be defined as in Exercise F.2.1.12. Show that the Riemann Hypothesis is equivalent to the estimate

$$\sum_{n=1}^{N} \delta_n^2 \ll_\varepsilon Q^{-1+\varepsilon}.$$

(h) (Landau 1924) Show that the Riemann Hypothesis is equivalent to the estimate

$$\sum_{n=1}^{N} |\delta_n| \ll_\varepsilon Q^{1/2+\varepsilon}.$$

14. Let b be an integer > 1, and suppose that the representation of x in base b is $x = 0.a_1a_2a_3\ldots$ where $0 \le a_n < b$ for all n. Suppose that $c_1, c_2, \ldots, c_K$ are integers such that $0 \le c_k < b$ for all k. We say that x is *normal base* b if

$$\lim_{N\to\infty} \frac{1}{N}\operatorname{card}\{n \in [1,N] : a_{n+k} = c_k\ (1 \le k \le K)\} = \frac{1}{b^K}$$

for each $K \ge 1$ and each of the b^K admissible choices of the c_k.

(a) Show that x is normal base b if and only if the sequence $\{xb^n\}$ is uniformly distributed (mod 1).

(b) Show that the numbers normal to base b form a set of first Baire category (i.e., the set can be expressed as a countable union of nowhere dense sets).

(c) Show that

$$\int_0^1 \left|\sum_{n=1}^{N} e(xb^n)\right|^2 dx = N.$$

(d) Let $D^\star(N;\boldsymbol{u}_x)$ denote the discrepancy of the sequence $\{xb^n\}_{n=1}^N$. Show that

$$\int_0^1 D^\star(N;\boldsymbol{u}_x)\,dx \ll N^{1/2}\log N.$$

(e) Show that almost all real numbers x are normal base b, in the sense of Lebesgue measure theory. This is interesting since as a set of first Baire category one might expect it to be small.

15. *Grössencharaktere* for $\mathbb{Q}(\sqrt{-1})$, continued from Exercise 11.3.14. Show that the number of pairs (a, b) of integers such that $a^2 + b^2 \le x$, $a^2 + b^2$ is prime, and $0 \le \arg(a + ib) \le \theta$ is

$$\frac{2\theta}{\pi}\,\mathrm{li}(x) + O\big(x \exp(-c\sqrt{\log x})\big)$$

uniformly for $0 \le \theta \le 2\pi$.

F.3 Kronecker's Theorem

We now generalize Theorem F.2 to m dimensions. We describe the distribution of the points $(\{qr_1\}, \{qr_2\}, \ldots, \{qr_m\})$ in $[0, 1)^m$. Since it is a nuisance to have to take the fractional part of real numbers, we simply consider $\boldsymbol{p}_q = (qr_1, qr_2, \ldots, qr_m)$ modulo $\mathbb{Z}^m$, or, equivalently, we consider $\boldsymbol{p}_q$ to represent a member of the m-dimensional circle group $\mathbb{T}^m = (\mathbb{R}/\mathbb{Z})^m$.

For $1 \le i \le m$, let $I_i = [\alpha_i, \beta_i]$ be an arc of $\mathbb{T}$ with $0 \le \beta_i - \alpha_i \le 1$, so that $\mathscr{B} = I_1 \times I_2 \times \cdots \times I_m$ is a box in $\mathbb{T}^m$. In the same way that we write a sum or product of numbers as $\sum_{i=1}^n a_i$ or $\prod_{i=1}^n a_i$, we may sometimes write a Cartesian product of a sequence of sets as $\mathsf{X}_{i=1}^n S_i$. Thus $\mathscr{B} = \mathsf{X}_{i=1}^m I_i$. For a given sequence $\boldsymbol{u}_1, \boldsymbol{u}_2, \ldots$ of points in $\mathbb{T}^m$, let

$$Z(N, \mathscr{B}) = \operatorname{card}\{n \in [1, N] : \boldsymbol{u}_n \in \mathscr{B}\}$$

and set

$$D(N, \mathscr{B}) = Z(N, \mathscr{B}) - N\prod_{i=1}^{m}(\beta_i - \alpha_i).$$

We say that the points $\boldsymbol{u}_n$ are *uniformly distributed* in $\mathbb{T}^m$ if $D(N, \mathscr{B}) = o(N)$ as $N \to \infty$ for every such box $\mathscr{B} \subseteq \mathbb{T}^m$. This is all in parallel with our treatment of the case $m = 1$, and we can also define a discrepancy function,

$$D(N) = \sup_{\mathscr{B} \subseteq \mathbb{T}^m} |D(N, \mathscr{B})|,$$

where the supremum is over all boxes as described above. Weyl's criterion extends to this situation in an obvious manner.

Theorem F.5 *Let $\boldsymbol{u}_1, \boldsymbol{u}_2, \ldots$ be a given sequence of points in $\mathbb{T}^m$. Then the following assertions are equivalent.*

(i) *The sequence $\{\boldsymbol{u}_n\}$ is uniformly distributed in $\mathbb{T}^m$.*

(ii) *If* $\boldsymbol{k}$ *is a non-zero lattice point (i.e.,* $\boldsymbol{k} \in \mathbb{Z}^m$, $\boldsymbol{k} \neq \mathbf{0}$*), then*

$$\sum_{n=1}^{N} e(\boldsymbol{k} \cdot \boldsymbol{u}_n) = o(N) \qquad (N \to \infty).$$

(iii) *If* f *is properly Riemann-integrable on* $\mathbb{T}^m$*, then*

$$\lim_{N\to\infty} \frac{1}{N} \sum_{n=1}^{N} f(\boldsymbol{u}_n) = \int_{\mathbb{T}^m} f(\boldsymbol{x})\, d\boldsymbol{x}.$$

(iv) $D(N) = o(N)$ *as* $N \to \infty$.

Proof The arguments of §F.1 carry over to the present context without change, except for the issue of constructing trigonometric majorants and minorants in several dimensions. We consider the majorants first. For $1 \le i \le m$, suppose that $T_i(x)$ is a trigonometric polynomial such that $\chi_{I_i}(x) \le T_i(x)$ for all x, and that $\int_0^1 T_i(x)\, dx \le \beta_i - \alpha_i + \varepsilon$. If we set

$$T_+(\boldsymbol{x}) = \prod_{i=1}^{m} T_i(x_i),$$

then $\chi_{\mathscr{B}}(\boldsymbol{x}) \le T_+(\boldsymbol{x})$ and

$$\int_{\mathbb{T}^m} T_+(\boldsymbol{x})\, d\boldsymbol{x} \le \prod_{i=1}^{m} (\beta_i - \alpha_i + \varepsilon) \le \operatorname{vol}(\mathscr{B}) + \big((1+\varepsilon)^m - 1\big).$$

This suffices as a majorant. As for minorants, we observe that I_i and its complement I_i^c partition $\mathbb{T}$ into two subsets. Hence the Cartesian products of the I_i and their complements partition $\mathbb{T}^m$ into 2^m boxes, say $\mathscr{B}_1, \mathscr{B}_2, \dots, \mathscr{B}_{2^m}$ where we take $\mathscr{B}_1 = \mathscr{B}$. Thus

$$\sum_{k=1}^{2^m} \chi_{\mathscr{B}_k}(\boldsymbol{x}) \equiv 1.$$

If $\chi_{\mathscr{B}_k}(x) \le T_k(\boldsymbol{x})$ for all $\boldsymbol{x}$ and $\int_{\mathbb{T}^m} T_k(\boldsymbol{x})\, d\boldsymbol{x} \le \operatorname{vol}(\mathscr{B}_k) + \varepsilon$, then

$$\chi_{\mathscr{B}}(\boldsymbol{x}) = 1 - \sum_{k=2}^{2^m} \chi_{\mathscr{B}_k}(\boldsymbol{x}) \ge 1 - \sum_{k=2}^{2^m} T_k(\boldsymbol{x}) = T_-(\boldsymbol{x}),$$

say, and $\int_{\mathbb{T}^m} T_-(\boldsymbol{x})\, d\boldsymbol{x} \ge \operatorname{vol}(\mathscr{B}) - \big(2^m - 1\big)\varepsilon$. This suffices to construct the required trigonometric minorant. □

We consider several forms of Kronecker's theorem, the simplest being a natural extension of Theorem F.2.

Theorem F.6 *Let $r_1, r_2, \ldots, r_m$ be real numbers. If the points*

$$\boldsymbol{p}_q = (qr_1, qr_2, \ldots, qr_m)$$

are dense in $\mathbb{T}^m$, then $1, r_1, r_2, \ldots, r_m$ are linearly independent over $\mathbb{Q}$. Conversely, if $1, r_1, r_2, \ldots, r_m$ are linearly independent over $\mathbb{Q}$, then the points $\boldsymbol{p}_q$ are not only dense in $\mathbb{T}^m$ but are uniformly distributed in $\mathbb{T}^m$.

Proof We first show that if the numbers $1, r_1, \ldots, r_m$ are linearly dependent, then the points $\boldsymbol{p}_q$ are not dense. Suppose that

$$u_0 + u_1 r_1 + u_2 r_2 + \cdots + u_m r_m = 0,$$

where the u_i are integers, not all 0. Clearly at least one of $u_1, u_2, \ldots, u_m$ is non-zero; without loss of generality, we may suppose that $u_m \ne 0$. If $\|qr_i\| \le \varepsilon$ for $1 \le i < m$, then

$$\|u_m q r_m\| = \left\| \sum_{i=1}^{m-1} u_i q r_i \right\| \le \sum_{i=1}^{m-1} \|u_i q r_i\| \le \sum_{i=1}^{m-1} |u_i| \|qr_i\| \le \varepsilon \sum_{i=1}^{m-1} |u_i|.$$

Suppose that ε is so small that the last quantity above is $\le |5u_m|^{-1}$. Then the box

$$\|x_1\| \le \varepsilon, \|x_2\| \le \varepsilon, \ldots, \|x_{m-1}\| \le \varepsilon, \left\| x_m - \frac{1}{2u_m} \right\| \le \frac{1}{5|u_m|}$$

contains no point $\boldsymbol{p}_q$, so the $\boldsymbol{p}_q$ are not dense.

Suppose now that $1, r_1, r_2, \ldots, r_m$ are linearly independent over $\mathbb{Q}$. Hence if $\boldsymbol{k} \in \mathbb{Z}^m$, $\boldsymbol{k} \ne \boldsymbol{0}$, then $\boldsymbol{k} \cdot \boldsymbol{r}$ is not an integer. Consequently, by (16.4) it follows that

$$\left| \sum_{q=1}^{Q} e(\boldsymbol{k} \cdot \boldsymbol{p}_q) \right| = \left| \sum_{q=1}^{Q} e(q\boldsymbol{k} \cdot \boldsymbol{r}) \right| \le \frac{1}{2\|\boldsymbol{k} \cdot \boldsymbol{r}\|} = O(1),$$

where the implicit constant depends on $\boldsymbol{k}$ and on $\boldsymbol{r}$. This is $o(Q)$ as $Q \to \infty$, so condition (ii) of Theorem F.5 is satisfied, and hence the $\boldsymbol{p}_q$ are uniformly distributed in $\mathbb{T}^m$. □

Let $\boldsymbol{x}_1, \boldsymbol{x}_2, \ldots$ be a sequence of points in $\mathbb{T}^m$. We may define a probability measure μ_N by placing a mass $1/N$ at each of the points $\boldsymbol{x}_n$ for $1 \le n \le N$, and put

$$\widehat{\mu}(\boldsymbol{k}) = \int_{\mathbb{T}^m} e(-\boldsymbol{k} \cdot \boldsymbol{x})\, d\mu(\boldsymbol{x}) = \frac{1}{N} \sum_{n=1}^{N} e(-\boldsymbol{k} \cdot \boldsymbol{x}_n). \tag{F.7}$$

Then Theorem F.5 could be formulated in terms of the μ_N, and indeed both the theorem and its proof apply equally to any sequence of probability measures. That is, the following assertions are equivalent.

(i) If $\mathscr{B} = \mathsf{X}_{i=1}^m I_i$ is a box in $\mathbb{T}^m$, then $\lim_{N\to\infty} \mu_N(\mathscr{B}) = \operatorname{vol} \mathscr{B}$.
(ii) If $\boldsymbol{k} \in \mathbb{Z}^m$, $\boldsymbol{k} \neq \boldsymbol{0}$, then $\lim_{N\to\infty} \widehat{\mu}_N(\boldsymbol{k}) = 0$.
(iii) If f is properly Riemann-integrable on $\mathbb{T}^m$, then

$$\lim_{N\to\infty} \int_{\mathbb{T}^m} f(\boldsymbol{x})\, d\mu_N(\boldsymbol{x}) = \int_{\mathbb{T}^m} f(\boldsymbol{x})\, d\boldsymbol{x}.$$

(iv) $\lim_{N\to\infty} \sup_{\mathscr{B}} |\mu_N(\mathscr{B}) - \operatorname{vol} \mathscr{B}| = 0$ where the supremum is over all boxes $\mathscr{B} = \mathsf{X}_{i=1}^m I_i \subseteq \mathbb{T}^m$.

In particular, if $\boldsymbol{p}(t)$ is the position vector of a continuous curve in $\mathbb{T}^m$, then we can define a probability measure

$$\mu_T(\mathscr{S}) = \frac{1}{T} \operatorname{meas}\{t \in [0, T] : \boldsymbol{p}(t) \in \mathscr{S}\}.$$

Thus if f is Riemann-integrable over $\mathbb{T}^m$, then

$$\int_{\mathbb{T}^m} f(\boldsymbol{x})\, d\mu_T(\boldsymbol{x}) = \frac{1}{T} \int_0^T f(\boldsymbol{p}(t))\, dt,$$

and hence we see that the curve $\boldsymbol{p}(t)$ is uniformly distributed in $\mathbb{T}^m$ if and only if

$$\int_0^T e(\boldsymbol{k} \cdot \boldsymbol{p}(t))\, dt = o(T) \qquad (T \to \infty)$$

for every non-zero lattice point $\boldsymbol{k} \in \mathbb{Z}^m$. A situation of this kind arises in our second formulation of Kronecker's theorem.

Theorem F.7 *Suppose that $r_1, r_2, \ldots, r_m$ are real numbers and let*

$$\boldsymbol{p}(t) = (tr_1, tr_2, \ldots, tr_m) \in \mathbb{T}^m,$$

where t is a real parameter. If the set $\mathscr{P} = \{\boldsymbol{p}(t) : t \in \mathbb{R}\}$ is dense in $\mathbb{T}^m$, then $r_1, r_2, \ldots, r_m$ are linearly independent over $\mathbb{Q}$. Conversely, if $r_1, r_2, \ldots, r_m$ are linearly independent over $\mathbb{Q}$, then $\mathscr{P}$ is not only dense in $\mathbb{T}^m$ but also uniformly distributed in the sense that

$$\lim_{T\to\infty} \sup_{\mathscr{B}} \left| \frac{1}{T} \operatorname{meas}\{t \in [0, T] : \boldsymbol{p}(t) \in \mathscr{B}\} - \operatorname{vol} \mathscr{B} \right| = 0,$$

where the supremum is taken over all boxes $\mathscr{B} = \mathsf{X}_{i=1}^m I_i \subseteq \mathbb{T}^m$.

It is easy to demonstrate by elementary reasoning that Theorems F.6 and F.7 are equivalent (see Exercise F.3.1.1). Thus it is possible to present Theorem F.7 as a consequence of Theorem F.6, but we find it instructive to derive it independently.

Proof Suppose that the r_i are linearly dependent, say $\boldsymbol{u} \cdot \boldsymbol{r} = 0$ where $\boldsymbol{u} \in \mathbb{Z}^m$ and $u_m > 0$. Hence if $\boldsymbol{p} \equiv t\boldsymbol{r} \pmod{\mathbb{Z}^m}$ for some t, then $\boldsymbol{u} \cdot \boldsymbol{p} = 0$. Let $\mathscr{B}$ denote the box of points $\boldsymbol{x}$ for which $|x_i| \le \varepsilon$ for $1 \le i < m$, and $|x_m - 1/(2u_m)| \le 1/(5u_m)$. If $\boldsymbol{x} \in \mathscr{B}$, then

$$|\boldsymbol{u} \cdot \boldsymbol{x} - 1/2| \le |u_m x_m - 1/2| + \sum_{i=1}^{m-1} |u_i x_i| \le \frac{1}{5} + \varepsilon \sum_{i=1}^{m-1} |u_i|.$$

Fix $\varepsilon > 0$ so that the last term above is $\le 1/5$. Then $\boldsymbol{u} \cdot \boldsymbol{x} \ge 1/10$, so $\boldsymbol{u} \cdot \boldsymbol{x} \ne 0$. Thus $\mathscr{B}$ contains no point of the curve $t\boldsymbol{r}$, and so the curve is not dense in $\mathbb{T}^m$.

Now suppose that the r_i are linearly independent over $\mathbb{Q}$. On defining the measure μ_T as above, we see that if $\boldsymbol{k} \in \mathbb{Z}^m$, $\boldsymbol{k} \ne \boldsymbol{0}$, then

$$\widehat{\mu}_T(\boldsymbol{k}) = \frac{1}{T} \int_0^T e(-\boldsymbol{k} \cdot t\boldsymbol{r})\, dt = \frac{1}{T} \cdot \frac{1 - e(-\boldsymbol{k} \cdot T\boldsymbol{r})}{2\pi i(\boldsymbol{k} \cdot \boldsymbol{r})} \longrightarrow 0$$

as $T \to \infty$. Hence the curve $t\boldsymbol{\theta}$ is uniformly distributed in $\mathbb{T}^m$. □

Theorem F.7 provides useful information concerning the values taken by exponential polynomials, as follows.

Corollary F.8 *Suppose that $f(t) = \sum_{r=1}^R a_r e(\lambda_r t)$ where the λ_r are real numbers. Suppose also that $|a_1| \ge |a_2| \ge \cdots \ge |a_R|$. If $|a_1| \le \sum_{r=2}^R |a_r|$, then the values of $f(t)$ for $t \in \mathbb{R}$ lie in the disk $|z| \le \sum_{r=1}^R |a_r|$. If $|a_1| > \sum_{r=2}^R |a_r|$, then the values of $f(t)$ lie in the annulus $|a_1| - \sum_{r=2}^R |a_r| \le |z| \le |a_1| + \sum_{r=2}^R |a_r|$. If the λ_r are linearly independent over $\mathbb{Q}$, then the values of $f(t)$ are dense in this disc (or annulus).*

Proof It is clear that the disc (or annulus) described is the set of points z that can be written in the form $z = \sum_{r=1}^R a_r e(\alpha_r)$. If the λ_r are linearly independent over $\mathbb{Q}$, then for any $\varepsilon > 0$ there exist real numbers t such that $\|\lambda_r t - \alpha_r\| \le \varepsilon$ and hence $|f(t) - z| \le C\varepsilon$ where $C = 2\pi \sum_{r=1}^R |a_r|$. □

We have found that if the numbers r_i are linearly independent over $\mathbb{Q}$, then the curve $t\boldsymbol{r}$ in $\mathbb{T}^m$ passes through any given box $\mathscr{B}$ infinitely many times. But we can actually prove a little more, namely that the gaps between returns to $\mathscr{B}$ are uniformly bounded. This is critical to our discussion of almost-periodic functions in §F.4.

Corollary F.9 *Suppose that the real numbers $r_1, r_2, \ldots, r_m$ are linearly independent over $\mathbb{Q}$, and let $\varepsilon > 0$ be given. Then there is a number $H > 0$, depending only on ε and the numbers $r_1, r_2, \ldots, r_m$, such that for any real number T and any $\boldsymbol{\alpha} \in \mathbb{T}^m$ there is a real number t, $T \le t \le T + H$, such that $\|tr_i - \alpha_i\| \le \varepsilon$ for $1 \le i \le m$.*

In order to clarify the relation of this new result to our earlier ones, we provide two proofs.

First Proof By Theorem F.7 we know that if H is sufficiently large, then

$$\left|\frac{1}{H}\operatorname{meas}\{t \in [0, H] : t\boldsymbol{r} \in \mathscr{B}\} - \operatorname{vol}\mathscr{B}\right| \le \varepsilon^m$$

for all boxes $\mathscr{B}$ in $\mathbb{T}^m$. Thus if $\operatorname{vol}\mathscr{B} > \varepsilon^m$, then there is a t, $0 \le t \le H$, such that $t\boldsymbol{r} \in \mathscr{B}$. Let $\mathscr{B}_0 = [-\varepsilon, \varepsilon]^m$; this is a box centred at $\mathbf{0}$ whose volume is $(2\varepsilon)^m > \varepsilon^m$. Then $\boldsymbol{c} + \mathscr{B}$ is a box of the same size, centred at $\boldsymbol{c}$. By taking $\mathscr{B} = \boldsymbol{c} + \mathscr{B}_0$, we see that for every $\boldsymbol{c}$ there is a t, $0 \le t \le H$, such that $t\boldsymbol{r} \in \boldsymbol{c} + \mathscr{B}_0$, which is to say that $\|tr_i - c_i\| \le \varepsilon$ for all i. Now take $c_i = \alpha_i - Tr_i$. Then $\|(t+T)r_i - \alpha_i\| \le \varepsilon$ for all i, and $T \le T + t \le T + H$. □

Second Proof Let $\mathscr{B}_0 = [-\varepsilon, \varepsilon]^m$ be a small box in $\mathbb{T}^m$ centred at $\mathbf{0}$, and let $T_-(\boldsymbol{x})$ be a trigonometric polynomial in m variables such that

$$T_-(\boldsymbol{x}) \le \chi_{\mathscr{B}_0}(\boldsymbol{x})$$

for all $\boldsymbol{x}$, and

$$\int_{\mathbb{T}^m} T_-(\boldsymbol{x})\, d\boldsymbol{x} > 0.$$

Then

$$\begin{aligned}\int_T^{T+H} \chi_{\mathscr{B}_0}(t\boldsymbol{r} - \boldsymbol{\alpha})\, dt &\ge \int_T^{T+H} T_-(t\boldsymbol{r} - \boldsymbol{\alpha})\, dt \\ &= \sum_{\boldsymbol{k}} \widehat{T}_-(\boldsymbol{k}) e(-\boldsymbol{k}\cdot\boldsymbol{\alpha}) \int_T^{T+H} e(t\boldsymbol{k}\cdot\boldsymbol{r})\, dt.\end{aligned}$$

Now if $\theta \ne 0$, then

$$\left|\int_T^{T+H} e(t\theta)\, dt\right| = \left|\frac{e((T+H)\theta) - e(T\theta)}{2\pi i\theta}\right| = \left|\frac{\sin \pi H\theta}{\pi\theta}\right| \le \frac{1}{\pi|\theta|}.$$

By hypothesis the r_i are linearly independent over $\mathbb{Q}$, which is to say that $\boldsymbol{k}\cdot\boldsymbol{r} \ne 0$ when $\boldsymbol{k} \in \mathbb{Z}^m$, $\boldsymbol{k} \ne \mathbf{0}$. Hence

$$\int_T^{T+H} \chi_{\mathscr{B}_0}(t\boldsymbol{r} - \boldsymbol{\alpha})\, dt \ge \widehat{T}_-(\mathbf{0})H - \sum_{\boldsymbol{k}\ne\mathbf{0}} \frac{|\widehat{T}_-(\boldsymbol{k})|}{\pi|\boldsymbol{k}\cdot\boldsymbol{r}|}.$$

Here the sum has finitely many summands because T_- is a trigonometric polynomial. Since $\widehat{T}_-(\mathbf{0}) > 0$, we see that if H is large enough, then the right-hand side above is positive, and so there is a t, $T \le t \le T + H$, such that $\|tr_i - \alpha_i\| \le \varepsilon$ for $1 \le i \le m$. □

In the proof of Theorem F.5, we constructed our minorant by rather inefficient means, which would be quantitatively inferior in higher dimensions. When quantitative precision is desired, the following construction may be useful.

Theorem F.10 *For $1 \le i \le m$, let I_i be intervals of $\mathbb{R}$ or arcs of $\mathbb{T}$, and let $\mathcal{B}$ be their Cartesian product. Suppose that $S_i^-(x) \le \chi_{I_i}(x) \le S_i^+(x)$ are respectively minorants and majorants of the characteristic function of I_i. Set $\boldsymbol{x} = (x_1, x_2, \ldots, x_m)$. Then*

$$S^-(\boldsymbol{x}) = \prod_{i=1}^m S_i^+(x_i) - \sum_{i=1}^m \left(S_i^+(x_i) - S_i^-(x_i)\right) \prod_{\substack{1\le j\le m\\ j\ne i}} S_j^+(x_j)$$

is a minorant of the characteristic function of $\mathcal{B}$.

Proof Suppose first that there is a k such that $S_k^-(x_k) \le 0$. Then

$$\begin{aligned} S^-(\boldsymbol{x}) &\le \prod_{i=1}^m S_i^+(x_i) - \left(S_k^+(x_k) - S_k^-(x_k)\right) \prod_{\substack{1\le j\le m\\ j\ne k}} S_j^+(x_j) \\ &= S_k^-(x_k) \prod_{\substack{1\le j\le m\\ j\ne k}} S_j^+(x_j) \\ &\le 0 \end{aligned}$$

since the first factor is ≤ 0 and all factors in the product are ≥ 0. If $\boldsymbol{x} \notin \mathcal{B}$, then there is a k such that $x_k \notin I_k$, so $S_k^-(x_k) \le 0$, and hence $S^-(\boldsymbol{x}) \le 0$.

Suppose now that $S_i^-(x_i) > 0$ for all i. Hence $x_i \in I_i$ for all i, so $\boldsymbol{x} \in B$. By induction on m we show that in this case,

$$S^-(\boldsymbol{x}) \le \prod_{i=1}^m S_i^-(x_i). \tag{F.8}$$

This is obvious when $m = 1$. We observe that

$$S^-(\boldsymbol{x}) = S_1^-(x_1) \prod_{i=2}^m S_i^+(x_i) - \sum_{i=2}^m \left(S_i^+(x_i) - S_i^-(x_i)\right) \prod_{\substack{j=1\\ j\ne i}}^m S_j^+(x_j).$$

Since $i \ge 2$ in the sum, the factor $S_1^+(x_1)$ always occurs in the second product. If we replace $S_1^+(x_1)$ by $S_1^-(x_1)$, then the product is made smaller, and the overall contribution larger. Hence the above is

$$\le S_1^-(x_1) \left(\prod_{i=2}^m S_i^+(x_i) - \sum_{i=2}^m \left(S_i^+(x_i) - S_i^-(x_i)\right) \prod_{\substack{j=2\\ j\ne i}}^m S_j^+(x_j) \right).$$

By the inductive hypothesis, the quantity inside the large parentheses is

$$\le \prod_{i=2}^{m} S_i^-(x_i),$$

so we have (F.8). Since $0 < S_i^-(x_i) \le 1$ for all i, it follows from (F.8) that $S^-(\boldsymbol{x}) \le 1 = \chi_B(\boldsymbol{x})$ so the proof is complete. □

F.3.1 Exercises

1. (a) Apply Theorem F.7 to the $m+1$ numbers $1, r_1, \ldots, r_m$. Deduce that there exist real numbers t such that $\|t - c_0\| < \varepsilon$, $\|tr_i - c_i\| < \varepsilon$ for $1 \le i \le m$. Take $c_0 = 0$, and hence deduce that t is near some integer, say q. Show that the numbers $\|qr_i - c_i\|$ are small, and hence deduce Theorem F.6.
   (b) Suppose that $r_1, \ldots, r_m$ are linearly independent numbers, and choose α to be linearly independent of them. Thus $1, r_1/\alpha, r_2/\alpha, \ldots, r_m/\alpha$ are linearly independent over $\mathbb{Q}$. Apply Theorem F.6 to obtain Theorem F.7 with $t = q/\alpha$.
2. Extend Theorem F.6 to allow for the possibility of linear dependances among 1 and the r_i, as follows. Let $r_1, r_2, \ldots, r_m$ and $\alpha_1, \alpha_2, \ldots, \alpha_m$ be real numbers. Show that the following two assertions are equivalent.
   (a) For every $\varepsilon > 0$ there is an integer q such that $\|qr_i - \alpha_i\| < \varepsilon$ for $i = 1, 2, \ldots, m$.
   (b) Let $u_1, u_2, \ldots, u_m$ be integers. If $\sum_{i=1}^m u_i r_i \in \mathbb{Z}$, then $\sum_{i=1}^m u_i \alpha_i \in \mathbb{Z}$.
3. Extend Theorem F.7 to allow for the possibility of linear dependances among the r_i, as follows. Let $r_1, r_2, \ldots, r_m$ and $\alpha_1, \alpha_2, \ldots, \alpha_m$ be real numbers. Show that the following two assertions are equivalent.
   (a) For every $\varepsilon > 0$ there exists a real number t such that $\|tr_i - \alpha_i\| < \varepsilon$ for $i = 1, 2, \ldots, m$.
   (b) If $u_1, u_2, \ldots, u_m$ are integers such that $\sum_{i=1}^m u_i r_i = 0$, then
   $$\sum_{i=1}^{m} u_i \alpha_i \in \mathbb{Z}.$$
4. Explain how we know that the numbers $\log p$ are linearly independent over the field of rational numbers.
5. Let $f(t) = \sum_{r=1}^R a_r \cos(\lambda_r t + \theta_r)$ where the a_r, the λ_r, and the θ_r are real numbers. Show that if the λ_r are linearly independent over $\mathbb{Q}$, then $(-A, A) \subseteq \operatorname{range} f \subseteq [-A, A]$ where $A = \sum_{r=1}^R |a_r|$.

6. (Selberg, oral communication) In this exercise we develop an alternative to the construction of Theorem F.10. We suppose that $0 \le A_i(x) \le 1$ for all x, that $P_i(x) \ge 0$ for all x, and that $A_i(x) - P_i(x) \le 0$ when $x \notin I_i$. (We think of A_i as being an approximation to the characteristic function of I_i, and of $P_i(x)$ as a peak function that compensates for the error in this approximation.) Put

$$S^-(\boldsymbol{x}) = \prod_{i=1}^{m} A_i(x_i) - \sum_{i=1}^{m} P_i(x_i) \prod_{\substack{1 \le j \le m \\ j \ne i}} A_j(x_j).$$

   (a) Show that $S^-(\boldsymbol{x}) \le 1$ for all $\boldsymbol{x}$.
   (b) Suppose that $x_k \notin I_k$. Show that

$$\begin{aligned} S^-(\boldsymbol{x}) &\le \prod_{i=1}^{m} A_i(x_i) - P_k(x_k) \prod_{\substack{1 \le j \le m \\ j \ne k}} A_j(x_j) \\ &= \big(A_k(x_k) - P_k(x_k)\big) \prod_{\substack{1 \le j \le m \\ j \ne k}} A_j(x_j) \le 0. \end{aligned}$$

   (c) Conclude that $S^-(\boldsymbol{x})$ minorizes the characteristic function of the box $\mathscr{B} = \times_{k=1}^{m} I_k$.

F.4 Almost Periodicity

The definition of almost periodicity is governed by our desire to characterize those functions $f(x)$ of a real variable that can be uniformly approximated by exponential polynomials, i.e., by finite sums of the form

$$P(x) = \sum_{m=1}^{M} a_m e(\lambda_m x). \tag{F.9}$$

To this end we call t an ε *almost-period* if $|f(x+t) - f(x)| < \varepsilon$ for all real x. The appropriate definition of almost periodicity is a little elusive because the mere existence of large almost-periods does not ensure that $f(x)$ can be uniformly approximated by exponential polynomials. The little bit more that is required is suggested by Corollary F.9.

Definition F.1 Suppose that $f(x)$ is a continuous function of a real variable. Then $f(x)$ is *almost periodic* if for every $\varepsilon > 0$ there exists a number H (depending on f and ε) such that every interval $[T, T+H]$ contains an ε almost-period of f.

We use Corollary F.9 to show that an exponential polynomial of the form (F.9) is almost periodic.

Theorem F.11 *Let $P(x)$ be defined as in* (F.9). *Then $P(x)$ is an almost-periodic function.*

Proof Let $\tau_1, \tau_1, \ldots, \tau_R$ be a maximal linearly independent subset of the λ_m. Thus there is an $M \times R$ matrix A with rational elements such that $\boldsymbol{\lambda} = A\boldsymbol{\tau}$. Let q be the least common denominator of the elements of A. Put $B = qA$, $\boldsymbol{\theta} = \frac{1}{q}\boldsymbol{\tau}$. Then the θ_r are linearly independent over $\mathbb{Q}$, and $\boldsymbol{\lambda} = B\boldsymbol{\theta}$, which is to say that each λ_m is an integral linear combination of the θ_r. That is, there is a trigonometric polynomial

$$T(\boldsymbol{x}) = \sum_{\boldsymbol{k}} c(\boldsymbol{k}) e(\boldsymbol{k} \cdot \boldsymbol{x})$$

in R variables such that $P(x) = T(x\boldsymbol{\theta})$. By Corollary F.9 we know that for any $\varepsilon > 0$ there is an H such that for any T there is a t, $T \le t \le T + H$, such that $\|t\theta_r\| \le \varepsilon$. Then

$$\begin{aligned}
|P(x+t) - P(x)| &= |T((x+t)\boldsymbol{\theta}) - T(x\boldsymbol{\theta})| \\
&= \left| \sum_{\boldsymbol{k}} c(\boldsymbol{k}) e(\boldsymbol{k} \cdot x\boldsymbol{\theta})(e(\boldsymbol{k} \cdot t\boldsymbol{\theta}) - 1) \right| \\
&\le 2\pi \sum_{\boldsymbol{k}} |c(\boldsymbol{k})| \, \|\boldsymbol{k} \cdot t\boldsymbol{\theta}\| \le 2\pi\varepsilon \sum_{\boldsymbol{k}} |c(\boldsymbol{k})| \sum_{r=1}^{R} |k_r| \\
&= C\varepsilon,
\end{aligned}$$

say. Thus t is a $C\varepsilon$ almost period of P. □

F.4.1 Exercises

1. Show that if $f(x)$ is almost periodic, then $f(x)$ is uniformly bounded.
2. Bohl (1906) Suppose that $f_1(x), f_2(x), \ldots, f_R(x)$ are periodic continuous functions. Show that $f_1(x) + f_2(x) + \cdots + f_R(x)$ is almost-periodic.
3. Let $f(x) = \sum_{m=1}^{\infty} a_m e(\lambda_m x)$ where $\sum_{m=1}^{\infty} |a_m| < \infty$ and the λ_m are distinct real numbers.

   (a) Show that $f(x)$ is almost-periodic.
   (b) Show that

$$\lim_{T\to\infty} \frac{1}{T} \int_0^T f(x) e(-\lambda x)\, dx = \begin{cases} a_m & \text{if } \lambda = \lambda_m \text{ for some } m, \\ 0 & \text{otherwise.} \end{cases}$$

(c) Show that

$$\lim_{T\to\infty} \frac{1}{T}\int_0^T |f(x)|^2\,dx = \sum_{m=1}^{\infty} |a_m|^2.$$

4. Let $\alpha(s) = \sum_{n=1}^{\infty} a_n n^{-s}$ be a Dirichlet series with abscissa of absolute convergence σ_a. Let σ be fixed, $\sigma > \sigma_a$, and put $f(t) = \alpha(\sigma + it)$.
   (a) Show that the preceding exercise applies to $f(t)$.
   (b) Let $\mathcal{L}(f)$ denote the set of those λ for which $c(\lambda) \neq 0$. Show that $\mathscr{L}(f) = \{\frac{-1}{2\pi}\log n \colon a_n \neq 0\}$.
5. (a) Suppose that f is an almost-periodic function, and that there is a $\delta > 0$ such that $|f(x)| \geq \delta$ for all real x. Show that $1/f(x)$ is an almost-periodic function.
   (b) Let $p_1, p_2, \ldots$ denote the prime numbers in increasing order. Put

$$f(t) = \sum_{r=1}^{\infty} \frac{p_r^{-it}}{r(r+1)}.$$

   Show that there is no real t such that $f(t) = 0$, but that $1/f(t)$ is not almost-periodic.

F.5 Notes

Section F.1. Weyl's Criterion originates in Weyl (1916).

For an extended discussion of uniform distribution, see Kuipers & Niederreiter (1974). There are many other applications of the techniques developed here, and the related approximation theorems of Appendix E, in analytic number theory, and even beyond this in areas such as dynamical systems. See, for example, Baker (1986), Brüdern & Fouvry (1996), Cochran (1988), Gonek & Montgomery (2016a,b), Harman (1993), Schwarz (1994), and Turán (1960).

Section F.2. Theorem F.3, with somewhat larger constants, was proved by Erdős & Turán (1948). Theorem F.4 is found in Vaaler (1985, Corollary 21) and Montgomery (1994, p. 8).

Section F.3. Kronecker (1884) achieved his general theorem using only the simplest algebraic and arithmetic tools. Many proofs of our Theorems F.6 and F.7 have been published; see for example Chen (2000). For a survey of these proofs as well as a sharp quantitative treatment, see Gonek & Montgomery (2016b).

The quantities $qr_1, qr_2, \ldots, qr_m$ of Theorem F.6 are linear forms in the single variable q. Kronecker considered linear forms in n variables $q_1, q_2, \ldots, q_n$, so his linear forms were $\sum_{j=1}^{n} r_{ij}q_j$ for $i = 1, 2, \ldots, m$. His full results are therefore as follows.

Theorem A *Let $R = [r_{ij}]$ be an $m \times n$ matrix with real entries, and suppose that $\boldsymbol{\alpha} \in \mathbb{R}^m$. The following two assertions are equivalent.*

(i) *For every $\varepsilon > 0$ there is a $\boldsymbol{t} \in \mathbb{R}^n$ such that*

$$\left\| \sum_{j=1}^{n} r_{ij} t_j - \alpha_i \right\| < \varepsilon$$

for $1 \le i \le m$.

(ii) *If $\boldsymbol{u} \in \mathbb{Z}^m$, and*

$$\sum_{i=1}^{m} u_i r_{ij} = 0$$

for $1 \le j \le n$, then $\sum_{i=1}^{m} u_i \alpha_i \in \mathbb{Z}$.

Theorem B *Let $R = [r_{ij}]$ be an $m \times n$ matrix with real entries, and suppose that $\boldsymbol{\alpha} \in \mathbb{R}^m$. The following two assertions are equivalent.*

(i) *For every $\varepsilon > 0$ there is a $\boldsymbol{q} \in \mathbb{Z}^n$ such that*

$$\left\| \sum_{j=1}^{n} r_{ij} q_j - \alpha_i \right\| < \varepsilon$$

for $1 \le i \le m$.

(ii) *If $\boldsymbol{u} \in \mathbb{Z}^m$, and*

$$\sum_{i=1}^{m} u_i r_{ij} \in \mathbb{Z}$$

for $1 \le j \le n$, then $\sum_{i=1}^{m} u_i \alpha_i \in \mathbb{Z}$.

As was the case with our Theorems F.6 and F.7, it is easy to show that Theorems A and B are equivalent. See Koksma (1936, pp. 83–86) for a review of Kronecker's Theorem up to 1936. Cassels (1957, pp. 53–59) gives a proof of the general $m \times n$ theorem, along classical lines. Siegel (1989, pp. 43–63) develops the theory of vector groups, from which Kronecker's theorem follows easily. For a quantitatively precise version of Kronecker's Theorem (in the case $n = 1$) see Gonek & Montgomery (2016b).

Theorem F.10 is from Barton, Montgomery, & Vaaler (2001, Theorem 7). Selberg pointed out the relations of Exercise F.3.1.6 to Jeff Vaaler in 1982, and remarked that they are also useful for forming the composition of two or more sieves.

Section F.4. The result of Exercise F.4.1.2 is due to Bohl (1906) (see p. 279 of his paper). Later, Bohr (1925) created an extensive theory of almost-periodic functions, and in the course of this demonstrated (cf. pp. 119–121) that Bohl's theorem is equivalent to the localized form of Kronecker's theorem, i.e., to our Corollary F.9.

Bohr (1925, 1932) defined almost-periodic functions and studied their properties in the hope that by applying his theory to Dirichlet series, a proof of RH would emerge. Others, such as Stepanov, Besicovitch (see Besicovitch 1932), Weyl, Bochner, von Neumann, and Turing generalized the concept.

We now state without proof a number of outstanding properties of almost-periodic functions. If $f(x)$ is almost periodic, then for every real number λ the limit

$$c(\lambda) = \lim_{T\to\infty} \frac{1}{T}\int_0^T f(x)e(-\lambda x)\,dx$$

exists. Let $\mathscr{L}(f)$ denote the set of those λ for which $c(\lambda) \neq 0$. The set $\mathscr{L}(f)$ is at most countable, and indeed there is a sort of Parseval identity:

$$\lim_{T\to\infty} \frac{1}{T}\int_0^T |f(x)|^2\,dx = \sum_{\lambda\in\mathscr{L}(f)} |c(\lambda)|^2.$$

If $f(x)$ is almost periodic, then for every $\varepsilon > 0$ there is an almost-periodic polynomial $T(x)$ of the shape (F.9) such that $|f(x) - T(x)| < \varepsilon$ for all x, and indeed such a $T(x)$ can be constructed so that $\lambda_m \in \mathscr{L}(f)$ for all m. Hence the sum or product of two almost-periodic functions is again almost-periodic.

Ingham (1962) used elementary tools of complex analysis to show that if

$$f(s) = \sum_{n=0}^{\infty} a_n e^{-\lambda_n s}$$

for $\sigma > 0$ where $\sum_{n=0}^{\infty} |a_n| < \infty$, $\lambda_0 = 0$, $\lambda_n > 0$ for $n > 0$, the λ_n are distinct, $\mathscr{E} = \{f(s) : \sigma > 0\}$, $\mathscr{D}$ is a neighbourhood in $\overline{\mathscr{E}}$, and ϕ is analytic and bounded in $\mathscr{D}$, then

$$g(s) = \phi(f(s)) = \sum_{n=0}^{\infty} b_n e^{-\mu_n s}$$

with $\sum_{n=0}^{\infty} |b_n| < \infty$, and the μ_n are linear combinations with positive integer coefficients of a finite collection of the λ_n. Hewitt & Williamson (1957) and Edwards (1957) used tools from functional analysis to establish the same thing in the special case $\phi(z) = 1/z$.

The notion of almost-periodicity that we have described here is known as *uniform almost periodicity* because it is based on the uniform norm. The function $f(y) = (\psi(e^y) - e^y)/e^{y/2}$ is not uniformly almost-periodic, but it can be shown that it is mean-square almost-periodic if the Riemann Hypothesis is true.

F.6 References

Baker, R. C. (1986). *Diophantine Inequalities*, Oxford: Oxford Clarendon Press.

Barton, J. T., Montgomery, H. L., & Vaaler, J. D. (2001). Note on a Diophantine inequality in several variables, *Proc. Amer. Math. Soc.* **129**, 337–345.

Besicovitch, A. S. (1932). *Almost Periodic Functions*. Cambridge: Cambridge University Press, xii+180 pp; New York: Dover, 1954.

Bohl, P. (1906). Über eine Differentialgleichung der Störungstheorie, *J. Reine Angew. Math.* **131**, 268–321.

Bohr, H. (1925). Zur Theorie der fast peiodischen Funktionen, I. Eine Verallgemeinerung der Theorie der Fourierreihen. *Acta Math.* **45**, 29–127.

(1932). *Fastperiodische Funktionen*, Ergebnisse der Mathematik und ihrer Grenzgebiete. 1. Bd., 5, Berlin: Springer; English Translation, H. Cohn, F. Steinhardt, New York: Chelsea, 1947, 1951.

Brüdern, J. & Fouvry, É. (1996). Le crible à vecteurs, *Compositio Math.* **102**, 337–355.

Cassels, J. W. S. (1957). *An Introduction to Diophantine Approximation*, Cambridge Tracts in Math. 45, London: Cambridge University Press.

Chen, Y.-G. (2000). The best quantitative Kronecker's theorem, *J. London Math. Soc.* (2) **61**, 691–705.

Cochran, T. (1988). Trigonometric approximation and uniform distribution modulo one, *Proc. Amer. Math. Soc.* **103**, 695–702.

Edwards, D. A. (1957). On absolute convergence of Dirichlet series, *Proc. Amer. Math. Soc.* **8**, 1067–1074.

Erdős, P. & Turán, P. (1948). On a problem in the theory of uniform distribution, I, *Nederl. Akad. Wetensch. Proc.* **10**, 1146–1154; II, 1262–1269 (= *Indag. Math.* **10**, 370–378; 406–413).

Franel, J. (1924). Les suites de Farey et le problème des nombres premiers. *Nachr. Akad. Wiss. Göttingen*, 198–201.

Gonek, S. M. & Montgomery, H. L. (2016a). Extreme values of the zeta function at critical points, *Q. J. Math.* **67**, 483–505.

(2016b). Kronecker's approximation theorem, *Indag. Math. (N. S.)* **27**, 506–523.

Harman, G. (1993). Small fractional parts of additive forms, *Philos. Trans. Roy. Soc. Londons Ser. A* **345**, 327–338.

Hewitt, E. & Williamson, H. (1957). Note on absolutely convergent Dirichlet series, *Proc. Amer. Math. Soc.* **8**, 863–868.

Ingham, A. E. (1962). On absolutely convergent Dirichlet series. In *Studies in Mathematical Analysis and Related Topics*, Stanford Studies in Mathematics and Statistics IV, Stanford: Stanford University Press, pp. 156–164.

Koksma, J. F. (1936). *Diophantische Approximationen*, Ergebnisse Mathematik 4, Berlin: Springer.

Kronecker, L. (1884). Näherungsweise ganzzahlige Auflösung lindearer Gleichungen, *Monatsb. König. Preuss. Akad. Wiss. Berlin* **1884**, 1179–1193, 1271–1299; *Werke*, K. Hensel, Ed., Vol. III, Leipzig: Teubner, 1899, pp. 47–110.

Kuipers, L. & Niederreiter, H. (1974). *Uniform Distribution of Sequences*, New York: Wiley & Sons.

Landau, E. (1924). Bemerkung zu der vorstehenden Arbeit von Herrn Franel, *Nachr. Akad. Wiss. Göttingen* **1924**, 202–206.

Montgomery, H. L. (1994). *Ten Lectures on the Interrace between Analytic Number Theory and Harmonic Analysis*, CBMS 84, Providence: Amer. Math. Soc.

Schwarz, W. (1994). *Arithmetical Functions: An Introduction to Elementary and Analytic Properties of Arithmetic Functions and to Some of Their Almost-Periodic Properties*, London Math. Soc. Lecture Notes 184, Cambridge: Cambridge University Press

Siegel, C. L. (1989). *Lectures on the Geometry of Numbers*, Berlin: Springer.

Turán, P. (1960). A theorem on diophantine approximation with application to Riemann zeta-function, *Acta Sci. Math. Szeged* **21**, 277–311; *Collected Papers*, Vol. 2, Budapest: Akad. Kiadó, 1990, pp. 1142–1163.

Vaaler, J. D. (1985). Some extremal functions in Fourier analysis, *Bull. Amer. Math. Soc.* **12**, 183–216.

Weyl, H. (1916). Über die Gleichverteilung von Zahlen mod. Eins, *Math. Ann.* **77** (3), 313–352.

Appendix G
Bounds for Bilinear Forms

G.1 The Operator Norm of a Matrix

In various situations we are confronted with a problem of bounding a bilinear form – namely an expression of the general shape

$$\sum_{m=1}^{M}\sum_{n=1}^{N} a_{mn}x_n y_m.$$

In applications, the x_n and y_m may have considerable arithmetic structure, but we can often obtain a serviceable estimate using only the mean square sizes of the variables. Thus we seek an inequality of the sort

$$\left|\sum_{m,n} a_{mn}x_n y_m\right| \le \Delta\Big(\sum_n |x_n|^2\Big)^{1/2}\Big(\sum_m |y_m|^2\Big)^{1/2}. \tag{G.1}$$

Here Δ depends on the coefficient matrix $A = [a_{mn}]$, but is independent of the vectors $\boldsymbol{x}, \boldsymbol{y}$.

Let $A = [a_{mn}]$ be an $M \times N$ matrix with complex entries. Then A determines a linear map $\boldsymbol{x} \mapsto \boldsymbol{y} = A\boldsymbol{x}$ from $\mathbb{C}^N$ to $\mathbb{C}^M$. The norm of A, as a linear operator, is the maximum of the ratio $\|\boldsymbol{y}\|/\|\boldsymbol{x}\|$ as $\boldsymbol{x}$ runs over all non-zero members of $\mathbb{C}^N$,

$$\|A\| = \max_{\boldsymbol{x}\neq \boldsymbol{0}} \frac{\|A\boldsymbol{x}\|}{\|\boldsymbol{x}\|},$$

where $\|\boldsymbol{x}\| = \big(\sum |x_n|^2\big)^{1/2}$ denotes the usual Euclidean norm. By homogeneity we may instead write

$$\|A\| = \max_{\|\boldsymbol{x}\|=1} \|A\boldsymbol{x}\|.$$

We now show that $\|A\|$ is the optimal constant in the inequality (G.1).

Theorem G.1 (Duality) *Let $A = [a_{mn}]$ be a fixed $M \times N$ matrix. The following three assertions concerning the positive constant Δ are equivalent.*

(i) *For any $\boldsymbol{x} \in \mathbb{C}^N$,*

$$\sum_{m=1}^{M}\left|\sum_{n=1}^{N} a_{mn}x_n\right|^2 \le \Delta^2 \sum_{n=1}^{N} |x_n|^2.$$

(ii) *For any $\boldsymbol{x} \in \mathbb{C}^N$ and any $\boldsymbol{y} \in \mathbb{C}^M$,*

$$\left|\sum_{m=1}^{M}\sum_{n=1}^{N} a_{mn}x_n y_m\right| \le \Delta \left(\sum_{n=1}^{N} |x_n|^2\right)^{1/2} \left(\sum_{m=1}^{M} |y_m|^2\right)^{1/2}.$$

(iii) *For any $\boldsymbol{y} \in \mathbb{C}^M$,*

$$\sum_{n=1}^{N}\left|\sum_{m=1}^{M} a_{mn}y_m\right|^2 \le \Delta^2 \sum_{m=1}^{M} |y_m|^2.$$

In terms of linear maps and inner products, these inequalities assert that

(i) $||A\boldsymbol{x}|| \le \Delta||\boldsymbol{x}||$;
(ii) $|(A\boldsymbol{x}, \boldsymbol{y})| \le \Delta||\boldsymbol{x}||||\boldsymbol{y}||$;
(ii) $||A^*\boldsymbol{y}|| \le \Delta||\boldsymbol{y}||$.

Here A^* is the *adjoint* of A. That is, $A^* = (\overline{A})^{\mathrm{T}}$ is the $N \times M$ matrix $A^* = [\overline{a_{nm}}]$. In terms of inner products, A^* is characterized by the property that $(A\boldsymbol{x}, \boldsymbol{y}) = (\boldsymbol{x}, A^*\boldsymbol{y})$ for all $\boldsymbol{x}$ and $\boldsymbol{y}$. Since (i) and (iii) are equivalent, we deduce that

$$||A|| = ||A^*||.$$

Proof We show that (i) and (ii) are equivalent. Then, by interchanging the roles of m and n it is clear that (ii) and (iii) are equivalent.

(i) $\Longrightarrow$ (ii). By Cauchy's inequality

$$\left|\sum_m\Big(\sum_n a_{mn}x_n\Big)y_m\right| \le \left(\sum_m\Big|\sum_n a_{mn}x_n\Big|^2\right)^{1/2}\left(\sum_m |y_m|^2\right)^{1/2}.$$

In the first factor on the right we insert the bound provided by (i) to obtain (ii).

(ii) $\Longrightarrow$ (i). Set

$$y_m = \sum_{n=1}^{N} a_{mn}x_n,$$

and let S denote the left-hand side of (i). Then $S = \sum_n a_{mn} x_n \overline{y_m}$, and by (ii) we see that

$$S \le \Delta \left(\sum_{n=1}^{N} |x_n|^2 \right)^{1/2} \left(\sum_{m=1}^{M} |y_m|^2 \right)^{1/2} = \Delta \left(\sum_{n=1}^{N} |x_n|^2 \right)^{1/2} S^{1/2}.$$

If $S = 0$, then (i) is obviously satisfied. Otherwise $S > 0$, and we may square both sides above and divide by S to obtain (i). □

Corollary G.2 *For any $M \times N$ matrix A,*

$$\|A\| = \|A^*\| \le \|A^* A\|^{1/2}.$$

By using Corollary G.11 below it will become apparent that the inequality here may be replaced by equality.

Proof The identity represents the equivalence of (i) and (iii). To obtain the inequality, let $\boldsymbol{x}$ be a unit vector for which $\|A\boldsymbol{x}\| = \|A\|$. Then

$$\|A\|^2 = \|A\boldsymbol{x}\|^2 = (A\boldsymbol{x}, A\boldsymbol{x}) = (A^* A\boldsymbol{x}, \boldsymbol{x}).$$

By (ii) with $\boldsymbol{y} = \boldsymbol{x}$ we see that this last expression is $\le \|A^* A\|$. □

As a first upper bound for $\|A\|$ we establish the following theorem.

Theorem G.3 *Let A be an $M \times N$ matrix. Then*

$$x\|A\| \le \left(\max_m \sum_{n=1}^{N} |a_{mn}| \right)^{1/2} \left(\max_n \sum_{m=1}^{M} |a_{mn}| \right)^{1/2}.$$

Proof By Cauchy's inequality,

$$\left| \sum_{m,n} a_{mn} x_n y_m \right| \le \left(\sum_{m,n} |a_{mn}| |x_n|^2 \right)^{1/2} \left(\sum_{m,n} |a_{mn}| |y_m|^2 \right)^{1/2}.$$

The first sum on the right-hand side is

$$\sum_n |x_n|^2 \sum_m |a_{mn}| \le \left(\max_n \sum_m |a_{mn}| \right) \sum_n |x_n|^2.$$

We treat the second sum similarly, and thus obtain the situation of Theorem G.1(ii) with

$$\Delta = \left(\max_n \sum_m |a_{mn}| \right)^{1/2} \left(\max_m \sum_n |a_{mn}| \right)^{1/2}.$$

Thus $\|A\| \le \Delta$ by Theorem G.1. □

In general, Theorem G.3 provides a useful bound only if the a_{mn} are non-negative and approximately the same size, or if the matrix is nearly diagonal. Otherwise the bound for $||A||$ may be weak because it takes no account of possible cancellation. We apply this to the matrix A^*A and appeal to Corollary G.2 to obtain

Corollary G.4 *Let $A = [a_{mn}]$ be an $M \times N$ matrix. Then*

$$||A|| \le \left(\max_{n_1} \sum_{n_2=1}^{N} \left| \sum_{m=1}^{M} \overline{a_{mn_1}} a_{mn_2} \right| \right)^{1/2}.$$

If the columns of A are nearly orthonormal, then A^*A is nearly the identity matrix, and by the above $||A||$ is not much more than 1. We may use columns rather than rows, by applying the above to A^{T} instead of A. If the columns are far from orthonormal, then the above bound will, in general, be weak. In some instances greater precision can be obtained by introducing a type of weighting factor.

Theorem G.5 *Let $A = [a_{mn}]$ be an $M \times N$ matrix, but suppose that the a_{mn} are defined for all integral values of m. Let w_m be non-negative and suppose that $w_m \ge 1$ for $1 \le m \le M$. Then*

$$||A|| \le \left(\max_{n_1} \sum_{n_2=1}^{N} \left| \sum_{m=-\infty}^{\infty} \overline{a_{mn_1}} a_{mn_2} w_m \right| \right)^{1/2}$$

provided that the inner sum converges for all n_1, n_2.

Proof Let $\boldsymbol{x}$ be a unit vector for which $||A\boldsymbol{x}|| = ||A||$. Then by the properties of the w_m we see that

$$||A\boldsymbol{x}||^2 = \sum_{m=1}^{M} \left| \sum_{n=1}^{N} a_{mn} x_n \right|^2 \le \sum_{m=-\infty}^{\infty} w_m \left| \sum_{n=1}^{N} a_{mn} x_n \right|^2.$$

We expand and take the sum over m inside to see that this is

$$\sum_{n_1} \overline{x_{n_1}} \sum_{n_2} x_{n_2} \sum_{m} w_m \overline{a_{mn_1}} a_{mn_2} = (B\boldsymbol{x}, \boldsymbol{x}),$$

where B is the matrix with entries

$$b_{n_1 n_2} = \sum_{m=-\infty}^{\infty} w_m \overline{a_{mn_1}} a_{mn_2}.$$

By Theorem G.1(ii) we know that $|(B\boldsymbol{x}, \boldsymbol{x})| \le ||B||$, so $||A|| \le ||B||^{1/2}$. Then by applying Theorem G.3 to B we obtain the stated result. □

If $w_m = 1$ for $1 \le m \le M$ and $w_m = 0$ otherwise, then the argument above reduces to the proof of Corollary G.4. If the a_{mn} are oscillatory and random in appearance, then the upper bounds for $||A||$ that we might derive from the theorems above are likely to be much larger than the true order of magnitude. In such a situation, the following lower bound may be closer to the truth.

Theorem G.6 *Let A be an $M \times N$ matrix. Then*

$$||A||^2 \ge \frac{\sum_{m,n} |a_{mn}|^2}{\min(M, N)}.$$

Proof We consider the size of $||A\boldsymbol{x}||$ with $x_n = e(n\theta)$, and average over θ. By the orthogonality of the functions $e(n\theta)$ we see that

$$\int_0^1 \sum_{m=1}^{M} \left| \sum_{n=1}^{N} a_{mn} e(n\theta) \right|^2 d\theta = \sum_{m,n} |a_{mn}|^2.$$

We choose a θ for which the integrand is at least as large as the right-hand side. Since $||\boldsymbol{x}|| = N^{1/2}$ for any θ, we conclude that

$$||A|| \ge \left(\frac{1}{N} \sum_{m,n} |a_{mn}|^2 \right)^{1/2}.$$

By applying this argument to A^{T} instead of A we obtain this lower bound with N replaced by M. Thus the proof is complete. □

G.1.1 Exercises

1. Let A be an $m \times n$ matrix, and let $C \subseteq \mathbb{C}^n$ denote the column space of $A^\star$, which is to say the set of all vectors of the form $A^\star \boldsymbol{y}$ for $\boldsymbol{y} \in \mathbb{C}^m$. Let Δ be the optimal constant in Theorem G.1(i). Suppose that $\boldsymbol{y} \in \mathbb{C}^m$ is chosen so that $||\boldsymbol{y}|| = 1$ and $||A^\star \boldsymbol{y}|| = \Delta$. Put $\boldsymbol{x} = A^\star \boldsymbol{y}$.
   (a) Show that $||A\boldsymbol{x}|| = \Delta ||\boldsymbol{x}||$.
   (b) Deduce that
   $$\max_{\substack{\boldsymbol{x} \in \mathbb{C}^n \\ \boldsymbol{x} \ne \boldsymbol{0}}} \frac{||A\boldsymbol{x}||}{||\boldsymbol{x}||} = \max_{\substack{\boldsymbol{x} \in C \\ \boldsymbol{x} \ne \boldsymbol{0}}} \frac{||A\boldsymbol{x}||}{||\boldsymbol{x}||}.$$
   (When seeking a bound for the norm of a matrix A, it is sometimes useful to know that it suffices to consider $\boldsymbol{x}$ of the form $A^\star \boldsymbol{y}$.)
2. For $\boldsymbol{x} \in \mathbb{C}^N$ and real $p > 1$, put $||\boldsymbol{x}||_p = \left(\sum |x_n|^p \right)^{1/p}$. Similarly, put $||\boldsymbol{x}||_\infty = \max |x_n|$. Suppose that p and q are real numbers, $1 \le p \le \infty$, $1 \le q \le \infty$, and that p' and q' are determined by the relations $1/p + 1/p' = 1$, $1/q + 1/q' = 1$. Let A be an $M \times N$ matrix. Show that the following assertions concerning the constant Δ are equivalent.

(a) For all $\boldsymbol{x} \in \mathbb{C}^N$,

$$\|A\boldsymbol{x}\|_p \le \Delta \|\boldsymbol{x}\|_q.$$

(b) For all $\boldsymbol{x} \in \mathbb{C}^N$ and $\boldsymbol{y} \in \mathbb{C}^M$,

$$\left|\sum a_{mn} x_n y_m\right| \le \Delta \|\boldsymbol{x}\|_q \|\boldsymbol{y}\|_{p'}.$$

(c) For all $\boldsymbol{y} \in \mathbb{C}^M$,

$$\|A^* \boldsymbol{y}\|_{q'} \le \Delta \|\boldsymbol{y}\|_{p'}.$$

3. Let B and C be rectangular matrices, and put

$$A = \begin{bmatrix} B & 0 \\ 0 & C \end{bmatrix}.$$

Show that $\|A\| = \max(\|B\|, \|C\|)$.

4. Suppose that $|a_{mn}| \le b_{mn}$ for all m and n. Show that $\|A\| \le \|B\|$.

5. Let A be an $M \times N$ matrix, and suppose that there are positive numbers C, $D, u_1, \ldots, u_N, v_1, \ldots, v_M$ such that

$$\sum_{m=1}^{M} |a_{mn}| v_m \le C u_n$$

for $1 \le n \le N$, and also that

$$\sum_{n=1}^{N} |a_{mn}| u_n \le D v_m$$

for $1 \le m \le M$.

(a) Show that if $\|\boldsymbol{x}\| = \|\boldsymbol{y}\| = 1$, then

$$|(A\boldsymbol{x}, \boldsymbol{y})|^2 \le \left(\sum_{m,n} |a_{mn}| v_m / u_n\right)\left(\sum_{m,n} |a_{mn}| u_n / v_m\right).$$

(b) Deduce that $\|A\| \le (CD)^{1/2}$.

6. Let A be an $M \times N$ matrix with $a_{mn} = 1$ for all m and n. Show that $\|A\| = (MN)^{1/2}$.

7. Let A be an $M \times N$ matrix with real entries. Show that

$$\max_{\substack{\boldsymbol{x} \in \mathbb{R}^N \\ \|\boldsymbol{x}\|=1}} \|A\boldsymbol{x}\| = \max_{\substack{\boldsymbol{x} \in \mathbb{C}^N \\ \|\boldsymbol{x}\|=1}} \|A\boldsymbol{x}\|.$$

8. Suppose that p and q are real numbers, $p > 1$, $q > 1$, and that $\frac{1}{p} + \frac{1}{p'} = 1$ and $\frac{1}{q} + \frac{1}{q'} = 1$. Let $A = [a_{mn}]$ be an $M \times N$ matrix. Show that the following three assertions concerning the positive constant Δ are equivalent.

(a) For any $\boldsymbol{x} \in \mathbb{C}^N$,

$$\left(\sum_{m=1}^{M} \left| \sum_{n=1}^{N} a_{mn} x_n \right|^{q'} \right)^{1/q'} \leq \Delta \left(\sum_{n=1}^{N} |x_n|^p \right)^{1/p}.$$

(b) For any $\boldsymbol{x} \in \mathbb{C}^N$ and any $\boldsymbol{y} \in \mathbb{C}^M$,

$$\left| \sum_{m=1}^{M} \sum_{n=1}^{N} a_{mn} x_n y_m \right| \leq \Delta \left(\sum_{n=1}^{N} |x_n|^p \right)^{1/p} \left(\sum_{m=1}^{M} |y_m|^q \right)^{1/q}.$$

(c) For any $\boldsymbol{y} \in \mathbb{C}^M$,

$$\left(\sum_{n=1}^{N} \left| \sum_{m=1}^{M} a_{mn} y_m \right|^{p'} \right)^{1/p'} \leq \Delta \left(\sum_{m=1}^{M} |y_m|^q \right)^{1/q}.$$

G.2 Square Matrices

The operator norm is defined for an arbitrary rectangular matrix, but if A is square, say $N \times N$, then further numbers can be associated with it. In the first place, A has N eigenvalues λ_n, which are the roots of the polynomial $\det(zI - A)$, and we define the *spectral radius* of A to be

$$\rho(A) = \max_n |\lambda_n|.$$

We also consider the *numerical radius* of A,

$$\nu(A) = \max_{\|\boldsymbol{x}\|=1} \left| \sum_{m,n} a_{mn} x_n \overline{x_m} \right| = \max_{\|\boldsymbol{x}\|=1} |(A\boldsymbol{x}, \boldsymbol{x})|.$$

These quantities are related to the operator norm $\|A\|$ in the following simple manner.

Theorem G.7 *Let A be an arbitrary $N \times N$ matrix. Then*

$$\rho(A) \leq \nu(A) \leq \|A\|.$$

Proof Let λ be an eigenvalue of A, and let $\boldsymbol{x} \neq \boldsymbol{0}$ be an associated eigenvector, so that $A\boldsymbol{x} = \lambda\boldsymbol{x}$. Without loss of generality we may suppose that $\|\boldsymbol{x}\| = 1$. For this vector, $(A\boldsymbol{x}, \boldsymbol{x}) = (\lambda\boldsymbol{x}, \boldsymbol{x}) = \lambda$, so that $\nu(A) \geq |\lambda|$, and hence $\nu(A) \geq \rho(A)$.

By Theorem G.1(ii),

$$\|A\| = \max_{\|\boldsymbol{x}\|=\|\boldsymbol{y}\|=1} |(A\boldsymbol{x}, \boldsymbol{y})|.$$

Thus $\nu(A) \leq \|A\|$, and the proof is complete. □

The first inequality in Theorem G.7 cannot be reversed in general since $\nu(A)$ may be large even when all the eigenvalues vanish. (Consider a matrix A for which $a_{mn} = 0$ whenever $m \ge n$.) However, $\nu(A)$ and $\|A\|$ are always comparable.

Theorem G.8 *Let A be an $N \times N$ matrix. Then*

$$\frac{1}{2}\|A\| \le \nu(A) \le \|A\|,$$

and if A is Hermitian (i.e., if $A^* = A$), then $\nu(A) = \|A\|$.

In Corollary G.11 it will also be established that if A is Hermitian, then $\rho(A) = \|A\|$.

Proof We establish the last assertion first. The hypothesis that A is Hermitian is equivalent to saying that $(A\boldsymbol{x}, \boldsymbol{y}) = (\boldsymbol{x}, A\boldsymbol{y})$ for all $\boldsymbol{x}$ and $\boldsymbol{y}$. Put $\boldsymbol{u} = \boldsymbol{x} + \boldsymbol{y}$ and $\boldsymbol{v} = \boldsymbol{x} - \boldsymbol{y}$. It is easily verified that if A is Hermitian, then

$$4\,\mathrm{Re}(A\boldsymbol{x}, \boldsymbol{y}) = (A\boldsymbol{u}, \boldsymbol{u}) - (A\boldsymbol{v}, \boldsymbol{v}).$$

By Theorem G.1(ii) we can choose unit vectors $\boldsymbol{x}$ and $\boldsymbol{y}$ so that $(A\boldsymbol{x}, \boldsymbol{y}) = \|A\|$. Then

$$4\|A\| = (A\boldsymbol{u}, \boldsymbol{u}) - (A\boldsymbol{v}, \boldsymbol{v}) \le \nu(A)(\|\boldsymbol{u}\|^2 + \|\boldsymbol{v}\|^2).$$

But $\|\boldsymbol{u}\|^2 = 2 + 2\,\mathrm{Re}(\boldsymbol{x}, \boldsymbol{y})$ and $\|\boldsymbol{v}\|^2 = 2 - 2\,\mathrm{Re}(\boldsymbol{x}, \boldsymbol{y})$, so that $\|\boldsymbol{u}\|^2 + \|\boldsymbol{v}\|^2 = 4$, and hence $\|A\| \le \nu(A)$.

The second displayed inequality follows trivially from Theorem G.1(ii) and the definition of $\nu(A)$. To establish an inequality in the reverse direction, suppose that A is an arbitrary $N \times N$ matrix. Write $A = B + iC$ where $B = (A + A^*)/2$ and $C = (A - A^*)/(2i)$. The triangle inequality holds for the operator norm $\|\cdot\|$, so $\|A\| \le \|B\| + \|C\|$. But B and C are Hermitian, so this latter quantity is $\nu(B) + \nu(C)$. For any $\boldsymbol{x} \in \boldsymbol{C}^N$, we see that $(B\boldsymbol{x}, \boldsymbol{x}) = \mathrm{Re}(A\boldsymbol{x}, \boldsymbol{x})$ and $(C\boldsymbol{x}, \boldsymbol{x}) = \mathrm{Im}(A\boldsymbol{x}, \boldsymbol{x})$. Hence $\nu(B) \le \nu(A)$, $\nu(C) \le \nu(A)$, and we conclude that $\|A\| \le 2\nu(A)$. □

We now consider the possibility that a square matrix A might be converted to a diagonal matrix by means of a suitable change of basis. In general, if S is non-singular, so that $\boldsymbol{x} = S\boldsymbol{u}$ expresses a linear change of variables, then the linear transformation $\boldsymbol{x} \mapsto A\boldsymbol{x}$ is computed as $\boldsymbol{u} \mapsto B\boldsymbol{u}$ in the new coordinate system, where $B = S^{-1}AS$. In this case we say that A and B are *similar*. An easy calculation reveals that if A and B are similar, then $\mathrm{tr}\,A = \mathrm{tr}\,B$, $\det A = \det B$, and indeed A and B have the same characteristic polynomial. Hence A and B have the same eigenvalues, so that $\rho(A) = \rho(B)$. On the other hand, the norm of a matrix is a metric quantity, and in general $\|A\| \neq \|B\|$. In order that $\|A\|$

should be invariant we restrict our attention to those similarity transformations that preserve distances. Let U be an $N \times N$ matrix. Then it is easy to verify that the following assertions are equivalent:

(i) U is unitary (i.e., $U^* = U^{-1}$);
(ii) the columns of U are orthonormal vectors;
(iii) the rows of U are orthonormal vectors;
(iv) the map $\boldsymbol{x} \mapsto U\boldsymbol{x}$ is an isometry of $\mathbb{C}^N$ (i.e., $\|U\boldsymbol{x}\| = \|\boldsymbol{x}\|$ for all $\boldsymbol{x} \in \mathbb{C}^N$);
(v) $(U\boldsymbol{x}, U\boldsymbol{y}) = (\boldsymbol{x}, \boldsymbol{y})$ for all $\boldsymbol{x}, \boldsymbol{y} \in \mathbb{C}^N$.

Thus a unitary transformation maps one orthonormal basis to another, and conversely, if two orthonormal bases are given, then there is a unitary transformation that takes one to the other. In the analogous situation of linear maps from $\mathbb{R}^N$ to itself, we would find that the orthogonal matrices have corresponding properties. (A matrix X is *orthogonal* if $X^{\mathrm{T}} = X^{-1}$.) If $A = U^{-1}BU$ where U is unitary, then we say that A and B are *unitarily similar*. In this case it is clear that $\|A\| = \|B\|$, and that $\nu(A) = \nu(B)$. Moreover, we note that A is Hermitian ($A^* = A$) if and only if B is, that A is *normal* ($AA^* = A^*A$) if and only if B is, and that A is unitary ($A^* = A^{-1}$) if and only if B is. We now produce a unitarily similar canonical form for A.

Theorem G.9 (Schur's triangularization theorem) *For any $N \times N$ matrix A there is an upper triangular matrix T that is unitarily similar to A, $T = U^{-1}AU$. The diagonal entries of T are the eigenvalues of A.*

Proof We prove the first assertion by induction on N. For $N = 1$ there is nothing to show. Suppose we have the result for $N - 1$. Let λ_1 be an eigenvalue of A, with $\boldsymbol{v}_1$ an associated unit eigenvector. Choose $\boldsymbol{v}_2, \ldots, \boldsymbol{v}_N$ so that the $\boldsymbol{v}_n$ form an orthonormal basis for $\mathbb{C}^N$, and let V be the matrix whose columns are the $\boldsymbol{v}_n$. Then V is unitary, and V^*AV has the form

$$V^*AV = \begin{bmatrix} \lambda_1 & * \\ \boldsymbol{0} & B \end{bmatrix}.$$

By the inductive hypothesis there is a unitary matrix W such that $W^{-1}BW$ is upper-triangular. Put

$$X = \begin{bmatrix} 1 & \boldsymbol{0} \\ \boldsymbol{0} & W \end{bmatrix}.$$

Then

$$X^*V^*AVX = \begin{bmatrix} \lambda_1 & * \\ \boldsymbol{0} & W^*BW \end{bmatrix}$$

is upper-triangular, and we take $U = VX$.

The second assertion is obvious since

$$\text{char poly } A = \text{char poly } T = \prod_{n=1}^{N}(x - t_{nn}). \qquad \square$$

If D is a diagonal matrix, then clearly $D^*D = DD^*$, so that D is normal. Conversely, suppose that T is a normal upper-triangular matrix. On comparing the diagonal entries of T^*T with those of TT^*, we see that

$$\sum_{m=1}^{n} |t_{mn}|^2 = \sum_{m=n}^{N} |t_{nm}|^2$$

for $1 \le n \le N$. On taking $n = 1$, we deduce that $t_{1m} = 0$ for $m > 1$. Then we set $n = 2$ to show that $t_{2m} = 0$ for $m > 2$. Hence by induction we find that $t_{mn} = 0$ for $m \ne n$, so that T is diagonal. Thus we have a corollary.

Corollary G.10 *A square matrix A is unitarily similar to a diagonal matrix, $U^*AU = D$, if and only if A is normal.*

If D is diagonal, then clearly $\rho(D) = \|D\|$. Thus we deduce another corollary.

Corollary G.11 *If A is normal, then $\rho(A) = \nu(A) = \|A\|$.*

We note that if A is Hermitian or unitary, then A is normal, and both corollaries apply. We consider again Corollary G.2, whose proof amounted to observing that

$$\|A\|^2 = \nu(A^*A) \le \|A^*A\|.$$

Since A^*A is Hermitian, we know by Theorem G.8 that equality holds here. By Corollary G.11 we can add the further observation that

$$\|A\|^2 = \rho(A^*A).$$

G.2.1 Exercises

1. (Schur 1909) Let $A = [a_{mn}]$ be an $N \times N$ matrix.
   (a) Show that
   $$\text{tr } AA^\star = \sum_{1 \le m,n \le N} |a_{mn}|^2.$$
   (b) Let U be a unitary matrix such that $UAU^* = T = [t_{mn}]$ is upper triangular. Show that
   $$\text{tr } TT^* = \sum_{1 \le m \le n \le N} |t_{mn}|^2 = \sum_{1 \le m,n \le N} |a_{mn}|^2.$$

(c) Let $\lambda_1, \lambda_2, \ldots, \lambda_N$ be the eigenvalues of A (e.g., $\lambda_n = t_{nn}$). Show that

$$\sum_{n=1}^{N} |\lambda_n|^2 \leq \sum_{1\leq m,n\leq N} |a_{mn}|^2,$$

and that equality holds if and only if A is normal.

2. Let A be an $M \times N$ matrix, and let $\lambda_1, \ldots, \lambda_N$ be the eigenvalues of A^*A. Show that the λ_n are non-negative, and that

$$\sum_{n=1}^{N} \lambda_n = \sum_{m,n} |a_{mn}|^2.$$

Use this to give a second proof of Theorem G.6.

3. Let A be an $M \times N$ matrix.

   (a) Show that $A(A^*A - zI)^{-1}A^* = I + z(AA^* - zI)^{-1}$ for any complex number z for which either of the inverses exists.

   (b) Show that the non-zero eigenvalues of A^*A coincide with those of AA^*.

4. Let A be an $N \times N$ matrix, and let $C, u_1, \ldots, u_N$ be positive numbers such that

$$\sum_{n=1}^{N} |a_{mn}|u_n \leq Cu_m \tag{G.2}$$

for $1 \leq m \leq N$.

   (a) Show that $\rho(A) \leq C$. (Suggestion: Let $\boldsymbol{x}$ be an eigenvector, and consider that m for which $|x_m|/u_m$ is maximal.)

   (b) Show that if $a_{mn} > 0$ for all m and n, and if C is chosen minimally, then equality holds in (G.2) for all m, so that $\rho(A)$ is an eigenvalue and $\boldsymbol{u}$ is an associated eigenvector with positive coordinates.

5. Show that an $N \times N$ matrix A is normal if and only if its eigenvectors form an orthogonal basis for $\mathbb{C}^N$.

6. Show that the following are equivalent:

   (a) U is unitary;

   (b) U is normal and all its eigenvalues are unimodular.

7. Show that the following are equivalent:

   (a) X is Hermitian;

   (b) X is normal and all its eigenvalues are real.

8. Let A be an $N \times N$ matrix. The *field of values* of A is the set of complex numbers $\{(A\boldsymbol{x}, \boldsymbol{x}) : \|\boldsymbol{x}\| = 1\}$.

(a) Show that if A and B are unitarily similar, then they have the same field of values.
(b) Show that if A is normal, then its field of values is the convex hull of its eigenvalues.
(c) Show that the field of values of A is an interval on the real line if and only if A is Hermitian.
(d) The field of values is a convex set that contains the eigenvalues of A.
(e) If B is an $M \times N$ matrix, then the field of values of B^*B is the same as the field of values of BB^*.

9. Let A be a Hermitian matrix for which $(A\boldsymbol{x},\boldsymbol{x}) \geq 0$ for all $\boldsymbol{x}$. Show that $|(A\boldsymbol{x},\boldsymbol{y})| \leq (A\boldsymbol{x},\boldsymbol{x})(A\boldsymbol{y},\boldsymbol{y})$. (Suggestion: Consider $(A(\lambda\boldsymbol{x} + \mu\boldsymbol{y}), \lambda\boldsymbol{x} + \mu\boldsymbol{y})$.)
10. Suppose that $A_1, \ldots, A_K$ are commuting normal matrices. Show that there is a unitary matrix U such that all the matrices U^*A_kU are diagonal.
11. (Watkins 1980) Suppose that A and B are real square matrices that are similar over $\mathbb{C}$, say $A = S^{-1}BS$ where S has complex entries. Write $S = P + iQ$ where P and Q have real entries.

    (a) Show that $PA = BP$ and that $QA = BQ$.
    (b) Deduce that $(P + rQ)A = B(P + rQ)$ for any real number r.
    (c) Let $p(z) = \det(P + zQ)$. Explain why $p(i) \neq 0$.
    (d) Explain why there is a real number r such that $p(r) \neq 0$.
    (e) Conclude that there is a non-singular square matrix R with real entries for which $A = R^{-1}BR$.

12. Let A be a real symmetric matrix. Show that any number of the form $(A\boldsymbol{x},\boldsymbol{x})$ where $\boldsymbol{x}$ is a unit vector in $\mathbb{C}^N$ can also be written in this form with $\boldsymbol{x}$ a unit vector in $\mathbb{R}^N$.
13. Let

$$A = \begin{bmatrix} 0 & 2 \\ 0 & 0 \end{bmatrix}.$$

Show that $\rho(A) = 0$, that $\nu(A) = 1$, and that $||A|| = 2$. (Thus the constant $1/2$ in the lower bound in Theorem G.8 is best possible.)

14. (a) Let

$$A = \begin{bmatrix} 2 & 0 & 0 \\ 0 & 0 & 1 \\ 0 & 0 & 1 \end{bmatrix}.$$

Show that the eigenvalues of A are $0, 1, 2$, that the eigenvalues of A^*A are $0, 2, 4$, that A is not normal, and that $\rho(A) = \nu(A) = ||A|| = 2$.

(b) Show that the converse of Corollary G.11 is true for $N \le 2$, but false for $N > 2$.

15. Let A be a normal matrix, λ a complex number, and $\boldsymbol{x}$ a vector. Put $\boldsymbol{e} = A\boldsymbol{x} - \lambda\boldsymbol{x}$. Show that A has an eigenvalue in the disc $|z - \lambda| \le \|\boldsymbol{e}\|/\|\boldsymbol{x}\|$. (Hint: If $A - \lambda I$ is singular, then this is obvious. Otherwise, argue that $\rho((A - \lambda I)^{-1}) = \|(A - \lambda I)^{-1}\| \ge \|\boldsymbol{x}\|/\|\boldsymbol{e}\|$.)
16. (a) Let C be an $N \times N$ Hermitian matrix such that $(C\boldsymbol{x}, \boldsymbol{x}) \ge 0$ for all $\boldsymbol{x} \in \mathbb{C}^N$. Show that there is an $N \times N$ matrix B such that $B^*B = C$.
    (b) Suppose that A is an $M \times N$ matrix, and put $\Delta = \|A^*A\|^{1/2}$. Show that there is an $N \times N$ matrix B such that $A^*A + B^*B = \Delta^2 I$.
    (c) Suppose that A is an $M \times N$ matrix for which condition (i) of Theorem G.1 holds. Show that there is an $N \times N$ matrix B such that

$$\sum_{m=1}^{M}\left|\sum_{n=1}^{N} a_{mn}x_n\right|^2 + \sum_{m=1}^{N}\left|\sum_{n=1}^{N} b_{mn}x_n\right|^2 = \Delta^2 \sum_{n=1}^{N} |x_n|^2$$

for all $\boldsymbol{x} \in \mathbb{C}^N$.
17. (Toeplitz 1910) Suppose that $f \in L^\infty(\mathbb{T})$ has Fourier coefficients

$$\widehat{f}(k) = \int_0^1 f(x)e(-kx)\,dx,$$

and put

$$S(x) = \sum_{n=1}^{N} x_n e(nx), \qquad T(x) = \sum_{m=1}^{N} y_m e(mx).$$

(a) Show that

$$\int_0^1 f(x)S(-x)T(-x)\,dx = \sum_{m=1}^{N}\sum_{n=1}^{N} \widehat{f}(m+n)x_n y_m.$$

(b) Show that

$$\int_0^1 f(x)S(-x)T(x)\,dx = \sum_{m=1}^{N}\sum_{n=1}^{N} \widehat{f}(m-n)x_n y_m.$$

(c) Explain why

$$\int_0^1 |S(-x)T(-x)|\,dx \le \left(\int_0^1 |S(x)|^2\,dx\right)^{1/2}\left(\int_0^1 |T(x)|^2\,dx\right)^{1/2}.$$

(d) Explain why

$$\int_0^1 |S(x)|^2\,dx = \sum_{n=1}^{N} |x_n|^2, \qquad \int_0^1 |T(x)|^2\,dx = \sum_{m=1}^{N} |y_m|^2.$$

(e) Show that

$$\left|\sum_{m=1}^{N}\sum_{n=1}^{N}\widehat{f}(m+n)x_n y_m\right| \le \|f\|_{L^\infty}\left(\sum_{n=1}^{N}|x_n|^2\right)^{1/2}\left(\sum_{m=1}^{N}|y_m|^2\right)^{1/2}.$$

(f) Show that

$$\left|\sum_{m=1}^{N}\sum_{n=1}^{N}\widehat{f}(m-n)x_n y_m\right| \le \|f\|_{L^\infty}\left(\sum_{n=1}^{N}|x_n|^2\right)^{1/2}\left(\sum_{m=1}^{N}|y_m|^2\right)^{1/2}.$$

18. Let $s(x)$ be the sawtooth function as defined in (F.6). Thus s has period 1, $s(0) = 0$, and $s(x) = x - \frac{1}{2}$ for $0 < x < 1$.

(a) Show that

$$\widehat{s}(k) = \begin{cases} \frac{i}{2\pi k} & \text{if } k \ne 0, \\ 0 & \text{if } k = 0. \end{cases}$$

(b) Show that

$$\left|\sum_{m=1}^{N}\sum_{n=1}^{N}\frac{x_n y_m}{m+n-1}\right| \le \pi\left(\sum_{n=1}^{N}|x_n|^2\right)^{1/2}\left(\sum_{m=1}^{N}|y_m|^2\right)^{1/2}.$$

(c) Show that

$$\left|\sum_{m=1}^{\infty}\sum_{n=1}^{\infty}\frac{x_n y_m}{m+n-1}\right| \le \pi\left(\sum_{n=1}^{\infty}|x_n|^2\right)^{1/2}\left(\sum_{m=1}^{\infty}|y_m|^2\right)^{1/2}.$$

(d) Show that

$$\left|\sum_{\substack{1\le m,n\le N\\ m\ne n}}\frac{x_n y_m}{m-n}\right| \le \pi\left(\sum_{n=1}^{N}|x_n|^2\right)^{1/2}\left(\sum_{m=1}^{N}|y_m|^2\right)^{1/2}.$$

19. Let $s(x)$ denote the sawtooth function, and suppose that $0 < \delta \le 1/2$.

(a) Show that

$$s(1-\delta+x) + s(1-\delta-x) + 2\delta = \begin{cases} 1 & (-\delta < x < \delta), \\ \frac{1}{2} & (x = \delta), \\ 0 & (\delta < x < 1-\delta). \end{cases}$$

(b) Suppose that the function U is even, has period 1, and is properly Riemann-integrable over bounded intervals. Show that

$$\begin{aligned}&\int_0^1 U(x+\delta)s(x)\,dx\\&\quad = (1/2-\delta)\int_{1-2\delta}^1 U(x+\delta)\,dx - \delta\int_0^{1-2\delta} U(x+\delta)\,dx\\&\quad = (1/2-\delta)\int_{-\delta}^{\delta} U(x)\,dx - \delta\int_\delta^{1-\delta} U(x)\,dx\\&\quad = (1/2-\delta)\int_0^1 U(x)\,dx - \frac{1}{2}\int_\delta^{1-\delta} U(x)\,dx.\end{aligned}$$

(c) Take $U(x) = \left|\sum_{n=1}^N e(nx)\right|^2$. Show that $\int_0^1 U(x)\,dx = N$, and use (16.4) to show that $\int_\delta^{1-\delta} U(x)\,dx \le (2\delta)^{-1}$.

(d) Show that if $\delta = 1/(2\sqrt{N})$, then

$$\int_0^1 U(x+\delta)s(x)\,dx \ge \frac{1}{2}N - \sqrt{N}.$$

(e) In Exercise G.2.1.17 set $x_n = e(-n\delta)$ and $y_m = e(m\delta)$. Note that $S(-x)T(x) = U(x+\delta)$.

(f) Show that in Exercise G.2.1.18(d), the number π cannot be replaced by $\pi - 2\pi/\sqrt{N}$.

20. (a) Let U be the $q \times q$ matrix with coefficients $u_{mn} = e(mn/q)/\sqrt{q}$. Show that U is unitary.

(b) Let $f(n)$ be an arithmetic function that is periodic with period q, and let C be the $q \times q$ matrix with coefficients $c_{mn} = f(m-n)$. (Such a matrix is called a *circulant*.) Show that U^*CU is diagonal.

(c) Let

$$\widehat{f}(k) = \frac{1}{q}\sum_{h=1}^{q} f(h)e(-hk/q)$$

be the discrete Fourier transform of f, as discussed in §4.1. Show that

$$\sum_{m=1}^{q}\sum_{n=1}^{q} f(m-n)x_m\overline{x_n} = \sum_{k=1}^{q}\widehat{f}(k)\left|\sum_{n=1}^{q} x_n e(kn/q)\right|^2.$$

(d) Put

$$\Delta = \max_k |\widehat{f}(k)|.$$

Show that

$$\left|\sum_{m=1}^{q}\sum_{n=1}^{q} f(m-n)x_m\overline{x_n}\right| \le \Delta \sum_{n=1}^{q} |x_n|^2$$

for arbitrary numbers x_n, and that the constant is best possible.

21. Let A be a square matrix.

 (a) By using the Schur triangularization theorem, or otherwise, show that the eigenvalues of A^2 are the squares of those of A.

 (b) Let $A = \begin{bmatrix} 2 & 1 \\ 1 & 3 \end{bmatrix}$, $B = \begin{bmatrix} 1 & 2 \\ -1 & 4 \end{bmatrix}$, and set $C = AB$. Show that there is no way to order the eigenvalues of A and of B so that their pairwise products form the eigenvalues of C.

22. (Schur 1921) For a given positive integer q let E be the $q \times q$ matrix $E = [e(mn/q)]$. This is the *Schur matrix*. Let $P = [p_{mn}]$ be the $q \times q$ permutation matrix with $p_{mn} = 1$ when $m \equiv n + 1 \pmod{q}$, and $p_{mn} = 0$ otherwise. Put $E_0 = PEP^t$.

 (a) Show that $E_0 = [e((m-1)(n-1)/q)]$.

 (b) Note that E_0 is a Vandermonde matrix. Deduce that

$$\det E = \prod_{0\le j<k<q} \big(e(k/q) - e(j/q)\big).$$

 (c) Deduce that

$$\det E = \prod_{0\le j<k<q} (2ie((j+k)/(2q)(\sin(\pi(k-j)/q)))\,.$$

 (d) Prove that

$$\sum_{0\le j<k<q} (j+k) = \frac{q(q-1)^2}{2}.$$

 (e) Conclude that

$$\det E = i^{q(q-1)/2} e\big(((q-1)/2)^2\big) \prod_{0\le j<k<q} \big(2\sin(\pi(k-j)/q)\big). \quad \text{(G.3)}$$

 Note that $e\big(((q-1)/2)^2\big) = 1$ if q is odd, and that it is $= i$ if q is even.

23. (Carlitz 1959) Let E be the Schur matrix, as in the preceding exercise, and let $\lambda_1, \lambda_2, \ldots, \lambda_q$ be the eigenvalues of E. Note that

$$\sum_{n=1}^{q} \lambda_n = \operatorname{tr} E = \sum_{n=1}^{q} e(n^2/q) = G(q),$$

say. Recall that in Corollary 9.16 that it was shown that $G(q)$ takes the values $(1+i)\sqrt{q}, \sqrt{q}, 0, i\sqrt{q}$ accordingly as $q \equiv 0, 1, 2, 3 \pmod 4$.

(a) Let $E^2 = B = [b_{mn}]$. Show that $b_{mn} = q$ if $m + n \equiv 0 \pmod q$, and that $b_{mn} = 0$ otherwise.
(b) Deduce that $\sum_{n=1}^{q} \lambda_n^2 = \operatorname{tr} B = q$ or $2q$ accordingly as q is odd or even.
(c) Show that $E^4 = B^2 = q^2 I$.
(d) Deduce that $|\det E| = q^{q/2}$.
(e) Deduce that every eigenvalue of E is of the form $i^a\sqrt{q}$ for some a.
(f) For $a = 0, 1, 2, 3$ let m_a be the number of eigenvalues equal to $i^a\sqrt{q}$.
(g) Explain why

$$m_0 + m_1 + m_2 + m_3 = q.$$

(h) Show that

$$m_0 + im_1 - m_2 - im_3 = \operatorname{tr} E/\sqrt{q} = G(q)/\sqrt{q},$$

and that

$$m_0 - im_1 - m_2 + im_3 = \overline{G}(q).$$

(i) Show that

$$m_0 - m_1 + m_2 - m_3 = \operatorname{tr} E^2/q = \begin{cases} 1 & (q \text{ odd}), \\ 2 & (q \text{ even}). \end{cases}$$

(j) Solve the equations in parts (f)–(h) to obtain the values of the multiplicities m_a: given in Table G.1.

	a			
q	0	1	2	3
0	$\frac{1}{4}q + 1$	$\frac{1}{4}q$	$\frac{1}{4}q$	$\frac{1}{4}q - 1$
1	$\frac{1}{4}(q+3)$	$\frac{1}{4}(q-1)$	$\frac{1}{4}(q-1)$	$\frac{1}{4}(q-1)$
2	$\frac{1}{4}(q+2)$	$\frac{1}{4}(q-2)$	$\frac{1}{4}(q+2)$	$\frac{1}{4}(q-2)$
3	$\frac{1}{4}(q+1)$	$\frac{1}{4}(q+1)$	$\frac{1}{4}(q+1)$	$\frac{1}{4}(q-3)$

Table G.1 *Multiplicity of the eigenvalue $i^a\sqrt{q}$, depending on q* (mod 4).

24. Let E be as in the preceding exercise, and suppose that q is an odd prime. Let $\boldsymbol{x}$ be the vector with coordinates $x_n = \left(\frac{n}{q}\right)$. Show that $\boldsymbol{x}$ is an eigenvector of E.

25. Let A be the $\varphi(q) \times \varphi(q)$ matrix $A = [\tau(\chi\overline{\psi})/\varphi(q)]$, where the rows are indexed by the Dirichlet character $\chi \pmod q$ and the columns are indexed by the Dirichlet character $\psi! \pmod q$.
   (a) Show that A is unitary.
   (b) Show that the vector $\boldsymbol{x}$ with coordinates $x_\psi = \psi(a)$ is an eigenvalue of A with eigenvalue $e(a/q)$.
   (c) Show that
$$\sum_{\chi,\psi} \tau(\chi\overline{\psi}) x_\chi \overline{x_\psi} = \sum_{\substack{a=1 \\ (a,q)=1}}^{q} \Big| \sum_\chi \chi(a) x_\chi \Big|^2 e(a/q).$$
   (d) Show that
$$\Big| \sum_{\chi,\psi} \tau(\chi\overline{\psi}) x_\chi \overline{x_\psi} \Big| \le \varphi(q) \sum_\chi |x_\chi|^2$$
for arbitrary complex numbers x_χ, and that the constant is best possible.
26. Let $f(n)$ be an arithmetic function with period q, and let $A = [f(mn)]$ be the $\varphi(q) \times \varphi(q)$ matrix whose rows m and columns n are indexed by the reduced residue classes (mod q).
   (a) Show that $\|A\| = \Delta$ where
$$\Delta = \max_\chi \Big| \sum_{n=1}^{q} f(n)\chi(n) \Big|.$$
   (b) Show that for arbitrary complex numbers x_n, y_m,
$$\Big| \sum_{\substack{m=1 \\ (mn,q)=1}}^{q} \sum_{n=1}^{q} f(mn) x_n \overline{y_m} \Big| \le \Delta \Bigg(\sum_{\substack{n=1 \\ (n,q)=1}}^{q} |x_n|^2 \Bigg)^{1/2} \Bigg(\sum_{\substack{m=1 \\ (m,q)=1}}^{q} |y_m|^2 \Bigg)^{1/2},$$
and that the constant Δ is best possible.
27. Let $\mathcal{S}$ be a set of N distinct Dirichlet characters modulo q.
   (a) Show that
$$\sum_{n=1}^{q} \Big| \sum_{s \in \mathcal{S}} \chi(n) \Big|^2 = N\varphi(q).$$
   (b) Show that
$$\sum_{n=1}^{q} \Big| \sum_{s \in \mathcal{S}} \chi(n) \Big|^4 \le N^3 \varphi(q).$$

(c) Deduce that

$$\sum_{n=1}^{q}\left|\sum_{s\in\mathcal{S}}\chi(n)\right| \ge \varphi(q).$$

(d) Suppose that q is prime, that $q \equiv 1 \pmod{N}$, and that $\mathcal{S}$ consists of the N characters χ modulo q for which $\chi^N = \chi_0$. Show that in this situation, equality holds in the lower bound above.

28. (a) Let f be an arithmetic function, and set $F(n) = \sum_{d|n} f(d)$. Let $R = [r_{mn}]$ be an $N \times N$ matrix with $r_{mn} = 1$ if $n|m$, and $r_{mn} = 0$ otherwise. Let Φ be an $N \times N$ diagonal matrix whose diagonal entries are $f(1), f(2), \dots, f(N)$. Let $A = [a_{mn}]$ be the $N \times N$ matrix whose entries are $F((m,n))$. Show that $A = R\Phi R^t$.

(b) (Smith 1876) Let $A = [a_{mn}]$ be the $N \times N$ matrix with $a_{mn} = (m,n)$. Show that

$$\det A = \prod_{n=1}^{N}\varphi(n).$$

This is the *Smith determinant.*

29. Let A_N denote the least number such that

$$\left|\sum_{pq\le N} x_p\overline{x_q}\right| \le A_N \sum_{p\le N}\frac{|x_p|^2}{p}$$

for all complex numbers x_p where p and q are to take only prime values. Show that $A_N \asymp N(\log N)^{-1/2}$.

30. Let B_N denote the least number such that

$$\left|\sum_{\substack{p<N,q\le N\\ pq\ge N}}\frac{x_p\overline{x_q}}{pq}\right| \le B_N \sum_{p\le N}\frac{|x_p|^2}{p}$$

for all complex numbers x_p where p and q are to take only prime values. Show that $B_N = 1 + O(1/\log N)$.

31. Let C_N be the least positive number such that

$$\left(\sum_{p\le N}\frac{x_p}{\sqrt{p}}\right)\left(\sum_{p\le N} x_p\sqrt{p}\right) \le C_N \sum_{p\le N}|x_p|^2$$

for all choices of the complex numbers x_p where p and q are to take only prime values. Show that $C_N \asymp \frac{N}{\log N}$.

32. Let A be the $N \times N$ matrix with coefficients

$$a_{mn} = \begin{cases} \Lambda(n/m)(m/n)^{1/2} & \text{if } m|n, \\ \Lambda(m/n)(n/m)^{1/2} & \text{if } n|m, \\ 0 & \text{otherwise.} \end{cases}$$

Show that $\|A\| = \log N + O(1)$. (Suggestion: Consider the vector $\boldsymbol{x}$ with coordinates $x_n = n^{-1/2}$.)

33. The object of this exercise is to show that if $f \in L^2[0,1]$ and $F(x) = \int_0^x f(u)\,du$ for $0 \le x \le 1$, then

$$\int_0^1 |F(x)|^2\,dx \le \frac{4}{\pi^2}\int_0^1 |f(x)|^2\,dx. \tag{G.4}$$

(a) Explain why it is enough to prove the above when $f(x) \ge 0$.
(b) Let $K(u,v) = \min(1-u, 1-v)$. Show that

$$\int_0^1 F(x)^2\,dx = \int_0^1\int_0^1 K(u,v)f(u)f(v)\,du\,dv.$$

(c) By a judicious application of the arithmetic–geometric inequality, show that

$$f(u)f(v) \le \frac{1}{2}f(u)^2\frac{\cos\frac{\pi}{2}v}{\cos\frac{\pi}{2}u} + \frac{1}{2}f(v)^2\frac{\cos\frac{\pi}{2}u}{\cos\frac{\pi}{2}v}$$

for $0 \le u, v < 1$.
(d) Show that if $0 \le v \le 1$, then

$$\int_0^1 K(u,v)\cos\frac{\pi}{2}u\,du = \frac{4}{\pi^2}\cos\frac{\pi}{2}v.$$

(e) Deduce (G.4).
(f) Show that if $f(u) = \cos\frac{\pi}{2}u$, then equality holds in (G.4).

G.3 Bessel's Inequality

Bessel's inequality asserts that if $\boldsymbol{\phi}_1, \boldsymbol{\phi}_2, \ldots, \boldsymbol{\phi}_R$ are orthonormal vectors in an inner product space V, then

$$\sum_{r=1}^{R} |(\boldsymbol{\xi}, \boldsymbol{\phi}_r)|^2 \le \|\boldsymbol{\xi}\|^2 \tag{G.5}$$

for all $\boldsymbol{\xi} \in V$. The proof of this is quite simple. For arbitrary y_r,

$$0 \le \left\| \boldsymbol{\xi} - \sum_{r=1}^{R} y_r \boldsymbol{\phi}_r \right\|^2 = \|\boldsymbol{\xi}\|^2 - 2\,\mathrm{Re} \sum_{r=1}^{R} \overline{y_r} (\boldsymbol{\xi}, \boldsymbol{\phi}_r) + \left\| \sum_{r=1}^{R} y_r \boldsymbol{\phi}_r \right\|^2$$

$$= \|\boldsymbol{\xi}\|^2 - 2\,\mathrm{Re} \sum_{r=1}^{R} \overline{y_r} (\boldsymbol{\xi}, \boldsymbol{\phi}_r) + \sum_{r=1}^{R} |y_r|^2. \tag{G.6}$$

Set $y_r = (\boldsymbol{\xi}, \boldsymbol{\phi}_r)$. Then the expression (G.6) is $\|\boldsymbol{\xi}\|^2 - \sum_r |(\boldsymbol{\xi}, \boldsymbol{\phi}_r)|^2$, so the proof is complete. However, in analytic number theory we often need to estimate a sum such as the one in (G.5) but with vectors $\boldsymbol{\phi}_r$ that are not quite orthogonal. It is therefore fortunate that we can extend Bessel's inequality to arbitrary vectors $\boldsymbol{\phi}_r$ with a constant that we can characterize in terms of the extent that the inner product matrix $[(\boldsymbol{\phi}_r, \boldsymbol{\phi}_s)]$ resembles the identity matrix.

Theorem G.12 *Let $\boldsymbol{\phi}_1, \boldsymbol{\phi}_2, \ldots, \boldsymbol{\phi}_R$ be arbitrary vectors in an inner product space V over the field $\mathbb{C}$ of complex numbers. For non-negative real numbers Δ, the following three assertions are equivalent.*

(i) *For every vector $\boldsymbol{\xi} \in V$,*

$$\sum_{r=1}^{R} |(\boldsymbol{\xi}, \boldsymbol{\phi}_r)|^2 \le \Delta^2 \|\boldsymbol{\xi}\|^2. \tag{G.7}$$

(ii) *For every vector $\boldsymbol{\xi} \in V$ and every vector $\mathbf{y} \in \mathbb{C}^R$,*

$$\left| \sum_{r=1}^{R} (\boldsymbol{\xi}, \boldsymbol{\phi}_r) y_r \right| \le \Delta \|\boldsymbol{\xi}\| \left(\sum_{r=1}^{R} |y_r|^2 \right)^{1/2}. \tag{G.8}$$

(iii) *For every vector $\mathbf{y} \in \mathbb{C}^R$,*

$$\sum_{r=1}^{R} \sum_{s=1}^{R} (\boldsymbol{\phi}_r, \boldsymbol{\phi}_s) y_r \overline{y_s} \le \Delta^2 \sum_{r=1}^{R} |y_r|^2. \tag{G.9}$$

Theorem G.12 contains Bessel's inequality as a special case, for if the $\boldsymbol{\phi}_r$ happen to be orthonormal, then the inequality (G.9) holds as an identity with $\Delta = 1$, and then (G.7) is Bessel's inequality. The coefficient matrix $C = [(\boldsymbol{\phi}_r, \boldsymbol{\phi}_s)]$ in (G.9) is Hermitian, and so by Corollary G.11 the best constant Δ for which the inequality (G.9) holds is the modulus of the largest eigenvalue of C. This quantity is known as the *spectral radius* of C; in symbols, $\Delta = \rho(C)$.

Proof (i) $\Longrightarrow$ (ii). By Cauchy's inequality,

$$\left| \sum_{r=1}^{R} (\boldsymbol{\xi}, \boldsymbol{\phi}_r) y_r \right| \le \left(\sum_{r=1}^{R} |(\boldsymbol{\xi}, \boldsymbol{\phi}_r)|^2 \right)^{1/2} \left(\sum_{r=1}^{R} |y_r|^2 \right)^{1/2}.$$

We apply the bound (G.7) to the first sum on the right above to obtain (G.8).

(ii) $\Longrightarrow$ (iii). Take $\boldsymbol{\xi} = \sum_{r=1}^{R} y_r \boldsymbol{\phi}_r$. Then

$$\sum_{r=1}^{R}\sum_{s=1}^{R}(\boldsymbol{\phi}_r, \boldsymbol{\phi}_s) y_r \overline{y_s} = \|\boldsymbol{\xi}\|^2. \tag{G.10}$$

But

$$\|\boldsymbol{\xi}\|^2 = \sum_{s=1}^{R}(\boldsymbol{\xi}, \boldsymbol{\phi}_s)\overline{y_s} \le \Delta\|\boldsymbol{\xi}\|\left(\sum_{s=1}^{R}|y_s|^2\right)^{1/2}$$

by (G.8). If $\boldsymbol{\xi} = \mathbf{0}$, then the left-hand side of (G.9) is 0, so there is nothing to prove. Otherwise, $\|\boldsymbol{\xi}\| > 0$, so we may cancel $\|\boldsymbol{\xi}\|$ from both sides above, and then square both sides. This gives (G.9), in view of (G.10).

(iii) $\Longrightarrow$ (i). We take the proof of Bessel's inequality as a model. For arbitrary y_r,

$$0 \le \left\|\boldsymbol{\xi} - \sum_{r=1}^{R} y_r \boldsymbol{\phi}_r\right\|^2 = \|\boldsymbol{\xi}\|^2 - 2\operatorname{Re}\sum_{r=1}^{R}\overline{y_r}(\boldsymbol{\xi}, \boldsymbol{\phi}_r) + \left\|\sum_{r=1}^{R} y_r \boldsymbol{\phi}_r\right\|^2.$$

Here the last term is

$$\sum_{r=1}^{R}\sum_{s=1}^{R}(\boldsymbol{\phi}_r, \boldsymbol{\phi}_s) y_r \overline{y_s}.$$

Thus by (iii) we see that

$$0 \le \|\boldsymbol{\xi}\|^2 - 2\operatorname{Re}\sum_{r=1}^{R}\overline{y_r}(\boldsymbol{\xi}, \boldsymbol{\phi}_r) + \Delta^2\sum_{r=1}^{R}|y_r|^2.$$

By taking $y_r = (\boldsymbol{\xi}, \boldsymbol{\phi}_r)/\Delta^2$ we find that

$$0 \le \|\boldsymbol{\xi}\|^2 - \frac{1}{\Delta^2}\sum_{r=1}^{R}|(\boldsymbol{\xi}, \boldsymbol{\phi}_r)|^2,$$

which gives (i). □

If the $\boldsymbol{\phi}_r$ are unit vectors that are nearly orthogonal so that the inner product matrix C is nearly the identity matrix, then we would expect that (G.7) holds with a constant not much larger than 1. The most immediate observation in this direction is as follows.

Theorem G.13 *The inequalities of Theorem* G.12 *hold with*

$$\Delta^2 = \max_{1\le r\le R}\sum_{s=1}^{R}|(\boldsymbol{\phi}_r, \boldsymbol{\phi}_s)|. \tag{G.11}$$

Proof By the arithmetic–geometric mean inequality we know that $|y_r y_s| \le \frac{1}{2}|y_r|^2 + \frac{1}{2}|y_s|^2$. Thus

$$\begin{aligned}\sum_{r=1}^{R}\sum_{s=1}^{R}(\boldsymbol{\phi}_r,\boldsymbol{\phi}_s)y_r\overline{y_s} &\le \sum_{r=1}^{R}\sum_{s=1}^{R}|(\boldsymbol{\phi}_r,\boldsymbol{\phi}_s)y_r\overline{y_s}| \le \sum_{r=1}^{R}|y_r|^2\sum_{s=1}^{R}|(\boldsymbol{\phi}_r,\boldsymbol{\phi}_s)| \\ &\le \left(\max_{1\le r\le R}\sum_{s=1}^{R}|(\boldsymbol{\phi}_r,\boldsymbol{\phi}_s)|\right)\sum_{r=1}^{R}|y_r|^2.\end{aligned}$$

Thus (G.9) holds with Δ as in (G.11), so the proof is complete. □

G.3.1 Exercises

1. The object of this exercise is to derive Theorem G.12 as an application of the duality principle of Theorem G.1. Let $\boldsymbol{e}_1, \boldsymbol{e}_2, \dots, \boldsymbol{e}_K$ be an orthonormal basis for $W = \text{span}(\boldsymbol{\phi}_1, \boldsymbol{\phi}_2, \dots, \boldsymbol{\phi}_R)$, and write

$$\boldsymbol{\phi}_r = \sum_{k=1}^{K} a_{rk}\boldsymbol{e}_k.$$

(a) Show that

$$\sum_{r=1}^{R}\sum_{s=1}^{R}(\boldsymbol{\phi}_r,\boldsymbol{\phi}_s)y_r\overline{y_s} = \left\|\sum_{r=1}^{R} y_r\boldsymbol{\phi}_r\right\|^2 = \sum_{k=1}^{K}\left|\sum_{r=1}^{R} a_{rk}y_r\right|^2.$$

(b) Put $v_k = (\boldsymbol{\xi}, \boldsymbol{e}_k)$, and define $\boldsymbol{\zeta}$ so that

$$\boldsymbol{\xi} = \boldsymbol{\zeta} + \sum_{k=1}^{K} v_k\boldsymbol{e}_k.$$

Thus $\boldsymbol{\zeta} \in W^{\perp}$. Show that

$$(\boldsymbol{\xi}, \boldsymbol{\phi}_r) = \sum_{k=1}^{K}\overline{a_{rk}}\, v_k.$$

(c) Deduce that

$$\sum_{r=1}^{R}(\boldsymbol{\xi}, \boldsymbol{\phi}_r)y_r = \sum_{r=1}^{R}\sum_{k=1}^{K}\overline{a_{rk}}y_r v_k,$$

and also that

$$\sum_{r=1}^{R}|(\boldsymbol{\xi}, \boldsymbol{\phi}_r)|^2 = \sum_{r=1}^{R}\left|\sum_{k=1}^{K}\overline{a_{rk}}v_k\right|^2.$$

(d) Show that

$$\sum_{k=1}^{K} |v_k|^2 \le \|\boldsymbol{\xi}\|^2.$$

(e) Use Theorem G.1 to prove Theorem G.12.

2. The object of this exercise is to use Theorem G.12 to prove Theorem G.1. Let $\boldsymbol{\xi} = (x_1, x_2, \ldots, x_N) \in \mathbb{C}^N$, and for $r = 1, 2, \ldots, R$ take $\boldsymbol{\phi}_r = (a_{r1}, a_{r2}, \ldots, a_{rN})$.

(a) Explain why

$$\sum_{r=1}^{R} |(\boldsymbol{\xi}, \boldsymbol{\phi}_r)|^2 = \sum_{r=1}^{R} \left| \sum_{n=1}^{N} a_{rn} x_n \right|^2.$$

(b) Show that

$$\sum_{r=1}^{R} (\boldsymbol{\xi}, \boldsymbol{\phi}_r) y_r = \sum_{r=1}^{R} \sum_{n=1}^{N} a_{rn} x_n y_r.$$

(c) Explain why

$$\left\| \sum_{r=1}^{R} y_r \boldsymbol{\phi}_r \right\|^2 = \sum_{n=1}^{N} \left| \sum_{r=1}^{R} a_{rn} y_r \right|^2.$$

(d) Use Theorem G.12 to prove G.1.

3. (Halász) Let $\boldsymbol{\xi}, \boldsymbol{\phi}_1, \boldsymbol{\phi}_2, \ldots, \boldsymbol{\phi}_R$ be arbitrary vectors in an inner product space V over the field $\mathbb{C}$ of complex numbers.

(a) Let c_r be chosen, $|c_r| = 1$, so that $c_r(\boldsymbol{\xi}, \boldsymbol{\phi}_r) = |(\boldsymbol{\xi}, \boldsymbol{\phi}_r)|$. Show that

$$\sum_{r=1}^{R} |(\boldsymbol{\xi}, \boldsymbol{\phi}_r)| = \left(\boldsymbol{\xi}, \sum_{r=1}^{R} \overline{c}_r \boldsymbol{\phi}_r \right).$$

(b) Explain why the right-hand side above is

$$\le \|\boldsymbol{\xi}\| \left\| \sum_{r=1}^{R} \overline{c}_r \boldsymbol{\phi}_r \right\|.$$

(c) Show that

$$\left\| \sum_{r=1}^{R} \overline{c}_r \boldsymbol{\phi}_r \right\|^2 = \sum_{1 \le r, s \le R} \overline{c}_r c_s (\boldsymbol{\phi}_r, \boldsymbol{\phi}_s).$$

(d) Conclude that

$$\sum_{r=1}^{R} |\boldsymbol{\xi}, \boldsymbol{\phi}_r)| \le \left(\sum_{1 \le r, s \le R} |(\boldsymbol{\phi}_r, \boldsymbol{\phi}_s)| \right)^{1/2} \|\boldsymbol{\xi}\|.$$

4. (Selberg, unpublished; cf. Bombieri 1971) Let $\boldsymbol{\xi}, \boldsymbol{\phi}_1, \boldsymbol{\phi}_2, \ldots, \boldsymbol{\phi}_R$ be elements of an inner product space V.

   (a) Explain why

$$0 \le \left\| \boldsymbol{\xi} - \sum_{r=1}^{R} c_r \boldsymbol{\phi}_r \right\|^2$$
$$= \|\boldsymbol{\xi}\|^2 - 2\,\mathrm{Re} \sum_{r=1}^{R} \overline{c}_r (\boldsymbol{\xi}, \boldsymbol{\phi}_r) + \sum_{1 \le r,s \le R} c_r \overline{c}_s (\boldsymbol{\phi}_r, \boldsymbol{\phi}_s).$$

   (b) Deduce that

$$2\,\mathrm{Re} \sum_{r=1}^{R} \overline{c}_r (\boldsymbol{\xi}, \boldsymbol{\phi}_r) \le \|\boldsymbol{\xi}\|^2 + \sum_{r=1}^{R} |c_r|^2 \sum_{s=1}^{R} |(\boldsymbol{\phi}_r, \boldsymbol{\phi}_s)|.$$

   (c) Take

$$c_r = (\boldsymbol{\xi}, \boldsymbol{\phi}_r) \left(\sum_{s=1}^{R} |(\boldsymbol{\phi}_r, \boldsymbol{\phi}_s)| \right)^{-1}$$

   and thus conclude that

$$\sum_{r=1}^{R} |(\boldsymbol{\xi}, \boldsymbol{\phi}_r)|^2 \left(\sum_{s=1}^{R} |(\boldsymbol{\phi}_r, \boldsymbol{\phi}_s)| \right)^{-1} \le \|\boldsymbol{\xi}\|^2.$$

   (d) Use the above to derive Theorem G.13.

5. Let $\boldsymbol{\phi}_1, \boldsymbol{\phi}_2, \ldots, \boldsymbol{\phi}_R$ and $\boldsymbol{\psi}_1, \boldsymbol{\psi}_2, \ldots, \boldsymbol{\psi}_S$ be any members of an inner product space V. Show that

$$\sum_{r=1}^{R} \sum_{s=1}^{S} |(\boldsymbol{\phi}_r, \boldsymbol{\psi}_s)|^2 \le \left(\sum_{r_1=1}^{R} \sum_{r_2=1}^{R} |(\boldsymbol{\phi}_{r_1}, \boldsymbol{\phi}_{r_2})|^2 \right)^{1/2} \left(\sum_{s_1=1}^{S} \sum_{s_2=1}^{S} |(\boldsymbol{\psi}_{s_1}, \boldsymbol{\psi}_{s_2})|^2 \right)^{1/2}.$$

G.4 Hilbert's Inequality

In classical analysis, the term 'Hilbert's inequality' refers to one of the bilinear form inequalities

$$\left| \sum_{m=1}^{\infty} \sum_{n=1}^{\infty} \frac{x_n y_m}{m+n-1} \right| \le \pi \left(\sum_{n=1}^{\infty} |x_n|^2 \right)^{1/2} \left(\sum_{m=1}^{\infty} |y_m|^2 \right)^{1/2}, \tag{G.12}$$

$$\left| \sum_{\substack{1 \le m,n < \infty \\ m \ne n}} \frac{x_n y_m}{m-n} \right| \le \pi \left(\sum_{n=1}^{\infty} |x_n|^2 \right)^{1/2} \left(\sum_{m=1}^{\infty} |y_m|^2 \right)^{1/2}. \tag{G.13}$$

These inequalities are easily proved (for the case of finite sums; see Exercise G.2.1.18). The constant π is best possible in both of the above, but equality is attained only when $x_n = 0$ for all n or $y_m = 0$ for all m. For our purposes, Hilbert's inequality is a bound for a bilinear form of the shape

$$\sum_{\substack{1\le m,n\le N\\ m\ne n}} \frac{x_n y_m}{\lambda_m - \lambda_n},$$

where the λ_n are distinct real numbers. Of course the bound we obtain for such a bilinear form depends on the extent to which the λ_n are well spaced.

Theorem G.14 *Let $\lambda_1, \lambda_2, \dots, \lambda_N$ be distinct real numbers, and let $\delta > 0$ have the property that $|\lambda_m - \lambda_n| \ge \delta$ whenever $m \ne n$. Then*

$$\left| \sum_{\substack{1\le m,n\le N\\ m\ne n}} \frac{x_n y_m}{\lambda_m - \lambda_n} \right| < \frac{\pi}{\delta}\left(\sum_{n=1}^N |x_n|^2\right)^{1/2}\left(\sum_{m=1}^N |y_m|^2\right)^{1/2} \tag{G.14}$$

for arbitrary real or complex numbers x_n and y_m.

On taking $y_m = \overline{x_m}$, we see in particular that

$$\left| \sum_{\substack{1\le m,n\le N\\ m\ne n}} \frac{x_n \overline{x_m}}{\lambda_m - \lambda_n} \right| \le \frac{\pi}{\delta}\sum_{n=1}^N |x_n|^2 \tag{G.15}$$

for arbitrary real or complex x_n.

Proof By Cauchy's inequality, the left-hand side above has absolute value not exceeding

$$\left(\sum_{m=1}^N |y_m|^2\right)^{1/2}\left(\sum_{m=1}^N \left| \sum_{\substack{1\le n\le N\\ n\ne m}} \frac{x_n}{\lambda_m - \lambda_n}\right|^2\right)^{1/2}.$$

Thus it suffices to show that

$$\sum_{m=1}^N \left| \sum_{\substack{1\le n\le N\\ n\ne m}} \frac{x_n}{\lambda_m - \lambda_n}\right|^2 \le \frac{\pi^2}{\delta^2}\sum_{n=1}^N |x_n|^2. \tag{G.16}$$

Indeed, by Theorem G.1, this inequality is equivalent to (G.14). Let $A = [a_{mn}]$ be the matrix with elements

$$a_{mn} = \begin{cases} \dfrac{1}{\lambda_m - \lambda_n} & \text{if } m \ne n, \\ 0 & \text{if } m = n. \end{cases}$$

We note that $A^* = -A$. Such a matrix is said to be *skew-Hermitian*. Since A is normal, by Corollary G.11 we know that $\rho(A) = \|A\|$, so in proving (G.16) we may assume that $\boldsymbol{x}$ is an eigenvector of A. Now $-iA$ is Hermitian, so an eigenvalue λ of $-iA$ is real, and $-iA\boldsymbol{x} = \lambda\boldsymbol{x}$ is equivalent to $A\boldsymbol{x} = i\lambda\boldsymbol{x}$. That is, any eigenvalue of A is of the form $i\lambda$ where λ is real. Thus as we continue, we assume that there is a real number λ such that

$$\sum_{\substack{1\le n\le N\\ n\ne m}} \frac{x_n}{\lambda_m - \lambda_n} = i\lambda x_m \tag{G.17}$$

for all m. In passing we note that since A is normal, it follows by Corollary G.11 that the special case (G.15) of (G.14) is equivalent to (G.14).

We square out the left-hand side of (G.16) and take the sum over m inside to see that this expression is

$$= \sum_{r=1}^{N} x_r \sum_{s=1}^{N} \overline{x_s} \sum_{\substack{1\le m\le N\\ m\ne r\\ m\ne s}} \frac{1}{(\lambda_m - \lambda_r)(\lambda_m - \lambda_s)}.$$

The terms with $r = s$ contribute

$$\sum_{n=1}^{N} |x_n|^2 \sum_{\substack{1\le m\le N\\ m\ne n}} \frac{1}{(\lambda_m - \lambda_n)^2}. \tag{G.18}$$

The terms with $r \ne s$ contribute

$$\sum_{\substack{1\le r,s\le N\\ r\ne s}} x_r\overline{x_s} \sum_{\substack{1\le m\le N\\ m\ne r\\ m\ne s}} \frac{1}{(\lambda_m - \lambda_r)(\lambda_m - \lambda_s)}. \tag{G.19}$$

Since $r \ne s$, we may write

$$\frac{1}{(\lambda_m - \lambda_r)(\lambda_m - \lambda_s)} = \frac{1}{\lambda_r - \lambda_s}\left(\frac{1}{\lambda_m - \lambda_r} - \frac{1}{\lambda_m - \lambda_s}\right).$$

On inserting this in (G.19), we find that the expression is

$$\sum_{\substack{1\le r,s\le N\\ r\ne s}} \frac{x_r\overline{x_s}}{\lambda_r - \lambda_s}\left(\sum_{\substack{1\le m\le N\\ m\ne r\\ m\ne s}} \frac{1}{\lambda_m - \lambda_r} - \sum_{\substack{1\le m\le N\\ m\ne r\\ m\ne s}} \frac{1}{\lambda_m - \lambda_s}\right).$$

In the first inner sum the summand is finite if we were to allow m to take the value s, so we drop the constraint $m \ne s$. Similarly, in the second inner sum we drop the constraint $m \ne r$. After accounting for the effect of these alterations, we find that the expression above is

$$
\begin{aligned}
&= 2 \sum_{\substack{1 \le r,s \le N \\ r \ne s}} \frac{x_r \overline{x_s}}{(\lambda_r - \lambda_s)^2} + \sum_{\substack{1 \le r,s \le N \\ r \ne s}} \frac{x_r \overline{x_s}}{\lambda_r - \lambda_s} \sum_{\substack{1 \le m \le M \\ m \ne r}} \frac{1}{\lambda_m - \lambda_r} \\
&\qquad - \sum_{\substack{1 \le r,s \le N \\ r \ne s}} \frac{x_r \overline{x_s}}{\lambda_r - \lambda_s} \sum_{\substack{1 \le m \le M \\ m \ne s}} \frac{1}{\lambda_m - \lambda_s} \\
&= 2T_1 + T_2 - T_3, \qquad \text{(G.20)}
\end{aligned}
$$

say.

Since $|x_r \overline{x_s}| \le \frac{1}{2}|x_r|^2 + \frac{1}{2}|x_s|^2$, it follows that

$$
|T_1| \le \sum_{r=1}^{N} |x_r|^2 \sum_{\substack{1 \le s \le N \\ s \ne r}} \frac{1}{(\lambda_r - \lambda_s)^2}. \qquad \text{(G.21)}
$$

We note that

$$
T_2 = \sum_{r=1}^{N} x_r \left(\sum_{\substack{1 \le s \le N \\ s \ne r}} \frac{\overline{x_s}}{\lambda_r - \lambda_s} \right) \left(\sum_{\substack{1 \le m \le N \\ m \ne r}} \frac{1}{\lambda_m - \lambda_r} \right).
$$

On taking complex conjugates of both sides of (G.17), and then setting $m = r$, we find that the first inner sum above is $= -i\lambda \overline{x_r}$ since λ is real. Thus

$$
T_2 = -i\lambda \sum_{r=1}^{N} |x_r|^2 \sum_{\substack{1 \le m \le N \\ m \ne r}} \frac{1}{\lambda_m - \lambda_r}.
$$

Similarly,

$$
T_3 = \sum_{s=1}^{N} \overline{x_s} \left(\sum_{\substack{1 \le r \le N \\ r \ne s}} \frac{1}{\lambda_r - \lambda_s} \right) \left(\sum_{\substack{1 \le m \le N \\ m \ne s}} \frac{1}{\lambda_m - \lambda_s} \right).
$$

By multiplying both sides of (G.17) by -1, and taking $m = s$, we find that the first inner sum above is $= -i\lambda x_s$. Thus

$$
T_3 = -i \sum_{s=1}^{N} |x_s|^2 \sum_{\substack{1 \le m \le N \\ m \ne s}} \frac{1}{\lambda_m - \lambda_s}.
$$

Hence $T_2 = T_3$, so the contributions of these terms in (G.20) cancel. Thus the expression (G.19) is precisely T_1. On combining (G.18) with (G.21) we deduce that the left-hand side of (G.16) does not exceed

$$3\sum_{n=1}^{N}|x_n|^2 \sum_{\substack{1\le m\le N\\ m\ne n}} \frac{1}{(\lambda_m-\lambda_n)^2}.$$

We may assume that the λ_n are in increasing order, so that $|\lambda_m - \lambda_n| \ge \delta|m-n|$. Hence the inner sum above does not exceed

$$\frac{1}{\delta^2}\sum_{\substack{1\le m\le N\\ m\ne n}} \frac{1}{(m-n)^2} \le \frac{2\zeta(2)}{\delta^2} = \frac{\pi^2}{3\delta^2}.$$

Thus we have (G.16), and the proof is complete. □

We now use Theorem G.14 to derive a trigonometric variant, which is useful when we work modulo 1.

Theorem G.15 *Let $\alpha_1, \alpha_2, \ldots, \alpha_R$ be distinct modulo 1, and let $\delta > 0$ have the property that $\|\alpha_r - \alpha_s\| \ge \delta$ whenever $r \ne s$. Then*

$$\left|\sum_{\substack{1\le r,s\le R\\ r\ne s}} \frac{u_r v_s}{\sin \pi(\alpha_r-\alpha_s)}\right| \le \frac{1}{\delta}\left(\sum_{r=1}^{R}|u_r|^2\right)^{1/2}\left(\sum_{s=1}^{R}|v_s|^2\right)^{1/2} \tag{G.22}$$

for arbitrary real or complex u_r and v_s.

On setting $v_s = \overline{u_s}$ we see, in particular, that

$$\left|\sum_{\substack{1\le r,s\le R\\ r\ne s}} \frac{u_r \overline{u_s}}{\sin \pi(\alpha_r-\alpha_s)}\right| \le \frac{1}{\delta}\sum_{r=1}^{R}|u_r|^2 \tag{G.23}$$

for arbitrary real or complex u_r.

Proof We recall that the Weierstrass product formula for the sine function asserts that

$$\sin \pi z = \pi z \prod_{k=1}^{\infty}\left(1-\frac{z}{k}\right)\left(1+\frac{z}{k}\right).$$

On taking logarithmic derivatives, it follows that

$$\pi \cot \pi z = \frac{1}{z} + \sum_{k=1}^{\infty}\left(\frac{1}{z-k}+\frac{1}{z+k}\right).$$

Now

$$\frac{1}{\sin \pi z} = \frac{1}{2}\cot\frac{\pi z}{2} - \frac{1}{2}\cot\frac{\pi(z+1)}{2},$$

so

$$\frac{\pi}{\sin \pi z} = \frac{1}{z} + \sum_{k=1}^{\infty}(-1)^k\left(\frac{1}{z-k} + \frac{1}{z+k}\right)$$
$$= \lim_{K\to\infty} \sum_{k=-K}^{K} \frac{(-1)^k}{z-k}. \tag{G.24}$$

We apply Theorem G.14 with doubly-indexed variables x_{nr}, y_{ms}, and λ_{nr}. Thus

$$\left| \sum_{\substack{r,s,m,n \\ (n,r)\neq(m,s)}} \frac{x_{nr} y_{ms}}{\lambda_{nr} - \lambda_{ms}} \right| \le \frac{\pi}{\delta} \left(\sum_{n,r} |x_{nr}|^2 \right)^{1/2} \left(\sum_{m,s} |y_{ms}|^2 \right)^{1/2}.$$

We now take $x_{nr} = (-1)^n u_r$, $y_{ms} = (-1)^m v_s$, and $\lambda_{nr} = n + \alpha_r$ for $1 \le m, n \le K$. Thus

$$\left| \sum_{(n,r)\neq(m,s)} \frac{(-1)^{n-m} u_r v_s}{n - m + \alpha_r - \alpha_s} \right| \le \frac{K\pi}{\delta} \left(\sum_{r=1}^{R} |u_r|^2 \right)^{1/2} \left(\sum_{s=1}^{R} |v_s|^2 \right)^{1/2}.$$

As

$$\sum_{\substack{1\le m,n\le K \\ m\neq n}} \frac{(-1)^{n-m}}{n-m} = 0,$$

we may replace the condition $(n,r) \neq (m,s)$ by the simpler condition $r \neq s$. We put $k = m - n$ and divide by K to see that

$$\left| \sum_{r\neq s} u_r v_s \sum_{k=-K}^{K} \frac{(-1)^k (1 - |k|/K)}{\alpha_r - \alpha_s - k} \right| \le \frac{\pi}{\delta} \left(\sum_{r=1}^{R} |u_r|^2 \right)^{1/2} \left(\sum_{s=1}^{R} |v_s|^2 \right)^{1/2}.$$

From (G.24) we see that the left-hand side above tends to

$$\pi \left| \sum_{r\neq s} \frac{u_r v_s}{\sin \pi(\lambda_r - \lambda_s)} \right|$$

as $K \to \infty$, so the proof is complete. □

Suppose that

$$\lambda_1 < \lambda_2 < \cdots < \lambda_N, \tag{G.25}$$

and let

$$\delta_n = \min_{\substack{m \\ m\neq n}} |\lambda_m - \lambda_n|. \tag{G.26}$$

Thus in Theorem G.14 we may take $\delta = \min_n \delta_n$. When some of the λ_n are more widely spaced from their neighbours than others, it is advantageous to work with the δ_n rather than with δ, as it is possible to derive a weighted form of the Hilbert inequality.

Theorem G.16 *Let $\lambda_1, \lambda_2, \ldots, \lambda_N$ be distinct real numbers, and let the numbers δ_n be defined as in* (G.26). *Then*

$$\left| \sum_{\substack{1\le m,n\le N \\ m\ne n}} \frac{x_m y_n}{\lambda_m - \lambda_n} \right| \le \frac{3}{2}\pi \left(\sum_{n=1}^{N} \frac{|x_n|^2}{\delta_n} \right)^{1/2} \left(\sum_{m=1}^{N} \frac{|y_m|^2}{\delta_m} \right)^{1/2} \tag{G.27}$$

for arbitrary real or complex numbers x_n and y_m.

This includes Theorem G.14 apart from the factor $3/2$. It is unknown whether the above is true with the constant π. On taking $y_m = \overline{x_m}$ we see, in particular, that

$$\left| \sum_{\substack{1\le m,n\le N \\ m\ne n}} \frac{x_n \overline{x_m}}{\lambda_m - \lambda_n} \right| \le \frac{3}{2}\pi \sum_{n=1}^{N} \frac{|x_n|^2}{\delta_n} \tag{G.28}$$

for arbitrary real or complex numbers x_n.

To prepare for the proof of Theorem G.16 we establish some useful inequalities.

Lemma G.17 *Let the λ_n and δ_n be as in* (G.25) *and* (G.26). *Suppose that f is defined on $(0, \infty)$, that f is positive, decreasing, and convex upwards, and that $\int_\delta^\infty f(x)\,dx < \infty$ for $\delta > 0$. Then*

$$\sum_{n>r} \delta_n f(\lambda_n - \lambda_r) \le (\lambda_{r+1} - \lambda_r) f(\lambda_{r+1} - \lambda_r) + \int_{\lambda_{r+1}-\lambda_r}^{\infty} f(x)\,dx \tag{G.29}$$

for $r < N$, and

$$\sum_{n<r} \delta_n f(\lambda_r - \lambda_n) \le (\lambda_r - \lambda_{r-1}) f(\lambda_r - \lambda_{r-1}) + \int_{\lambda_r-\lambda_{r-1}}^{\infty} f(x)\,dx \tag{G.30}$$

for $r > 1$.

Proof The contribution of $n = r + 1$ in (G.29) is

$$\delta_{r+1} f(\lambda_{r+1} - \lambda_r) \le (\lambda_{r+1} - \lambda_r) f(\lambda_{r+1} - \lambda_r)$$

since $\delta_{r+1} \le \lambda_{r+1} - \lambda_r$ and f is positive. For $n > r + 1$ we set $\mathscr{M}_n = [\lambda_n - \frac{1}{2}\delta_n, \lambda_n + \frac{1}{2}\delta_n]$, and observe that

$$\delta_n f(\lambda_n - \lambda_r) \le \int_{\mathscr{M}_n} f(x - \lambda_r)\,dx$$

by the convexity of f. The intervals $\mathscr{M}_n$ are disjoint and lie in the interval $[\lambda_{r+1}, \infty)$, so

$$\sum_{n>r+1} \delta_n f(\lambda_n - \lambda_r) \le \int_{\lambda_{r+1}}^{\infty} f(x - \lambda_r)\,dx$$

since f is non-negative. Thus we have (G.29), and (G.30) is proved similarly. □

Corollary G.18 *Let the λ_n and δ_n be as in* (G.25) *and* (G.26). *Then*

$$\sum_{\substack{n=1\\ n\neq r}}^{N} \frac{\delta_n}{(\lambda_n-\lambda_r)^2} \le \frac{4}{\delta_r} \tag{G.31}$$

and

$$\sum_{\substack{n=1\\ n\neq r}}^{N} \frac{\delta_n}{(\lambda_n-\lambda_r)^4} \le \frac{8}{3\delta_r^3} \tag{G.32}$$

for $1 \le r \le N$.

Proof By taking $f(x) = 1/x^2$ in (G.29) we find that

$$\sum_{r<n\le N} \frac{\delta_n}{(\lambda_n-\lambda_r)^2} \le \frac{2}{\lambda_{r+1}-\lambda_r} < \frac{2}{\delta_r},$$

and the corresponding sum over $n < r$ is bounded similarly using (G.30). This gives (G.31), and (G.32) is proved similarly by taking $f(x) = 1/x^4$. □

Lemma G.19 *Let the λ_n and δ_n be as in* (G.25) *and* (G.26). *If* $1 \le r, s \le N$, *and* $r \neq s$, *then*

$$\sum_{\substack{1\le n\le N\\ n\neq r\\ n\neq s}} \frac{\delta_n}{(\lambda_n-\lambda_r)^2(\lambda_n-\lambda_s)^2} \le \frac{4}{(\lambda_r-\lambda_s)^2}\left(\frac{1}{\delta_r}+\frac{1}{\delta_s}\right). \tag{G.33}$$

Proof Let $f(x) = (x-\lambda_r)^{-2}(x-\lambda_s)^{-2}$. We first show that f is concave upwards. By taking logarithmic derivatives we see that

$$\frac{f'}{f}(x) = \frac{-2}{x-\lambda_r} + \frac{-2}{x-\lambda_s}.$$

By differentiating both sides of this we find that

$$\left(\frac{f'}{f}\right)'(x) = \frac{2}{(x-\lambda_r)^2} + \frac{2}{(x-\lambda_s)^2}.$$

Here the left-hand side is $(f''(x)f(x) - f'(x)^2)/f(x)^2$, so by multiplying both sides of the above by $f(x)^2$ and then adding $f'(x)^2$ we deduce that

$$f''(x)f(x) = f'(x)^2 + f(x)^2\left(\frac{2}{(x-\lambda_r)^2} + \frac{2}{(x-\lambda_s)^2}\right) > 0.$$

Since $f(x) > 0$, it follows that $f''(x) > 0$.

Let $\mathscr{M}_n = [\lambda_n - \delta_n/2, \lambda_n + \delta_n/2]$. Since f is convex upwards, it follows that

$$\delta_n f(\lambda_n) \le \int_{\mathscr{M}_n} f(x)\,dx,$$

and on summing this over n we find that

$$\sum_{\substack{1\le n\le N\\ n\ne r\\ n\ne s}} \delta_n f(\lambda_n) \le \sum_{\substack{1\le n\le N\\ n\ne r\\ n\ne s}} \int_{\mathscr{M}_n} f(x)\,dx. \tag{G.34}$$

Now

$$\begin{aligned}\lambda_n + \frac{1}{2}\delta_n &= \lambda_{n+1} - (\lambda_{n+1} - \lambda_n) + \frac{1}{2}\delta_n \le \lambda_{n+1} - \frac{1}{2}(\lambda_{n+1} - \lambda_n)\\ &\le \lambda_{n+1} - \frac{1}{2}\delta_{n+1},\end{aligned}$$

so the intervals $\mathscr{M}_n$ are pairwise disjoint. Let $\mathscr{R} = \mathbb{R} \setminus \big(\mathscr{M}_r \cup \mathscr{M}_s\big)$. Since $f(x) > 0$ for all x, it follows that the right-hand side of (G.34) is

$$\le \int_{\mathscr{R}} f(x)\,dx. \tag{G.35}$$

We note that

$$\frac{1}{(x-\lambda_r)(x-\lambda_s)} = \frac{1}{\lambda_r - \lambda_s}\left(\frac{1}{x-\lambda_r} - \frac{1}{x-\lambda_s}\right).$$

On squaring both sides of this, and then expanding the right-hand side, we deduce that

$$\begin{aligned} f(x) &= \frac{1}{(\lambda_r-\lambda_s)^2(x-\lambda_r)^2} - \frac{2}{(\lambda_r-\lambda_s)^2(x-\lambda_r)(x-\lambda_s)}\\ &\qquad + \frac{1}{(\lambda_r-\lambda_s)^2(x-\lambda_s)^2}\\ &= f_1(x) + f_2(x) + f_3(x),\end{aligned}$$

say.

Since $f_1(x) \ge 0$ for all x, it follows that

$$\begin{aligned}\int_{\mathscr{R}} f_1(x)\,dx \le \int_{\mathscr{M}_r^c} f_1(x)\,dx &= \frac{2}{(\lambda_r-\lambda_s)^2}\int_{\delta_r/2}^{\infty} x^{-2}\,dx\\ &= \frac{4}{\delta_r(\lambda_r-\lambda_s)^2},\end{aligned} \tag{G.36}$$

and similarly

$$\int_{\mathscr{R}} f_3(x)\,dx \le \frac{4}{\delta_s(\lambda_r-\lambda_s)^2}. \tag{G.37}$$

It remains to treat $\int_{\mathcal{R}} f_2(x)\,dx$. We observe that

$$\begin{aligned} f_2(x) &= \frac{-2}{(\lambda_r-\lambda_s)^3}\left(\frac{1}{x-\lambda_r}-\frac{1}{x-\lambda_s}\right) \\ &= \frac{-2}{(\lambda_r-\lambda_s)^3(x-\lambda_r)}+\frac{2}{(\lambda_r-\lambda_s)^3(x-\lambda_s)} \\ &= f_{21}(x)+f_{22}(x), \end{aligned}$$

say. Let $\mathcal{I}(X)=[-X,X]$. Then

$$\begin{aligned} \int_{\mathcal{R}} f_2(x)\,dx &= \lim_{X\to\infty}\int_{\mathcal{R}\mathcal{I}(X)} f_2(x)\,dx \\ &= \lim_{X\to\infty}\int_{\mathcal{R}\mathcal{I}(X)} f_{21}(x)\,dx+\lim_{X\to\infty}\int_{\mathcal{R}\mathcal{I}(X)} f_{22}(x)\,dx. \end{aligned} \tag{G.38}$$

Suppose that X is large. Then

$$\int_{\mathcal{R}\mathcal{I}(X)} f_{21}(x)\,dx=\int_{\mathcal{M}_r^c\mathcal{I}(X)} f_{21}(x)\,dx-\int_{\mathcal{M}_s} f_{21}(x)\,dx<\int_{\mathcal{M}_r\mathcal{I}(X)} f_{21}(x)\,dx$$

since $f_{21}(x)>0$ for $x\in\mathcal{M}_s$. The remaining integral above is $-2/(\lambda_r-\lambda_s)^3$ times

$$\int_{\lambda_r+\frac{1}{2}\delta_r}^{X}\frac{dx}{x-\lambda_r}+\int_{-X}^{\lambda_r-\frac{1}{2}\delta_r}\frac{dx}{x-\lambda_r}=\log\frac{X-\lambda_r}{X+\lambda_r},$$

which tends to 0 as $X\to\infty$. Thus the first limit in (G.38) is negative. As for the second limit, we note that

$$\int_{\mathcal{R}\mathcal{I}(X)} f_{22}(x)\,dx=\int_{\mathcal{M}_s^c\mathcal{I}(X)} f_{22}(x)\,dx-\int_{\mathcal{M}_r} f_{22}(x)\,dx<\int_{\mathcal{M}_s^c\mathcal{I}(X)} f_{22}(x)\,dx$$

since $f_{22}(x)>0$ for $x\in\mathcal{M}_r$. The remaining integral above is $2/(\lambda_r-\lambda_s)^3$ times

$$\int_{\lambda_s+\frac{1}{2}\delta_s}^{X}\frac{dx}{x-\lambda_s}+\int_{-X}^{\lambda_s-\frac{1}{2}\delta_s}\frac{dx}{x-\lambda_s}=\log\frac{X-\lambda_s}{X+\lambda_s},$$

which tends to 0 as $X\to\infty$. Thus the second limit in (G.38) is also negative, so

$$\int_{\mathcal{R}} f_2(x)\,dx<0.$$

The stated result now follows by combining this with (G.36) and (G.37) in (G.35). □

Proof of Theorem G.16 Put $u_n = x_n/\sqrt{\delta_n}$ and $v_m = y_m/\sqrt{\delta_m}$. Thus we have to show that

$$\left|\sum_{\substack{1\le m,n\le N\\ m\ne n}} \frac{\sqrt{\delta_m\delta_n}}{\lambda_m-\lambda_n}u_n v_m\right| \le \frac{3}{2}\pi\left(\sum_{n=1}^N |u_n|^2\right)^{1/2}\left(\sum_{m=1}^N |v_m|^2\right)^{1/2}$$

for all u_n and v_m. By Cauchy's inequality, the left-hand side above is

$$\le \left(\sum_{m=1}^N |v_m|^2\right)^{1/2}\left(\sum_{m=1}^N\left|\sum_{\substack{n=1\\ n\ne m}}^N \frac{\sqrt{\delta_m\delta_n}}{\lambda_m-\lambda_n}u_n\right|^2\right)^{1/2}.$$

Thus it suffices to show that

$$\sum_{m=1}^N\left|\sum_{\substack{n=1\\ n\ne m}}^N \frac{\sqrt{\delta_m\delta_n}}{\lambda_m-\lambda_n}u_n\right|^2 \le \frac{9}{4}\pi^2\sum_{n=1}^N |u_n|^2 \tag{G.39}$$

for all u_n. Let $A = [a_{mn}]$ be the $N \times N$ matrix with coefficients

$$a_{mn} = \begin{cases} \dfrac{\sqrt{\delta_m\delta_n}}{\lambda_m-\lambda_n} & \text{if } m \ne n,\\ 0 & \text{if } m = n.\end{cases}$$

Thus $A^* = -A$, so that A is skew-Hermitian, and hence normal. Thus by Corollary G.11 we know that $\rho(A) = \|A\|$. Thus we may assume that $\boldsymbol{u}$ is an eigenvector of A. Since $-iA$ is Hermitian, any eigenvalue λ of $-iA$ is real, so that if $\boldsymbol{u}$ is an associated eigenvector, then $-iA\boldsymbol{u} = \lambda\boldsymbol{u}$. On multiplying both sides of this by i, we deduce that $A\boldsymbol{u} = i\lambda\boldsymbol{u}$. Thus the eigenvalues of A are of the form $i\lambda$ where λ is real. As we continue, we assume that $\boldsymbol{u}$ is such an eigenvector, so that

$$\sum_{\substack{n=1\\ n\ne m}}^N \frac{\sqrt{\delta_m\delta_n}}{\lambda_m-\lambda_n}u_n = i\lambda u_m \tag{G.40}$$

for all m. We may further assume that $\boldsymbol{u}$ is a unit vector, which is to say that

$$\sum_{n=1}^N |u_n|^2 = 1. \tag{G.41}$$

We expand the left-hand side of (G.39) and take the sum over m inside, to see that the expression is

$$= \sum_{r=1}^N \sqrt{\delta_r}u_r\sum_{s=1}^N \sqrt{\delta_s}\overline{u_s}\sum_{\substack{1\le m\le N\\ m\ne r\\ m\ne s}} \frac{\delta_m}{(\lambda_m-\lambda_r)(\lambda_m-\lambda_s)}.$$

The terms with $r = s$ contribute

$$\sum_{r=1}^{N} \delta_r |u_r|^2 \sum_{\substack{m=1 \\ m\neq r}}^{N} \frac{\delta_m}{(\lambda_m - \lambda_r)^2}. \tag{G.42}$$

The terms with $r \neq s$ contribute

$$\sum_{\substack{1\le r,s\le N \\ r\neq s}} \sqrt{\delta_r\delta_s}\, u_r\overline{u_s} \sum_{\substack{1\le m\le N \\ m\neq r \\ m\neq s}} \frac{\delta_m}{(\lambda_m - \lambda_r)(\lambda_m - \lambda_s)}. \tag{G.43}$$

Since $r \neq s$ in the above, we may write

$$\frac{\delta_m}{(\lambda_m - \lambda_r)(\lambda_m - \lambda_s)} = \frac{1}{\lambda_r - \lambda_s}\left(\frac{\delta_m}{\lambda_m - \lambda_r} - \frac{\delta_m}{\lambda_m - \lambda_s}\right).$$

On inserting this in (G.43), we find that the expression is

$$= \sum_{\substack{1\le r,s\le N \\ r\neq s}} \frac{\sqrt{\delta_r\delta_s}}{\lambda_r - \lambda_s} u_r\overline{u_s}\left(\sum_{\substack{1\le m\le N \\ m\neq r \\ m\neq s}} \frac{\delta_m}{\lambda_m - \lambda_r} - \sum_{\substack{1\le m\le N \\ m\neq r \\ m\neq s}} \frac{\delta_m}{\lambda_m - \lambda_s}\right).$$

In the first sum over m there is no need to exclude $m = s$, and in the second sum over m there is no need to exclude $m = r$. On inserting these terms we see that the above is

$$\begin{aligned} &= \sum_{\substack{1\le r,s\le N \\ r\neq s}} \frac{\sqrt{\delta_r\delta_s}(\delta_r + \delta_s)}{(\lambda_r - \lambda_s)^2} u_r\overline{u_s} + \sum_{\substack{1\le r,s\le N \\ r\neq s}} \frac{\sqrt{\delta_r\delta_s}}{\lambda_r - \lambda_s} u_r\overline{u_s} \sum_{\substack{m=1 \\ m\neq r}}^{N} \frac{\delta_m}{\lambda_m - \lambda_r} \\ &\quad - \sum_{\substack{1\le r,s\le N \\ r\neq s}} \frac{\sqrt{\delta_r\delta_s}}{\lambda_r - \lambda_s} u_r\overline{u_s} \sum_{\substack{m=1 \\ m\neq s}}^{N} \frac{\delta_m}{\lambda_m - \lambda_s} \\ &= T_1 + T_2 - T_3, \end{aligned} \tag{G.44}$$

say. By taking $m = r$ in (G.40) and then taking complex conjugates we find that

$$\sum_{\substack{s=1 \\ s\neq r}}^{N} \frac{\sqrt{\delta_r\delta_s}}{\lambda_r - \lambda_s}\overline{u_s} = -i\lambda\overline{u_r}$$

since λ is real. Thus

$$T_2 = -i\lambda \sum_{r=1}^{N} |u_r|^2 \sum_{\substack{m=1 \\ m\neq r}}^{N} \frac{\delta_m}{\lambda_m - \lambda_r}.$$

By taking $m = s$ in (G.40) and then multiplying by -1 we find that

$$\sum_{\substack{r=1 \\ r\neq s}}^{N} \frac{\sqrt{\delta_r \delta_s}}{\lambda_r - \lambda_s} u_r = -i\lambda u_s.$$

Thus

$$T_3 = -i\lambda \sum_{s=1}^{N} |u_s|^2 \sum_{\substack{m=1 \\ m\neq s}}^{N} \frac{\delta_m}{\lambda_m - \lambda_s}.$$

On comparing these formulæ we deduce that $T_2 = T_3$, so that these terms cancel in (G.40), so we now estimate T_1.

Clearly

$$|T_1| \le \sum_{\substack{1\le r,s\le N \\ r\neq s}} \frac{\sqrt{\delta_r \delta_s}(\delta_r + \delta_s)}{(\lambda_r - \lambda_s)^2} |u_r u_s| = U,$$

say. We note that

$$U = 2\sum_{r=1}^{N} |u_r| \sum_{\substack{s=1 \\ s\neq r}}^{N} \frac{\sqrt{\delta_r \delta_s}\delta_s}{(\lambda_r - \lambda_s)^2} |u_s|.$$

Thus by Cauchy's inequality,

$$(U/2)^2 \le \left(\sum_{r=1}^{N} |u_r|^2\right)\left(\sum_{r=1}^{N} \delta_r \left|\sum_{\substack{s=1 \\ s\neq r}}^{N} \frac{\delta_s^{3/2}|u_s|}{(\lambda_r - \lambda_s)^2}\right|^2\right).$$

Here the first factor on the right-hand side is 1 in view of (G.41). In the second factor we expand the modulus-squared and take the sum over r inside. Thus the right-hand side above is

$$= \sum_{1\le s,t\le N} \delta_s^{3/2}\delta_t^{3/2} |u_s u_t| \sum_{\substack{1\le r\le N \\ r\neq s \\ r\neq t}} \frac{\delta_r}{(\lambda_r - \lambda_s)^2(\lambda_r - \lambda_t)^2}.$$

By distinguishing those terms for which $s = t$ from those for which $s \neq t$, we see that the above is

$$= \sum_{s=1}^{N} \delta_s^3 |u_s|^2 \sum_{\substack{r=1 \\ r\neq s}}^{N} \frac{\delta_r}{(\delta_r - \delta_s)^4}$$

$$+ \sum_{\substack{1\le s,t\le N \\ s\neq t}} \delta_s^{3/2}\delta_t^{3/2} |u_s u_t| \sum_{\substack{1\le r\le N \\ r\neq s \\ r\neq t}} \frac{\delta_r}{(\lambda_r - \lambda_s)^2(\lambda_r - \lambda_t)^2}.$$

In the first term we use (G.32) to bound the sum over r. By (G.41) it follows that this term is $\le 8/3$. In the second term we use (G.33) to estimate the sum over r. The resulting bound is precisely $4U$. Since $(U/2)^2 \le 8/3 + 4U$, it follows that

$$|T_1| \le U \le 8 + 4\sqrt{14/3} < 16.641.$$

This is our upper bound for the expression (G.43). By (G.31) and (G.41) we see that the expression (G.42) is ≤ 4. On summing these estimates we see that the left-hand side of (G.39) is ≤ 20.641. The right-hand side is $9\pi^2/4 \ge 22.206$, so (G.39) holds, and the proof is complete. □

In the same way that we derived Theorem G.15 from Theorem G.14, we can derive a weighted inequality for use modulo 1 from the weighted Hilbert inequality (Theorem G.16).

Theorem G.20 *Let $\alpha_1, \alpha_2, \ldots, \alpha_R$ be distinct modulo 1, and put*

$$\delta_r = \min_{\substack{1\le s\le R\\ s\ne r}} \|\alpha_r - \alpha_s\|.$$

Then

$$\left| \sum_{\substack{1\le r,s\le R\\ r\ne s}} \frac{u_r v_s}{\sin \pi(\alpha_r - \alpha_s)} \right| \le \frac{3}{2} \left(\sum_{r=1}^R \frac{|u_r|^2}{\delta_r} \right)^{1/2} \left(\sum_{s=1}^R \frac{|v_s|^2}{\delta_s} \right)^{1/2} \tag{G.45}$$

for arbitrary real or complex u_r and v_s.

Proof As in the proof of Theorem G.15, we employ a doubly-indexed family of λ, namely $\lambda_{nr} = n + r$ for $1 \le n \le K$ and $1 \le r \le R$. Thus $|\lambda_{nr} - \lambda_{ms}| \ge \delta_r$ whenever $(m, s) \ne (n, r)$. We continue as in the proof of Theorem G.15, but with an appeal to Theorem G.16 in place of Theorem G.14. □

G.4.1 Exercise

1. Write $\cos \pi\alpha = (e^{i\pi\alpha} + e^{-i\pi\alpha})/2$, and apply Theorem G.15 twice to show that

$$\left| \sum_{\substack{1\le r,s\le R\\ r\ne s}} u_r v_s \cot \pi(\alpha_r - \alpha_s) \right| \le \frac{1}{\delta} \left(\sum_{r=1}^R |u_r|^2 \right)^{1/2} \left(\sum_{s=1}^R |v_s|^2 \right)^{1/2}$$

for arbitrary real or complex u_r and v_s.

G.5 Notes

For more material on bilinear forms and matrix inequalities, see Hardy, Littlewood, & Pólya (1952, pp. 196–259), Marcus & Minc (1964), Bellman (1970), and Beckenbach & Bellman (1965). For properties of integral matrices see Newman (1972).

Section G.1. Theorem G.1 is due to Hellinger & Toeplitz (1910), who also dealt with the convergence issues that arise when accepting infinite-dimensional matrices. The case $q = p'$ of Exercise G.1.1.1 is due to F. Riesz (1913), and the general case is due to his younger brother, M. Riesz (1927).

Section G.2. Theorem G.9, is Satz I of Schur (1909); it is one of the foundational results of linear algebra. The bound of Exercise G.2.1.17(f) can be sharpened, slightly, by replacing the supremum of $|f|$ by its *essential supremum* which is defined to be the supremum of the set of those numbers V for which $\{x : |f(x)| \ge V\}$ has positive measure. With this refinement, the bound is best possible since the argument of Exercise G.2.1.19 can be extended with $s(x)$ replaced by an arbitrary measurable function. Schur (1921) gave a simple proof that $|G(q)| = \sqrt{q}$ when q is odd, determined $\det E$, E^2, E^4 (as in Exercises G.2.1.22–23), and then deduced the multiplicities m_a, and hence the values of $G(q)$ when q is odd. His argument is reproduced in Landau (1958, pp. 207–212), except that Schur took for granted that the eigenvalues of A^2 are the squares of those of A. (This is an easy consequence of his triangularization, Theorem G.9.) Morton (1980) has constructed a set of q linearly independent eigenvectors of the Schur matrix. Balatoni (1969) has derived both upper and lower bounds for the largest and smallest eigenvalues of the matrix whose determinant is the Smith determinant.

Section G.3. The original Bessel inequality was published by the physicist/astronomer/mathematician F. W. Bessel in 1828. Boas (1941) and Bellman (1944) proposed generalizations of Bessel's inequality, in which the given vectors are close to orthonormal. Rényi (1949a,b,c, 1950, 1958, 1959) developed a number of principles along these lines, for purposes of improving the large sieve of Linnik. Heilbronn (1958) gave a further bound, which turns out to be a little weaker than the estimate of Halász in Exercise G.3.1.3. From the first two exercises at the end of this section we see that such extensions of Bessel's inequality are equivalent to consideration of bilinear forms, although sometimes (e.g., in §E.3.3) we still find it convenient to think in terms of Bessel's inequality. Discussion of the large sieve continued to be framed in terms of Bessel's inequality even after the seminal works of Roth (1965) and Bombieri (1965). Elliott (1971, 1973) and Matthews (1972a,b, 1973) were among the first to address the large sieve in terms of bilinear forms.

Section G.4. In lectures, Hilbert proved the inequalities (G.12), (G.13), but the latter with the constant 2π. His proof is reproduced in Hardy, Littlewood, & Pólya (1952, pp. 235–236). The inequalities were first proved with the optimal constant π by Schur (1911); for his proof see Schur (1911, p. 213). For an extended discussion of the original Hilbert inequalities see Hardy, Littlewood, & Pólya (1952, §8.12, Chapter IX, and Appendix III).

Atle Selberg wrote out Theorems G.14, G.16, and their proofs for the authors. He left us to deal with the problem of proving Theorems G.15 and G.20. We achieved this in Montgomery & Vaughan (1974) by inserting trigonometric functions in all of his formulæ, although, as the reader will see, it has now been found that these latter theorems are more easily derived directly from Selberg's original theorems, by exploiting the partial fraction expansion (G.24) of the cosecant function. Preissmann (1984) showed that the constant $\frac{3}{2}\pi$ in (G.27) can be replaced by $\frac{4}{3}\pi$. In conversations, Selberg reported that he had shown that the inequality holds with the constant 3.2, but it seems that no trace remains of the method he used to achieve this. Selberg (1991, pp. 220–225) later derived Theorem G.14 by a different method, but our proof above of Theorem G.15 follows Selberg's original unpublished argument. Let C_0 denote the best constant that could take the place of $\frac{3}{2}\pi$ in (G.27). By following Selberg's method as found in this section, one encounters the problem of establishing an inequality of the sort

$$\sum_{r=1}^{N} |u_r| \sum_{\substack{s=1 \\ s \neq r}}^{N} \frac{\sqrt{\delta_r \delta_s}(\delta_r + \delta_s)}{(\lambda_r - \lambda_s)^2} |u_s| \le C_1 \sum_{r=1}^{N} |u_r|^2. \tag{G.46}$$

This form is Hermitian and positive, so we would expect that it might be easy to estimate. From this approach we find that

$$C_0 \le \sqrt{\frac{\pi^2}{3} + C_1}.$$

If we could establish (G.46) with $C_1 = \frac{2}{3}\pi^2$, then we would have $C_0 = \pi$. However, Yangjit (2023) recently showed that the best constant C_1 in (G.46) is $\ge 0.70094\pi^2$. If Selberg reached 3.2 by estimating C_1, then his bound was very close to optimal, and that approach would never give $C_0 \le 3.19$. For more on this topic see Li (2005) and Preissmann & Lévêque (2013).

Montgomery & Vaaler (1998) introduced a still more general weighted form of Hilbert's inequality: Let the λ_n be as before, and suppose that $\beta_n \ge 0$ for all n. Then

$$\left| \sum_{\substack{m=1 \\ m \neq n}}^{N} \sum_{n=1}^{N} \frac{z_m \overline{z_n}}{\beta_m + \beta_n + i(\gamma_m - \gamma_n)} \right| \le 84 \sum_{n=1}^{N} \frac{|z_n|^2}{\delta_n}. \tag{G.47}$$

G.6 References

Balatoni, F. (1969). On the eigenvalues of the matrix of the Smith determinant (in Hungarian), *Mat. Lapok* **20**, 397–403.

Beckenbach, E. E. & Bellman, R. (1965). *Inequalities*, Ergebnisse der Mathematik und ihrer Grenzgebiete (N.F.) 30, Berlin: Springer-Verlag.

Bellman, R. (1944). Almost orthogonal series, *Bull. Amer. Math. Soc.* **50**, 517–519.

(1970). *Introduction to Matrix Analysis*, New York: McGraw-Hill.

Boas, R. P. (1941). A general moment problem, *Amer. J. Math.* **63**, 361–370.

Bombieri, E. (1965). On the large sieve, *Mathematika* **12**, 201–225.

(1971). A note on the large sieve, *Acta Arith.* **18**, 401–404.

Carlitz, L. (1959). Some cyclotomic matrices, *Acta Arith.* **5**, 293–308.

Elliott, P. D. T. A. (1971). On inequalities of large sieve type, *Acta Arith.* **18**, 405–422.

(1973). On connections between the Turán–Kubilius inequality and the large sieve: Some applications. In *Analytic Number Theory*, Proc. Sympos. Pure Math. (St. Louis, 1972), Vol. XXIV, Providence: Amer. Math. Soc., pp. 77–82.

Hardy, G. H., Littlewood, J. E., & Pólya, G. (1952). *Inequalities*, Cambridge: Cambridge University Press.

Heilbronn, H. (1958). On the averages of some arithmetical functions of two variables, *Mathematika* **5**, 1–7.

Hellinger, E. & Toeplitz, O. (1910). Grundlagen für eine Theorie der unendlichen Matrizen, *Math. Ann.* **69**, 289–330.

Landau, E. (1958). *Elementary Number Theory*, transl. J. E. Goodman, New York: Chelsea.

Li, X.-J. (2005). A note on the weighted Hilbert's inequality, *Proc. Amer. Math. Soc.* **133**, 1165–1173.

Marcus, M. & Minc, H. (1964). *A Survey of Matrix Theory and Matrix Inequalities*, Boston: Allyn & Bacon.

Matthews, K. R. (1972a). On an inequality of Davenport and Halberstam, *J. London Math. Soc.* (2) **4**, 638–642.

(1972b). On a bilinear form associated with the large sieve, *J. London Math. Soc.* **5**, 567–570.

(1973). Hermitian forms and the large and small sieves, *J. Number Theory* **5**, 16–23.

Montgomery, H. L. & Vaaler, J. D. (1998). A further generalization of Hilbert's inequality, *Mathematika* **45**, 35–39.

Montgomery, H. L. & Vaughan, R. C. (1974). Hilbert's inequality, *J. London Math. Soc.* (2) **8**, 73–82.

Morton, P. (1980). On the eigenvectors of Schur's matrix, *J. Number Theory* **12**, 122–127.

Newman, M. (1972). *Integral Matrices*, New York: Academic Press.

Preissmann, E. (1984). Sur une inégalité de Montgomery–Vaughan, *Enseign. Math.* (2) **30**, 95–113.

Preissmann, E. & Lévêque, O. (2013). On generalized weighted Hilbert matrices, *Pacific J. Math.* **265**, 199–219.

Rényi, A. (1949a). Un nouveau théorème concernant les fonctions indépendantes et ses applications à la théorie des nombres, *J. Math. Pures Appl.* (9) **28**, 137–149.

(1949b). On a theorem of the theory of probability and its application in number theory, *Časopis Pěst. Mat. Fys.* **74**, 167–175.

(1949c). Sur un théorème général de probabilité, *Ann. Inst. Fourier (Grenoble)* **1**, 43–52.

(1950). On the large sieve of Yu. V. Linnik, *Compositio Math.* **8**, 68–75.

(1958). On the probabilistic generalization of the large sieve of Linnik, *Magyar Tud. Akad. Mat. Kutató Int. Közl.* **3**, 199–206.

(1959). New version of the probabilistic generalization of the large sieve, *Acta Math. Acad. Sci. Hungar.* **10**, 441–451.

Riesz, F. (1913). *Les systèmes d'équations linéaires à une infinité d'inconnues*. Paris: Gauthier-Villars.

Riesz, M. (1927). Sur les maxima des formes bilinéaires et sur les fonctionnelles linéaires, *Acta Math.* **49**, 465–497.

Roth, K. F. (1965). On the large sieves of Linnik and Rényi, *Mathematika* **12**, 1–9.

Schur, I. (1909). Über die charakteristischen Wurzeln einer linearen Substitution mit einer Anwendung auf die Theorie der Integralgleichungen, *Math. Ann.* **66**, 488–510; *Gesammelte Abhandlungen I*, Berlin: Springer-Verlag, 1973, pp. 272–510.

(1911). Bemerkungen zur Theorie der beschraänkten Bilinearformen mit unendlich vielen Veränderlichen, *J. Reine Angew. Math.* **140**, 1–28; *Gesammelte Abhandlungen I*, Berlin: Springer-Verlag, 1973, pp. 464–510.

(1921). Über die Gaußschen Summen, *Nachr. könig. Gesell. Göttingen*, **1921**, 147–153; *Gesammelte Abhandlungen II*, Berlin: Springer-Verlag, 1973, pp. 327–333.

Selberg, A. (1991). *Collected Papers*, Vol. II, Berlin: Springer-Verlag.

Smith, H. J. S. (1876). On the value of a certain arithmetical determinant, *Proc. London Math. Soc.* **7**, 208–212.

Toeplitz, O. (1910). Zur Theorie der quadratischen Formen von unendlich vielen Veränderlichen, *Nach. Akad. Wiss. Göttingen* **1910**, 489–506.

Watkins, W. (1980). Similarity of matrices, *Amer. Math. Monthly* **87**, 300.

Yangjit, Wijit (2023). On the Montgomery–Vaughan weighted generalization of Hilbert's inequality, *Proc. Amer. Math. Soc.* Ser. B **10**, 439–454.

Appendix H
Linear Programming

H.1 Fundamental Theory

The following simple and intuitively obvious result is fundamental.

Theorem H.1 *Let $\mathscr{C}$ be a closed convex set in $\mathbb{R}^m$, and suppose that $\boldsymbol{b} \notin \mathscr{C}$. Then there is a hyperplane $\mathscr{H} = \{\boldsymbol{u} \in \mathbb{R}^m : \boldsymbol{n} \cdot \boldsymbol{u} = c\}$ that separates $\boldsymbol{b}$ from $\mathscr{C}$ in the sense that $\boldsymbol{b}\cdot\boldsymbol{n} < 0$ and $\boldsymbol{n}\cdot\boldsymbol{u} \geq 0$ for all $\boldsymbol{u} \in \mathscr{C}$.*

Proof Let $\boldsymbol{u}_0$ be a point of $\mathscr{C}$ whose distance from $\boldsymbol{b}$ is minimal. It is clear that this minimal distance is attained by some point $\boldsymbol{u}_0$ of $\mathscr{C}$, even if $\mathscr{C}$ is unbounded and therefore not compact, since we may restrict our attention to a sufficiently large compact subset of $\mathscr{C}$. Set $\boldsymbol{n} = \boldsymbol{u}_0 - \boldsymbol{b}$ and put $c = \boldsymbol{u}_0\cdot\boldsymbol{n}$. Then

$$\boldsymbol{b}\cdot\boldsymbol{n} = (\boldsymbol{u}_0 - \boldsymbol{n})\cdot\boldsymbol{n} = c - |\boldsymbol{n}|^2 < c$$

since $\boldsymbol{n} \neq \boldsymbol{0}$.

Suppose on the other hand that $\boldsymbol{u} \in \mathscr{C}$. The points $(1-t)\boldsymbol{u}_0 + t\boldsymbol{u}$, $0 \leq t \leq 1$, constitute the line segment joining $\boldsymbol{u}_0$ to $\boldsymbol{u}$. Since $\mathscr{C}$ is convex, these points are also members of $\mathscr{C}$. Consider the distance of such a point from $\boldsymbol{b}$. We note that

$$|(1-t)\boldsymbol{u}_0 + t\boldsymbol{u} - \boldsymbol{b}|^2 = |\boldsymbol{u} + t(\boldsymbol{u} - \boldsymbol{u}_0)|^2 = |\boldsymbol{n}|^2 + 2t(\boldsymbol{u} - \boldsymbol{u}_0)\cdot\boldsymbol{n} + |\boldsymbol{u} - \boldsymbol{u}_0|^2 t^2.$$

If it were the case that $(\boldsymbol{u} - \boldsymbol{u}_0)\cdot\boldsymbol{n} < 0$, then the above would be smaller than $|\boldsymbol{n}|^2$ if we took t sufficiently small and positive. Then we would have a point of $\mathscr{C}$ that is closer to $\boldsymbol{b}$ than $\boldsymbol{u}_0$. Since $\boldsymbol{u}_0$ was chosen to minimize this distance, we conclude that $(\boldsymbol{u} - \boldsymbol{u}_0)\cdot\boldsymbol{n} \geq 0$, which is to say that $\boldsymbol{u}\cdot\boldsymbol{n} \geq c$. □

Corollary H.2 *A closed convex set in $\mathbb{R}^m$ is the intersection of its supporting translated half-spaces.*

Suppose that $\boldsymbol{a}$ and $\boldsymbol{b}$ are vectors in $\boldsymbol{R}^m$. We say that $\boldsymbol{a} \geq \boldsymbol{b}$ if $a_i \geq b_i$ for all respective coordinates.

Theorem H.3 (Farkas' lemma, Farkas 1902) *Suppose that $\boldsymbol{b} \in \mathbb{R}^m$ and that A is an $m \times n$ real matrix. Then exactly one of the following is true.*

(i) *There is an $\boldsymbol{x} \in \mathbb{R}^n$ such that $A\boldsymbol{x} = \boldsymbol{b}$ and $\boldsymbol{x} \geq \boldsymbol{0}$.*
(ii) *There is a $\boldsymbol{y} \in \mathbb{R}^m$ such that $\boldsymbol{b} \cdot \boldsymbol{y} < 0$ and $A^{\mathrm{T}}\boldsymbol{y} \geq \boldsymbol{0}$.*

Proof That the alternatives are mutually exclusive is clear, for if both (i) and (ii) held, then we would have

$$0 > \boldsymbol{y}^{\mathrm{T}}\boldsymbol{b} = \boldsymbol{y}^{\mathrm{T}}(A\boldsymbol{x}) = (\boldsymbol{y}^{\mathrm{T}}A)\boldsymbol{x} = \left(A^{\mathrm{T}}\boldsymbol{y}\right)^{\mathrm{T}}\boldsymbol{x} \geq 0.$$

Let $\mathscr{C} = \{A\boldsymbol{x} \in \mathbb{R}^m : \boldsymbol{x} \geq \boldsymbol{0}\}$ be the closed convex cone generated by the columns of A. By Theorem H.1, either $\boldsymbol{b} \in \mathscr{C}$, in which case we are in case (i), or else there is a vector $\boldsymbol{n} \in \mathbb{R}^m$ and a real number c such that $\boldsymbol{n} \cdot \boldsymbol{b} < c$ but $\boldsymbol{n} \cdot \boldsymbol{u} \geq c$ for all $\boldsymbol{u} \in \mathscr{C}$. Since $\boldsymbol{0} \in \mathscr{C}$ it follows that $c \leq 0$. On the other hand, if there were an $\boldsymbol{u} = A\boldsymbol{x} \in \mathscr{C}$ such that $\boldsymbol{n} \cdot \boldsymbol{u} < 0$, then such numbers would be unbounded below, since $\boldsymbol{u}$ can be replaced by $\alpha\boldsymbol{u}$ with $\alpha \geq 0$ arbitrarily large. Thus $c = 0$. We take $\boldsymbol{y} = \boldsymbol{n}$, and observe that $\boldsymbol{y}^{\mathrm{T}}A\boldsymbol{x} \geq 0$ for all $\boldsymbol{x} \geq 0$ if and only if $\boldsymbol{y}^{\mathrm{T}}A \geq \boldsymbol{0}$. Thus the proof is complete. □

Theorem H.4 *Suppose that $\boldsymbol{b} \in \mathbb{R}^m$ and that A is an $m \times n$ matrix. Then exactly one of the following is true.*

(i) *There is an $\boldsymbol{x} \in \mathbb{R}^n$ such that $A\boldsymbol{x} \leq \boldsymbol{b}$ and $\boldsymbol{x} \geq \boldsymbol{0}$.*
(ii) *There is a $\boldsymbol{y} \in \mathbb{R}^m$ such that $\boldsymbol{b} \cdot \boldsymbol{y} < 0$, $A^{\mathrm{T}}\boldsymbol{y} \geq \boldsymbol{0}$, and $\boldsymbol{y} \geq \boldsymbol{0}$.*

Proof That the alternatives are mutually exclusive is clear, for if both (i) and (ii) held, then we would have

$$0 > \boldsymbol{y}^{\mathrm{T}}\boldsymbol{b} \geq \boldsymbol{y}^{\mathrm{T}}(A\boldsymbol{x}) = (\boldsymbol{y}^{\mathrm{T}}A)\boldsymbol{x} \geq 0.$$

We apply Theorem H.3 with n replaced by $m + n$, A replaced by $[A \mid I]$, and $\boldsymbol{x}$ replaced by $\left[\frac{\boldsymbol{x}}{\boldsymbol{w}}\right]$ where $\boldsymbol{w} \in \mathbb{R}^m$. In case (i) of Theorem H.3 we have

$$[A \mid I]\begin{bmatrix} \boldsymbol{x} \\ \text{---} \\ \boldsymbol{w} \end{bmatrix} = \boldsymbol{b},$$

where $\boldsymbol{x} \geq \boldsymbol{0}$ and $\boldsymbol{w} \geq \boldsymbol{0}$. That is, $A\boldsymbol{x} + \boldsymbol{w} = \boldsymbol{b}$, which is case (i) above. Alternatively, in case (ii) of Theorem H.3 there is a $\boldsymbol{y} \in \mathbb{R}^m$ such that $\boldsymbol{b} \cdot \boldsymbol{y} < 0$ and $\boldsymbol{y}^{\mathrm{T}}[A \mid I] \geq 0$. That is, $\boldsymbol{y}^{\mathrm{T}}A \geq \boldsymbol{0}$ and $\boldsymbol{y} \geq \boldsymbol{0}$. Thus (ii) holds and the proof is complete. □

We are now in a position to prove the Fundamental Duality Theorem of linear programming.

Theorem H.5 (Fundamental Duality Theorem) *Let A be an $m \times n$ matrix with real entries, and suppose that $\boldsymbol{b} \in \mathbb{R}^m$ and $\boldsymbol{c} \in \mathbb{R}^n$ are given. Put $\mathscr{X} = \{\boldsymbol{x} \in \mathbb{R}^n : \boldsymbol{x} \geq \boldsymbol{0},\ A\boldsymbol{x} \leq \boldsymbol{b}\}$ and $\mathscr{Y} = \{\boldsymbol{y} \in \mathbb{R}^m : \boldsymbol{y} \geq \boldsymbol{0},\ A^{\mathrm{T}}\boldsymbol{y} \geq \boldsymbol{c}\}$. If $\mathscr{X}$ and $\mathscr{Y}$ are both non-empty, then*

$$\max_{\boldsymbol{x} \in \mathscr{X}} \boldsymbol{c} \cdot \boldsymbol{x} = \min_{\boldsymbol{y} \in \mathscr{Y}} \boldsymbol{b} \cdot \boldsymbol{y}. \tag{H.1}$$

If $\mathscr{X}$ is non-empty, then $\sup_{\boldsymbol{x} \in \mathscr{X}} \boldsymbol{b} \cdot \boldsymbol{x} = +\infty$ if and only if $\mathscr{Y} = \emptyset$. Similarly, if $\mathscr{Y}$ is non-empty, then $\inf_{\boldsymbol{y} \in \mathscr{Y}} \boldsymbol{c} \cdot \boldsymbol{y} = -\infty$ if and only if $\mathscr{X} = \emptyset$.

It is possible that both $\mathscr{X}$ and $\mathscr{Y}$ are empty.

Proof Let L denote the left-hand side of (H.1) if $\mathscr{X}$ is non-empty, and let R denote the right-hand side of (H.1) if $\mathscr{Y}$ is non-empty. Since $\mathscr{X}$ is connected and $\boldsymbol{c} \cdot \boldsymbol{x}$ is a continuous function of $\boldsymbol{x}$, it follows that the values $\boldsymbol{c} \cdot \boldsymbol{x}$, for $\boldsymbol{x} \in \mathscr{X}$, form an interval on the real line, I_L, say. Similarly, the values $\boldsymbol{b} \cdot \boldsymbol{y}$, for $\boldsymbol{y} \in \mathscr{Y}$, form an interval I_R. Suppose that $\boldsymbol{x} \in \mathscr{X}$ and $\boldsymbol{y} \in \mathscr{Y}$. Since $\boldsymbol{c} \leq A^{\mathrm{T}}\boldsymbol{y}$ and $\boldsymbol{x} \geq \boldsymbol{0}$, it follows that

$$\boldsymbol{c} \cdot \boldsymbol{x} = \boldsymbol{c}^{\mathrm{T}}\boldsymbol{x} \leq \left(\boldsymbol{y}^{\mathrm{T}}A\right)\boldsymbol{x} = \boldsymbol{y}^{\mathrm{T}}\left(A\boldsymbol{x}\right).$$

On the other hand, $\boldsymbol{y} \geq \boldsymbol{0}$ and $A\boldsymbol{x} \leq \boldsymbol{b}$, so the above is

$$\leq \boldsymbol{y}^{\mathrm{T}}\boldsymbol{b} = \boldsymbol{b} \cdot \boldsymbol{y}.$$

Thus the interval I_L lies entirely to the left of the interval I_R. What is further asserted in (H.1) is that there is no gap between these intervals. From the above it is clear that if $\mathscr{X}$ and $\mathscr{Y}$ are both non-empty, then the interval I_L is bounded above and the interval I_R is bounded below. It remains to show that if $\mathscr{X} \neq \emptyset$ and $\mathscr{Y} = \emptyset$, then I_L extends to $+\infty$, and similarly that if $\mathscr{X} = \emptyset$ and $\mathscr{Y} \neq \emptyset$, then the interval I_R extends to $-\infty$.

Suppose that $\mathscr{X}$ is non-empty and that μ is a number chosen so large that $\mu > L$. Then there does not exist an $\boldsymbol{x} \geq \boldsymbol{0}$ such that

$$\begin{bmatrix} A \\ \hline -\boldsymbol{c}^{\mathrm{T}} \end{bmatrix} \boldsymbol{x} \leq \begin{bmatrix} \boldsymbol{b} \\ \hline -\mu \end{bmatrix}.$$

We apply Theorem H.4 to this situation, and see that case (i) is excluded. Thus there is a $\left[\frac{\boldsymbol{z}}{q}\right] \in \mathbb{R}^{m+1}$ such that

$$[\boldsymbol{b}^{\mathrm{T}} \mid -\mu] \begin{bmatrix} \boldsymbol{z} \\ \hline q \end{bmatrix} < 0, \qquad \left[A^{\mathrm{T}} \mid -\boldsymbol{c}\right] \begin{bmatrix} \boldsymbol{z} \\ \hline q \end{bmatrix} \geq \boldsymbol{0}, \qquad \boldsymbol{z} \geq \boldsymbol{0}, \qquad q \geq 0.$$

That is, $\boldsymbol{b} \cdot \boldsymbol{z} < \mu q$ and $A^{\mathrm{T}}\boldsymbol{z} \geq q\boldsymbol{c}$. We see that $q > 0$, for if $q = 0$, then we would have $\boldsymbol{b} \cdot \boldsymbol{z} < 0$, $A^{\mathrm{T}}\boldsymbol{z} \geq \boldsymbol{0}$, and $\boldsymbol{z} \geq \boldsymbol{0}$, which is case (ii) of Theorem H.4. But then case (i) would be excluded, which is to say that $\mathscr{X}$ would be empty, contrary to assumption. Thus $q > 0$ and we set $\boldsymbol{y} = \frac{1}{q}\boldsymbol{z}$. Then $\boldsymbol{y} \in \mathscr{Y}$ and $\boldsymbol{b}\cdot\boldsymbol{y} < \mu$. Thus we learn that if it is possible to choose a number μ larger than all members of I_R, then $\mathscr{Y}$ is non-empty, and also that $\mu > R$. Thus if $\mathscr{X}$ is non-empty, then $L = +\infty$ if and only if $\mathscr{Y}$ is empty. As we observed at the outset, if $\mathscr{X}$ and $\mathscr{Y}$ are non-empty, then $L \le R$. What we have now shown is that there is no number μ such that $L < \mu < R$. Thus $L = R$.

The proof is now complete except for the very final assertion. For this we can argue similarly, or we can exchange m and n, replace $\boldsymbol{b}$ by $-\boldsymbol{c}$, replace $\boldsymbol{c}$ by $-\boldsymbol{b}$, and replace A by $-A^{\mathrm{T}}$. Then $\mathscr{X}$ and $\mathscr{Y}$ are exchanged, L is replaced by $-R$, and R is replaced by $-L$. Through this reversal we see that the final assertion follows from the one immediately before it, so the proof is complete. □

H.1.1 Exercises

1. Let $\mathscr{C}$ be a closed convex set in $\mathbb{R}^m$, suppose that $\boldsymbol{b} \notin \mathscr{C}$, and let $\boldsymbol{u}_0$ be a member of $\mathscr{C}$ that is closest to $\boldsymbol{b}$, as in the proof of Theorem H.1. Show that $\boldsymbol{u}_0$ is unique.
2. Let $\mathscr{C}$ be a convex set in $\mathbb{R}^m$, and suppose that $\boldsymbol{b} \notin \mathscr{C}$. Show that there is a hyperplane $\mathscr{H} = \{\boldsymbol{u} \in \mathbb{R}^m : \boldsymbol{n}\cdot\boldsymbol{u} = c\}$ that separates $\boldsymbol{b}$ from $\mathscr{C}$ in the sense that $\boldsymbol{n}\cdot\boldsymbol{b} \le 0$ but $\boldsymbol{n}\cdot\boldsymbol{u} \ge 0$ for all $\boldsymbol{u} \in \mathscr{C}$.
3. Derive Theorem H.3 from Theorem H.4 by applying Theorem H.4 with m replaced by $2m$, and with A, $\boldsymbol{b}$, $\boldsymbol{y}$ replaced, respectively, by

$$\begin{bmatrix} A \\ \hline -A \end{bmatrix}, \quad \begin{bmatrix} \boldsymbol{b} \\ \hline -\boldsymbol{b} \end{bmatrix}, \quad \begin{bmatrix} \boldsymbol{y}_1 \\ \hline -\boldsymbol{y}_2 \end{bmatrix}.$$

H.2 The Application to Sieves

Let $\mathscr{A}$ be a sequence of non-negative numbers $a(k)$, let $\mathscr{P}$ be a finite set of primes, set $P = \prod_{p\in\mathscr{P}} p$, and for $\delta|P$ set

$$S_\delta = \sum_{\substack{k \\ (k,P)=\delta}} a(k)$$

so that

$$X_d = \sum_k a(kd) = \sum_{\substack{\delta \\ d|\delta|P}} S_\delta$$

for $d|P$. Let $n = 2^{\omega(P)}$. Thus n is the number of $\delta|P$, and the S_δ play the role of the x_j in Theorem H.5. The X_d are linear forms in the S_δ, which must obey bounds of the sort

$$|X_d - \rho(d)X| \le R_d \tag{H.2}$$

for $d|P$. Here $\rho(d)$ is a multiplicative function defined on the divisors of P with $0 \le \rho(p) \le 1$ for all $p|P$. This gives rise to $m = 2n$ linear inequalities,

$$\begin{aligned} X_d &\le \rho(d)X + R_d, \\ -X_d &\le -\rho(d)X + R_d \end{aligned}$$

for $d|P$. Let A be the $m \times n$ partitioned matrix

$$A = \begin{bmatrix} A^+_{d\delta} \\ \text{---} \\ A^-_{d\delta} \end{bmatrix},$$

where

$$A^+_{d\delta} = \begin{cases} 1 & \text{if } d|\delta, \\ 0 & \text{otherwise,} \end{cases} \qquad A^-_{d\delta} = \begin{cases} -1 & \text{if } d|\delta, \\ 0 & \text{otherwise.} \end{cases}$$

Let $\boldsymbol{S}$ be a column vector whose n coordinates are the numbers S_δ, and let $\boldsymbol{b}$ be a column vector whose $m = 2n$ coordinates are partitioned so that the first n coordinates are the numbers $b^+_d = \rho(d)X + R_d$, followed by n further coordinates $b^-_d = -\rho(d)X + R_d$. Thus the vectors $\boldsymbol{S}$ are subject to the condition $\boldsymbol{S} \ge \boldsymbol{0}$ and $A\boldsymbol{S} \le \boldsymbol{b}$. Let $\mathscr{X}$ denote the set of these admissible vectors $\boldsymbol{S}$. Let $\boldsymbol{c}$ be a vector in $\mathbb{R}^n$ whose coordinates are indexed by the $\delta|P$ with

$$c_\delta = \begin{cases} 1 & \text{if } \delta = 1, \\ 0 & \text{if } \delta > 1. \end{cases}$$

For an upper bound sieve, we would want to derive an upper bound for the size of $\boldsymbol{c} \cdot \boldsymbol{S}$. To construct the dual extremal problem we let $\boldsymbol{\lambda} \in \mathbb{R}^m$ have non-negative coordinates, partitioned into two halves, so that $\boldsymbol{\lambda} = [\lambda^+_d \,|\, \lambda^-_d]$. Note that in this situation, λ^+_d is not an upper bound sifting function, nor is λ^-_d a lower bound sifting function. Rather, they are building blocks which will be used to form an upper bound sifting function. In addition to $\boldsymbol{\lambda} \ge \boldsymbol{0}$, the $\boldsymbol{\lambda}$ are required to satisfy $\boldsymbol{\lambda} A \ge \boldsymbol{c}$. That is,

$$\sum_{\substack{d \\ d|\delta}} \lambda^+_d - \sum_{\substack{d \\ d|\delta}} \lambda^-_d \ge c_\delta = \begin{cases} 1 & \text{if } \delta = 1, \\ 0 & \text{if } \delta > 1. \end{cases}$$

Let $\mathscr{Y}$ denote the set of those $\boldsymbol{\lambda}$ that meet these requirements. For any $\boldsymbol{\lambda} \in \mathscr{Y}$, the quantity $\boldsymbol{b}\cdot\boldsymbol{\lambda}$ is an upper bound for $\boldsymbol{c}\cdot\boldsymbol{S}$ for all $\boldsymbol{S} \in \mathscr{X}$. That is,

$$S_1 \le [b_d^+ \mid b_d^-] \cdot [\lambda_d^+ \mid \lambda_d^-] = X \sum_{d|P} (\lambda_d^+ - \lambda_d^-)\rho(d) + \sum_{d|P} (\lambda_d^+ + \lambda_d^-) R_d.$$

Suppose that λ_d is given, and that λ_d^+ and λ_d^- take non-negative values so that $\lambda_d^+ - \lambda_d^- = \lambda_d$. If $\lambda_d \ge 0$, then the quantity $\lambda_d^+ + \lambda_d^-$ is minimized by taking $\lambda_d^+ = \lambda_d$ and $\lambda_d^- = 0$. Similarly, if $\lambda_d \le 0$, the quantity $\lambda_d^+ + \lambda_d^-$ is minimized by taking $\lambda_d^+ = 0$ and $\lambda_d^- = -\lambda_d$. Thus

$$S_1 \le X \sum_{d|P} \lambda_d \rho(d) + \sum_{d|P} |\lambda_d| R_d, \tag{H.3}$$

where

$$\sum_{d|\delta} \lambda_d \ge \begin{cases} 1 & \text{if } \delta = 1, \\ 0 & \text{if } \delta > 1. \end{cases}$$

Moreover, by Theorem H.5 we know that the minimum of the bound (H.3) over $\boldsymbol{\lambda} \in \mathscr{Y}$ is equal to the maximum of S_1 as $\boldsymbol{S}$ takes all possible values in $\mathscr{X}$, provided that both $\mathscr{X}$ and $\mathscr{Y}$ are non-empty. For $\mathscr{Y}$ this is easy: Just take $\lambda_1 = 1$ and $\lambda_d = 0$ for $d > 1$. To exhibit a point in $\mathscr{X}$, we note that if

$$S_\delta = \rho(\delta) \prod_{p|P/\delta} (1 - \rho(p))X$$

for all $\delta|P$, then $X_d = \rho(d)X$ for all $d|P$. Hence, in particular, any upper bound that can be derived from the hypotheses (H.2) must be at least as large as

$$X \prod_{p|P} (1 - \rho(p)),$$

and any lower bound cannot exceed this value.

To obtain a corresponding result for lower bound sieves, we let A, $\boldsymbol{b}$, and $\mathscr{X}$ be defined as above, but we now set $\boldsymbol{c} \in \mathbb{R}^n$ to be $\boldsymbol{c} = (c_\delta)$ with

$$c_\delta = \begin{cases} -1 & \text{if } \delta = 1, \\ 0 & \text{if } \delta > 1\ \delta|P. \end{cases}$$

That is, $\boldsymbol{c}$ is the negative of its former value. This results in a change in the definition of $\mathscr{Y}$. We still take $\mathscr{Y}$ to be the subset of $\mathbb{R}^{2n}$ consisting of partitioned vectors $\boldsymbol{\lambda} = [\lambda_d^+, \lambda_d^-]$ such that $A^T\boldsymbol{\lambda} \ge \boldsymbol{c}$, but this condition now reads

$$\sum_{\substack{d \\ d|\delta}} \lambda_d^+ - \sum_{\substack{d \\ d|\delta}} \lambda_d^- \ge \begin{cases} -1 & \text{if } \delta = 1, \\ 0 & \text{if } \delta > 1\ \delta|P \end{cases}$$

for $\delta|P$. Put $\lambda_d = \lambda_d^- - \lambda_d^+$. Then

$$\sum_{\substack{d \\ d|\delta}} \lambda_d \le \begin{cases} 1 & \text{if } \delta = 1, \\ 0 & \text{if } \delta > 1,\ \delta|P. \end{cases}$$

Hence

$$\begin{aligned} -S_1 \le [b_d^+ \,|\, b_d^-] \cdot [\lambda_d^+ \,|\, \lambda_d^-] &= X \sum_{d|P} (\lambda_d^+ - \lambda_d^-)\rho(d) + \sum_{d|P} (\lambda_d^+ + \lambda_d^-) R_d \\ &= -X \sum_{d|P} \lambda_d \rho(d) + \sum_{d|P} (\lambda_d^+ + \lambda_d^-) R_d. \end{aligned}$$

To minimize the value of $\lambda_d^+ + \lambda_d^-$, if $\lambda_d \ge 0$, set $\lambda_d^- = \lambda_d$, $\lambda_d^+ = 0$, and if $\lambda_d < 0$ set $\lambda_d^+ = -\lambda_d$ and $\lambda_d^- = 0$. On rearranging the inequality above, we find that

$$X \sum_{d|P} \lambda_d \rho(d) - \sum_{d|P} |\lambda_d| R_d \le S_1,$$

and that the maximum of the left-hand side over all $\boldsymbol{\lambda} \in \mathscr{Y}$ is equal to the minimum of the right-hand side over all $\mathbf{S} \in \mathscr{X}$. Note that $\mathscr{X}$ is non-empty because it is unchanged from our upper bound discussion, and that $\mathscr{Y}$ is non-empty because $\mathbf{0} \in \mathscr{Y}$. Indeed, it sometimes happens that the best lower bound for S_1 is 0, and in that case $\boldsymbol{\lambda} = \mathbf{0}$ is optimal. (For example, this happens when $\mathcal{A} = \{2, 3, 4\}$, $\mathscr{P} = \{2, 3\}$.)

It may seem that our considerations are not very useful because the number of variables grows exponentially as a function of n. However, in most practical situations, the X_d are well approximated only for d of limited size, which is to say for $d \le z$ for some parameter z. We then set $R_d = \infty$ for $d > z$, which has the effect of forcing the support of $\boldsymbol{\lambda}$ to lie in $[1, z]$.

H.2.1 Exercise

1. (USA Mathematical Olympiad 2011 Problem 6) Let $\mathscr{A}$ be a set of 225 integers, let $\mathscr{A}_1, \ldots, \mathscr{A}_{11}$ be subsets of $\mathscr{A}$ such that $\mathscr{A}_i = 45$ for $1 \le i \le 11$, and also such that $\operatorname{card} \mathscr{A}_i \cap \mathscr{A}_j = 9$ for $1 \le i < j \le 11$. Let $\mathscr{D} = \mathscr{A}_1 \cup \mathscr{A}_2 \cup \cdots \cup \mathscr{A}_{11}$, and set $\mathscr{R} = \mathscr{A} \setminus \mathscr{D}$. The object of this exercise is to show that $\operatorname{card} \mathscr{D} \ge 165$, and to show that this is best possible.

   (a) Note that the first part of the object is equivalent to showing that $\operatorname{card} \mathscr{R} \le 60$. This resembles an upper bound sieve problem, in which the $\mathscr{A}_i$ correspond to multiples of a prime p_i which are *deleted*, and the numbers in $\mathscr{R}$ *remain*. For $a \in \mathscr{A}$ define a weight

$$w(a) = 1 + \lambda_1 \sum_{\substack{1 \le i \le 11 \\ a \in \mathscr{A}_i}} 1 + \lambda_2 \sum_{\substack{1 \le i < j \le 11 \\ a \in \mathscr{A}_i \cap \mathscr{A}_j}} 1,$$

where λ_1 and λ_2 are yet to be chosen.

(b) Note that if $w(a) \ge 0$ for all $a \in \mathscr{A}$ and $w(a) \ge 1$ for all $a \in \mathscr{R}$, then

$$\operatorname{card} \mathscr{R} \le \sum_{a \in \mathscr{A}} w(a).$$

(c) Show that

$$\sum_{a \in \mathscr{A}} w(a) = 225 + 495(\lambda_1 + \lambda_2).$$

(d) For $a \in \mathscr{A}$, let $m = m(a)$ denote the number of i, $1 \le i \le 11$, for which $a \in \mathscr{A}_i$. Show that

$$w(a) = 1 + \binom{m}{1}\lambda_1 + \binom{m}{2}\lambda_2 = f(m),$$

say.

(e) Choose λ_1 and λ_2 so that $f(3) = f'(3) = 0$. With the λ_j chosen this way, show that $\lambda_1 + \lambda_2 = -1/3$, and that $f(m) = (m-3)^2/9$, with the result that $f(0) = 1$, $f(m) \ge 0$ for all m, and $f(m) = 0$ only when $m = 3$.

(f) Conclude that $\operatorname{card} \mathscr{R} \le 60$, which is to say that $\operatorname{card} \mathscr{D} \ge 165$.

(g) To achieve equality in the above argument, the $\mathscr{A}_i$ must be chosen so that $m(a) = 3$ for all $a \in \mathscr{D}$. Note that $\binom{11}{3} = 165$. Choose 165 distinct integers, and for each triple (i, j, k) with $1 \le i < j < k \le 11$ place one of these integers in $\mathscr{A}_i \cap \mathscr{A}_j \cap \mathscr{A}_k$. The $\mathscr{A}_i$ are to have no other members. Show that $\operatorname{card} \mathscr{D} = 165$, that $\operatorname{card} \mathscr{A}_i = 45$ for all i, and that $\operatorname{card} \mathscr{A}_i \cap \mathscr{A}_j = 9$ for all pairs $i < j$.

H.3 Notes

Section H.1. Theorem H.1 is also true in many other spaces, but it is important that the space in question be locally convex.

The history of linear programming can be traced back to Fourier (1826), who determined whether a polyhedron defined by linear inequalities is empty by projecting it to a space of dimension one less; thus he eliminated one variable at a time. This process is now known as Fourier–Motzkin elimination. See Dantzig & Eaves (1973) and Williams (1986) for accounts of this. The next major development was the discovery of Farkas (1902), but this had little impact until much later. De la Vallée Poussin (1911) devised an iterative procedure for solving a minimax problem, but this also seems not to have attracted attention. Retrospectively, Farebrother (2006) argues that with a few small adjustments, de la Vallée Poussin's procedure could have been converted to provide a linear programming algorithm. Economists made progress in the 1930s and 40s, and George Dantzig invented the simplex method in the summer of 1947, but it was on 3 October 1947, when Dantzig described his work to von Neumann at the Institute for Advanced Study, that von Neumann immediately conjectured the duality principle. Von Neumann's contention was that Dantzig's problem was essentially equivalent to a problem in the theory of games that had already been discussed in von Neumann & Morgenstern (1944). Dantzig was assigned the job of writing up rigorous proofs, a task he completed by 5 January 1948. However, he did not publish his paper because he considered it to be the work of von Neumann. In 1948, Gale, Kuhn, and Tucker started their work on non-linear programming, and soon discovered duality, independently of von Neumann. See Gale, Kuhn, & Tucker (1951). Further seminal papers from this era are found in the conference proceedings edited by Koopmans (1951). For details as to how and by whom such terms as 'primal', 'simplex method', and 'linear programming' were invented, see Dantzig (1982) and Dorfman (1984).

In most applications, the simplex method seems to run in a little more than linear time, but some artificial situations have been constructed in which it runs much slower. Shor (1970) proposed a different algorithm for linear programming, of a type called 'ellipsoidal'. Khachiyan (1979) modified it, and thus was able to prove that the linear programming problem runs in polynomial time. However, these algorithms are not so fast in practice. On the other hand, Karmarkar (1984), at Bell Labs, proposed a method that deals simultaneously with issues of projection and scaling, and is fast in practice. For more details, with instructive code fragments, see Chakraborty, Chandru, & Rao (2020).

Section H.2. Chebyshev advanced our understanding of the distribution of prime numbers by employing truncated versions of the Möbius function. The first person to modify the Möbius function to form a sifting function as we think of them today was a young French mathematician, Jean Merlin, who was killed in WWI. Thus we have from him only one brief announcement, Merlin (1911), communicated by Poincaré, and one posthumous paper, Merlin (1915), prepared by Hadamard. Viggo Brun was stimulated by these items, and developed an effective sieve method. Buchstab devised a method by which sieve estimates could be improved, but without any indication that sifting functions had the capability of delivering optimal results. In his Stony Brook lectures, Selberg

(1971) argued that sifting functions can deliver optimal bounds because they represent the supporting planes of a certain convex body. We have expressed the situation in terms of linear programming in order to make it more amenable to numerical explorations.

The published account of the Olympiad Problem includes three solutions, none of which treat the problem as one of linear programming. The number 225 was cunningly chosen so that $45 = 225/5$ and $9 = 225/5^2$, but this has no bearing on the solution. If 225 is replaced by any number $N > 165$, then $\operatorname{card} \mathscr{D} = 165$ and $\operatorname{card} \mathscr{R} = N - 165$. We understand that this problem was solved by exactly two contestants.

H.4 References

Chakraborty, A., Chandru, V., & Rao, M. R. (2020). A linear programming primer from Fourier to Karmarkar, *Ann. Oper. Res.* **287**, 593–616.

Dantzig, G. B. (1982). Reminiscences about the origins of linear programming, *Oper. Res. Lett.* **1**, 43–48.

Dantzig, G. B. & Eaves, B. C. (1973). Fourier–Motzkin elimination and its dual, *J. Combinatorial Theory Ser. A* **14**, 288–297.

Dorfman, R. (1984). The discovery of linear programming, *Ann. Hist. Comput.* **6**, 283–295.

Farebrother, R. W. (2006). A linear programming procedure based on de la Vallée Poussin's minimax estimation procedure, *Comput. Statist. Data Anal.* **51**, 453–456.

Farkas, J. (1902). Theorie der einfachen ungleichungen, *J. Reine Angew. Math.* **124**, 1–27. doi:10.1515/crll.1902.124.1

Fourier, J. B. J. (1826). Solution d'une question particulière du calcul des inégalités, *vol. II* Paris, pp. 317–328.

Gale, D., Kuhn, H. W., & Tucker, A. W. (1951). *Linear Programming and the Theory of Games*, Cowles Commission Monographs, No. 13, New York: Wiley, pp. 317–329.

Karmarkar, N. K. (1984). A new polynomial-time algorithm for linear programming, *Combinatorica* **4**, 373–395.

Khachiyan, L. G. (1979). A polynomial algorithm in linear programming, *Dokl. Akad. Nauk SSSR* **244**, 1093–1096; *Soviet Mathematics Doklady* **20**, 191–194.

Koopmans, T. C. (1951). *Activity Analysis of Production and Allocation*, New York: Wiley.

Merlin, J. (1911). Sur quelques théorèmes d'arithmétique et un énoncé qui les contient, *C. R. Acad. Sci. Paris* **153**, 516–518.

(1915). Un travail de Jean Merlin sur les nombres premiers, *Bull. Sci. Math.* (2) **39**, 121–136.

von Neumann, J. & Morgenstern, O. (1944). *Theory of Games and Economic Behavior*, Princeton: Princeton University Press.

Selberg, A. (1971). Sieve methods. In *Proc. Sympos. Pure Math.* (SUNY Stony Brook 1969), Vol. XX, pp. 311–351, Providence: Amer. Math. Soc.; *Collected Papers* I, pp. 568–608, Berlin: Springer-Verlag (1989).

Shor, N. Z. (1970). Convergence rate of the gradient descent method with dilation of the space, *Cybernetics* **6**, 102–108.

de la Vallée Poussin, C. J. E. G. N. (1911). Sur la méthode de l'approximation minimum, *Ann. Soc. Sci. Bruxelles* **35**, 1–16.

Williams, H. P. (1986). Fourier's method of linear programming and its dual, *Amer. Math. Monthly* **93**, 681–695. `https://doi.org/10.2307/2322281`.

Errata for Volume 1

page	line	Correction
6	15	The value given of $\mathrm{li}(10^{13})$ is incorrect; it should be 346065645809.01.
7	−10	For 'k' read 'K'.
17	−2	insert comma before 'then'.
23	−6	Replace '$n = 1$' by '$n = 2$'.
41	−10	The typeface in the first line under the first sum is too small.
42	14	insert parentheses: $c = \big(2C_0 - 1 - \zeta'(2)\big)/\zeta(2)$
57	7	For '1/d' read '$1/d$'.
64	12	After 'Show that' insert 'if $q > 1$, then'.
67	−1	In summation replace $f \le x$ by $f > x$.
70	14	The right-hand side of the inequality should read $\mathrm{li}(\log n) + O\big((\log n)\exp\big(-c\sqrt{\log\log n}\big)\big)$
70	16	The right-hand side of the inequality should read $\mathrm{li}(\log n) + O\big((\log n)\exp\big(-c\sqrt{\log\log n}\big)\big)$
88		In Exercise 6(d), replace '$\zeta(2) - 1/z$' by '$\zeta(2) - 1/(z-1)$'.
92	3	After 'Λ_1' insert '= 1'.
92	6	Replace '$g([d,e])$' by '$b([d,e])$'.
117	−9	Replace 'Lemma 4.2' by 'Lemma 4.3'.
122	8	For '$\frac{L'}{L}(s,\chi)$' read '$\frac{L'}{L}(s,\chi_0)$'.
126	1	The condition '$n \equiv a \pmod q$' should be '$p \equiv a \pmod q$'.
126	2	The condition '$n \equiv a \pmod q$' should be '$p \equiv a \pmod q$'.
126	3	The condition '$n \equiv a \pmod q$' should be '$p \equiv a \pmod q$'.
131	13	For $\left(1 - \frac{1}{N(\mathfrak{p})}\right)^{-1}$ read $\left(1 - \frac{1}{N(\mathfrak{p})^s}\right)^{-1}$

page	line	Correction
133	3	For 'be written' read 'may be written'.
133	−5	For '$L(1,\chi)$' read '$L(1,\chi)\neq 0$'.
138	−3	For 'x_0^{σ}' read 'x^{σ_0}'.
139	12, 13	'*si*' should be 'si'.
139	13	'$4^{\sigma_0}+x^{\sigma_0}$' should be '$(4x)^{\sigma_0}$'.
140	-7	'$4^{\sigma_0}+x^{\sigma_0}$' should be '$(4x)^{\sigma_0}$'.
147	5	Replace '$+\frac{1}{2}$' by '$+\frac{1}{4}$', '$-\frac{1}{2}$' by '$-\frac{1}{4}$', '$+\frac{i}{2}$' by '$+\frac{i}{4}$', and '$-\frac{i}{2}$' by '$-\frac{i}{4}$'.
158	−5	Replace $\sum_{s\to 0^+}$ by $\lim_{s\to 0^+}$.
193	3	For '$cos\theta$' read '$\cos\theta$'.
258	4	For 'contraction' read 'contradiction'.
258	8	For 'arithemtic' read 'arithmetic'.
284	7	For 'for some integer k.' read 'for some integer k, when $(n,p)=1$.'
286	−7	Replace '$e(a/q)$' by '$e(-a/q)$'.
310	−5	Replace $\chi(2)$ by $\overline{\chi}(2)$.
311	4, 6, 8	Replace N/q by q/N in three places.
318	3	For 'l' read 'ℓ'.
346	−9	For '$M_h(R)$' read $M_h(R)=\max_{\|z\|\le R}\|h(z)\|$'.
347	3	For 'K_2' read '$\mathscr{K}_2$'.
348	−13	For 'Exercise 10.1' read 'Exercise 10.2.1'.
348	−12	For 'Exercise 10.4' read 'Exercise 10.2.4'.
369	−9	For '(5.23)' read '(5.25)'.
369	−8	For '$e^{n/x}$' read '$e^{-n/x}$'.
374	3	Replace '(10)' by '(11.10)'.
377	8	Delete '(a)'.
377	12, 13	Delete all of part (b) of the exercise.
386	−5	Replace 'x' by 'n' in two places.
386	−7	Replace 'x' by 'n'.
389	7	Replace 'e^{C_0}' by 'e^{-C_0}'.
409	−10	close the space between $\left(\frac{\zeta'}{\zeta}\right)'$ and (0).
409	−2	Replace $+(-1+\cosh 1/z)\log z$ by $-(-1+\cosh 1/z)\log z$.
411	11	For '$\phi(s)$' read '$\Phi(s)$'.
423	−4	For '$\gamma_2<-\gamma_1$' read '$\gamma_2<-\gamma_1/2$'.
423	−3	Replace '$\gamma_2<-\gamma_1$' by '$\gamma_2<-\gamma_1/2$' in two places.
430	Ex. 2	Replace F by ψ_1 in five places.
434	5	Replace '(13.25)' by '(13.35)'.
434		In the last error term in (13.37), replace '$y^{1-\sigma}$' by '$x^{1-\sigma}$'.
434		In the last term of (13.38), replace '$y^{1-\sigma}$' by '$x^{1-\sigma}$'.

page	line	Correction
435	−2	Replace ‘Corollary 13.13’ by ‘Theorem 13.13’.
438	6	In the second sum, replace ‘$\frac{\Lambda(n)}{n\log n}$’ by ‘$\frac{1}{kp^k}$’.
442	−9	Between ‘Put’ and ‘σ_1' insert ‘$\sigma_0 = 1 + 1/\log x$,’.
442	−5	For ‘Theorem 13.22’ read ‘Theorem 13.23’.
442	−4	For ‘$\int_{\sigma_1} \sigma_0$’ read ‘$\int_{\sigma_1}^{\sigma_0}$’.
442	−2	For ‘Theorem 13.22’ read ‘Theorem 13.23’.
444	−1	Replace $\lvert L(s,\chi)\rvert$ by $\lvert\log L(s,\chi)\rvert$.
445	4	The displayed formula should read $\lvert\log L(s,\chi)\rvert \le \log\log\log q\tau + O(1)$.
445	7	Replace $\lvert L(s,\chi)\rvert$ by $\lvert\log L(s,\chi)\rvert$.
446	−5	For ‘$B_1(x/d^2)$’ read ‘$B_1(\{x/d^2\})$’.
461	2	For ‘$2\pi S(t) \le c\log\log T$’ read ‘$S(t) \le \frac{c}{\pi}\sqrt{\frac{1}{2}\log\log T}$’, and for ‘$\int_{\infty}^{c}$’ read ‘$\int_{-\infty}^{c}$’.
464	4	Replace ‘and’ by ‘and if $\Theta > 1/2$, then’
465	−6	Replace ‘since $\Theta \ge 1/2$, it follows that’ by ‘if $\Theta > 1/2$, then’.
465	−11	For ‘Lemma 1’ read ‘Lemma 15.1’.
475	−2	Replace ‘$e^{i\gamma_1\phi}/\rho_1^K$’ by ‘$e^{i\gamma_1\phi}/\rho_1^{K+1}$’.
476	10	Replace ‘no prime power’ by ‘no logarithm of a prime power’.
492	11	For ‘$\xi \le x \le b$’ read $a \le x \le \xi$’.
500	12	For ‘$\zeta(2) = \pi/6$’ read ‘$\zeta(2) = \pi^2/6$’.
501	6	For ‘(1)’ read ‘(B.1)’.
503	−8	For ‘N’ read ‘N’.
508	7	Replace ‘$B_1(x)$’ by ‘$B_1(\{x\})$’.
508	−5	For ‘c’ read ‘C’.
508	−2	Replace ‘$B_1(x)$’ by ‘$B_1(\{x\})$’.
508	−1	Replace ‘$B_2(x)$’ by ‘$B_2(\{x\})$’.
520	5	For ‘constatnt’ read ‘constant’.
530	−8	For ‘$\int_0^\infty$’ read ‘$\int_0^1$’.
535	−7	Replace ‘$f(k)$’ by ‘$\widehat{f}(k)$’.
536		In the figure, the label 0 is at 0.1, but it should be at 0.0 on the horizontal axis.
536	−2	For ‘Z’ read ‘$\mathbb{Z}$’.
539	2	For ‘$\widehat{f}(k)$’ read ‘$\widehat{F}(k)$’.
551	−22	For ‘powe series’ read ‘power series’.

Name Index

Subject Index

For EU product safety concerns, contact us at Calle de José Abascal, 56–1°,
28003 Madrid, Spain or eugpsr@cambridge.org.

www.ingramcontent.com/pod-product-compliance
Lightning Source LLC
LaVergne TN
LVHW011657200926
848484LV00004B/975
* 9 7 8 1 0 0 9 4 4 5 0 5 4 *